BCS: 50 Years

BCS: 50 Years

Edited by

Leon N Cooper
Dmitri Feldman
Brown University, USA

NEW JERSEY • LONDON • SINGAPORE • BEIJING • SHANGHAI • HONG KONG • TAIPEI • CHENNAI

Published by

World Scientific Publishing Co. Pte. Ltd.

5 Toh Tuck Link, Singapore 596224

USA office: 27 Warren Street, Suite 401-402, Hackensack, NJ 07601

UK office: 57 Shelton Street, Covent Garden, London WC2H 9HE

Library of Congress Cataloging-in-Publication Data

BCS : 50 years / edited by Leon N. Cooper and Dmitri Feldman.
p. cm.
Includes bibliographical references and index.
ISBN-13: 978-9814304641 (hardcover : alk. paper)
ISBN-10: 9814304646 (hardcover : alk. paper)
ISBN-13: 978-9814304658 (pbk. : alk. paper)
ISBN-10: 9814304654 (pbk. : alk. paper)
1. Superconductivity--History. 2. Superconductors--History. I. Cooper, Leon N. II. Fel'dman, D. E. (Dmitrii Eduardovich)
QC611.92.B37 2010
537.6'23--dc22

2010038551

British Library Cataloguing-in-Publication Data

A catalogue record for this book is available from the British Library.

The cover image was created by Richard Fishman and Xiangjun Shi.

PREFACE

More than 50 years have passed since the publication of John Bardeen, Leon Cooper and Robert Schrieffer's Theory of Superconductivity, known universally as BCS. In that time so many new experiments have been done, so much additional theory has been developed and so many papers and books have been written that it is hardly possible to begin to span the vast universe that this subject has become in a single volume.

Although the publication of BCS might have been thought to be the end of a fifty year long search for the solution to the problem of superconductivity, in fact, it appears to have been more like the beginning. Since 1957, a remarkable number of variations have been explored. There have been high temperature and other new superconductors, applications, from Josephson junctions and SQUIDS to type II superconducting magnets and new phenomena such as the superfluidity of ^{3}He and the BE/BCS crossover in cold atoms. In addition, the fundamental pairing idea has been applied to systems as diverse as nuclei, neutron stars and quark matter, and the mathematical methods including the concept of broken gauge symmetry, introduced by BCS, have been applied to the standard model.

This volume begins with historical chapters presenting the point of view of major players in the field. Although these somewhat Rashomon-like recollections do not agree in every detail, together, we believe they provide a fascinating, first hand account of what are now long-ago events.

It follows with chapters on SQUIDS, superconductivity in complex and disordered materials, high T_c and other new superconductors, and concludes with chapters on BCS beyond superconductivity including superfluid ^{3}He, the BE/BCS crossover in cold atoms and the application of ideas and methods introduced by BCS to neutron stars, quark matter and the standard model.

In all, we have tried to present an overview of the evolution of BCS from its origins to detailed discussions of some of the most interesting current topics and still unanswered questions.

Acknowledgments

We would like, above all, to express our appreciation to the authors who have contributed to this volume. They fit the task of writing their chapters into already tight schedules, and, in addition, endured some harassment from the editors, eager to finish before the BCS 100th anniversary.

We would like to thank Richard Fishman and Xiangjun Shi for creating the image that appears on the cover.

We would also like to express our appreciation to Lakshmi Narayanan, Yubing Zhai and the rest of the devoted editorial team of World Scientific Publishing Co. and last, but certainly not least, we would like to thank Pete Bilderback without whose editorial, administrative and organizational efforts this volume could not have come to be.

Leon N Cooper and *Dmitri Feldman*

CONTENTS

I. Historical Perspectives

REMEMBRANCE OF SUPERCONDUCTIVITY PAST

Leon N Cooper

Department of Physics and Institute for Brain and Neural Systems, Brown University, Box 1843, Providence, RI 02912, USA
Leon_Cooper@Brown.edu

Recollections of events that led to BCS.

Memory fades. Is it possible to recreate the events that led to BCS? Can we recapture the great difficulty (even conjectured insolubility) superconductivity presented more than half a century ago? Perhaps not. But yellowed notes, like tea-soaked crumbs of madeleine, have awakened for me, sometimes with startling clarity, images and memories from that somewhat distant past.

My interaction with superconductivity began when I met John Bardeen in April or May of 1955 at the Institute for Advanced Study in Princeton. John was on the East Coast looking for a postdoc to work with him on superconductivity. He had written to several physicists, among them T. D. Lee and Frank Yang, asking if they knew of some young fellow, skilled in the latest and most fashionable theoretical techniques (at that time, Feynman Diagrams, renormalization methods, and functional integrals) who might be diverted from the true religion of high energy physics (as it was then known) and convinced that it would be of interest to work on a problem of some importance in solid-state.

As far as I can recall, it was the first time I had even heard of superconductivity. (Columbia, where I earned my Ph.D. did not, to my memory, offer a course in modern solid-state physics.) Superconductivity is mentioned in the second edition of Zemansky's *Heat and Thermodynamics* (the text that was assigned for the course in Thermodynamics taught by Henry Boorse that I took in my last semester at Columbia College), but I suspect those pages were not assigned reading. Although John, no doubt, explained something about the problem, it is unlikely that I absorbed very much at the time.

The long and very imposing list of physicists (among them Bohr, Heisenberg and Feynman) who had tried or were trying their hand at superconductivity should have given me pause. Even Einstein, in 1922 — before the quantum theory of metals was in place — had attempted to construct a theory of superconductivity. Fortunately, I was unaware of these many unsuccessful attempts. So when John invited me to join him (he, somehow, neglected to mention these previous efforts), I decided to take the plunge.

In September of 1955, I arrived in Champaign-Urbana. I was assigned a wooden desk in John's office, the same desk that had been previously occupied by David Pines and later by Gerald Rickayzen. The atmosphere at Illinois was friendly and collegial. Down the hall, one could find Joe Weneser and Arnold Feingold who shared an office. Geoffrey Chew and Francis Low shared another. Among the experimentalists were John Wheatley and Charlie Slichter.

The department was led by Wheeler Loomis and Gerald Almy. Bob Schrieffer was one of John's graduate students. His desk was in an attic office above the third floor of the Physics building with a sign on the door that read "Institute for Retarded Studies." John had assigned, or Bob had chosen, superconductivity as a thesis problem. We would spend considerable time in that office commiserating with each other.

John's first suggestion was that I read Schoenberg's book. So, early in September of 1955, I began to consume Schoenberg and other books and articles on the facts and the theoretical attempts to solve the problem of superconductivity. I went through many of John's calculations, some classical attempts going back to Heisenberg and the arguments of H. and F. London. John at that time was writing an article for the *Encyclopedia of Physics* reviewing the theory of superconductivity; one of my assignments was to go over the arguments and to proofread the article. Also, sometime during that year, I gave a series of seminars on current methods in quantum field theory.

As Brian Pippard was to emphasize in a lecture he gave at Illinois, the facts of superconductivity appear to be simple. It seemed clear that superconductivity must be a fairly general phenomenon that does not depend on the details of metal structure, that there must be a qualitative change in the nature of the electron wave function that occurs in a wide variety of conditions in many metals and many crystal structures.

In many lectures we presented these simple facts more or less as follows:

In 1911, Kamerlingh Onnes found that the resistance of a mercury sample disappeared suddenly below a critical temperature
R
R
n
T
T
c

Fig. 1.

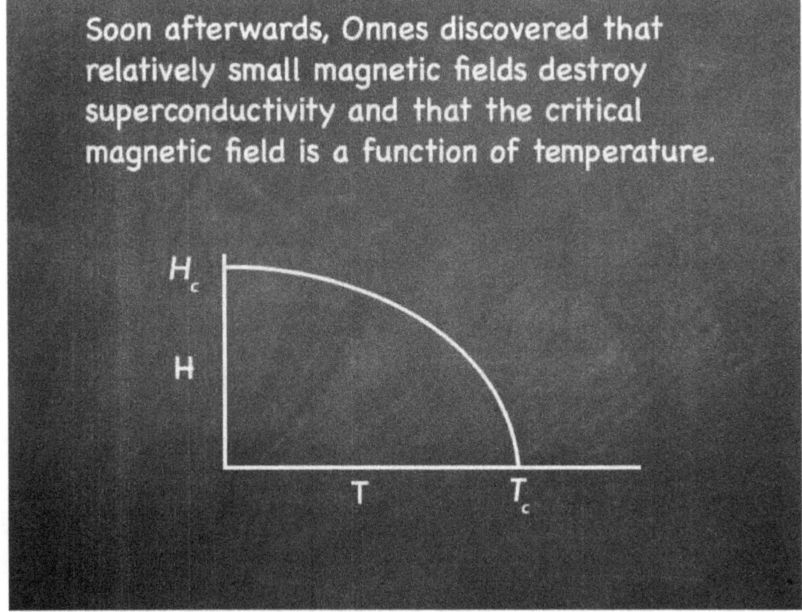
Soon afterwards, Onnes discovered that relatively small magnetic fields destroy superconductivity and that the critical magnetic field is a function of temperature.
H
c
H
T
T
c

Fig. 2.

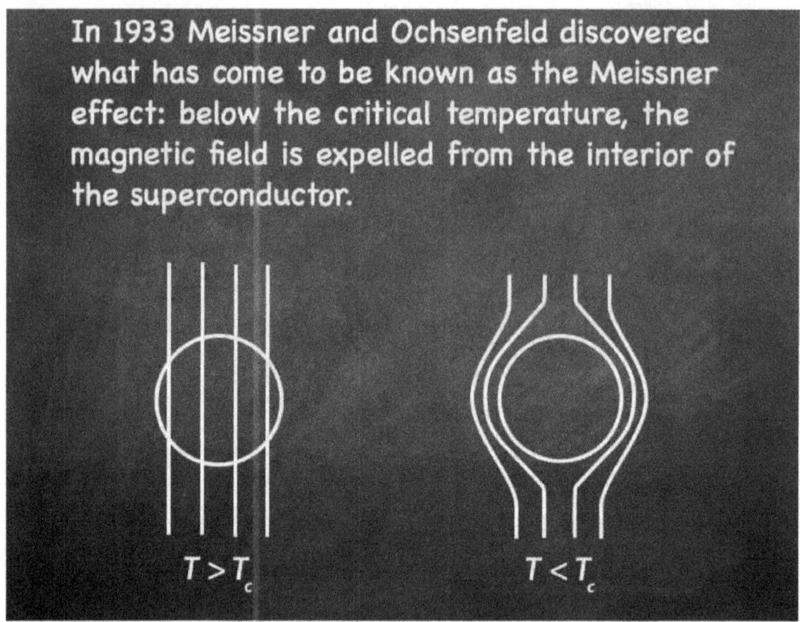
In 1933 Meissner and Ochsenfeld discovered what has come to be known as the Meissner effect: below the critical temperature, the magnetic field is expelled from the interior of the superconductor.
$T > T_c$
$T < T_c$

Fig. 3.

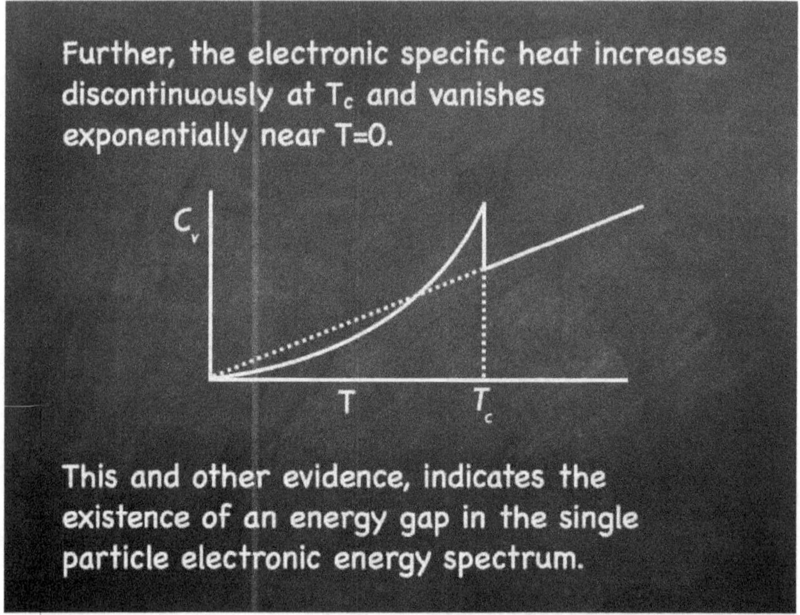
Further, the electronic specific heat increases discontinuously at T_c and vanishes exponentially near T=0.
C_v
T
T_c
This and other evidence, indicates the existence of an energy gap in the single particle electronic energy spectrum.

Fig. 4.

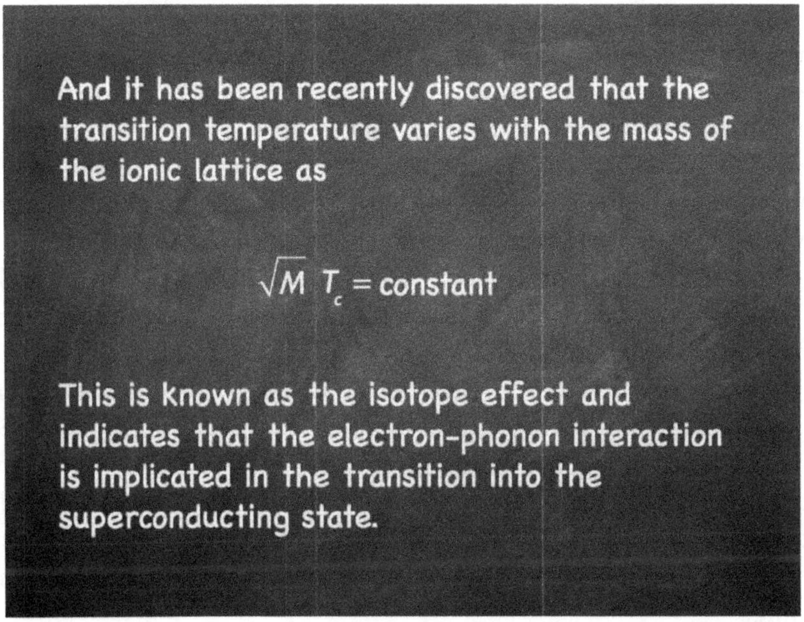

And it has been recently discovered that the transition temperature varies with the mass of the ionic lattice as

$$\sqrt{M}\, T_c = \text{constant}$$

This is known as the isotope effect and indicates that the electron-phonon interaction is implicated in the transition into the superconducting state.

Fig. 5.

I became convinced early that the essence of the problem was an energy gap in the single particle excitation spectrum. This, of course, was the prevailing view at Illinois and seemed very reasonable — almost necessitated by the exponentially decreasing electronic specific heat near $T = 0$. John had shown that such a gap would be likely to yield the Meissner effect. His argument had been criticized on grounds of lack of gauge invariance, but John's opinion was that longitudinal excitations that would be of higher energy due to long-range Coulomb forces would take care of such objections. This also seemed reasonable. In any case, a single particle energy spectrum so greatly reduced from that of a normal metal would no doubt lead to a state with radically different properties.

So I started to attack the many-electron system to see how such an energy gap might come about. I attempted to sum various sets of diagrams: ladders, bubbles and many others. I tried low energy theorems, Breuckner's method, functional integrals and so on. After more attempts than I care to mention, during the fall of 1955, I was no longer feeling so clever. So, when John left for a trip in December, I put my calculations aside and tried to think things through anew.

Paradoxically, the problem seemed conceptually simple. The Sommerfeld–Bloch individual particle model (refined as Landau's Fermi liquid theory) gives a fairly good description of normal metals but no hint of superconductivity. Superconductivity was thought to occur due to an interaction between electrons. The Coulomb interaction contributes an average energy of 1 eV/atom. However, the average energy in the superconducting transition as estimated from T_c is about 10^{-8} eV/atom. Thus, the enormous qualitative changes that occur at the superconducting transition involve an energy change orders of magnitude smaller than the Coulomb energy. This huge energy difference contributed to the great difficulty of the problem. Also, the Coulomb interaction is present in all metals, but only some metals become superconductors.

After the work of Fröhlich, Bardeen and Pines, it was realized that an electron–electron interaction, due to phonon exchange, under some conditions could produce an attractive interaction between electrons near the Fermi surface. But, so far, there was no suggestion of how this would give rise to the superconducting state. In total frustration, I decided to shelve diagrams and functional integrals and step through the problem from the beginning.

This began to come together for me on a trip to New York for a family reunion and Christmas party. The train ride took about seventeen hours. Johnson once remarked that the prospect of hanging powerfully concentrates the intellect. He might have added seventeen sleepless hours on a train.

I began to pose the question in a highly simplified fashion, introducing complications one by one. Where does the problem become insoluble? Consider first N electrons in a container of volume Ω. This would give a Fermi sphere and, at least qualitatively, some of the properties of normal metals. Now turn on an attractive electron–electron interaction, even one that is very weak (so weak that interacting electrons would be limited to a small region about the Fermi surface). Immediately one is faced with an extraordinarily difficult problem in which vast numbers of degenerate quantum states interact with each other. What would the properties of the solutions be? The more I thought about this the clearer it became that, for all of the sophistication of various approaches, we did not yet know the answer to what seemed to be a fairly elementary question.

On my return to Illinois in January of 1956, I recall asking Joe Weneser, who worked in an office down the hall and who was always kind enough to listen sympathetically, "How do you solve problems in quantum mechanics for very degenerate systems?" Joe thought a while and said, "Why don't you look it up in Schiff?" Well I looked it up in Schiff and realized that Professor Yukawa, from whom I learned quantum mechanics at

Columbia, had already taught me what was there and that it didn't help very much.

So I embarked on an analysis of problems in quantum mechanics involving degeneracy. This led me to think about all types of matrices that might result from highly degenerate quantum systems. Piled on my desk, I may have had every book related to this subject in the physics library. I learned obscure matrix theorems, studied properties of stochastic and other special matrices and began a variety of theoretical experiments.

I was no longer doing what John expected me to do, and was unable to communicate what it was that I was doing. (I was not sure myself.) So, for a time we worked separately.

It was early in the Spring of 1956 that I realized that matrices, of a type that were certain to arise as sub-matrices of the Hamiltonian of the many-electron system I was considering, characteristically would have among their solutions states that split from the rest which were linear combinations of large numbers of the original states. These new states had properties qualitatively different from the original states of which they were formed. They were typically separated from the rest by NV (the number of states multiplied by the interaction energy) and therefore, since $N \sim \Omega$ and $V \sim \Omega^{-1}$, the separation energy would be independent of the volume. It was soon clear that this was a natural, almost inevitable, property of the degenerate systems I was considering.

Among the easiest such sub-matrices to pick, because of two body interactions, conservation of momentum, symmetry of the wave function, and all of the arguments that have been made many times since, were those that came about due to transitions of zero-spin electron pairs of given total momentum. I then focused my attention on such pairs.

I should note that I came about these pairs in my attempt to solve the degeneracy problem. Schafroth, Butler and Blatt had considered a system of charged electron pair molecules whose size was less than the average distance between them so they could be treated as a charged Bose–Einstein gas. They had shown that such a system displayed a Meissner effect and a critical temperature condensation. Schafroth, as I recall, gave a colloquium at Illinois presenting his ideas. I am not sure when that colloquium was given: whether it was before or after my own pair idea. However I was aware of Schafroth's argument by the time I submitted my letter to *Physical Review* in September 1956. As far as I was concerned, pairs spreading over distances of the order of 10^{-4} cm so that 10^6 to 10^7 occupy the same volume bore little relation to what Schafroth had proposed. These extended pairs might have some Bose properties but it seemed unlikely that they would undergo a Bose condensation in the Schafroth sense.

It is mildly ironic that at present we can go continuously from BCS extended pairs to Schafroth molecule-like pairs in the BCS/Bose–Einstein crossover. I thought, from time to time, that it would be extraordinarily interesting if one could vary the strength of the interaction in order to observe this transition and had hoped for a while that the interaction would be strong enough in high temperature superconductors so that this transition might be seen as the temperature was lowered, a possibility I mentioned at a roundtable discussion in Stockholm at the time of the Nobel award for high T_c superconductivity. That has turned out, at least so far, not to be possible, but the Feshbach resonance has made it possible to vary the interaction strength between fermions continuously so that we now can actually see this transition. This is explained in more detail in Wolfgang Ketterle and Gordon Baym's chapters in this volume.

A page from my notes (Fig. 6) from that period shows the pair solutions as they first appeared to me. The last crossing on the left is the surprising coherent state, split from the continuum by an energy proportional to $\hbar\omega$ multiplied by the exponential factor displaying the now well-known essential singularity.

The energy of a ground state composed of such 'bound' pair states would be proportional to $(\hbar\omega)^2$ and therefore inversely proportional to the isotopic mass as expected. In addition, the exponential factor seemed to give a natural explanation of why the transition energy into the superconducting state was so small.

It seemed clear that if somehow the ground state could be composed of such pairs, one would have a state with qualitatively different properties from the normal state, with the ground state probably separated by an energy gap from single particle excited states and thus likely, following arguments that had already been given by Bardeen, to produce the qualitative properties of superconductors.

In addition, all of this could be accomplished in what was close to a variational solution of the many-electron Schrödinger equation with demonstrably lower energy than the state of independent particles from which the pairs had been formed. I could show, using one stochastic matrix of which I was particularly fond, a matrix with zero diagonal elements (zero kinetic energy — what I called the strong coupling limit, $N(0)V \gg 1$) that $N(0)\hbar\omega$ non-interacting pairs with average available phase space $N(0)\hbar\omega$ for transitions (this would result if we chose pairs and pair states in a shell $\hbar\omega$ about the

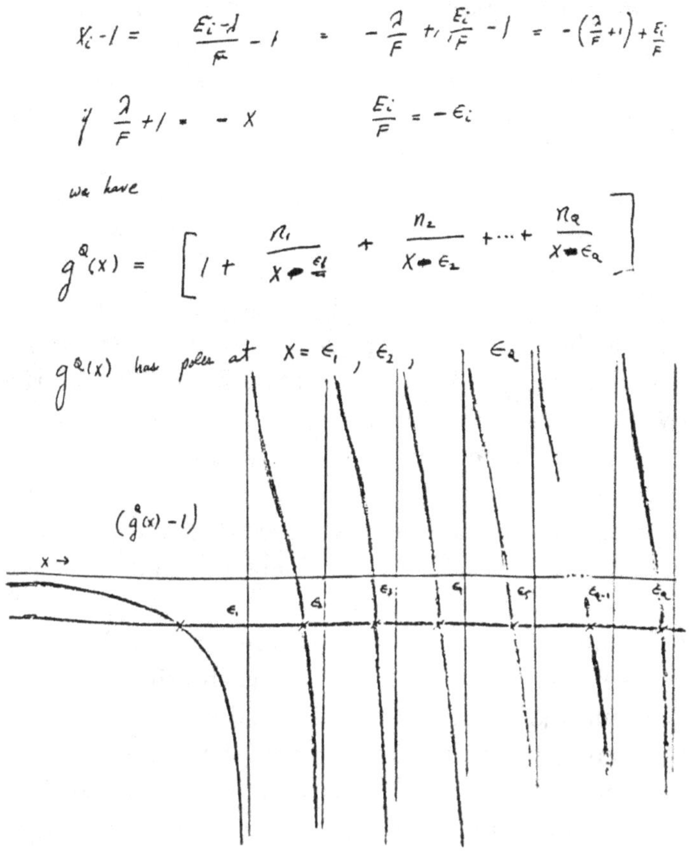

Fig. 6. Handwritten notes from Spring 1956.

Fermi surface) would lead to a condensation energy $-N(0)^2(\hbar\omega)^2V$. This, amusingly, turns out to be the BCS ground state energy difference in the same strong coupling limit. I did not know this last fact, of course, but these results were enough to convince me that this was the way to go. The solution of the next problem, to produce a wave function that incorporated the pairs and satisfied the exclusion principle with which one could calculate, took the rest of the year.

I was reasonably excited by these results but became aware, painfully, in the next months, how long the road still was. The way ahead was not as clear to everyone else as it appeared to me. It was not easy to see what the solution meant and it was not evident how to use it in calculations. I was asked over and over how one could have a bound state whose energy was larger than zero. (Going over my notes from that period, I read that

I was then calling it the "strange bound state"). At a seminar that I gave on the subject at Illinois some time during the spring of 1956, Francis Low asked how one could get a volume-independent energy shift if the interaction matrix elements went as Ω^{-1}.

In addition to trying to construct and calculate with a wave function composed of non-interacting pairs that satisfied the exclusion principle, I spent too much of the next six months answering questions of this type, showing that there could be developed a theory of Green's functions of electrons interacting above a Fermi sea, working on a pair algebra, making many attempts to prove that qualitative features revealed in the pair solutions would not be lost in the full many-body system. (I continued such efforts for several years afterward, long after everyone else regarded the problem as solved.) I thus spent much of my time engaged in missionary work: lecturing, trying to convince blank-eyed listeners and proving theorems. If I had been as mature then as I perhaps am today, I would have used this precious time to calculate and forge ahead. (As I now tell my graduate students: "Stop thinking and start working.")

Richard Feynman has related that at a meeting of the American Physical Society, likely sometime in 1956, he was chatting with Onsager when a wild-eyed young man came up to them and said that he had solved the problem of superconductivity. (There were, at the time, quite a few wild-eyed young — and not so young — men who were convinced they had solved the problem.) As Feynman relates, he could not understand what the young man was saying and concluded that the fellow was probably crazy. Onsager, on the other hand, according to Feynman, thought for a while and said, "I think he's right." Feynman believed that the wild-eyed young man was me. I am not at all sure whether or not this meeting actually occurred, but it might have. I certainly approached many people — trying not to be wild-eyed — while attempting to explain my ideas. But if it did happen, I sure wish Onsager had said something to me. I could have used the encouragement.

I spent the summer months of 1956 in California working at Ramo-Wooldridge (later Space Technology Laboratories). I do not remember what it was that I worked on there, but I did give a series of lectures on superconductivity — including, of course, the strange bound state.

Among other news items that greeted us in the Fall of 1956 when I returned to Illinois from California was a report of a talk by Feynman that was given at the Low Temperature Physics conference in Seattle. He concluded with the statement, "And when one works on it [the problem of superconductivity], I warn you before you start, one comes up finally to a terrible shock; one discovers that he is too stupid to solve the problem."

I had been working on a rather long paper describing possible methods for putting the pairs into a many-body wave function. Hearing that Feynman was hard at work on superconductivity, jolted me to move faster. So, I hurriedly put together the letter that described my pair results. This was submitted to *Physical Review* on September 18, 1956.

I continued working trying to find a way to calculate with some kind of a many-body pair wave function, but went sideways as much as forward. I kept wandering back to the full many-body problem, the problem of the changing signs of general matrix elements — just the problem that restricting oneself to pair transitions would avoid.

Then came the news that John had won the Nobel Prize for his part in the invention of the transistor. Before John left for Stockholm, Bob, nervous about the lack of progress towards his thesis, spoke to him about possibly changing his thesis subject. John offered Bob his now famous advice, "Give it [superconductivity] another month ... keep working ... maybe something will happen ..."

Then it happened. Bob finally made the crucial step of embodying the pairs in a wave function that satisfied the Pauli principle with which we could easily calculate — which is what we had been trying to do for the past six months. I believe he obtained this result in December 1956 or January 1957. It may have been while we were on the East Coast at the Stevens Institute of Technology for a conference on many body systems. If I recall correctly, Bob once told me that he worked this out on a New York City subway car. If so, he found more virtue in the New York subway system in this short sojourn than I discovered in all of the years that I rode it.

We met by accident in the Champaign airport, both on our way back from the East Coast. It was here that Bob first showed me his results. We were jumping up and down with excitement, oblivious to what others, who were in that airport on that cold winter evening, might have thought.

Bob's results finally convinced John. (As Bob said to me at the time: "We've turned the battleship around.") So one morning, late in January or early February of 1957, John asked me if I would agree to write a paper with Bob and him on the theory of superconductivity. I guess I must have agreed.

The next five months were a period of the most concentrated, intense and incredibly fruitful work I have experienced. We divided our efforts: Bob would focus on thermodynamic properties, I would focus on electro-dynamic

TABLE II. Matrix elements of single-particle scattering operator.

Wave functions[a] Initial, Ψ_i $(\mathbf{k}\uparrow, -\mathbf{k}\downarrow)$	$(\mathbf{k}'\uparrow, -\mathbf{k}'\downarrow)$	Final, Ψ_f $(\mathbf{k}\uparrow, -\mathbf{k}\downarrow)$	$(\mathbf{k}'\uparrow, -\mathbf{k}'\downarrow)$	Ground (+) or excited (−) $\mathbf{k}$	$\mathbf{k}'$	Energy difference $W_i - W_f$	Probability of initial state	Matrix elements $c_{\mathbf{k}'\uparrow}^* c_{\mathbf{k}\uparrow}$ or $c_{-\mathbf{k}'\downarrow}^* c_{\mathbf{k}\uparrow}$	$c_{-\mathbf{k}\downarrow}^* c_{-\mathbf{k}'\downarrow}$ or $-c_{-\mathbf{k}\downarrow}^* c_{\mathbf{k}'\uparrow}$
(a)				+	+	$E-E'$	$\frac{1}{2}s(1-s'-p')$	$[(1-h)(1-h')]^{1/2}$	$-(hh')^{1/2}$
$X0$	00	00	$X0$	−	−	$E'-E$	$\frac{1}{2}sp'$	$(hh')^{1/2}$	$-[(1-h)(1-h')]^{1/2}$
$X0$	XX	XX	$X0$	+	−	$E+E$	$\frac{1}{2}sp'$	$-[(1-h)h']^{1/2}$	$-[h(1-h')]^{1/2}$
				−	+	$-(E+E')$	$\frac{1}{2}s(1-s'-p')$	$-[h(1-h')]^{1/2}$	$-[(1-h)h']^{1/2}$
(b)				+	+	$E'-E$	$\frac{1}{2}s'(1-s-p)$	$(hh')^{1/2}$	$-[(1-h)(1-h')]^{1/2}$
XX	$0X$	$0X$	XX	−	−	$E-E'$	$\frac{1}{2}s'p$	$[(1-h)(1-h')]^{1/2}$	$-(hh')^{1/2}$
00	$0X$	$0X$	00	+	−	$-(E+E')$	$\frac{1}{2}s'(1-s-p)$	$[h(1-h')]^{1/2}$	$[h'(1-h)]^{1/2}$
				−	+	$E+E'$	$\frac{1}{2}s'p$	$[(1-h)h']^{1/2}$	$[h(1-h')]^{1/2}$
(c)				+	+	$E+E'$	$\frac{1}{4}ss'$	$[(1-h)h']^{1/2}$	$[h(1-h')]^{1/2}$
$X0$	$0X$	00	XX	−	−	$-(E+E')$	$\frac{1}{4}ss'$	$-[h(1-h')]^{1/2}$	$-[h'(1-h)]^{1/2}$
		XX	00	+	−	$E-E'$	$\frac{1}{4}ss'$	$[(1-h)(1-h')]^{1/2}$	$-(hh')^{1/2}$
				−	+	$E'-E$	$\frac{1}{4}ss'$	$-(hh')^{1/2}$	$[(1-h)(1-h')]^{1/2}$
(d)				+	+	$-(E+E')$	$(1-s-p)(1-s'-p')$	$[h(1-h')]^{1/2}$	$[(1-h)h']^{1/2}$
XX	00	$0X$	$X0$	−	−	$E+E'$	pp'	$-[(1-h)h']^{1/2}$	$-[h(1-h')]^{1/2}$
00	XX			+	−	$E'-E$	$(1-s-p)p'$	$-(hh')^{1/2}$	$[(1-h)(1-h')]^{1/2}$
				−	+	$E-E'$	$p(1-s'-p')$	$[(1-h)(1-h')]^{1/2}$	$-(hh')^{1/2}$

[a] For transitions which change spin, reverse designations of $(\mathbf{k}'\uparrow, -\mathbf{k}'\downarrow)$ in the initial and in the final states.

Fig. 7. Matrix elements as presented in *Physical Review*, 1957.

properties, with John focusing on transport and non-equilibrium properties. New results appeared almost every day. John's vast experience came to the fore. All of the calculations done previously with a normal metal (these John had in his head) could be turned on this new theory of superconductivity if one put in the appropriate matrix elements.

It is hard to believe that with the notation and techniques we used at that time that we could have ever obtained the correct results. As some may recall, we had excited singles, excited pairs, etc. How simple things seem now.

In less than a month, February 18, 1957, we communicated some of our earliest results in a letter to *Physical Review*. This letter describes the utility of the pairing approximation in solving the sign change problem that occurs for general many body matrix elements mentioned above and presents some very preliminary results; but there was enough to excite interest.

We communicated further results in two post-deadline papers to the American Physical Society meeting in March of 1957. John generously allowed Bob and me to present the papers. Bob made the trip to Philadelphia via New Hampshire (as I recall, to visit a friend) and I had the responsibility for bringing the slides for both of our talks to Philadelphia. It turned out that, for some reason I do not recall, Bob could not make it to Philadelphia on time. So, at the last minute, I had to give both of our talks. Although I missed having Bob there to support me, this was no particular problem;

either of us could have given either talk. The session audience overflowed the lecture room; the level of interest was incredible. Yet, I do not remember feeling overwhelmed. I was too inexperienced to fully appreciate what was happening, to appreciate how rare such an event was. In any case, even at that early presentation with only partial results, enthusiasm was enormous and acceptance was almost immediate.

In the course of working out the electromagnetic properties of our superconductor, I had many discussions with John and Bob on questions related to gauge invariance. There were two problems:

The first arose due to the momentum dependence of the electron–electron interaction. This approximation, taken from my pair argument, simplified the calculations. We could easily have employed a momentum independent interaction, but this would have made the calculations much more cumbersome. Our feeling that this was not very important was confirmed by Phil Anderson who communicated his result sometime that spring and is mentioned in a note added in proof in our paper.

The second problem arose due to the fact that our calculation of the Meissner effect was done in the radiation gauge. This calculation was modeled on the one done previously by John assuming an energy gap in the single particle energy spectrum and was based on an early calculation by O. Klein. In a general gauge, we would have to include longitudinal excitations. John and I had talked about this on and off since I first looked over his calculations in the fall of 1955. His opinion, as mentioned above, was that the longitudinal excitations would be more or less unaltered from what they were in the normal metal and that due to the long-range Coulomb interaction would be the very high energy plasma modes and so would not affect our results. Further, since we all believed that our theory was gauge invariant, we could calculate in whatever gauge we chose. This view (also mentioned in the footnote added in proof) was corroborated by Anderson and almost at the same time, in an elegant fashion, using Green's function methods first introduced by Gor'kov and Nambu. Further, Gor'kov was able to derive the Ginsburg–Landau equation from his formulation of BCS and this equation is explicitly gauge invariant.

On my own, at this point, I might have gone off on a side-track to prove that our theory was gauge invariant. Here John's experience played a crucial role. He encouraged me not to divert my attention and to continue my calculations of the electrodynamics. This time I took his advice.

We had some real problems. In my calculations of the Meissner effect, I was obtaining a penetration depth a factor of two too small. John and I went over these calculations and could not find anything wrong. We were almost ready to accept that, somehow, this might, in fact, be the case.

It was at a concert (probably in April of 1957) featuring the music of Harry Parch, brother of the *New Yorker* cartoonist Virgil Parch (a somewhat idiosyncratic composer who orchestrated his pieces with instruments of his own design — some of them huge wooden instruments — so that his music which, as I recall was not bad at all, was not playable except with the instruments shipped from his studio that required special training to play) thinking through the Meissner effect matrix elements for the thousandth time, I realized that between the initial and final state there existed another path, a path that did not occur in normal metal calculations. In the relevant limit the additional term would be of equal magnitude to the usual term but would possibly differ by a sign. The way we did the calculations then, there were about a half a dozen creation and annihilation operators to be manipulated. It was with some anxiety that I realized I could never, in the course of that concert, determine whether we would get the expected penetration depth or whether the Meissner effect would disappear. The family fortune was on the roulette table: double or nothing.

You will believe that I redid the calculations several times that weekend; by Monday morning there was no longer any doubt. Rushing to the office, possibly somewhat earlier than usual, I recall a vivid image: John in his chair, listening intently and absorbing every word. When I was finished I turned from the blackboard and said, "You see it comes out right." John was unusually loquacious that morning. He nodded in agreement and said: "Hmmm."

This new effect was immediately included in all of our calculations. It led to the now-famous coherence factors, with their surprising dependence on the behavior of the interaction under time reversal. Charlie Slichter, who was then doing the experiment with Chuck Hebel on the temperature dependence of the relaxation time for nuclear spins in superconductors, a quick study indeed, very soon was working along with us calculating the theoretical spin relaxation rate.

Somewhat later, Bob Morse communicated the results he had obtained with Bohm at Brown University on the temperature dependence of ultrasonic attenuation in superconductors. Nuclear spin relaxation and ultrasonic attenuation were compared in a note we added in proof. Our theory explained the remarkable and counterintuitive increase in the relaxation rate of the nuclear spins, which contrasted markedly with the unexpected sharp decrease in ultrasonic attenuation just below the transition temperature (see

Figs. 16 and 18 in Charlie Slichter's chapter in this volume). As we remarked in the note, this provided experimental confirmation of the effect of the coherence factors.

By July, exhausted after months of non-stop calculation, we knew that we had solved the problem. Our paper was submitted to *Physical Review* on July 8, 1957.

With a few exceptions, acceptance of our theory was almost immediate. There were some complaints (as expected, supposed lack of gauge invariance was one). As I have said, we had discussed this but, wisely, had not spent too much time on it.

There were also some regrets. One rather well known low temperature physicist (possibly hoping for a new law of nature) expressed his disappointment that "... such a striking phenomenon as superconductivity [was] ... nothing more exciting than a footling small interaction between atoms and lattice vibrations."

Sometime the next fall I received a preprint from Valatin, describing his method for constructing a much easier to use set of orthogonal excitations. What a relief. Gone were excited singles and pairs. We now could calculate more easily and much more rapidly. Some time later I received a preprint from Bogoliubov who, building on his previous liquid helium work, had arrived at the same excitations. And very soon thereafter I learned of the Green's function methods of Gor'kov and Nambu which made issues such as gauge invariance much more transparent. This is discussed in Nambu's chapter in this volume and the part of the story that occurred in the Soviet Union is related in Gor'kov's chapter.

As this volume makes evident, the simple facts of superconductivity are no longer so simple. Among other varieties, we now have superconductors that are resistive, that are gapless, and that display no Meissner effect. Superconductors now come in more flavors than Baskin-Robbins ice cream. We could not, in 1957, have been aware of all of these variations. But, more important, had we known all of this and worried about it, we might never have constructed anything. This is an example of what I believe is a general principle for the investigation of difficult scientific questions. One must make things as simple as possible ("but no simpler" as Einstein, supposedly, once said). In the case of superconductors, the key was to extract at least one qualitative feature of the superconducting state that was strikingly different from the normal state. For us it was the energy gap. By concentrating our attention on how such a gap could arise in a degenerate system of interacting electrons, we discovered the new ground state. An ironic consequence of the full theory is that a variation exists in which the energy gap may disappear.

Among the consequences of our theory is the startling non-zero vacuum expectation value of two-electron creation or annihilation operators: what is now known as spontaneously broken gauge symmetry. It has become fashionable, in some circles, to call superconductivity a manifestation of broken symmetry and to assert that once gauge symmetry is broken the properties of superconductors follow. For example, Steve Weinberg writes in this volume that "A superconductor of any kind is nothing more or less than a material in which ... electromagnetic gauge is spontaneously broken ... All of the dramatic exact properties of superconductors ... follow from the assumption that electromagnetic gauge invariance is broken ... with no need to inquire into the mechanism by which the symmetry is broken." This is not — strictly speaking — true, since broken gauge symmetry might lead to molecule-like pairs and a Bose–Einstein rather than a BCS condensation. But, more important, such statements turn history on its head. Although we would not have used these words in 1957, we were aware that what is now called broken gauge symmetry would, under some circumstances (an energy gap or an order parameter), lead to many of the qualitative features of superconductivity. This had been well-known since the Gorter–Casimir two fluid model and the work of the Londons. The major problem was to show how an energy gap, an order parameter or "condensation in momentum space" could come about — to show how, in modern terms, gauge symmetry could be broken spontaneously. We demonstrated — I believe for the first time, and again using current language — how the gauge-invariant symmetry of the Lagrangian could be spontaneously broken due to interactions which were themselves gauge invariant. It was as though we set out to build a car and, along the way, invented the wheel.

It is true that in 1957 we never mentioned this very important symmetry breaking property of our theory, or that it was analogous to the symmetry breaking that occurs in the ferromagnetic transition. Though even I was aware of the properties of the ferromagnetic transition (not to mention Bob or John), we never explicitly pointed out the connection. Perhaps, as consolation for our oversight, we might remind ourselves that the great James Clerk Maxwell, to my knowledge, never mentioned that his equations were invariant under Lorentz or gauge transformations.

In the early sixties I returned for a while to the effort to prove that the pairing wave function was, in fact, a good solution of the many-body Hamiltonian, that the terms in the interaction we had omitted would not

Fig. 8. Bardeen, Cooper and Schrieffer (BCS). Courtesy: AIP Emilio Segrè Visual Archives.

change the qualitative results. These efforts, I must say, met with very limited success. I recall that after a seminar on this subject, a young man asked me, "Why are you interested in this problem? Everyone knows there is a pair condensation."

How quickly impossibly difficult had become obvious.

This article is based in part on an account I gave of the origins of the BCS Theory at the University of Illinois at Urbana-Champaign's BCS@50 Conference, as well as various talks celebrating the 50th anniversary of the BCS Theory.

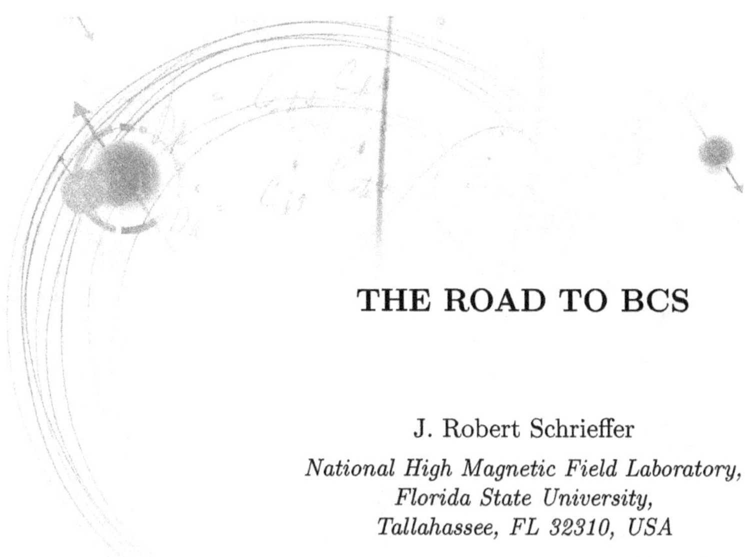

THE ROAD TO BCS

J. Robert Schrieffer

National High Magnetic Field Laboratory,
Florida State University,
Tallahassee, FL 32310, USA

This article is based on an interview with Bob Schrieffer. The questions are by Joan N. Warnow. The audio version is available on the AIP website at: http://www.aip.org/history/mod/superconductivity/02.html.

Q: Bob, Everybody Knows You Were Involved in What Turned Out to be the Explanation of Superconductivity. How Did that Come About?

I recall the second year I was at Urbana, that was '54, '55 ... and I had really hoped all the time when I went there that I would get to work on superconductivity .... I came and asked Bardeen for a real thesis problem, and I'm sure he had this in mind. And he said, "Come in and see me." Exactly how the discussion came, I don't quite recall, but he traditionally kept in his bottom drawer a list of problems. And I remember there were ten problems on this particular list and the tenth was superconductivity. He said, "Well, why don't you think about it?"

Q: Well, What Did You Do?

I went and chatted with Francis Low about this, because I felt that I could chat with him. He was very open. And I asked him what he thought about it, should I try this? He, I recall, asked, "How old are you?" and I told him. And he said, "Well, you can waste a year of your life and see how it goes."

The program really had been worked out in John's mind, I don't know, ten years before or what have you .... He had this thing so nailed down on every corner: he understood the experiments, he understood the general requirements of the theory. The whole thing was more or less jelled in his

Fig. 1. Robert Schrieffer.

mind. And then there was this stumbling block, and that was, you know, how to write down the wave function.

Q: Now, Just Where Were You Located — I Mean, Physically — at the University?

I was at what was called the "Institute for Retarded Study" — affectionately known — and it was on the third and a half floor of the building .... It was again a wonderful format. There were people all together in one large area. There were field theorists, there were nuclear physicists — all theorists came there. And if somehow you were able to move to the Institute for Retarded Study, you had made it. That was considered the greatest. And when there was a place open, a desk open, then everyone would sort of scramble around to see who could get in there .... There was a great blackboard, and there were always two or three people at the blackboard, arguing and discussing. So that was fun. They were all students there.

Fig. 2. University of Illinois at Urbana-Champaign Department of Physics circa 1957. The "Institute for Retarded Study" (indicated by arrow) was on the top floor on the right.

Q: That Seems Marvelous! Now — Let's See — John Bardeen and Leon Cooper Were in the Physics Building and You Were at the "Institute." How Did You All Interact?

Bardeen and Cooper shared an office, and that was very important. They could wheel around their chairs and talk to each other continually. But I would come down and say something to John and then Leon was there and we'd get together into a three way discussion — or if Leon was out, John and I would chat. But it was sort of a round robin where I think John and Leon probably didn't talk too much more than I chatted, but they were always together — and when they had a question, it would come up and they would discuss it. So, that was a very happy relationship which largely came about because there weren't enough offices for everyone. They were just squeezed in. I'd been working on the Brueckner theory, and Leon took very seriously the energy gap aspect and focused on that.

Leon's discovery, that a pair is unstable, suggested a direction we should move to understand which hunk of the Hamiltonian we should look at — which piece of the total interaction was important .... So we started thinking about how we could make a many-body theory which took into account many pairs at the same time .... We said, "OK, let's write down the problem

Fig. 3. Bardeen.

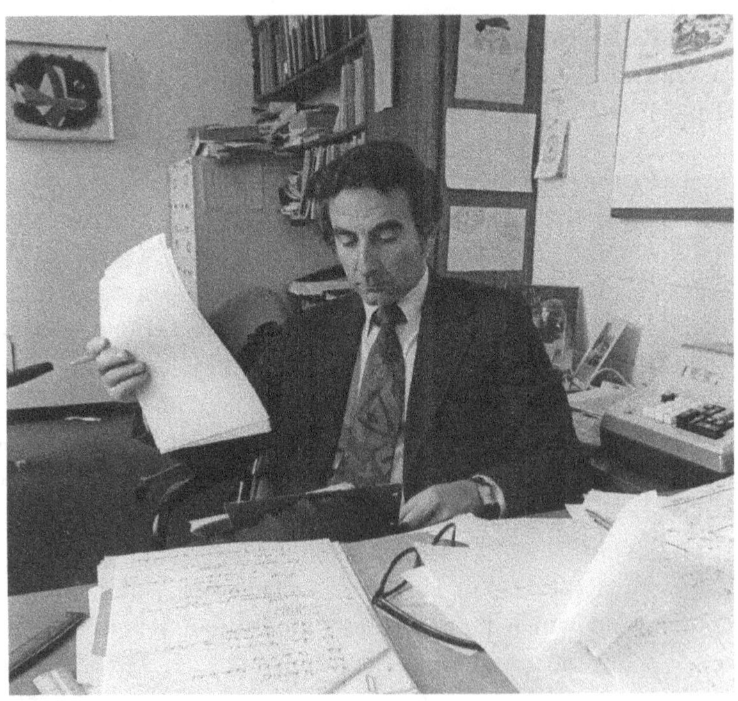

Fig. 4. Cooper.

where all electrons are treated, but we treat them in the second quantization formalism corresponding to pairs of zero momentum, and try and solve that problem" .... And the fact that we concentrated on the pairs of zero momentum, rather than trying to treat all momentum pairs simultaneously, was to a certain extent out of simplicity: do the simplest thing first, if it doesn't work, then go on to the next most complicated. That seemed obvious. But then the problem came, we couldn't even solve that simplest of problems.

We wrote down the Hamiltonian and looked at it and couldn't make any progress on it. We didn't know how to approach it — various ideas about variational methods, we thought — tried all sorts of approximate schemes .... It was very exciting, but it was very frustrating, needless to say. And it wasn't clear what was going to happen .... We also felt we were really hot. It was sort of this mixed feeling. We were really on the trail, and it was sort of almost a schizophrenia, you know — we're going to do it, and we're not.

Q: You All Knew You Were Looking for a Wave Function. Am I Right?

That's right.

Q: And This Went on for Several Months?

Yes.

Q: Weren't You All Feeling Somewhat Discouraged?

I personally had become somewhat discouraged at being able to make significant progress taking Leon's beautiful result and making a many-body theory out of it. I had started to quietly work on ferromagnetism. And I had mentioned to Bardeen that I thought perhaps I would like to change the thesis topic, because I didn't quite see that we were going anywhere.

Q: And What Was Bardeen's Reaction?

Well, I remember, just before John left for Stockholm, he said, "Give it another month or a month and a half, wait 'til I get back, and keep working, and maybe something will happen and then we can discuss it a little later."

In any event, we proceeded on, and then there was this meeting at Stevens and the New York meeting. And that was in the middle to end — I guess

Fig. 5. Bardeen at his desk.

the end of January. And, somehow, during that couple of days in New York — whether it was at the Stevens part of it or the APS meeting part, it was some time during that week — I started to think about the variational scheme associated with this Tomonaga wave function .... I wanted to use a variational scheme because there didn't seem to be any other scheme that was appropriate. One had to guess the answer, if you like, and then use some sort of a variational approach.

And I said, "Well, look, there are so many pairs around, that some sort of a statistical approach would be appropriate." That was sort of floating around in my mind — that there are so many pairs, they're overlapping — some sort of a statistical approach is appropriate. And, then the other one was this Tomonaga wave function — all sort of crystallized in saying, "Well, suppose I put an amplitude," I think I called it the square root of h "that the state is occupied, and a square root of 1 minus h, that it's unoccupied — the amplitude — and then let's product these over all states k." And that's just what Tomonaga did for that problem. I said, "Well, at least that allows the electrons to hop from — or the pairs to hop from — state to state, and that seemed like a reasonable guess." We were fooling around and that was one try.

So I set that down and then I looked at it, and I realized that that didn't conserve the number of electrons. It was a variable number of electrons and that had worried me, I remember. And so I decided, "Well, what I should do is multiply that wave function by a term involving e to minus

Fig. 6. Schrieffer as a graduate student.

the number of particles and — just like in the grand canonical ensemble in statistical mechanics — sort of extend that idea to the wave function in quantum mechanics." And I said, "Gee, I don't know if it's going to work, but it seems to me a reasonable approach. Let me try it."

So I guess it was on the subway, I scribbled down the wave function and I calculated the beginning of that expectation value and I realized that the algebra was very simple. I think it was somehow in the afternoon and that night at this friend's house I worked on it. And the next morning, as I recall, I did the variational calculation to get the gap equation and I solved the gap equation for the cutoff potential. It was just a few hours work. It was really exciting, it was fun. It was sort of beautiful and elegant — things worked out. It was all algebraic and I didn't have to go to a computer, or you know, there weren't terms I just threw away because I just couldn't handle them, but the whole thing was analytic. There were certain beauties, a simplicity, which — you might call it esthetics. I think that's — to my mind, that's a phony word, it implies more than that. But, it was sort of nice and I liked it.

Fig. 7. Cooper circa 1972.

Q: So, Now You Had It — That Wave Function. Did You Feel that Things Were Falling into Place?

The consequences, you know, weren't clear to me or weren't important.

Q: And You Were Also Very Young. 25?

Right. I'd seen a certain amount of physics. And I didn't have perspective. I didn't have any basis to judge right or wrong, so I assumed that this was, perhaps not wrong, but it was a beginning of another interesting idea. Like Leon had a very good idea and it worked to a certain extent. I assumed that this was perhaps a good idea and it would move one along, but this wasn't the solution to the problem. So people keep saying, "Nature is ultimately simple." I guess in some sense it depends upon the eyes. But this was so simple I didn't believe it. And that was sort of the other side.

It was an intuitive leap. And any intuitive leap, you have to justify it through a lot of tie points to experiment, and ultimately you hope there's a theoretical deductive way of getting there. But it was certainly far from that and I think even today we're not there. But I guess the main point I wanted to make was, I thought it was too simple and this just can't be the answer. It was exciting because it was fun to do, it worked out. And I met Leon then at the Champaign airport. Apparently he'd come in also from New York. Why we came there — I don't know — at the same time, but we appeared. I showed him this and he seemed very interested. He said, "Great, looks terrific," and "Let's go and talk to John in the morning" .... You know, we really worked as a team, and I can't imagine of any more cooperative feeling .... So the next morning we went and chatted with Bardeen and very quickly, as I recall, he looked at it and he said he thought that there was something really there. It was so fantastically exciting that we sort of worked 18 hours a day, because there was just so much to do .... So we were working on two levels. One was saying, "Isn't it fantastic? It's all breaking open." But on the other level we were having mechanical difficulties of doing all the calculations and working and checking, etc. So it was an intensive period of intellectual activity, but also just hard work.

Q: What Seemed to be the Biggest Problem at this Point?

We did the low temperature thermodynamics and we tried very hard to get the second order of phase transition — the jump in the specific heat — and that just didn't come out .... Then I think it was about three weeks to a month later — I'd been working very hard on it and Bardeen had — and I remember it was a Wednesday I thought I'd broken the problem. And I had made a slip of a sign .... But I think that Friday night, a distinguished Swedish scientist — Berelius, I believe — was visiting the Bardeens. And so, as I recall — again, my memory may not be accurate here — that John was somehow off on Cloud 7 that night. And there were long gaps in the conversation where John was staring into space, and the conversation was going on, but in a very strange sort of way. And it was clear that John was thinking hard about something. And what he was thinking about was how to get the second order phase transition and exactly how to write the wave function down. So the next morning — apparently that night he had cracked the problem and called up the next morning. He woke me up early in the morning ... and sort of said, "I've got it, I've got it. The whole thing's worked out." But I had to write the thesis. So I went off to New Hampshire

Fig. 8. Schrieffer, Bardeen and Cooper, 1972.

in — what? — the beginning or the middle of March, quietly getting the thing written out.

Then came — let's see — then Fred Seitz had called Eli Burstein (who was somehow in charge of, or at least related to, the March meeting of the American Physical Society here in Philadelphia, the solid state physics meeting) and said that a major break in the theory of superconductivity had occurred — or at least John believed so — and was it possible to have two post-deadline papers? So those were arranged, and John refused himself to come to speak about the theory because he wanted to make sure that the young people got the credit. And, you know, that's unbelievable, fantastic. So Leon was able to come and I got the word so late that I couldn't get on the plane to come. So he gave both papers together: he gave the one I was to give and the one he gave. This was a particularly interesting event not only because it was announcing the theory fairly early after its inception, if you like, and in a very raw form. It had only been — what? — a month and a half old, and the system responded to provide a possibility or a vehicle to get this out. But much more so, it was to my mind a remarkable insight into the personal character of John Bardeen, who, I think, in many ways has felt the intrinsic intellectual contribution he made through superconductivity in some ways superseded that which was made in the invention of the transistor.

Fig. 9. A last minute wave function adjustment. Heinrich Böll and Gerald Edelman oversee the proceedings.

He's said this on various occasions. And yet, for him, after struggling with the problem with a great amount of success, and having finally come to the pinnacle of achievement in his professional life, in a sense, steps aside for two young people — one of whom was a graduate student just sort of began in the field a year and a half before; the other wasn't from the field at all but was a postdoc brought in — and says, "OK, you go out and tell the world and I will stay here in Urbana." It's just beyond belief.

So, I think, to my mind, that's probably the most exciting message of the whole thing.

All photos courtesy AIP Emilio Segré Visual Archives except Fig. 2, courtesy Richard W. Vook and the University of Illinois at Urbana-Champaign with special thanks to Celia Elliot.

DEVELOPMENT OF CONCEPTS IN SUPERCONDUCTIVITY

John Bardeen

Department of Physics, University of Illinois,
1110 West Green Street, Urbana, IL 61801, USA

This is an excerpt from a talk that John Bardeen gave on the development of the theory of superconductivity in London, England on September 17, 1962 when he received the Fritz London award for his work developing the BCS theory of superconductivity. The talk was given at the *Eighth International Conference on Low Temperature Physics* at Queen Mary College in London and was reprinted in *Physics Today* in January of 1963.

My first attempt to construct a theory of superconductivity was made in the late '30s and was strongly influenced by London's picture .... I thought that it might be possible to extend the Bloch one-electron model to account for superconductivity. A periodic potential introduces Brillouin-zone boundaries in k space, with an energy gap at the boundary proportional to the Fourier coefficient of the potential. If one could produce zone boundaries at nearly all parts of the Fermi surface, one would get a lowering of energy of the electrons in states just inside the surface. No matter how complex the Fermi surface, it should be possible to accomplish this by introducing many small periodic distortions of the lattice corresponding to a very large complex unit cell. The attempt to construct a theory along these lines was not successful; various objections were raised. Further, more accurate estimates showed that this type of instability is unlikely to occur in real metals at low temperatures. The work was interrupted by the war and all that was published was an abstract of the talk. Much later, Fröhlich developed a far more complete theory for a one-dimensional model along similar lines.

After the war, my research interests turned to semiconductors, and it was not until May 1950, when I heard about the isotope effect from Serin, that I resumed work on superconductivity theory. Separated isotopes became available after the war, so that it was possible to determine whether or not there was a dependence of critical temperature on isotopic mass. Experiments were undertaken independently by Reynolds, Serin *et al.* at Rutgers and by Maxwell at the National Bureau of Standards, first on mercury. These showed, surprisingly at the time, that T_c varies inversely with the square root of the isotopic mass. The mass would not be an important parameter unless the motion of the ions is involved, which suggested that superconductivity must arise from some sort of interaction between the electrons and zero-point vibrations of the lattice. I attempted to develop a theory in which I suggested that the effect of the interaction would be such as to lower the energy of electrons near the Fermi surface, but as a result of dynamic interactions with the zero-point motion rather than by periodic lattice distortions.

About a week after I sent in a letter to the editor outlining these ideas. Fröhlich visited the Bell Telephone Laboratories where I was working at the time. He told me about his own work on a theory of superconductivity

Fig. 1. John Bardeen.

based on electron-phonon interactions, which he had done at Purdue in the spring of 1950. Fröhlich's work was done without knowledge of the isotope effect. He was greatly encouraged when he learned, just about the time he was ready to send his manuscript to *The Physical Review*, about this strong experimental confirmation of his approach. Although there were mathematical difficulties in both his approach and mine, primarily because of a use of perturbation theory in a region where it is not justified, we were both convinced that at last we were on the road to an explanation of superconductivity.

It did not take long to discover that the difficulties with these theories were basic and not easy to overcome. This was shown perhaps most clearly by a calculation of Schafroth, who contributed much to superconductivity theory. His untimely death cut short a promising career. Schafroth showed that a theory based on treating the electron-phonon interaction by perturbation theory could not account for the Meissner effect, even though the expansion is carried to arbitrarily high order.

These theories of Fröhlich and myself were based essentially on the self-energy of the electrons in the phonon field rather than on a true interaction between electrons. It became evident that all or nearly all of the self-energy is included in the normal state and is not much changed by the transition.

In Fig. 2, we have reproduced a slide made in 1955 to illustrate the status of the theory up to that time. The thermal properties gave evidence for an energy gap for excitation of a quasi-particle from the superconducting ground state. Further, I showed that if one assumed a reasonable energy-gap model, one could account for the Meissner effect, but with a nonlocal theory similar to that proposed by Pippard. The "derivation of the Meissner effect" which I gave at that time has been criticized by Buckingham and others on the grounds that the calculation is not gauge invariant, but I believe that the argument as given is essentially correct and is in accord with the present microscopic theory. The energy-gap model was the unifying theme of my review article which appeared in 1956 in *Handbuch der Physik*, Vol. XV. At that time there was no way to derive an energy-gap model from microscopic theory. While the Heisenberg–Koppe theory based on Coulomb interactions could be interpreted in terms of an energy-gap, it did not yield the isotope effect and was also subject to other difficulties. Thus, at that time, it appeared that the main problem of the microscopic theory was to show how electron-phonon interactions might yield an energy gap.

That electron-phonon interactions lead to an effective attractive interaction between electrons by exchange of virtual phonons was shown by Fröhlich

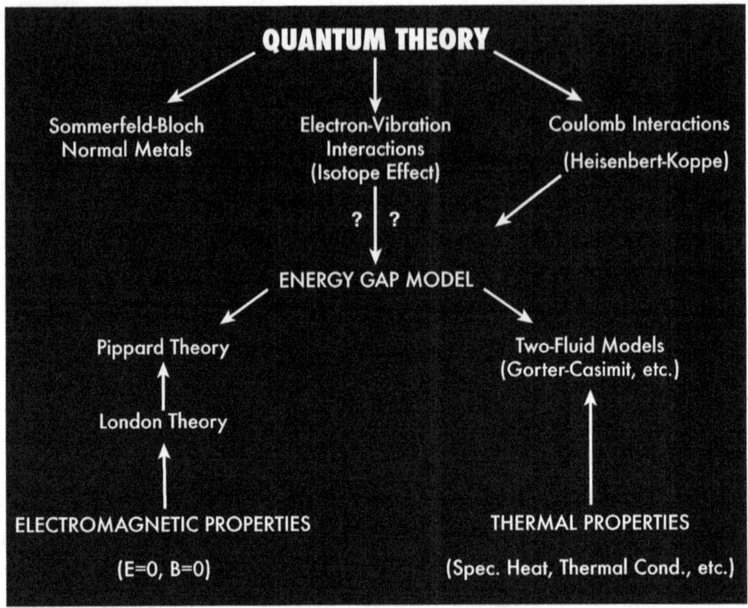

Fig. 2. Reproduction of a slide made in 1955 to illustrate the status of the theory at that time. Experiments on thermal properties gave evidence for an energy gap for excitation of electrons from the superconducting ground state. It was shown that a reasonable energy-gap model would most likely lead to Pippard's nonlocal modification of the phenomenological London equations to describe the electromagnetic properties. Thus, it seemed, the major problem was to see how an energy gap might follow from a microscopic theory based on interactions between electrons and phonons, as indicated by the isotope effect.

by use of field-theoretic techniques. His analysis was extended by Pines and myself to include Coulomb interactions. In second order, there is an effective interaction between the quasi-particle excitations of the normal state which is the sum of the attractive phonon-induced interaction and a screened Coulomb interaction. In the *Handbuch* article, I suggested that one should take the complete interaction, not just the diagonal self-energy terms, and use it as the basis for a theory of superconductivity.

The next major step was made by Cooper, who, following up this approach, showed that if there is an effective attractive interaction, a pair of quasi-particles above the Fermi sea will form a bound state no matter how weak the interaction. If the binding energy is of the order of kT_c, the size of the pair wave function is of the order of 10^{-5} to 10^{-4} cm. This calculation showed definitely that, in the presence of attractive interactions, the Fermi sea which describes the ground state of the normal metal is unstable against the formation of such bound pairs. However, one could not use this calculation immediately to construct a theory of superconductivity. If all of the electrons within $\sim kT_c$ of the Fermi surface form such bound pairs, the spacing between the pairs would be only $\sim 10^{-6}$ cm, a distance much smaller than the size of a pair. Because of the considerable overlap between the pairs, and because of the exclusion principle and required anti-symmetry of the wave functions, they cannot be regarded as moving independently. Thus, the picture proposed earlier by Schafroth (1955), and developed more completely in cooperation with Butler and Blatt of electron pairs as "localized entities (pseudo-molecules) whose center-of-gravity motion is essentially undisturbed", and which at low temperatures undergo an Einstein–Bose condensation is not valid. New methods were required to construct a theory of superconductivity, and this was first accomplished by the joint efforts of Cooper, Schrieffer, and myself. While the theory can be and has been developed by use of a variety of mathematical techniques, I believe that the variational method used in our original publications gives as good a picture as any of the ground-state wave functions and of the quasi-particle excitation spectrum with a gap.

One may describe the low-lying configurations for the normal phase of a metal by specifying the occupancy in k-space of the quasi-particles above the Fermi sea and of unoccupied states or holes below the sea. In accordance with the Landau Fermi-liquid model, the energy of one quasi-particle may depend on the distribution of the other quasi-particles. These quasi-particle configurations are not exact solutions of the Hamiltonian when Coulomb and phonon interactions are included, but are reasonably well defined if the excitation energies are not too high. The configurations are presumed to include correlation energies and quasi-particle self-energies characteristic of the normal phase. Superconductivity arises from residual attractive interactions between the normal quasi-particles.

Cooper, Schrieffer, and I took for the variational wave-function ground state of a superconductor a linear combination of normal configurations in which the quasi-particle states are occupied in pairs $(k_1 \uparrow, k_2 \downarrow)$ of opposite spin and the same total momentum, $k_1 + k_2 = q$, common to all pairs. In any configuration, the two states of a pair are either both occupied or both empty. Values of q different from zero describe current flow in the ground state, that for $q = 0$ for zero current has the lowest energy. We also worked out a quasi-particle excitation spectrum for a superconductor in one-to-one correspondence with that for a normal metal with a temperature-dependent energy gap for excitation of particles from the superconducting ground state.

A superconductor differs from a semiconductor in that the gap in the former is relative to the Fermi surface and is not fixed in k-space. The entire system with interactions can be displaced in momentum space to give a net current flow.[a] If v_s is the velocity of flow, the mass of the flow at $T = 0°$ K is ρv_s where $\rho = nm$ is the density of the electrons. At a finite temperature, quasi-particle excitations will reduce the current, but when a local equilibrium is established corresponding to a given v_s, a net flow $\rho_s v_s$ will remain. This defines the density of the superfluid component of the two-fluid model, ρ_s. With increasing temperature, ρ_s decreases from ρ at $T = 0°$ K to zero as $T \to T_c$. When the Fermi sea of a normal metal is displaced in momentum space, quasi-particle excitations soon reduce the current to zero, so that $\rho_s = 0$. A superfluid is characterized by a value of ρ_s different from zero. These considerations are analogous to those Landau used to account for the superfluidity of liquid helium.

The theory has been applied to a wide variety of properties such as specific heats, electromagnetic properties, thermal conductivity, ultrasonic attenuation, nuclear spin relaxation times, the Knight shift and electron spin paramagnetism, electron tunneling, critical fields and currents, boundary effects, and other problems. In nearly all cases excellent agreement between theory and experiment is found when the parameters of the theory are evaluated empirically. Difficulties associated with thermal conductivity for phonon scattering and with the Knight shift appear to be on the way to resolution through a combination of experimental and theoretical work.

The metastability of persistent currents does not occur because of lack of scattering. Quasi-particles are readily scattered, but such scattering does not change the common momentum of the pairs and thus v_s. It only results in fluctuations about the current corresponding to local quasi-particle equilibrium, $\rho_s v_s$.

An unexpected feature of the theory is the marked effect of coherence on the matrix elements for scattering of quasi-particles in a superconductor. It accounts for phenomena which would be inexplicable on the basis of any simple two-fluid model. In the early spring of 1957, when Cooper, Schrieffer, and I were first working out the details of the theory, Hebel and Slichter, also working at Illinois, made the first measurements of nuclear spin relaxation times in a superconductor by use of ingenious experimental techniques.

[a]To simplify the argument, we omit effects of the magnetic field on current paths, which is not valid except for flow in very thin films.

They found, surprisingly, a marked *decrease* in the relaxation time as the temperature dropped below T_c in the superconducting state, followed by an increase at still lower temperatures. Relaxation of the nuclear spins occurs from interaction with the conduction electrons in which there is a spin flip of the electron as well as the nucleus. The experiments indicated a *larger* interaction in the superconducting than in the normal state, even though specific heats and other experiments showed that there must be a marked decrease in the number of quasi-particle excitations as the temperature drops below T_c. For example, the attenuation of ultrasonic waves drops abruptly at T_c. These apparently contradictory experiments are accounted for by coherence effects. In calculating matrix elements for quasi-particle transitions in a superconductor, we found that it is necessary to add coherently the contributions from electrons of opposite spin and momentum in the various normal configurations which make up the quasi-particle states of a superconductor. For the case of a spin flip, the two contributions to the matrix element add constructively, and the larger transition probability in the superconducting state is a result of the increased density of states in energy. For an ordinary interaction such as occurs in ultrasonic attenuation, the contributions add destructively, giving a drop with an infinite slope at T_c as observed. The experimental check of these very marked effects of coherence provides one of the best confirmations of pairing in the wave functions.

In working out the properties of our simplified model and comparing with experimental results on real metals, we were continually amazed at the excellent agreement obtained. If there was serious discrepancy, it was usually found on rechecking that an error was made in the calculations. Everything fitted together neatly like the pieces of a jigsaw puzzle. Accordingly, we were unprepared for the skepticism with which the theory was greeted in some quarters. Those most skeptical had generally worked long and hard on superconductivity theory themselves, and had their own ideas of what the theory should be like. Most of the criticism centered on our derivation of the Meissner effect, because it was not carried out in a manifestly gauge-invariant manner. While our derivation is not mathematically rigorous, we gave what we believe are good physical arguments for our use of a transverse gauge, and our procedure has been justified in subsequent work. As we have seen, our model is exactly of the sort which should account for superconductivity according to London's ideas.

At the opposite extreme were some who felt that the explanation of superconductivity would mark the end of what had long been a puzzling and challenging scientific problem. On the contrary, the theory has stimulated much new experimental and theoretical work; it has helped put new life into the field. While some questions have been answered, many others have been raised as we probe more deeply, and plenty of problems remain, as is evident from the papers submitted to this meeting.

Since the original publications, the mathematical formulation of the theory has been developed considerably. Several different mathematical formulations have been given which have improved the rigor and have extended the theory so as to apply to a wider variety of problems. Particular mention should be made of the work of Bogoliubov and co-workers, who, along with Valatin, introduced the now famous transformation to quasi-particle variables, gave a much improved treatment of Coulomb interactions, provided a treatment of collective excitations, and made other noteworthy contributions. Independently of this work, Anderson gave a derivation based on an equation-of-motion approach which introduced collective excitations and allowed a manifestly gauge-invariant treatment of the Meissner effect. The approaches of Bogoliubov and of Anderson were extended by Rickayzen to give probably the most complete derivation of the Meissner effect to date. Green's-function methods, borrowed from quantum field theory, have been used widely and with great success, following the initial work of Gor'kov, Martin and Schwinger, Kadanoff, and others. Gor'kov, in particular, has used these methods to solve several difficult problems in superconductivity theory. Fröhlich was one of the pioneers in the use of field-theoretic methods in solid-state problems ...

We have seen that the development of our understanding of superconductivity has resulted from a close interplay of theory and experiment. Physical insight into the nature of the superconducting state gained from a study of the experimental findings has been essential to make progress in the theory. Increased theoretical understanding has suggested new experiments, new paths to explore, and has helped to understand better such seemingly unrelated fields as nuclear structure and elementary particles.

FAILED THEORIES OF SUPERCONDUCTIVITY

Jörg Schmalian

Department of Physics and Astronomy, and Ames Laboratory, Iowa State University, Ames, IA 50011, USA

Almost half a century passed between the discovery of superconductivity by Kamerlingh Onnes and the theoretical explanation of the phenomenon by Bardeen, Cooper and Schrieffer. During the intervening years the brightest minds in theoretical physics tried and failed to develop a microscopic understanding of the effect. A summary of some of those unsuccessful attempts to understand superconductivity not only demonstrates the extraordinary achievement made by formulating the BCS theory, but also illustrates that mistakes are a natural and healthy part of scientific discourse, and that inapplicable, even incorrect theories can turn out to be interesting and inspiring.

The microscopic theory of superconductivity was developed by Bardeen, Cooper and Schrieffer (BCS).[1] It was published in 1957, 46 years after the original discovery of the phenomenon by Kamerlingh Onnes.[2] During those intervening years numerous scientists tried and failed to develop an understanding of superconductivity, among them the brightest minds in theoretical physics. Before 1957 other correct descriptions of the phenomenon superconductivity were developed. Those include the work by Cornelius Gorter and Hendrik Casimir in 1934,[3] and most notably the phenomenological theories of Heinz and Fritz London in 1935,[4] and Vitaly Ginzburg and Lev Landau in 1950.[5] The triumph of the BCS theory was, however, that it gave an explanation that started from the basic interactions of electrons and atoms in solids, i.e. it was a microscopic theory, and, at the same time, explained essentially all of the complex properties of conventional superconductors. For a number of observables, the agreement between theory and experiment is of unparalleled precision in the area of interacting many body systems.

When discussing failed attempts to understand superconductivity, we must keep in mind that they are a natural and healthy part of scientific discourse. They are an important part of the process of finding the right answers. These notes are not written to taunt those who tried and did not succeed. On the contrary, it is the greatness that comes with names like Joseph John Thompson, Albert Einstein, Niels Bohr, Léon Brillouin, Ralph Kronig, Felix Bloch, Lev Landau, Werner Heisenberg, Max Born, and Richard Feynman that demonstrates the dimension of the endeavor undertaken by John Bardeen, Leon N. Cooper and J. Robert Schrieffer. Formulating the theory of superconductivity was one of the hardest problems in physics of the 20th century.

In light of the topic of this article, it is not without a sense of irony that the original discovery by Kamerlingh Onnes seems to have been motivated, at least in part, by an incorrect theory itself, proposed by another highly influential thinker. Lord Kelvin had argued, using the law of corresponding states, that the resistivity of all substances diverges at sufficiently low temperatures.[6] Kamerlingh Onnes was well aware of Kelvin's work, and while he might have been skeptical about it, the proposal underscored the importance to investigate the $T \to 0$ limit of the electric resistivity. For a discussion of Kelvin's role in motivating Kamerlingh Onnes' experiment, see Ref. 7.

We know now that superconductivity is a macroscopic quantum effect. Even though many elements of the new quantum theory were developed by Planck, Bohr, Sommerfeld, Einstein and others starting in 1900, it is clear in hindsight that it was hopeless to explain superconductivity before the formulation of quantum mechanics by Heisenberg[8] and Schrödinger[9] during 1925–1926. This was crisply articulated by Albert Einstein when he stated during the 40th anniversary of Kamerlingh Onnes' professorship in Leiden in 1922: "*With our far-reaching ignorance of the quantum mechanics of composite systems we are very far from being able to compose a theory out of these vague ideas.*"[10] The vague ideas resulted out of his own efforts to understand superconductivity using the concept of "*molecular conduction chains*" that carry supercurrents. What Einstein had in mind resembles to some extent the soliton motion that does indeed occur in one-dimensional conductors.[11,12] In his view "*supercurrents are carried through closed molecular chains where electrons undergo continuous cyclic exchanges.*" Even though no further justification for the existence of such conduction paths was given, the approach was based on the view that superconductivity is deeply rooted in the specific chemistry of a given material and based on the existence of a state that connects the outer electrons of an atom or molecule with those of its neighbors. Einstein also suggested an experiment to falsify his theory: bringing two different superconducting materials in contact, no supercurrent should go from one superconductor to the other, as the molecular conducting chains would be interrupted at the interface. Again, it was Kamerlingh Onnes who performed the key experiment and showed that supercurrents pass through the interface between lead and tin, demonstrating that Einstein's theory was incorrect, as was stated in a *post scriptum* of Einstein's original paper.[10] For historical accounts of Einstein's work on superconductivity and his impact on condensed matter physics in general, see Refs. 13 and 14, respectively.

While Einstein's concept of molecular conduction chains did not turn out to be the right one, he was correct in insisting that a theory of superconductivity cannot be based on the concept of non-interacting electrons in solids. It is also remarkable to read the introductory paragraph of his paper on superconductivity, where he states that "*...nature is a merciless and harsh judge of the theorist's work. In judging a theory, it never rules 'Yes' in best case says 'Maybe', but mostly 'No'. In the end, every theory will see a 'No'.*" Just like Einstein, most authors of failed theories of superconductivity were well aware that their proposals were sketchy and their insights of

Fig. 1. Einstein, Bohr, Kronig, Landau, Bloch and Brillouin made proposals for microscopic theories of superconductivity prior to the groundbreaking experiment by Meissner and Ochsenfeld in 1934.

preliminary character at best. Einstein's theory should not be considered as a singular intuition, on the spur of the moment. Joseph John Thompson, who discovered the electron in his famous cathode ray experiment,[15] had already made a proposal to explain superconductivity in terms of fluctuating electric dipole chains in 1915.[16] In 1921, Kamerlingh Onnes proposed a model of superconducting filaments.[17] Below the superconducting transition temperature, conduction electrons would "*slide, by a sort of association, through the metallic lattice without hitting the atoms.*" In judging these early ideas about superconductivity, one must appreciate that even the normal state transport properties of metals were only poorly understood. In fact the bulk of Einstein's paper is concerned with a discussion of the normal state electric and heat conductivities.

After the formulation of quantum mechanics, motivated to a large extent by the properties of single electrons and atomic spectra, it soon became clear that this new theory explained phenomena that were much more complex. Heisenberg's important contributions to the theory of magnetism[18] and Felix Bloch's lasting work on the theory of electrons in crystals[19] were already published during 1928. Soon after, an understanding of ordinary electrical conductors and of the peculiar thermal properties of solids was developed using the new quantum theory (see Ref. 20 for an historical account). The

Fig. 2. Between 1941 and the formulation of the BCS theory, unsuccessful attempts to formulate microscopic theories of superconductivity were made by Bardeen, Heisenberg, London, Born, Fröhlich and Feynman.

main tools needed to formulate the theory of superconductivity were now or would soon become available. Given the swift success in other areas of solid state physics, the inability to formulate a theory of superconductivity demonstrated the need for conceptually new insights that were required to solve this problem.

Of the "failures" to explain superconductivity during the early post quantum mechanics days, two theories are noteworthy for their elegance and the distinguished participants involved. Those are the spontaneous current approach independently proposed by Felix Bloch and Lev Landau[21] and the theory of coherent electron-lattice motion by Niels Bohr and, around the same time, by Ralph de Laer Kronig.[22,23] It may be said that the authors of these ideas worked on wrong theories, but certainly not that they proposed uninteresting ideas. Both concepts had some lasting impact or relevance.

In the Drude formula for electric transport[24] the conductivity is given as,

$$\sigma = \frac{e^2 n}{m}\tau, \tag{1}$$

with electron charge e, mass m, density n and scattering time τ, respectively. This result led to the frequently proposed view that superconductors are perfect conductors, i.e. $\sigma \to \infty$, due to a vanishing scattering rate τ^{-1}. In his 1933 paper, Landau gave a very compelling argument that superconductors should not be considered as perfect conductors.[21] His reasoning was based upon the fact that the resistivity right above T_c is still finite, i.e. τ is finite and electrons must be undergoing scattering events. Landau stressed that a mechanism based on the notion of a perfect conductor requires that τ^{-1} jumps discontinuously from a finite value to zero. He argued that it is highly implausible that all interactions are suddenly switched off at the transition temperature. The remarkable aspect of this argument is that it was almost certainly made without knowledge of the crucial experiment by Meissner and Ochsenfeld[25] that ruled against the notion of superconductors as perfect conductors. In addition, Landau's formulation of the theory in 1933, *albeit* wrong, contained the first seeds of the eventually correct Ginzburg–Landau theory of superconductivity.[5] Having rejected the idea of superconductivity due to an infinite conductivity, he analyzed the possibility of equilibrium states with finite current. Landau proposed to expand the free energy of the system in powers of the current $\mathbf{j}$:[21]

$$F(\mathbf{j}) = F(\mathbf{j} = \mathbf{0}) + \frac{a}{2}\mathbf{j}^2 + \frac{b}{4}\mathbf{j}^4 . \tag{2}$$

No odd terms occurred in $F(\mathbf{j})$ as the energy should not depend on the current direction. The equilibrium current $\langle \mathbf{j} \rangle$ is given by the value that minimizes $F(\mathbf{j})$. Landau argued that $b > 0$, to ensure a continuous transition, and that the coefficient of the quadratic term changes sign at the transition temperature $a \propto T - T_c$, to allow for a finite equilibrium current, $\langle \mathbf{j} \rangle \neq \mathbf{0}$, below T_c. He pointed out that the resulting heat capacity jump agrees with experiment, but cautioned that the temperature variation of the current, $|\mathbf{j}| \propto (T_c - T)^{1/2}$, seems to disagree with observations. The approach was inspired by the theory of ferromagnetism and already used the much more general concept of an order parameter that discriminates between different states of matter. Landau's first order parameter expansion of the free energy of antiferromagnets was published in 1933,[26] after his manuscript on superconductivity. The widely known Landau theory of phase transition was only published in 1937.[27] It is amazing that these lasting developments seem to have been inspired by an incorrect theory of superconductivity. Landau expansions have since been successfully used to describe the critical point of water, liquid crystals, the properties of complex magnets, the inflationary cosmic evolution right after the big bang, and many other systems.

Felix Bloch did not publish his ideas about coupled spontaneous currents, which were, just like Landau's theory, motivated by the theory of ferromagnetism. Yet, through his efforts he became highly knowledgeable about the status of the experimental observations in the field. During the early 1930's

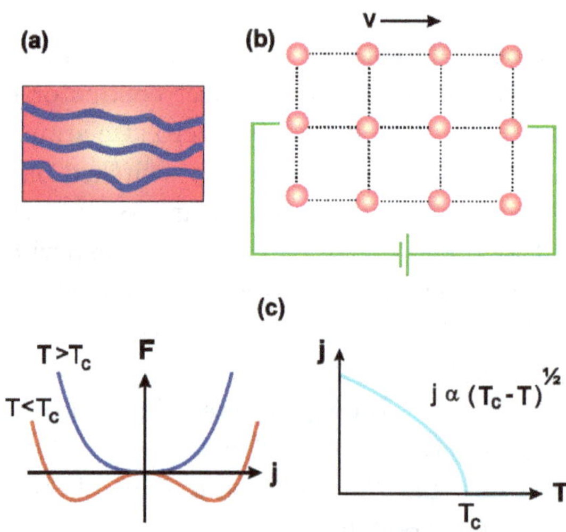

Fig. 3. (a) Sketch of Einstein's molecular conduction chains. (b) Kronig's electron crystal that was supposed to slide as a whole in an infinitesimal external electric field. (c) Landau's expansion of the free energy with respect to the equilibrium current.

Bloch's infamous, somewhat cynical, second theorem, that every theory of superconductivity can be disproved, was frequently cited among theorists.[20] It reflected the degree of despair that must have existed in the community. His first theorem, which was, in contrast to the second one, perfectly serious, was concerned with the energy of current carrying states. The theorem contained the proof that Bloch's own theory of coupled spontaneous currents was in fact false. David Bohm gave a summary of Bloch's first theorem in 1949:[28] suppose a finite momentum $\langle \mathbf{P} \rangle = \langle \Psi | \mathbf{P} | \Psi \rangle \neq \mathbf{0}$ in the ground state Ψ of a purely electronic system, which leads to a finite current $\langle \mathbf{j} \rangle = e \langle \mathbf{P} \rangle / m$. Let the Hamiltonian

$$H = \sum_i \left(-\frac{\hbar^2 \nabla^2}{2m} + U(\mathbf{r}_i) \right) + \sum_{i<j} V(\mathbf{r}_i - \mathbf{r}_j) \tag{3}$$

consist of the kinetic energy, the potential $U(\mathbf{r}_i)$ due to the ion lattice and the electron–electron interaction $V(\mathbf{r}_i - \mathbf{r}_j)$. One then finds that the wave function $\Phi = \exp(i\delta\mathbf{p} \cdot \sum_i \mathbf{r}_i / \hbar)\Psi$ has a lower energy than Ψ if the variational parameter $\delta\mathbf{p}$ points opposite to $\langle \mathbf{P} \rangle$. Thus, Ψ cannot be the ground state unless $\langle \mathbf{j} \rangle = \mathbf{0}$ for the purely electronic problem, Eq. (3). This result clearly invalidated Landau's and Bloch's proposals.

The second idea proposed in 1932 by Bohr and Kronig was that superconductivity would result from the coherent quantum motion of a lattice of electrons. Given Bloch's stature in the field, theorists like Niels Bohr where eager to discuss their own ideas with him. In fact Bohr, whose theory for superconductivity was already accepted for publication in the July 1932 issue of the journal "Die Naturwissenschaften", withdrew his article in the proof stage, because of Bloch's criticism (see Ref. 20). Kronig was most likely also aware of Bloch's opinion when he published his ideas.[22] Only months after the first publication he responded to the criticism made by Bohr and Bloch in a second manuscript.[23] It is tempting to speculate that his decision to publish and later defend his theory was influenced by an earlier experience: in 1925 Kronig proposed that the electron carries spin, i.e. possesses an internal angular momentum. Wolfgang Pauli's response to this idea was that it was interesting but incorrect, which discouraged Kronig from publishing it. The proposal for the electron spin was made shortly thereafter by Samuel Goudsmit and George Uhlenbeck.[29] Kronig might have concluded that it is not always wise to follow the advice of an established and respected expert.

In his theory of superconductivity, Kronig considered the regime where the kinetic energy of electrons is sufficiently small such that they crystallize to minimize their Coulomb energy. Given the small electron mass, he estimated that the vibrational frequencies of this electron crystal were much higher than those of crystal lattice vibrations, amounting to its strong rigidity. In the presence of an electric field, this electron crystal was then supposed to slide as a whole, as the rigidity of the electron crystal would suppress scattering by individual electrons. The superconducting transition temperature would correspond to the melting point of the electron crystal. Furthermore, the suppression of superconductivity with magnetic field could, in Kronig's view, be explained due to the interference of field induced circular orbits with the crystalline state. Clearly, the constructed approach was a sophisticated version of the idea of a perfect conductor. The flaw of the approach was that it overestimated quantum zero-point fluctuations that were supposed to prevent the crystal from getting pinned. Still, it is noteworthy that the proposal used the rigidity of a macroscopic state — the electron crystal — to avoid single electron scattering. As Kronig pointed out, the notion of an electron lattice and its potential importance for electron transport and superconductivity were already voiced by Frederick Lindemann in 1915,[30] and refined by J. J. Thompson in 1922.[31] It is nevertheless remarkable that Kronig used the concept of an electron crystal within a quantum mechanical approach two years prior to Eugene Wigner's pioneering work on the subject.[32]

Léon Brillouin, who made key contributions to quantum mechanics, solid state physics, and information theory, proposed his own theory of superconductivity during the spring of 1933.[33] He assumed an electronic band structure $\varepsilon(\mathbf{p})$ with a local maximum at some intermediate momentum $\mathbf{p}_0$. $\mathbf{p}_0$ is neither close to $\mathbf{p} = \mathbf{0}$ nor, to use the contemporary terminology, at the Brillouin zone boundary. He showed that the equilibrium current of such a system vanishes, but that non-equilibrium populations $n(\mathbf{p})$ of momentum states, may give rise to a net current. He then argued that due to the local maximum in $\varepsilon(\mathbf{p})$ such non-equilibrium states have to overcome a barrier to relax towards equilibrium, leading to metastable currents. At higher temperatures, equilibration becomes possible, causing those supercurrents to relax to zero. Brillouin realized that his scenario naturally implied a critical current. Since Brillouin argued that superconductivity was a metastable state, his proposal was at least not in conflict with Bloch's first theorem. In 1934, Gorter and Casimir gave strong evidence for the fact that superconductivity is an equilibrium phenomenon,[3] which ruled out Brillouin's approach.

Before discussing further examples of "failures", the breakthrough experiment by Walter Meissner and Robert Ochsenfeld,[25] followed by the pioneering theory of Heinz and Fritz London,[4] must be mentioned. Meissner and Ochsenfeld demonstrated that the magnetic flux is expelled from a superconductor, regardless of its state in the distant past. The phenomenological theory that naturally accounted for this finding was soon after proposed by the London brothers in their 1934 paper in the *Proceedings of the Royal Society*.[4] To appreciate some of the later proposals for microscopic theories of superconductivity, we deviate from the theme of this manuscript and briefly summarize this correct and lasting phenomenological theory: London and London discussed that in a perfect conductor, Ohm's law, $\mathbf{j} = \sigma \mathbf{E}$, is replaced by the acceleration equation

$$\frac{\partial \mathbf{j}}{\partial t} = \frac{ne^2}{m} \mathbf{E}, \tag{4}$$

appropriate for the frictionless motion of a charged particle in an electric field $\mathbf{E}$. Combining this relation with Faraday's law yields

$$\nabla \times \frac{\partial \mathbf{j}}{\partial t} = \frac{ne^2}{mc} \frac{\partial \mathbf{B}}{\partial t}, \tag{5}$$

with magnetic induction $\mathbf{B}$. From Ampere's law follows then that the magnetic field decays at a length scale λ_L to its initial value $\mathbf{B}_0$ where $\lambda_L^{-2} = 4\pi ne^2/(mc^2)$. London and London concluded that in case of a perfect conductor "*the field* $\mathbf{B}_0$ *is to be regarded as 'frozen in' and represents a permanent memory of the field which existed when the metal was last cooled below the transition temperature.*" If correct, cooling from the normal state to the superconducting state in an external field would keep the magnetic induction at its finite high temperature value $\mathbf{B}_0 \neq 0$, directly contradicting the Meissner–Ochsenfeld experiment. In addition, it would imply that superconductors are not in thermodynamic equilibrium as their state depends on the system's history in the distant past. These facts led London and London to abandon the idea of a perfect conductor. They further realized that the history dependence of the field value is a direct consequence of the presence of time derivatives on both sides of Eq. (5), while the desired decay of the field on the length scale λ_L follows from the curl operator on the left-hand side of Eq. (5). Instead of using Eq. (5), they simply proposed to drop the time derivatives, leading to a new material's equation for superconductors:

$$\nabla \times \mathbf{j} = \frac{ne^2}{mc} \mathbf{B}. \tag{6}$$

This phenomenological London equation guarantees that the field decays to zero on the length λ_L, referred to as the London penetration depth. The authors concluded that "*in contrast to the customary conception that in a supraconductor a current may persist without being maintained by an electric or magnetic field, the current is characterized as a kind of diamagnetic volume current, the existence of which is necessarily dependent upon the presence of a magnetic field.*" In other words, London and London recognized that superconductors are perfect diamagnets. With the Nazi regime coming into power in 1933 a sudden shift of the research efforts from Germany to the United States and England took place, changing the priorities of numerous researchers. While the theory of the London brothers is a beautiful phenomenological account of the Meissner effect, a microscopic theory was not immediately inspired by this key experiment.

After the Second World War, Werner Heisenberg, one of the creators of modern quantum mechanics, took serious interest in formulating a theory of superconductivity.[34] His theory was based on the assumption that strong Coulomb interactions dramatically alter the character of electrons. Instead of forming plane waves, electrons near the Fermi energy would localize. The model was treated using a variational single electron wave function. Heisenberg realized that a crucial challenge for any theory of superconductivity was to derive the tiny observed energy advantage of the superconducting state, compared to other possible states. His search for new bound states in the vicinity of the Fermi energy was quite original and clearly pointed in the right direction. His confidence regarding his own results is nevertheless impressive: "*If ... condensation takes place through the Coulomb forces, one can scarcely think of any other mechanism which reduces ordinary Coulomb energies to so small values.*"[35] The shortcoming in the approach was that Heisenberg did not accept that the Meissner effect is at odds with the infinite conductivity approach to superconductivity. He stated: "*The essential difference from several more recent attempts is the assumption that the perfect conductivity rather than the diamagnetism is the primary feature of the phenomenon.*"[35] Heisenberg claimed to have derived the London equation from his theory. Following his calculation one finds that the derivation is based on the implicit assumption that the initial magnetic field value vanishes, i.e. $\mathbf{B}_0 = \mathbf{0}$.

Max Born, in collaboration with Kai Chia Cheng, proposed a theory of superconductivity in 1948.[36] One goal of the theory was to explain why some metals are superconducting while others are not. The authors gave empirical evidence for the fact that superconductors tend to have a Fermi surface located in the near vicinity of the Brillouin zone boundary, which suggested to them that ionic forces, i.e. those due to $U(\mathbf{r}_i)$ in Eq. (3) would play an important role. In their analysis, which is essentially a density functional theory, they found that the electron–electron interaction causes changes in the occupation of momentum states near the Brillouin zone boundary.

Apparently unaware of the fact that Bloch's first theorem is valid even in the presence of the ionic potential, they claimed, without giving further details, to have found asymmetric occupations with a net momentum and therefore a spontaneous current. Born and Cheng were correct to point out that interactions affect the distribution in momentum space, and become important for systems with large portions of the Fermi surface close to the Brillouin zone boundary. Yet, a correct analysis of their own formalism should have led them to the conclusion that asymmetric occupations do not correspond to the lowest energy state.

Fritz London strongly disagreed with Heisenberg and Born. Being challenged by those ideas, he formulated his own microscopic theory of superconductivity.[37] In a 1948 paper, he first demonstrated that Heisenberg's theory cannot yield superconductivity as a matter of principle. He then pointed out that Heisenberg missed the leading term in his treatment of the Coulomb interaction, which, in London's view, is the exchange interaction, discovered by Heisenberg himself many years before.[18] London then proposed that this exchange interaction, when it favors ferromagnetism, is responsible for superconductivity as it can lead to an "*attraction in momentum space.*" In materials known then, superconductivity is not caused by this effect, but rather by vibrations of the crystal lattice, i.e. sound excitations. On the other hand, the superfluid state of ^{3}He,[38–40] and superconductivity in a number of strongly correlated electron systems have been discussed in terms of an exchange of collective degrees of freedoms, such as ferromagnetic, antiferromagnetic or current fluctuations,[40–44] that bear some resemblance to what London had in mind. Of course, our understanding of these materials is based on the BCS theory. London's ideas, while interesting, remained vague. The most remarkable aspect of this 1948 paper by Fritz London was however that it emphasized clearly that a superconductor is a macroscopic object in a coherent quantum state: "*The superconductor as a pure quantum mechanism of macroscopic scale.*" This view was also one of the key elements of the elegant phenomenological theory by Ginzburg and Landau that was proposed soon thereafter.[5]

If we talk about the creators of unsuccessful theories of superconductivity we should not ignore that Bardeen himself belonged, at least temporarily, to this honorable club. One attempt to solve the problem, briefly mentioned in an abstract in Physical Review, goes back to 1941 and is based on the idea that electrons couple to small periodic distortions of the lattice.[45] In his scenario, the electron-lattice interaction would cause small lattice distortions with large unit cells containing approximately 10^6 atoms, leading to 10^6 bands in a tiny Brillouin zone. Bardeen estimated that approximately 10^4 of those bands were located in the vicinity of $\pm k_B T_c$ of the Fermi energy. Since some bands would have a very small mass, he expected them to contribute to a large diamagnetic response. He further estimated that a fraction of approximately 10^{-6} of the electrons would participate in this diamagnetic shielding. As the reasoning is based upon lattice distortions, Bardeen concluded that the electron-lattice coupling of good superconductors should be large, a conclusion that also followed from the BCS theory, yet for very different reasons.

Another key experiment that elucidated the origin of superconductivity was the measurement of the isotope effect at Rutgers University in the group of Bernard Serin[46] and by Emanuel Maxwell[47] at the National Bureau of Standards, Washington, D.C. in 1950. These experiments found a change in the transition temperature upon changing the ion mass via isotope substitution, and demonstrated that vibrations of the crystalline lattice were closely tied to the emergence of superconductivity. In fact, Herbert Fröhlich predicted the isotope effect theoretically.[48] In 1950 Bardeen and, independently, Fröhlich worked on the problem and came to the conclusion that vibrations of the crystalline lattice lead to a net attraction between electrons and are a likely cause of superconductivity.[48,49] This attraction is present even if one includes the electron–electron Coulomb interaction as shown by Bardeen and Pines.[50] For an historical account of this period of time, see Ref. 51. Starting in 1950 one should not refer to Bardeen's or Fröhlich's ideas as failed, rather as incomplete. The link between the established attraction between electrons due to sound excitations and superconductivity was still unresolved. For example, in 1953 Fröhlich developed a beautiful theory for one-dimensional system with electron lattice coupling[52] and obtained in his calculation a gap in the electron spectrum, as was in fact seen in a number of experiments. The theory is a perfectly adequate and very interesting starting point for the description of sliding charge density waves. We know now that the one-dimensionality makes the model somewhat special and that the gap is incomplete in three dimensions. Most importantly, the theory did not explain superconductivity and did not account for the Meissner effect. Still, there is strong evidence that sliding charge density waves, as proposed by Fröhlich to explain superconductivity, have in fact been observed in quasi-one-dimensional conductors; see, for example, Refs. 53 and 54.

Richard Feynman is another scientist who battled with the problem of superconductivity. When he was asked in 1966 about his activities in the '50s he answered: "*There is a big vacuum at that time, which is my attempt to solve the superconductivity problem — which I failed to do.*"[55,56] Feynman's views, shortly before the BCS theory appeared, are summarized in the proceedings of a meeting that took place the fall of 1956.[57] He stated that the Fröhlich–Bardeen model was, in his view, the correct approach; it only needed to be solved adequately. Even by performing a very accurate analysis using the method of perturbation theory, i.e. by including the vibrations of the crystal without fundamentally changing the properties of the electrons, Feynman was still unable to obtain a superconducting state. From this calculation, he concluded that the solution must be beyond the scope of perturbation theory, which turned out to be correct. The formulation of such a theory was accomplished by John Bardeen, Leon N. Cooper and J. R. Schrieffer.

After one goes over these unsuccessful attempts to understand superconductivity, there seem to be a few natural conclusions that one can draw: first, the development of the BCS theory is among the most outstanding intellectual achievements in theoretical physics. Second, theories emerge quite differently from how they are taught in the classroom or presented in textbooks, and, finally, an inapplicable theory, even a wrong theory, can be interesting, inspiring and may turn out to be useful in another setting.

Acknowledgments

The author acknowledges Paul C. Canfield, Rafael M. Fernandes, Jani Geyer and Vladimir G. Kogan for a critical reading of the manuscript.

References

1. J. Bardeen, L. N. Cooper and J. R. Schrieffer, *Phys. Rev.* **106**, 162 (1957).
2. H. Kamerlingh Onnes, *Commun. Phys. Lab. Univ. Leiden* **29**, 1 (1911).
3. C. S. Gorter and H. Casimir, *Zeitschrift für Technische Physik* **15**, 539 (1934).
4. H. London and F. London, *Proc. Roy. Soc. A* **149**, 71 (1935).
5. V. L. Ginzburg and L. D. Landau, *Soviet Phys. JETP* **20**, 1064 (1950).
6. L. Kelvin, *Philos. Mag.* **3**, 257 (1902).
7. P. H. E. Mejier, *American J. Phys.* **62**, 1105 (1994).
8. W. Heisenberg, *Zeitschrift für Physik* **33**, 879 (1925); *ibid.* **39**, 499 (1926).
9. E. Schrödinger, *Annalen der Physik* **79**, 361 (1926); *ibid.* **79**, 489 (1926); *ibid.* **81**, 109 (1926).
10. A. Einstein, Theoretische Bemerkungen zur Supraleitung der Metalle, in *Het Natuurkundig Laboratorium der Rijksuniversiteit te Leiden*, in *de Jaren*, Vol. 429, 1904–1922 (Eduard Ijdo, Leiden) (1922).
11. W. P. Su, J. R. Schrieffer and A. J. Heeger, *Phys. Rev. B* **22**, 2099 (1980).

12. A. J. Heeger, S. Kivelson, J. R. Schrieffer and W.-P. Su, *Rev. Mod. Phys.* **60**, 781 (1988).
13. T. Sauer, *Archive for History of Exact Sci.* **61**, 159 (2007).
14. M. Cardona, Albert Einstein as the father of solid state physics, in 100 anys d'herència Einsteiniana, ed. P. González-Marhuenda, pp. 85–115, Universitat de València (2006) [arXiv:physics/0508237].
15. J. J. Thomson, *Philos. Mag.* **44**, 293 (1897).
16. J. J. Thompson, *Philos. Mag.* **30**, 192 (1915).
17. H. Kamerlingh Onnes, *Commun. Phys. Lab. Univ. Leiden. Supplement* **44a**, 30 (1921).
18. W. Heisenberg, *Zeitschrift für Physik* **49**, 619 (1928).
19. F. Bloch, *Zeitschrift für Physik* **52**, 555 (1928).
20. L. Hoddeson, G. Baym and M. Eckert, *Rev. Mod. Phys.* **59**, 287 (1987).
21. L. D. Landau, *Physikalische Zeitschrift der Sowjetunion* **4**, 43 (1933).
22. R. Kronig, *Zeitschrift für Physik* **78**, 744 (1932).
23. R. Kronig, *Zeitschrift für Physik* **80**, 203 (1932).
24. P. Drude, *Annalen der Physik* **306**, 566 (1900); *ibid.* **308**, 369 (1900).
25. W. Meissner and R. Ochsenfeld, *Naturwissenschaften* **21**, 787 (1933).
26. L. D. Landau, *Physikalische Zeitschrift der Sowjetunion* **4**, 675 (1933).
27. L. D. Landau, *Sov. Phys. JETP* **7**, 19 (1937); *ibid.* **7**, 627 (1937).
28. D. Bohm, *Phys. Rev.* **75**, 502 (1949).
29. S. Goudsmit and G. E. Uhlenbeck, *Physica* **6**, 273 (1926); G. E. Uhlenbeck and S. Goudsmit, *Naturwissenschaften* **47**, 953 (1925).
30. F. A. Lindemann, *Philos. Mag.* **29**, 127 (1915).
31. J. J. Thompson, *Philos. Mag.* **44**, 657 (1922).
32. E. P. Wigner, *Phys. Rev.* **46**, 1002 (1934).
33. L. Brillouin, *Comptes Rendus Hebdomadaires des Seances de L Academie des Sciences* **196**, 1088 (1933); *J. Phys. Radium* **4**, 333 (1933); *ibid.* **4**, 677 (1933).
34. W. Heisenberg, *Zeitschrift f. Naturforschung* **2a**, 185 (1947); *ibid.* **3a**, 65 (1948).
35. *The Electron Theory of Superconductivity*, in Two Lectures by W. Heisenberg (Cambridge Univ. Press, 1949).
36. M. Born and K. C. Chen, *Nature* **161**, 968 (1948); *ibid.* **161**, 1017 (1948).
37. F. London, *Phys. Rev.* **74**, 562 (1948).
38. A. J. Leggett, *Rev. Mod. Phys.* **47**, 331 (1975).
39. J. C. Wheatley, *Rev. Mod. Phys.* **47**, 415 (1975).
40. D. Vollhardt and P. Wölfle, *The Superfluid Phases of Helium 3* (Taylor & Francis, 1990).
41. D. J. Scalapino, *J. Low Temp. Phys.* **117**, 179 (1999).
42. T. Moriya and K. Ueda, *Rep. Progr. Phys.* **66**, 1299 (2003).
43. A. V. Chubukov, D. Pines and J. Schmalian, in *Physics of Superconductors 1*, eds. K. H. Bennemann and J. B. Ketterson (Springer, 2003).
44. Y. Yanasea, T. Jujob, T. Nomurac, H. Ikedad, T. Hotta and K. Yamada, *Phys. Rep.* **387**, 1 (2003).
45. J. Bardeen, *Phys. Rev.* **59**, 928 (1941).
46. C. A. Reynolds, B. Serin, W. H. Wright and L. B. Nesbitt, *Phys. Rev.* **78**, 487 (1950).

47. E. Maxwell, *Phys. Rev.* **78**, 477 (1950); *ibid.* **79**, 173 (1950).
48. H. Fröhlich, *Proc. Phys. Soc. A* **63**, 778 (1950); *Phys. Rev.* **79**, 845 (1950); *Proc. Phys. Soc. A* **64**, 129 (1951).
49. J. Bardeen, *Phys. Rev.* **79**, 167 (1950); *ibid.* **80**, 567 (1950); *ibid.* **81**, 829 (1951).
50. J. Bardeen and D. Pines, *Phys. Rev.* **99**, 1140 (1955).
51. L. Hoddeson, *J. Statist. Phys.* **103**, 625 (2001).
52. H. Fröhlich, *Proc. Roy. Soc. A* **223**, 296 (1954).
53. N. Harrison, C. H. Mielke, J. Singleton, J. Brooks and M. Tokumoto, *J. Phys.: Condensed Matter* **13**, L389 (2001).
54. G. Blumberg, P. Littlewood, A. Gozar, B. S. Dennis, N. Motoyama, H. Eisaki and S. Uchida, *Science* **297**, 584 (2002).
55. Interview of Richard F. Feynman, by Charles Weiner, June 28, 1966, p. 168, American Institute of Physics, College Park, MD.
56. D. Goodstein and J. Goodstein, *Phys. Perspective* **2**, 30 (2000).
57. R. P. Feynman, *Rev. Mod. Phys.* **29**, 205 (1957).

NUCLEAR MAGNETIC RESONANCE AND THE BCS THEORY

Charles P. Slichter

Department of Physics, University of Illinois,
1110 West Green Street, Urbana, IL 61801, USA

The author describes the inspiration for the experiment by Hebel and Slichter to measure the nuclear spin–lattice relaxation time in superconductors, the design considerations for the experiment, the surprising experimental results, their theoretical treatment using the Bardeen–Cooper–Schrieffer theory, and how comparing the nuclear relaxation results with those for ultrasound absorption confirmed the central idea of the BCS theory, the BCS pair wave function.

The year 2007 was the 50th anniversary of the famous Bardeen–Cooper–Schrieffer (BCS) theory of superconductivity. To celebrate it, we held a four-day symposium in Urbana. I was one of the speakers. I talked about the experiment my remarkable student Chuck Hebel and I did,[1] inspired by a talk given by John Bardeen. It gave me an opportunity to set our experiment in an historical context, describe a bit about how John's thinking was evolving, and present some of the thoughts and motivation behind what we did that are not in our published papers. That talk and all the others of the Symposium are available on the web at: http://www.conferences.uiuc.edu/bcs50/video.html.

In 1951, John Bardeen joined the faculty of the University of Illinois with an appointment as Professor, half in the Department of Electrical Engineering (EE), half in the Department of Physics. As an undergraduate at the University of Wisconsin, he had majored in EE, but had built a strong background in other areas with teachers such as Warren Weaver in mathematics and J. H. Van Vleck in physics. His PhD at Princeton had been in Mathematics, but his thesis advisor was the physicist Eugene Wigner. He was in fact Wigner's second PhD student, and the second one to work on a problem in solid-state physics. Fred Seitz, Wigner's first PhD student

Fig. 1. John Bardeen in his Physics Department office in Urbana, early 1950s.

Fig. 2. Fred Seitz, early 1950s.

had been one of John's first friends at Princeton. Through this friendship, Illinois had learned that John was unhappy at Bell Laboratories and was able to persuade John to come to Illinois with this joint appointment.

John established an experimental program in semiconductors in EE, with post docs supervising the laboratory activity. In physics, he set up a program in theoretical physics, focusing on the subject of his long time interest, superconductivity.

Fig. 3. A happy J. H. Van Vleck receiving early morning congratulations on receiving the 1977 Nobel Prize in Physics. In his typical self-deprecating way, he joked with friends saying, "I thought I was probably ruled out by the Statute of Limitations."

I had come to Illinois right out of graduate school in September 1949. My undergraduate advisor was J. H. Van Vleck, who had moved from the University of Wisconsin to Harvard in the early 1930s. He had recommended to me to do graduate work at Harvard and then later suggested I ask Purcell to be my thesis advisor to study paramagnetic salts, a long time interest of Van Vleck, by electron spin resonance (ESR). I was Purcell's third student after Nico Bloembergen and George Pake and the only one doing electron spin resonance. I built an x-band rig to measure crystal field splittings of paramagnetic salts, and a second one with high power to measure their spin lattice relaxation times.[2]

Erwin Hahn had just completed his PhD in Urbana that June doing pulsed nuclear magnetic resonance (NMR), and over the summer had discovered spin echoes. We had a wonderful year together before he left for Stanford as an NRC Postdoctoral Fellow.

By the time John arrived, my students and I had been doing experimental NMR studies of metals (Dick Norberg, Don Holcomb),[3] ESR studies of conduction electrons (Tom Carver, Bob Schumacher)[4] and double resonance studies that confirmed Overhauser's theory[5] of dynamic polarization of nuclei in metals (Tom Carver).[6]

Fig. 4. Photo of Erwin Hahn in his naval uniform, submitted as part of his application to graduate school in physics at the University of Illinois. During the war, Erwin was a radar technician.

The year 1950 had been a big year for superconductivity with the discovery of the isotope effect, the publication of the Ginsberg–Landau theory, and Pippard's proposal of his non-local version of the London equations. In 1954, evidence was found that the heat capacity of superconductors depended exponentially on inverse temperature.[7] Bardeen[8] proposed that a simple model of a superconductor, the conventional free electron theory of metals with a gap inserted in the density of states in energy, would lead to the Meissner effect and a non-local version of the London equations.

In 1954 he gave a talk at Illinois in which he tried to show how an electron-lattice interaction would lead to an energy gap (this was not the BCS theory). I was in the audience. I did not really understand most of what John was saying, but as a result of all the work my students had been doing, I immediately realized that a gap in the density of states at the Fermi energy would have a big effect on the nuclear spin-lattice relaxation time, T_1. Nuclear relaxation arises by a process in which the nuclear magnetic moment scatters a conduction electron to absorb or emit the energy to flip the nuclear spin. The small size of the nuclear magnetic moment means that only a small amount of energy is exchanged in such a scattering event. The

Fig. 5. Charlie Slichter, Urbana, 1950.

only electrons that could participate in such a process are those near the Fermi energy where there are empty states into which an electron can be scattered.

The idea that an energy gap would have a strong effect on the nuclear spin-lattice relaxation time was very exciting until I suddenly realized that a superconductor is perfectly diamagnetic and would therefore exclude the magnetic field (this was before the discovery of Type II superconductors). How could one do NMR in a substance that excluded magnetic fields? By this time in John's talk I had quit listening and was focused on how to do NMR in a superconductor.

Then an idea came to me: use a time dependent "static" magnetic field, cycling between a strong field that suppressed the superconductivity of the sample and zero magnetic field that rendered the sample superconducting (Fig. 6). Initiate the experiment by applying a magnetic field H_o, much stronger than the superconducting critical field H_c so that the superconductivity of the sample was suppressed, and leave it on long enough for the nuclear magnetization to be established. Use an NMR apparatus operating at a frequency that places the resonance at a field H_{reson} less than H_o but

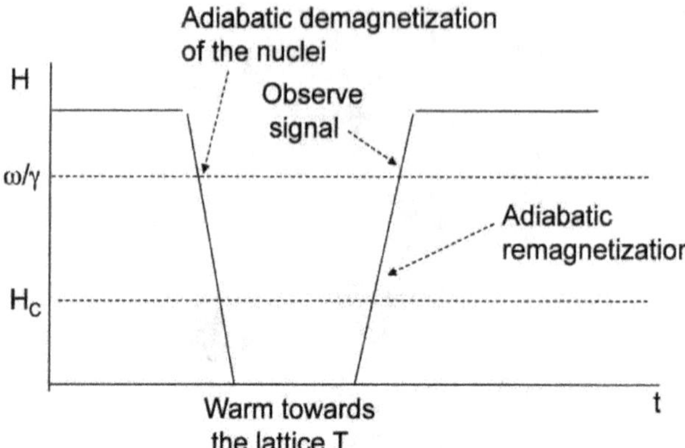

Fig. 6. Schematic picture of the cycle of magnetic field, H, versus time, t, to make possible measurement of nuclear relaxation in a material that excludes "static" magnetic fields when the sample is in the superconducting state. As long as H is greater than the superconducting critical field, H_c, the sample is in the normal state, permitting static and alternating fields to penetrate the sample. The adiabatic demagnetization leaves the nuclear spin temperature far below the lattice temperature and the sample in the superconducting state. The spin-lattice relaxation process warms the nuclear spins in the superconducting state. The extent of the warming is monitored by observing the resonance on the fly as the sample is adiabatically remagnetized.

stronger than H_c. Turn the magnetic field to zero, sufficiently slowly to be an adiabatic demagnetization. This step would leave the nuclear spins at a very low spin temperature compared with the lattice and the sample superconducting. The nuclei would then start to warm towards the lattice temperature. After a time t_{warm}, turn the magnetic field back on to its initial value H_o, inspecting the NMR signal on the fly as the field passes through the value H_{reson}. Then study the strength of the NMR signal as a function of t_{warm}, thus deducing the nuclear spin-lattice relaxation time in the superconducting state.

The nuclear spin-lattice relaxation time, T_1, is the length of time it takes the nuclei to establish their thermal equilibrium magnetization, M_o, in an applied magnetic field. The first theory of T_1 in a metal was actually made by Heitler and Teller[9] in 1936, long before the invention of magnetic resonance, in a paper in which they explored using the nuclear magnetization of a metal to achieve cooling by adiabatic demagnetization of the nuclear magnetism. The nuclear spins would then cool the solid. They predicted that the nuclear spin lattice relaxation time in a metal would vary inversely with $1/T$.

In 1950, Korringa[10] worked out a detailed general theory of the nuclear spin-lattice relaxation time in a metal, showing that it was closely related to the Knight shift. Korringa had worked out the theory for metals in a strong magnetic field. In this case, the nuclear energy levels are just those of the Zeeman effect. But in zero field, the Hamiltonian is just the dipolar coupling. The eigenstates and energy levels were not known in this case. For my PhD thesis I had used magnetic resonance to measure the T_1 of electron spins in paramagnetic salts.[11] I had worked out a general expression describing the relaxation of a spin system that at all times during the relaxation process could be described by a spin temperature. I give the derivation below. Using it, we found that we could express the quantum mechanical aspects of the theory as a trace, making it unnecessary to solve the zero field Hamiltonian to determine the magnetic field dependence of the relaxation time.

In many cases the relaxation process is described by the equation for the time dependence of the magnetization, M:

$$\frac{dM}{dt} = -\frac{M - M_o}{T_1}, \tag{1}$$

where the thermal equilibrium magnetization, M_o, obeys Curie's law relating magnetization of N nuclei of spin I, with gyromagnetic ratio γ_n, to the lattice temperature T_L, the Boltzmann constant k_B, and the Curie constant, C[12]

$$M_o = \frac{CH_o}{T_L}, \tag{2}$$

where

$$C = N\frac{\gamma_n^2 \hbar^2 I(I+1)}{3k_B}. \tag{3}$$

Notice that if H_o is zero, the thermal equilibrium magnetization is zero for any value of the lattice temperature. In our experiment, we turn the external magnetic field, H_o, to zero. I have called it an "adiabatic demagnetization" since we take pains to employ a reversible process that maintains the order of the system. From Curie's law, we see that the magnetization is zero in zero applied magnetic field. In the strong field, the order of the system is manifested in the magnetization. Where has it gone when the magnetization is zero in zero applied field? Clearly, it now must reside in the alignment of the nuclear spins along the local magnetic fields of their neighbors. Since these local fields are randomly oriented, no net magnetization results.

In such a process, the entropy of the system, σ, remains constant during the demagnetization process. Consequently, if the system were perfectly isolated from the lattice, one could then turn the magnetic field back to its original value, thereby recovering the initial magnetization without further need for a T_1 process.

For a system of N spins, the entropy when the Zeeman energy is lower than the thermal energy at temperature θ is given by

$$\sigma = Nk_B \ln[2I+1] - \frac{C(H_o^2 + H_{\text{loc}}^2)}{2\theta^2}, \tag{4}$$

$$CH_{\text{loc}}^2 = \frac{1}{(2I+1)^N} \text{Tr}[H_D^2]s, \tag{5}$$

$$H_D = \frac{1}{2}\sum_{i,j} \frac{(\gamma_n \hbar)^2}{R_{ij}^3}\left[I_i \cdot I_j - 3\frac{(I_i \cdot R_{ij})(I_j \cdot R_{ij})}{R_{ij}^2}\right], \tag{6}$$

and H_D is the Hamiltonian of the nuclear dipole-dipole coupling, and R_{ij} is the distance between nuclei i and j. This relationship shows that if one starts with a spin system in a strong magnetic field H_{oo}, in thermal equilibrium with the lattice at temperature θ_L, lowering the applied field to zero, cools the spin system to a temperature, θ, of

$$\theta = \theta_L \frac{H_{\text{loc}}}{H_{oo}}. \tag{7}$$

To describe the spin-lattice process warming the system in zero magnetic field, it is simplest to talk about the temperature of the spin system since the magnetization is zero as long as the applied field vanishes. The concept of spin temperature goes back to Casimir and du Pre[13] in the 1930s when adiabatic demagnetization was invented as a method of obtaining low temperature. Gorter utilized the spin temperature concept extensively in his monograph on paramagnetic relaxation[14] with which I had become familiar since Gorter taught a course using it while a visiting professor at Harvard in the summer of 1947. The trace methods, involving spin temperature, that we utilize below were first introduced to calculations of relaxation rates by Van Vleck in his paper on paramagnetic relaxation in Ti and Cr alums.[15] We then describe the relaxation process in terms of the energy of the spin system. It is convenient to introduce the notation

$$\beta = \frac{1}{k_B \theta}. \tag{8}$$

For all values of applied magnetic field, the nuclear spin system is described by the Hamiltonian

$$H = H_z + H_D, \tag{9}$$

where H_z is the Zeeman energy of all N nuclear spins,

$$H_z = -\gamma_n \hbar H_o \sum_{j=1}^{N} I_{zj} \,. \tag{10}$$

We label the eigenvalues of the Hamiltonian, **H**, as E_n. Then, the mean energy, E, of the system is given by

$$E = \sum_n p_n E_n \,, \tag{11}$$

where p_n is the probability that state n is occupied.

Using the fact that

$$p_n = \frac{\exp(-\beta E_n)}{(2I+1)^N} \,, \tag{12}$$

we find

$$E = -\beta \left[\frac{\sum\limits_n E_n^2}{(2I+1)^N} \right] . \tag{13}$$

The spin-lattice relaxation process changes the populations of the energy states. Then, assuming that we have simple rate processes

$$\frac{dp_n}{dt} = \sum_m p_m W_{mn} - p_n W_{nm} \,, \tag{14}$$

where W_{mn} is the rate of transitions induced from state m to state n by the external world (the thermal reservoir) so that

$$\frac{dE}{dt} = \sum_n E_n \frac{dp_n}{dt} \sum_{n,m} E_N [p_m W_{mn} - p_n W_{nm}] \,. \tag{15}$$

If the system is described by a spin temperature as the energy change takes place, then we also have

$$\frac{dE}{dt} = -\frac{d\beta}{dt} \frac{\sum\limits_n E_n^2}{(2I+1)^N} \,. \tag{16}$$

To guarantee that the transitions will lead to a population corresponding to the lattice temperature, we invoke the principle of detailed balance:[16]

$$W_{mn} = W_{nm} \exp[(E_m - E_n)\beta_L] \,. \tag{17}$$

Combining Eqs. (14)–(16), we find

$$\frac{d\beta}{dt} = \frac{\beta - \beta_L}{T_1}$$

$$\frac{1}{T_1} = \frac{1}{2} \frac{\sum_{n,m} W_{mn}(E_m - E_n)^2}{\sum_n E_n^2} . \tag{18}$$

Equation (14) is a normal modes problem with many real exponential roots. The imposition of the assumption that the spins are described by spin temperature at all stages of the relaxation leads to a process described by a single exponential relaxation as described by Eq. (18). Experimentally, we find a single exponential, providing an experimental justification of the spin temperature assumption.

In a metal, the dominant source of nuclear relaxation is coupling between the magnetic moments of the nuclei and the conduction electrons. The process may be viewed as a scattering process in which the nuclei in an initial state m scatter an electron from a state k, s into a final state k', s', while making a transition to a final nuclear state n. (Note that these are the eigenstates of the many-nucleus system, not those of a single nucleus.) For many metals, the scattering interaction is the Fermi contact term between an electron at position r_i and a nucleus at position R_j,

$$H_{\text{Fermi}} = \left(\frac{8\pi}{3}\right) \gamma_e \gamma_n \hbar^2 S_i \cdot I_j \delta(r_i - R_j) . \tag{19}$$

Then, the probability per second of the transition from initial state m, k, s to final state n, k', s' is

$$P_{mks,nk's'} = \left(\frac{2\pi}{\hbar}\right) |\langle mks|H_{\text{Fermi}}|nk's'\rangle|^2 \delta(E_{mks} - E_{nk's'}) . \tag{20}$$

To have the initial state occupied and the final state empty, satisfying the Pauli principle, for the normal state we have

$$W_{mn} = \sum_{ks,ks'} P_{mks,nk's'} f(E_{ks})(1 - f(E_{k's'})) , \tag{21}$$

where $f(E)$ is the Fermi function.

Assuming that the electron wave functions are products of an electron spin function and a Bloch function $\psi_k(r_i) = u_k(r_i)e^{ikr_i}$, where u_k has the periodicity of the lattice, and assuming a spherical Fermi surface, we showed in the appendix of our 1958 paper that

$$W_{mn} = \sum_{i,j,\alpha(=x,y,z)} a_{ij} \langle n|I_{i\alpha}|m\rangle\langle m|I_{j\alpha}|n\rangle\,, \tag{22}$$

where

$$a_{ij} = \frac{64\pi^3\hbar^3}{9}\gamma_n^2\gamma_e^2|u(0)^4|_{\text{Fermi}}\frac{\sin^2(k_F R_{ij})}{(k_F R_{ij})^2}\int \rho(E_i)\rho(E_f)f(E_i)[1-f(E_f)]dE_i\,. \tag{23}$$

We have pulled the wave function out of the integral, replacing it by its value averaged over the Fermi surface. The temperature dependence arises from its effect on the Fermi functions.

It is useful to define the temperature dependent integral $\Sigma_n(T)$ for later use. It determines the temperature variation of the spin lattice relaxation rate, $R(T)$, for a metal in the normal state,

$$\Sigma_n(T) = \int \rho(E_i)\rho(E_f)f(E_i)[1-f(E_f)]dE_i\,. \tag{24}$$

Utilizing Eqs. (18) and (22), we find that

$$R \equiv 1/T_1 = \mathrm{Tr}\{[H, I_{i\alpha}][I_{j\alpha}, H]\}/2\,\mathrm{Tr}(H^2)\,. \tag{25}$$

If we keep just the on-site terms $i = j$, we find for the magnetic field dependence of the relaxation rate

$$R(T) = a_{00}(T)\frac{\mathrm{Tr}\,H_z^2 + 2\,\mathrm{Tr}\,H_{dd}^2}{\mathrm{Tr}\,H_z^2 + \mathrm{Tr}\,H_{dd}^2}\,. \tag{26}$$

In a strong magnetic field, this gives $R(T) = a_{00}(T)$ which is just the Korringa result. It is easy to show that the rate then is proportional to the absolute temperature, T.

In zero applied field, the rate is predicted to be twice as fast as the rate in strong applied field. Experimentally, this is found to be true for the alkali metals whose relaxation rates are slow enough for the field cycling method to work. For Al, we found that the experimental ratio is 3.35 ± 0.35. Although this differed from the predicted ratio, the data fit a similar mathematical form

$$R(H,T)/R(\infty,T) = (H^2 + A')/(H^2 + A)\,. \tag{27}$$

We do not know why Al differs from the case of the alkali metals.

Dick Norberg and Don Holcomb had both gone to college at DePauw in Greencastle, Indiana. They told me of another very bright Illinois graduate student from there, Chuck Hebel, who was at that time working as a

Fig. 7. Chuck Hebel in 1953.

research assistant at the Betatron and was thinking of switching fields and urged me to try to recruit him for the experiment. I told Chuck about the idea for the experiment and, to my delight, he decided to undertake it for his PhD thesis. As I explain below, this experiment was quite unconventional, pushing the boundaries of NMR work. It was a very challenging project, requiring substantial building of apparatus and overcoming difficult experimental challenges.

Almost everything my students had been doing was quite novel and unconventional. Norberg[17] studied the famous H-Pd system, using ^{1}H NMR spin echoes from a short piece of Pd wire containing H. Norberg and Holcomb's study of the alkali metals introduced use of the boxcar integrator, giving pulse techniques the ability to signal average analogous to the use of lock-in amplifiers for steady state NMR. Our experiments on the alkali metals were the first to measure self diffusion in solids by NMR pulse techniques. In the Overhauser effect experiments, we did magnetic resonance at magnetic fields of only a few tens of gauss, performed the first electron-nuclear double resonance, made the first measurements of the spin lattice relaxation time of conduction electrons, and made the first demonstration of dynamic nuclear polarization. Our experiments on static magnetic susceptibility showed for the first time how, using magnetic resonance, one could directly measure the spin susceptibility of conduction electrons and were

Table 1. Properties of some candidate metals.

	^{207}Pb	^{199}Hg	^{115}In	^{27}Al
$T_C(K)$	7.19	4.13	3.40	1.17
Abundance	21%	16%	95%	100%
ν (MHz/T)	8.9	7.6	9.3	11.1
H_C (gauss)	803	412	293	105
T_1 at 1 K (msec)	$\approx$ 10, too fast	$\approx$ 10, too fast	$\approx$ 10, too fast	450

the very first measurements of the famous Pauli electron spin susceptibility. My students generated an atmosphere of excitement, were seeking new challenges, and eager to do things that had never been done before. What a joy to be around them! That was the atmosphere Chuck stepped into and rapidly enhanced.

The key issue in planning was the selection of the sample material since its properties determined all the other experimental parameters. There were four possible candidates, Hg, Pb, In and Al. Their properties are listed in Table 1.

We had to be able to turn the magnet on and off in a time fast compared with the nuclear spin-lattice relaxation times. Al had a spin-lattice relaxation time of about a half second, a very favorable number, whereas the others had expected T_1's in the millisecond time range, too short to be workable. However, the superconducting transition temperature of Al was only 1.17 K, a difficult temperature to achieve. At that time, there were two choices to reach such a temperature: pump on ^{4}He or adiabatic demagnetization. This experiment was the first one my group had attempted in the liquid helium range, so we consulted our colleagues Dillon Mapother and John Wheatley. It is very hard to get much below about 2 K by pumping on ^{4}He because of the Rollin film, but the alternative of cooling by adiabatic demagnetization would have added very great complexity to what was already a challenging experiment. We decided to pump on ^{4}He using a special design of He chamber Chuck developed based on their advice. We utilized three Dewars. An outer liquid nitrogen Dewar enclosed a liquid He Dewar containing liquid He at 4.2 K and an inner liquid He Dewar connected to a oil booster pump backed by a mechanical pump. The inner Dewar contained the sample that sat in a ^{4}He bath that was sealed beneath two one-millimeter constrictions placed for the purpose of cutting down the flow of the Rollin film.

We needed a magnet that we could switch sufficiently rapidly yet produced a strong enough magnetizing field to give an observable NMR signal. Conventional NMR used iron magnets that generated 10,000 gauss, putting the NMR frequency at about 10 MHz. The time to turn such a magnet off or on is measured in minutes, not fractions of a second. At lower magnetic fields the NMR signals are weaker, but our experience from our work on the Overhauser effect where we used 50 kHz for the NMR frequency with magnetic field of order 30 gauss (generated by an air core solenoid) gave us confidence that we could work at magnetic fields much lower than 10,000 gauss. Since H_c of Al is 105 gauss, Chuck and I decided to use a polarizing field, H_o, of 500 gauss, and magnetic field to detect the resonance, H_{reson}, of 360 gauss, putting the NMR frequency at 400 kHz. From his work at the Betatron, Chuck knew that there were some thin sheets of magnet iron left over from the recent construction of the Betatron, and proposed using them to construct a magnet we could pulse off and on. It consisted of 0.014-in thick silicon steel sheets with adjustable laminated sections near the magnet gap as shims to obtain favorable magnetic field homogeneity at the position of the sample. Figure 8 shows the magnet and Dewar assembly Chuck designed.

Fig. 8. Our apparatus, showing the Dewar system hanging vertically into the magnet gap. The magnet, made of sheets of Betatron iron, and the energizing coils are evident on either side of the bottom of the Dewar.

It was important to control the time history of the magnet when we turned the magnetic field off or on in order to satisfy the adiabatic condition. Think of the individual nuclear spins as being quantized along the magnetic field each nucleus experiences, either parallel or anti parallel to the effective magnetic field. The crucial time period is when the values of the applied field were comparable to the local fields, since that was when the nuclear moments that are quantized along external magnetic field change to quantization along the local field. If one turns the field to zero too rapidly, the system arrives at zero field with the high field magnetization. As Eq. (2) shows, the system then is not described by a spin temperature since in zero magnetic field, no matter what the spin temperature may be, the magnetization must be zero. The transition from strong field to zero magnetic field needs to happen sufficiently slowly that the nuclear spin quantization remains at all times along the vector resultant of the applied and local fields at each nuclear site. During the turn-off of the magnet, adjustment of the time constant of the turn-off easily enabled us to satisfy the condition. If we simply reversed the process when we turned on the magnetic field, the fastest rise would occur at low magnetic fields, violating the adiabatic conditions. Turning on the magnet required that we have a process that was slow at low fields, but then became fast to avoid relaxation during the turn-on. Figure 9 shows the scheme we used.

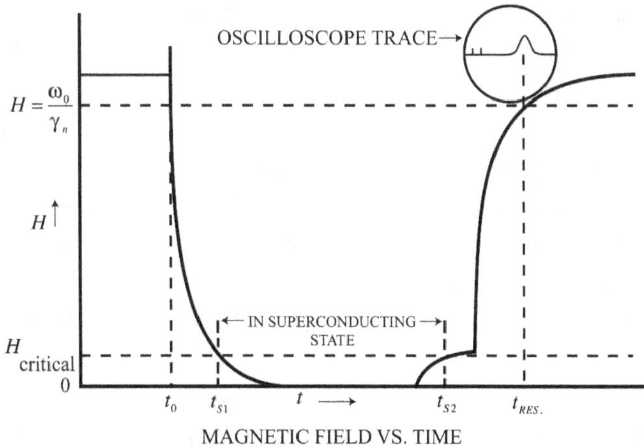

Fig. 9. The actual magnetic field cycle. The demagnetization followed an exponential decay. The remagnetization began with a small exponential to carry the H slowly through values comparable to the local magnetic field, from the neighboring nuclei, then rose rapidly to high values. The resonance was observed on the fly as it passed through the NMR resonance condition.

Our sample was "atomized" particles of 99.9% purity obtained from Alcoa and sieved through a 325-mesh sieve. The mean particle size determined by microscope was 10 microns. The sample was annealed after many measurements were made and several of the measurements repeated to look for effects of strain. DeBye-Scherrer x-ray photographs were taken of the annealed and unannealed samples to look for the presence of dislocations, but no difference in the two photographs could be detected. However, we did find that, after annealing, the sample showed pronounced supercooling that led to substantial loss in signal because the transition to the superconducting state was precipitous rather than adiabatic. With this apparatus, we were able to cool to 0.9 K (a T/T_c of 0.8) and we could turn the magnetic field to zero in about 1 msec.

The resonance was observed with a bridgeless system similar to that used for the thesis of Schumacher. A 400 kHz oscillator of the Pound–Watkins type[18] fed the rf through a high impedance to the sample coil, which was resonated with a parallel capacitance, and fed directly to the input of an rf amplifier. As used, the apparatus was sensitive to the imaginary part of the nuclear magnetic susceptibily, χ''. As mentioned above, the rf from the 400 kHz was on continuously, but the NMR resonance condition was satisfied only for the short time that the time-dependent magnetic field passed through the resonance value. Therefore, the NMR signal was a transient pulse. The signal was then amplified and displayed on an oscilloscope. Figure 10 shows an oscilloscope photo of the signal following a single switching cycle at 1 K. In addition, following the method introduced by Norberg and Holcomb, we recorded the signal on a gated detector (a boxcar integrator) that enabled us to signal average multiple acquisitions of the signal. We were able to improve the signal-to-noise ratio by nearly a factor of 3. Because we had an rf oscillator on during this acquisition, this was phase sensitive detection, to my knowledge in fact the first use of phase sensitive detection of pulsed NMR signals.[19]

Meanwhile there were important theoretical developments. In 1955, David Pines and John Bardeen[20] worked out the theory of the interaction of the conduction electrons and the lattice and how that modified the electron–electron coupling. In 1956, Leon Cooper, who succeeded David as John's postdoc, worked out the theory of a pair of electrons interacting above a filled Fermi sea,[21] introducing the important ideas of electron pairing and the important role of degeneracy. And Bob Schrieffer had started work as John's graduate student working on the problem of the origin of superconductivity.

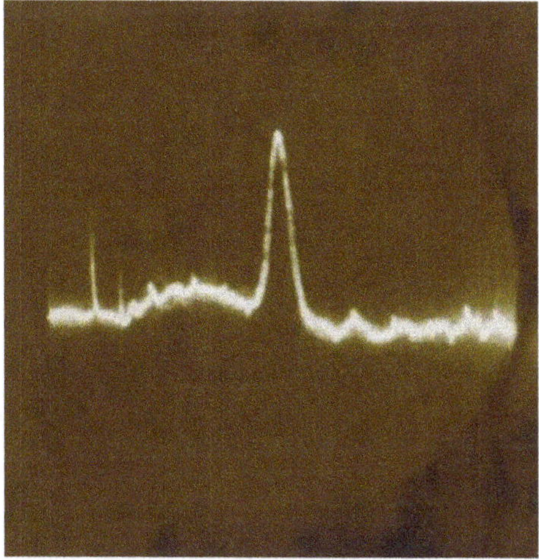

Fig. 10. Photo of the transient NMR signal observed on the screen of an oscilloscope as the magnetic field passed through resonance condition. Such signals were also recorded on a "boxcar integrator" circuit to improve signal to noise ratio.

Fig. 11. David Pines while a postdoc with Bardeen.

We finished building our apparatus in mid 1956. For our first experiments we studied the relaxation time in zero field in the normal state of Al. We confirmed experimentally the general form of our theoretical prediction of the magnetic field dependence of the nuclear spin-lattice relaxation time in

Fig. 12. Leon Cooper while a postdoc with Bardeen.

Fig. 13. Bob Schrieffer when entering the University of Illinois for graduate study in Physics.

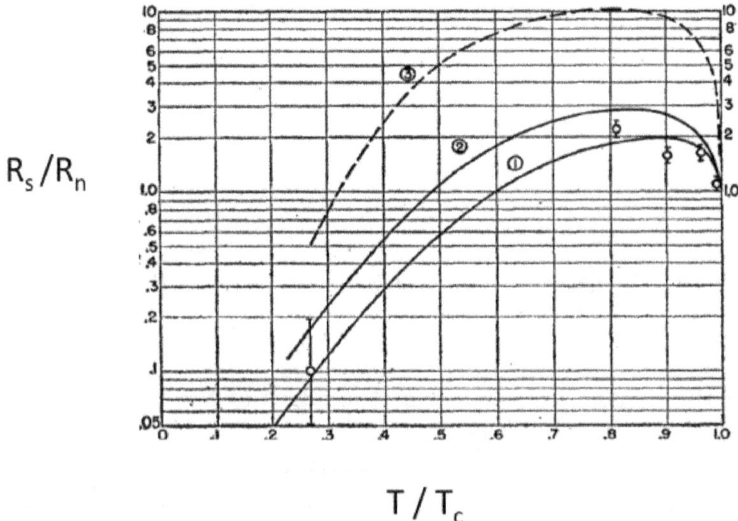

Fig. 14. Our data: Ratio of the relaxation rate in the superconducting state, R_s, to its value in the normal state, R_n, versus T/T_c. The lines are theoretical curves added later after the creation of the BCS theory.

the normal state and that at all values of applied static magnetic field, the nuclear spin lattice relaxation rate $1/T_1$, was proportional to T.

We got our first experimental results for the superconducting state in the early fall of 1956. Of course, since there was no broadly accepted theory of superconductivity, we had no formal prediction of what we would find. But there were approximate ways to think about the problem. A two fluid model in which there was a normal fluid and a super fluid was popular at that time. It would appear to imply that the nuclear relaxation would arise from the normal fluid only and thus the nuclear spin-lattice relaxation rate would be slower in the superconducting state than in the normal state. We felt that the same would be true if there were an energy gap since a gap would inhibit excitations of the electrons need to relax the nuclei.

We were therefore greatly surprised to find that the nuclear relaxation rate (compared with what it would have been for the normal state at that temperature) *increased by a factor of two* as we cooled below T_c. Our results are shown in Fig. 14. Several theoretical curves are also shown, but they came much later.

Chuck and I puzzled about this. We had one possible explanation based on the Korringa theory and the idea that an energy gap opened up at the Fermi surface by pushing some states up and others down in energy,

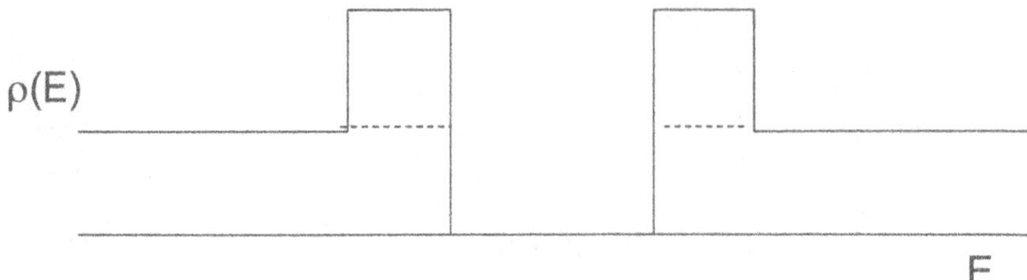

Fig. 15. Schematic density of states showing a pile-up at the edge of the energy gap.

increasing the density of states near the edges of the gap. The density of states then might look like Fig. 15.

As we see from Eqs. (23) and (24), the relaxation rate is determined by an integral over energy. The energy to flip a nuclear spin is very small compared with kT. Therefore, we can treat the electron initial and final energies as nearly equal. The factor $f(1-f)$ peaks at the Fermi energy, and is zero for energies more than a few kT away from the Fermi energy. So the relaxation rate is determined by the integral of the square of the density of states over this region. If the peaks in Fig. 15 fall within the region, the integral will be bigger than if the density of states had not been gapped and peaked. The argument is similar to the inequality $2^2 + 0^2 > 1^2 + 1^2$.

We went to John to show him our experimental results and asked what he thought of our explanation. He kindly said that at least it was not foolish! As we shall see, it turned out later from the BCS theory that this argument is part of the story, but omits what is the most exciting part of the explanation, the role of the pairing coherence of the BCS wave function on the matrix elements!

In October 1956, everyone at Illinois was wildly excited to learn that John, Walter Brattain, and Shockley would receive the Nobel Prize in Physics for the invention of the transistor. In December, after his return from Sweden, John very sweetly told us all about the events surrounding the Prize, even bringing in the gold medal for us all to see. This was all done without conceit, but rather the way a dear friend includes one in something meaningful.

Then one morning in early February, I walked out of my office on the second floor of the Physics Building and encountered John. He clearly had something on his mind and wanted to say something to me. John was not much of a talker. I realized that if I spoke up, I would preempt what he wanted to say. So we just stood there while I waited. It seemed like forever, but was probably just a few seconds. Then he said "Well, I think we have figured out superconductivity." What a message! That was no doubt the most exciting event in my life as a scientist. Evidently the night before John, Bob, and Leon had become convinced that they had the solution in hand and he was coming down the hall to tell me.

In the next few days they sent off a Letter to Physical Review describing their concepts.[22]

As yet, there were many predictions to be worked out, but they had enough results to realize that they had identified the essence of what caused superconductivity. In the next few days, they showed Chuck and me their theory: the Hamiltonian, the wave function, the density of states, how to write excited states. Chuck and I decided to try to use the theory to calculate the nuclear spin-lattice relaxation time. The BCS theory was written using annihilation and creation operators. Neither Chuck nor I had ever used them. So first we set out to use them to solve the relaxation in the normal state where we knew the answer (the Korringa result). We each did the calculation. Then we each repeated the calculation using the BCS wave functions, energy levels, and density of states. To our delight, our answers agreed with each other! Our result predicted that indeed the relaxation rate should increase in the temperature region of our data.

We give the details of our calculation in Hebel's thesis[23] and in our 1959 paper. I will not repeat it here. The form of the result is an integral very similar to the conventional result shown in Eq. (23). (Of course, the answer Hebel and I found was the same as that found by Bardeen, Cooper, and Schrieffer.) However, the integral Σ_n is replaced by Σ_s

$$\Sigma_s(T) = \int_0^\infty \rho_s(E_i)\rho_s(E_f)C_{nmr}(E_i, E_f, T)f(E_i, T)[1-f(E_f, T)]dE_i\,, \quad (28)$$

where the superconducting density of states ρ_s obeys

$$\begin{aligned} \rho_s(E) &= 0 \quad \text{for } |E| < \varepsilon_0(T) \\ &= \rho_n(E)\sqrt{E^2/[E^2 - \varepsilon_0^2(T)]} \quad \text{for } |E| > \varepsilon_0(T)\,, \qquad (29) \\ \text{and } C_{nmr}(E_i, E_f, T) &= 1 + [\varepsilon_0^2(T)/E_iE_f]\,, \end{aligned}$$

with $\varepsilon_0(T)$ the energy gap at temperature T, and $\rho_n(E)$ the density of states of the normal metal.

There is an energy gap. The density of states in the superconductor outside of the gap is seen to be larger than in the normal state, as in Fig. 15. However, at the edges of the gap, ρ_s has an inverse square root infinity. Thus, the resulting relaxation rate would blow up were we to neglect the nuclear energy change associated with relaxation and simply set $E_i = E_f$. In fact, we predicted a tenfold increase in relaxation rate. To bring the theory into agreement with experiment, we postulated that the electronic energy levels had a natural width. We postulated that the superconducting states had a lifetime broadening, and replaced ρ_s by ρ'_s

$$\begin{aligned} \rho'_s &= \int \rho_s(E',\theta)\Delta(E'-E)dE' \\ &\int \Delta(E'-E)dE' = 1\,. \end{aligned} \tag{30}$$

Hebel used a simple rectangular function for Δ. Expressions derived in this manner are plotted on the data plot. Also included is a point obtained by Reif[24] from T_1 measurements on colloidal mercury using the saturation method.

In 1957, Morse and Bohm measured sound absorption in superconducting In and Sn.[25] Like NMR, absorption of ultrasound involves scattering of conduction electrons to absorb (emit) the energy when a sound quantum is emitted (absorbed). In contrast to the NMR result, they found that the sound absorption rate drops precipitously as one cools below the superconducting transition. In any one-electron theory, sound absorption and NMR relaxation should differ primarily in the scale of the interaction strength, and thus would have the same temperature dependence. Figure 16, based on a figure from Cooper's Nobel Prize lecture,[26] shows the ultrasound data. Clearly, the data differ drastically from our NMR result.

It is a major triumph of the Bardeen, Cooper, and Schrieffer theory that it successfully predicts the ultrasound result and thus resolves this paradox. The essential change is in the integral Σ_s. The ultrasound absorption rate $R_{us}(T)$ obeys

$$R_{us}(T) \propto \int_0^\infty \rho_s(E_i)\rho_s(E_f)C_{us}(E_i,E_f,T)f(E_i,T)[1-f(E_f,T)]dE_i\,, \tag{31}$$

where

$$C_{us}(E_i,E_f,T) = 1 - [\varepsilon_o^2(T)/E_iE_f]\,. \tag{32}$$

For ultrasound, the factor C_{us} goes to zero as the $E_{i,f}$ energies approach the gap ε_o, canceling the infinities from the density of states, whereas for NMR the factor C_{nmr} goes to 2, giving infinities in the density of states full sway.

If one goes back to the origin of the C factors, one finds that they are an expression of the pair nature of the wave function and of an interference between the state k, spin up and −k, spin down that form the Cooper pairs.

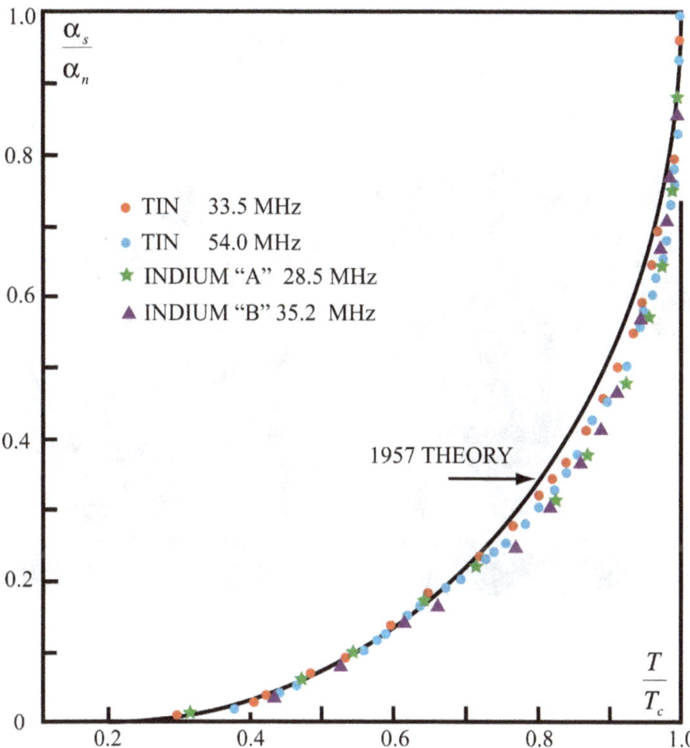

Fig. 16. Data of Morse and Bohm,[25] for tin and indium, for the temperature variation of the ratio of the ultrasound absorption in the superconducting state to its value in the normal state compared. The solid line is the prediction of the BCS theory.

Thus it is the pair nature of the BCS wave function that contributes two terms. For NMR the terms are added; for sound absorption one term is subtracted from the other. Therefore the different temperature dependence of the two experiments is a direct result of the pair nature of the BCS wave function. The detailed explanation is a feature of Leon Cooper's Nobel Prize Lecture.

Although the BCS theory of nuclear relaxation had a peak just below the superconducting transition, at lower temperatures it predicted that the relaxation rate would drop exponentially with inverse temperature. We could not reach such temperatures.

Al Redfield at the IBM Watson Laboratory had become interested in studying nuclear relaxation in superconductors and had independently invented the idea of using magnetic field cycling for the experiment. He

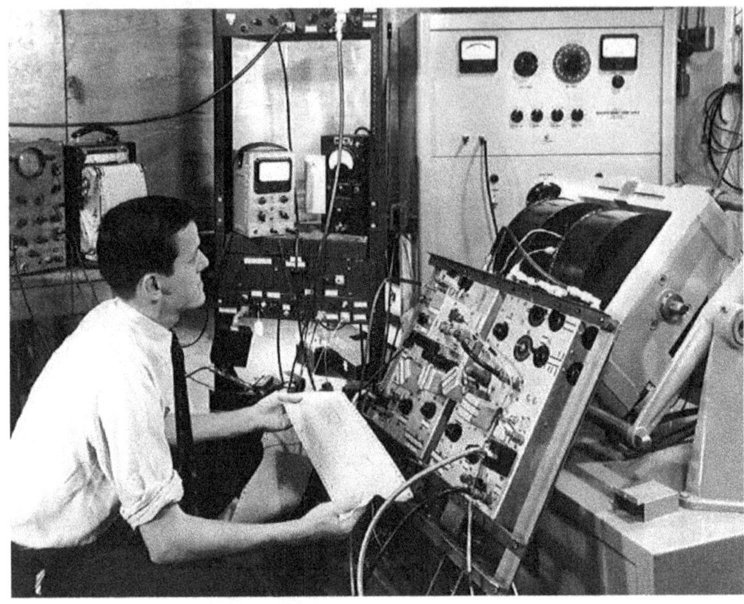

Fig. 17. Al Redfield at the controls of his NMR rig while a postdoc with Bloembergen at Harvard.

learned about our work after Chuck and I were well along on our experiment. Al and I had been good friends for many years. We met during the war when I worked at the Woods Hole Oceanographic Institution on a war project. At that time he was in his mid teens. Later we saw each other at Harvard when I was in graduate school and he an undergraduate. I had urged him to come to Illinois for graduate study. He did so and in fact did his PhD thesis using one of the two magnets of my group, measuring the Hall effect in diamonds. Bob Maurer was his thesis advisor. He then had gone to postdoc with Bloembergen at Harvard where he worked on NMR, then to the IBM Watson Lab at Columbia.

He decided to try to extend the NMR data to lower temperatures using adiabatic demagnetization. He had some results by this method. Then a major event happened: the price of ^{3}He dropped tenfold. Seidel and Keesom proposed using it as a cryogenic fluid since it did not display superfluidity and thus did not suffer from the problems arising from the Rollin films. They demonstrated pumping on ^{3}He to cool well below 1 K.[27] Redfield then switched to this method and obtained the NMR data well below the temperatures of the peak, verifying the low temperature prediction of our calculation. His data and ours were displayed in Fig. 18 based on a figure taken from

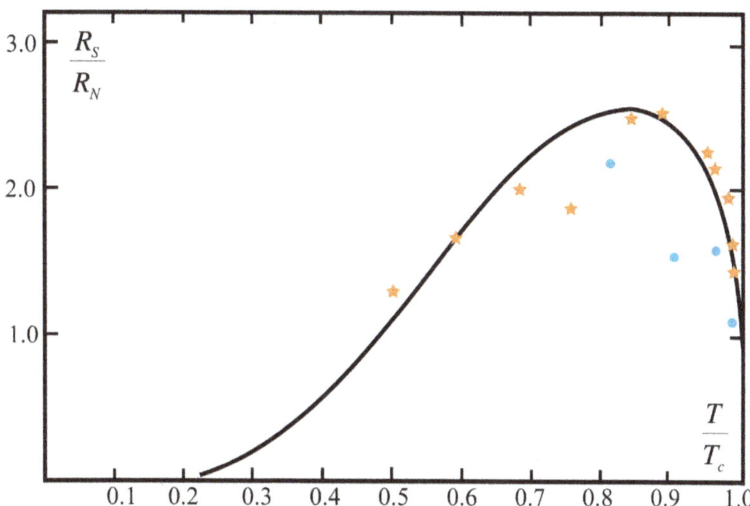

Fig. 18. The combined data taken from Cooper's Nobel Prize Lecture. Al NMR data of Hebel and Slichter (blue circles) and Redfield (orange stars) for R_s/R_n versus T/T_c, showing the peak just below the superconducting transition arising from the coherence of the electron spin pairing, and the low temperature fall-off resulting from the presence of the energy gap.

Leon's talk in Stockholm. Later, Redfield made substantial further studies with his students Anderson and Masuda.[28]

Reflections

Al Redfield, Chuck Hebel, and I look back on our experiments with very special pleasure. It was a time of great excitement. Explanation of superconductivity had eluded scientists for many years. To feel that one might be close to a solution was a great impetus for our experiments. Even before the breakthrough by John, Leon, and Bob, there was an air of excitement since so many new results were arising. When John, Leon, and Bob announced their theory everyone was transfixed! Experimenters rapidly accepted it since it explained so much experimental data. Some theorists were more cautious in accepting the theory.

For Chuck and me, the most remarkable thing was to be instructed in the theory at the very time that its authors were developing their calculations. What remarkable openness! What remarkable kindness to take the time to help us when they were racing to explore the consequences of their discovery!

This experience has put John, Leon, and Bob in a very special place in our affection and our admiration. It is a great joy to reflect on those days.

Looking back, I am especially impressed at how quickly John in collaboration with his three young colleagues got the solution after leaving Bell Labs and his work on transistors. His recognition of the consequences of an energy gap, based on a very simple modification of the free electron theory of metals in his paper of 1955 illustrates how closely he followed experiment. It also shows what an intuitive feel he had for the nature of the superconducting state.

Another special lesson derives from the fact that our work was launched by John's thinking before BCS had solved the problem. In his talk that inspired me to try to measure the nuclear relaxation in a superconductor, he was presenting a theoretical approach he later abandoned. But he had already such a depth of understanding that his thinking had tremendous validity. That is why our experiment turned out to have an interesting result. It has made me realize how important it is to listen with respect to the thinking of other scientists, even when there is great scientific controversy, because even an incorrect or an incomplete theory may contain crucial elements of truth.

John was remarkable in his ability either to work by himself or to work cooperatively with others. The BCS theory is a triumph of both of these strengths. He identified and attracted the wonderful postdocs David Pines and Leon Cooper. And he was the magnet who drew Bob Schrieffer, who could have gone anywhere he wanted for graduate work, to Illinois. John's kindness, his openness, and his friendliness were ever manifest. This spirit attracted those brilliant colleagues who shared these same qualities. Chuck and I will forever be grateful for our time with John, Leon, Bob, and David.

Acknowledgments

For me, a key part of this paper is the collection of images of the participants. They bring back vivid memories of these marvelous people and my association with them. I am indebted to the Harvard University Gazette and the Harvard University Archives for their gracious help in providing the image of Professor Van Vleck, and to Alfred Redfield for the image of him with his rig in his lab at Harvard. Sid Drell provided my picture dated 1950. Above all, I thank Celia Elliott for her invaluable help and warm support. She devoted many hours to finding, scanning, and, when necessary, "photoshopping" all the other images dating to the 1950s of the people who are important to me. Her work gives life to the story.

References

1. L. C. Hebel and C. P. Slichter, *Phys. Rev.* **107**, 901 (1957); L. C. Hebel and C. P. Slichter, *Phys. Rev.* **113**, 1504 (1959).
2. C. P. Slichter and E. M. Purcell, *Phys. Rev.* **76**, 466 (1949).
3. R. E. Norberg and C. P. Slichter, *Phys. Rev.* **83**, 1074 (1951); D. F. Holcomb and R. E. Norberg, *Phys. Rev.* **98**, 1074 (1955).
4. R. T. Schumacher, T. R. Carver and C. P. Slichter, *Phys. Rev.* **95**, 1089 (1954); R. T. Schumacher and C. P. Slichter, *Phys. Rev.* **101**, 58 (1956).
5. A. W. Overhauser, *Phys. Rev.* **92**, 411 (1953).
6. T. R. Carver and C. P. Slichter, *Phys. Rev.* **92**, 212 (1953); T. R. Carver and C. P. Slichter, *Phys. Rev.* **102**, 975 (1956).
7. W. S. Corak, B. B. Goodman, C. B. Satterthwaite and A. Wexler, *Phys. Rev.* **96**, 1442 (1954).
8. J. Bardeen, *Phys. Rev.* **97**, 1724 (1955).
9. W. Heitler and E. Teller, *Proc. Roy. Soc.* **A155**, 629 (1936).
10. J. Korringa, *Physica* **16**, 601 (1960).
11. C. P. Slichter, PhD thesis, Harvard University (1949).
12. For detailed background on topics in this article, such as derivation of Curie's law, see C. P. Slichter, *Principles of Magnetic Resonance*, 3rd edition (1996), Springer, New York. Referred to the present article as "*Principles*".
13. H. G. B. Casimir and F. K. du Pré, *Physica* **5**, 507 (1938).
14. C. J. Gorter, *Paramagnetic Relaxation* (Elsevier Publishing Company, Inc., New York, 1947).
15. J. H. Van Vleck, *Phys. Rev.* **57**, 426 (1940).
16. See *Principles*, Chapter 1.
17. R. E. Norberg, *Phys. Rev.* **93**, 638 (1952).
18. G. D. Watkins and R. V. Pound, *Phys. Rev.* **82**, 343 (1951).
19. We later extended the method to obtain phase sensitive detection for signal averaging spin echoes. J. J. Spokas and C. P. Slichter, *Phys. Rev.* **113**, 1462 (1959).
20. J. Bardeen and D. Pines, *Phys. Rev.* **99**, 1140 (1955).
21. L. N. Cooper, *Phys. Rev.* **104**, 1189 (1956).
22. J. Bardeen, L. N. Cooper and J. R. Schrieffer, *Phys. Rev.* **106**, 162 (1957).
23. L. C. Hebel, PhD thesis, University of Illinois, 1957.
24. F. Reif, *Phys. Rev.* **102**, 1417 (1956); *Phys. Rev.* **106**, 208 (1957).
25. R. W. Morse and H. V. Bohm, *Phys. Rev.* **108**, 1094 (1957).
26. L. N. Cooper, *Nobel Lectures: Physics 1971–1980*, ed. Stig Lundqvist (World Scientific Publishing Co., Singapore, 1992).
27. G. Seidel and P. H. Keesom, *Rev. Sci. Instr.* **29**, 606 (1958).
28. A. G. Redfield and A. G. Anderson, *Physica* **24**, S150 (1958); A. G. Redfield, *Phys. Rev. Lett.* **3**, 85 (1959); A. G. Anderson and A. G. Redfield, *Phys. Rev.* **116**, 583 (1959).

SUPERCONDUCTIVITY: FROM ELECTRON INTERACTION TO NUCLEAR SUPERFLUIDITY

David Pines

Physics Department and ICAM, UC Davis, Davis, CA 94103, USA

I present an expanded version of a talk given at the Urbana symposium that celebrated the fiftieth anniversary of the publication of the microscopic theory of superconductivity by Bardeen, Cooper, and Schrieffer — BCS. I recall at some length, the work with my Ph.D. mentor, David Bohm, and my postdoctoral mentor, John Bardeen, on electron interaction in metals during the period 1948–55 that helped pave the way for BCS, describe the immediate impact of BCS on a small segment of the Princeton physics community in the early spring of 1957, and discuss the extent to which the Bardeen–Pines–Frohlich effective electron-electron interaction provided a criterion for superconductivity in the periodic system. I describe my lectures on BCS at Niels Bohr's Institute of Theoretical Physics in June 1957 that led to the proposal of nuclear superfluidity, discuss nuclear and cosmic superfluids briefly, and close with a tribute to John Bardeen, whose birth centennial we celebrated in 2008, and who was my mentor, close colleague, and dear friend.

1. Introduction

In 1948, when I was beginning my Ph.D. thesis research with David Bohm at Princeton, nearly 40 years had passed since the discovery of superconductivity, yet the prospects for developing a microscopic theory that explained it continued to be grim. While it could be argued that developing a theory of superconductivity was the outstanding challenge in all of physics, the failed attempts to solve it by the giants in the field, from Einstein, Bohr, Heisenberg, Born, Bloch, Landau and Fritz London to the young John Bardeen (see J. Schmalian, this volume), had led most theorists to look elsewhere for promising problems on which to work, despite the fact that considerable progress had been achieved experimentally and phenomenologically on the properties of the superconducting state.

Superconductivity was obviously an amazing emergent electronic phenomenon, in which the transition to the superconducting state must involve a fundamental change in the ground and excited states of electron matter, but efforts to understand how an electron interaction could bring this about had come to a standstill. A key reason was that one had so little fundamental understanding of the influence of electron interaction on normal state metallic properties, much less how it could bring about superconductivity. In other words, the success of a nearly free electron model for most metallic properties offered no clues for how some metals could be turned into superconductors by an electron interaction that seemed barely able to affect their normal state behavior.

Landau famously said "You cannot repeal Coulomb's law," but as far as the normal metals were concerned, it appeared that the best way to deal with it was to ignore it. For example, when John Bardeen sought in 1936 to go beyond the free electron model in a calculation of corrections to the electronic specific heat coming from electron interaction, he ran into serious trouble. Because of the long range of the Coulomb interaction, which varies as $1/r$, the lowest-order correction term to the electron self-energy (coming from the Hartree–Fock exchange energy) is such that the specific heat develops a term that varies as $T/\ln T$, a variation that was not seen experimentally. Similar difficulties awaited those who attempted to carry out first-principles calculations that went beyond the Hartree–Fock approximation in calculating the influence of electron interaction on the cohesive energy, the bandwidth, etc. Assuming that the interaction between electrons was screened, as a simple static calculation by Fermi and Thomas had suggested, provided a way out for calculations of the perturbative corrections of the specific heat, but one then could no longer obtain a qualitative, much less quantitative account, of the ground state and cohesive energy of simple metals.

In 1948, Bohm was in his second year of an assistant professorship at Princeton. He was widely regarded as one of the very best of Robert Oppenheimer's students to have come out of Berkeley and as perhaps one of the three best of the post-WW2 generation of theoretical physicists (the others being Richard Feynman and Julian Schwinger). He had become deeply involved in the issue of the transition from a quantum to a classical description of measurement theory in the course of working on what became his superb text on quantum mechanics. He was also interested in ways to deal with the divergences in quantum electrodynamics and in extending his work with Eugene Gross on the interplay of collective and single particle behavior in classical plasmas to the quantum domain in the hope that by so doing one might get an improved understanding of electron interaction in metals and thereby make a start on developing a microscopic theory of superconductivity.

I had met Bohm much earlier in 1943, when, as an undergraduate in Berkeley, I had heard him give a guest lecture on plasmas in 1943 in Joe Weinberg's course in electrodynamics, and I came to know him reasonably well before going off to the Navy in July, 1944. But it was through Robert Oppenheimer that I came to be Bohm's graduate student. When I returned to Berkeley in 1946 as a graduate student, it was my hope and expectation to do a Ph.D. thesis with Oppenheimer, who was as inspiring as a teacher of physics as he had been as the leader of the Manhattan Project. When Opje (his Dutch nickname that we all used) decided to leave Berkeley to become Director of the Institute for Advanced Study in Princeton the following year, I very much wanted to follow him to Princeton so I could do a thesis with him there. Opje liked the idea and arranged for me to receive late admission to the Physics Department at the university. I then wrote to Bohm to inquire about housing, and he wrote back that his then roommate, S. Kusaka, had tragically died in a boating accident, and that I was welcome to share his room in a house on Nassau St.

During those immediate post-war years, it was not unusual for graduate students to know junior faculty members well, because the world of physics was still quite small, and many graduate students were in their mid-twenties, having spent some three to five years in the armed services or government laboratories before returning to academic life. I shared a room with Dave Bohm for some nine months, during which I learned a very great deal of physics through our many informal conversations over shared meals. Moreover, since Opje had become far too busy with directing the Institute for Advanced Study and providing advice to the government on atomic weapons to take on any more students, I decided that if I passed my qualifying examination in the spring of 1948, I would very much like to do my Ph.D. research with Bohm. I passed the prelim, and Dave took me on as his student. He invited me to share his office, which I did for the next two years while I carried out the research for my 1950 Ph.D. thesis, "The Role of Plasma Oscillations in Electron Interactions."

2. My Ph.D. Thesis

Bohm's ideas for my research were based on his knowledge of classical plasmas, ionized gases made up of freely moving electrons and positive ions, in which one found high frequency collective longitudinal modes, plasma oscillations. These are collisionless modes (like the zero sound mode subsequently proposed by Landau) whose origin is not in the frequent collisions between electrons that bring about hydrodynamic sound at long wavelengths, but rather in the influence of the mean electric fields produced by their mutual Coulomb interaction. Because of the long-range of that interaction, the frequency of a plasma oscillation in the limit of long wavelengths is finite, being $\omega_p = (4\pi \mathrm{N}e^2/m)^{1/2}$, where N is the electron density and m its mass.

A second aspect of the emergent behavior Bohm had studied in classical plasmas was the screening of low energy electronic motion; as a given electron moves it polarizes the plasma in such a way that its interaction with another electron is screened, with the screening length being that first calculated by Debye, $[\langle v\rangle/\omega_p]$, where $\langle v\rangle$ is an average electron velocity. With Eugene Gross, my predecessor as his student, Bohm had studied in detail the interplay between the resulting collective and individual particle behavior in classical plasmas. The task Bohm set me was to extend his work with Gross to the quantum domain, where it might be applicable to a simple model of a metal, electrons moving in a uniform background of positive charge, a state of matter that Conyers Herring subsequenty dubbed "jellium".

Bohm suggested that I attempt to develop a Hamiltonian approach in which plasma oscillations appeared explicitly, and use this to explore the extent to which that this would take into account the major consequence of the long range part of the Coulomb interaction. As a warm-up exercise, I studied the "transverse" analogue of the Coulomb interaction, the magnetic interaction between electrons produced by their coupling to electromagnetic waves. I developed a classical and quantum Hamiltonian approach in which I used a series of canonical transformations to calculate both the modification in the electromagnetic wave dispersion relation coming from their coupling to the electron plasma and the modifications in the Biot–Savart law that described the magnetic interaction between electrons arising from that coupling.

I then developed a corresponding classical Hamiltonian approach to the Coulomb interaction by working in a longitudinal gauge in which subsidiary conditions linked the vector potential to the charge density fluctuations. By carrying out a series of canonical transformations on the starting Hamiltonian, I could obtain a new Hamiltonian in which plasma oscillations appeared explicitly, while their coupling to each other and to the electrons could be calculated. I found that within what is now known as the random phase approximation (RPA) this coupling produced only small corrections to Debye's effective electron interaction while the subsidiary conditions could easily be satisfied. The long-range part of the Coulomb interaction was completely redescribed in terms of the fields describing plasma oscillation, leaving in its wake, a short-ranged screened interaction between electrons with a screening radius, $[\langle v \rangle / \omega_p]$. I was thus able to provide a microscopic justification for the physical approach taken by Bohm and Gross for the classical plasma.

It seemed straightforward to extend this Hamiltonian approach to the quantum domain, but at the time Bohm and I did not see an easy way to sort out the quantum influence of the subsidiary conditions on electron motion. Therefore in my thesis, I confined myself to a derivation of the dispersion relation for the quantum plasma in two different ways: a self-consistent field approach and an equation of motion approach for its density fluctuations, $\rho_k = \Sigma_i \exp[ik \cdot r_i] = \Sigma_p c_{p-k} * c_p$, on using the second quantized operators that create and destroy individual electrons. In both approaches, I found that on making the RPA, the plasma dispersion relation could be written as:

$$1 = [4\pi e^2/m]\Sigma_i 1/[(\omega - k \cdot P_i/m)^2 - h^2 k^4/16(m\pi)^2]$$

where ω is the oscillation frequency for a collective mode of momentum k, and one is summing over all electron momenta, P_i, from which it follows that at long wavelengths, $\omega = \omega_p$. Moreover, ρ_k and its time derivative could be taken to be the field coordinates, Q_k and P_k, for quantized plasma waves, in that when suitably renormalized, these obeyed the canonical quantum commutation relations.

I concluded in my thesis that a proper quantum treatment of a starting Hamiltonian identical to that considered classically, with due account of the subsidiary conditions, would lead to a new Hamiltonian in which the long range part of the Coulomb interaction was replaced by that for the plasma waves, while the remaining short range part would resemble closely a screened interaction with the quantum Fermi–Thomas screening length, and expressed the hope that in this way, one could begin a systematic study of the influence of electron interaction on metallic properties.

3. Physics, Politics, and Collaboration

My thesis work and its subsequent publication (which I will describe shortly) became entangled in politics when I was less than half-way through my research, as Bohm became a target of the House Committee on Un-American Activities as part of its efforts to taint his mentor, Robert Oppenheimer, with a Communist brush. Bohm received a notice to appear before that now infamous group in late April 1949, and in his subsequent appearances (in late

May and early June) refused to testify about his one-time Berkeley Radiation Laboratory colleagues or his own involvement with the Communist Party (of which one subsequently learned he had been a member for some nine months during 1942–43) on the ground that such testimony would interfere with his First Amendment right of free speech and his Fifth Amendment right to avoid self-incrimination.

Dave was by then sharing a house near Roosevelt, NJ with my wife and myself and another graduate student couple, and needless to say our conversations for some months were more about politics than physics. He was, however, supported by his colleagues and the Princeton administration, so that despite the distractions occasioned by his trips to Washington, he was able to continue teaching and directed my thesis research until its completion in July 1950.

However Bohm's support by the Princeton administration disappeared in December 1950, when he was indicted by Congress and arrested for his refusal to testify. Without waiting for the outcome of his trial, Harold Dodds, the then President of Princeton University, suspended Bohm from his position on the Princeton faculty, while proposing to continue his salary as long as he did not set foot on the campus. Despite his 5/31/51 acquittal in a federal court, when his initial university contract expired in July 1951, in a decision made by Dodds without input from the Physics faculty, Bohm was not rehired by the university. Despite efforts to find him another faculty position or research appointment in the US by those in the physics community who knew him well, from Albert Einstein to colleagues at Bell Laboratories and RCA Labs, and especially by Lyman Spitzer, the leader of Princeton's Matterhorn Project that was directed toward harnessing the fusion energy of the sun, the political atmosphere in the US was at that time so poisonous that he could not obtain any physics-related position in the US. He became arguably our most distinguished scientific political exile when, not long after his seminal text on quantum theory had been published, he accepted a position at the University of São Paolo. He went there in October, 1951, beginning an exile from the US scientific community that continued, apart from occasional visits, until his death in London 41 years later.

Bohm's scientific exile from the US was enormously costly to both Dave and to the United States. Dave, who was described once by Eugene Gross as a "professional thinker", typically had some five new ideas a day. He relied in no small part on daily conversations with his students and colleagues and his frequent contact with experimentalists to decide which of these to pursue. Once he was no longer in a major center of physics research, he focused more and more on his unsuccessful hidden variables theory and, apart from a brilliant paper with Aharonov, gradually became lost to the world of physics as he made the transition from physics to philosophy of physics to philosophical observations about the brain and the universe, etc.

Moreover, at the time Dave left the US, he arguably understood plasmas better than any other scientist then alive, and his advice and scientific input on work at Princeton and elsewhere on attempts at fusion energy would have been invaluable. In retrospect, had he been able to actively participate in the program, through his deep understanding of plasma instabilities and plasma turbulence he would arguably have saved the US at least five years and countless millions of dollars.

As a result of Bohm's problems with Princeton and his move to Sao Paulo, it took some time to publish the major findings in my thesis. After receiving my Ph.D. I had gone, in September 1950 to the University of Pennsylvania as an Instructor in Physics. Since soon thereafter Bohm was indicted and forbidden to work on the Princeton campus, he would come to our apartment in Philadelphia so that we could begin writing up our work for publication. Before he left for Brazil we managed to complete the first two in our planned series of papers describing and expanding on my thesis.

In the first[1] we described the warm-up problem in my thesis — a collective Hamiltonian approach to the modification of transverse electromagnetic waves by a quantum plasma and the description of electron interactions in a quantum plasma that are induced by their coupling to transverse electromagnetic fields.

In the second[2] we used an equation of motion approach to examine the interplay between collective and individual particle motion in mainly classical plasmas. We introduced the term, random phase approximation (RPA), to describe the principal approximation made in our approach, and argued that within the RPA for many metals quantized plasma oscillations would be lightly damped. Importantly, we noted that there was experimental verification of our prediction of quantized plasma oscillations in metals, a verification that increased my confidence in our ability to develop a satisfactory microscopic theory of their behavior.

Conyers Herring had brought to our attention the pre-WWII characteristic energy loss experiments of Ruthemann and Lang who had studied the behavior of fast (keV) electrons passing through thin metallic films. For some metals, instead of finding the expected featureless distribution of energy losses, they found a series of discrete peaks. For Be and Al these were at 19 eV and 14.7 eV, amazing close to the energy, $[h\omega_p/2\pi]$, contained in a single quantum of plasma oscillation for these materials, [18.8 eV and 15.9 eV, respectively], computed by assuming that all the valence electrons outside the closed shells of the atoms in these metals were free, while the mean free path for producing such losses was also quite close to that calculated under this assumption.

It was while at Penn that I also realized that the RPA might represent a quite general way of identifying potential collective modes in any many-body system, and suggested to my first and only student there, Mel Ferentz, that it be applied to the nuclear many-body problem, with particular attention to explaining the nuclear giant dipole resonance as a collisionless collective mode brought about in nuclear matter by hadron-hadron interaction. In work that was published in collaboration with Murray Gell-Mann, who suggested how to deal with its behavior in a finite nucleus, we showed that the resulting energy was in good accord with the experimental findings.

However, completion of the third paper with Bohm, in which we planned to provide the microscopic justification for the central results on electron interaction in metals obtained in my Ph.D. thesis, was delayed substantially by his departure for Brazil and mine to Urbana in July 1952. In the interest of full disclosure, my move from Penn to Urbana was triggered by the quite unexpected and initially dismaying news that I was not going to be recommended for promotion in the Penn Physics Department. As I look back on that decision, made just after I had finished the above papers and received my first national recognition (an invitation to give an invited paper at the APS meeting in March 1952), I cannot decide whether it was simply because I was too impatient with my departmental elders or whether my connection to Bohm played a role. Either way, it was for me a most fortunate turn of events, as when I sought the advice of Fred Seitz on what to do next, he responded that he would talk with John Bardeen, whom he had recently attracted to the Illinois faculty. John responded by inviting me to come to Urbana as his first post-doctoral research associate.

4. John Bardeen and Superconductivity: The Early Years

Let me set the stage for my time in Urbana by sharing with the reader parts of a recent memoir about John Bardeen and his early efforts to understand superconductivity.[3] "John Bardeen, whose 100th birthday we celebrated on May 23, 2008, was arguably the most influential scientist/inventor of the latter part of the twentieth century. Through his scientific discoveries, his instinct for invention, and his impact on his colleagues, he made possible the electronics revolution and the information explosion that have changed dramatically our every day lives.

Bardeen was an authentic American genius. Scientific genius is no easier to pinpoint than artistic genius. It derives from a combination of factors, including — but not limited to — intuition, imagination, far-reaching vision and exceptional native gifts that blossom into significant technical skills, and the willingness and ability to challenge conventional wisdom. Perhaps even more importantly, scientific genius depends on an instinct for invention, an ability to focus on the problem at hand, to juggle multiple approaches, and a fierce determination to pursue that problem to a successful conclusion.

The inventor of the transistor and leader of the team that developed the microscopic theory of superconductivity, John Bardeen possessed all of these qualities. But what made him different from so many fellow scientific geniuses in physics of the 20th century — Einstein, Bohr, Dirac, Feynman, Landau, Pauli, and Oppenheimer? The answer lies not only in his two Nobel Prizes for Physics (in 1956 and 1972), but also his remarkable modesty, his deep interest in the application of science, and his genuine ability to collaborate easily with experimentalist and theorist alike."

Early on in his scientific career, Bardeen had become deeply interested in superconductivity. "In an abstract of a talk he gave at the Spring (Washington) meeting of the American Physical Society in May 1941, Bardeen summarized what might be described as his first "wrong" explanation for superconductivity — suggesting that superconductivity might originate in a small periodic distortion of the lattice that produced a fine grained zone structure in momentum space such that the energy gain from the resulting discontinuities would outweigh the cost of producing the distortion. He argued that one could achieve perfect diamagnetism (the ability of a superconductor to screen out external magnetic fields) in this way with only a fraction of the electrons near their ground state being involved and used his earlier calculations of the resistivity of simple metals to estimate the strength of the electron-lattice required to bring this about. He concluded that a high density of valence electrons and a strong electron-phonon interaction were favorable to superconductivity, a remarkably prescient argument of which it could be said that he was right for the wrong reasons."

Bardeen returned to superconductivity in early 1950, "when he learned in a phone call from Bernard Serin of the discovery of the isotope effect (the dependence of the superconducting transition temperature on the inverse square root of the isotopic mass) on the superconducting transition temperature of lead, he essentially dropped everything else to return to his work of a decade earlier on a connection between lattice vibrations and superconductivity. ... Bardeen, and independently (and in advance of the Serin–Maxwell result) the British theorist, Herbert Frohlich, decided that the key physical effect must be a phonon-induced change in the self energy of a fraction of the electrons near their ground state configuration, and both published their ideas in the early part of 1950. However, neither was able to demonstrate how this could lead to the formation of a coherent state of matter that could flow without resistance and screen out external magnetic fields ("perfect" diamagnetism), the key physical idea introduced by Fritz London to explain, at a phenomenological level, superconductivity. So this approach turned out to be Bardeen's second "failed" effort to develop a microscopic theory.

It was during this period, in the spring of 1950, that I first met Bardeen, who had come to Princeton to teach a seminar on the physics of semiconductors. Bardeen was already a legend among those of us who were Princeton graduate students — as an exceptionally gifted young theorist who had been Wigner's best student and gone on to invent the transistor. Bardeen's lectures on semiconductors were not memorable, but were typical of his lecturing style — clear, informative, low key, softly spoken, with little in the way of emphasis. A highlight of those weekly visits were his discussions about superconductivity with my office-mate, David Bohm, and the excitement he conveyed that a solution might be on the way ...

Although it became increasingly clear to Bardeen that a change in the electron self-energy would not in itself lead to superconductivity, he continued at Bell Labs to work on ways the coupling of electrons to lattice vibrations could play a significant role. He was, however, working in a highly unsatisfactory atmosphere (he was still a member of Shockley's group, and Shockley gave him no latitude to work with experimentalists) and began to explore opportunities elsewhere. When he told his old friend, Fred Seitz, who had moved in 1949 to lead a group in solid-state physics at the University of Illinois in Urbana-Champaign, that he was ready to leave Bell Laboratories, Seitz quickly put together an offer for Bardeen to join him there, with a joint appointment in the Department of Electrical Engineering and the Department of Physics, which Bardeen accepted, and moved to Urbana in the fall of 1951."

5. With Bardeen in Urbana: 1952–1955

When I arrived in Urbana in July 1952, to work with Bardeen as the Physics Department's first Research Assistant Professor, I was given a desk in a corner of John's office. This gave me a unique opportunity to work with John on a variety of problems and to learn from him a way of doing physics that proved invaluable in my subsequent scientific career.

I arrived with the unfinished business of continuing at long distance the collaboration with David Bohm on extending our collective description to the quantum domain, but was eager to take advantage of the opportunity that working with Bardeen provided to move in some new directions. I asked John what I might work on, and he suggested I look at the polaron problem — the motion of slow electrons that are strongly coupled to the optical modes found in polar crystals — beginning with papers by Frohlich and Pekar. John suggested I do this in the hope that an improved understanding of strong coupling might provide insight into the role that electron-phonon coupling played in bringing about superconductivity.

It was part of what I subsequently realized was John's multi-pronged approach to solving the problem. These ranged from immersing himself in the experimental data and developing a phenomenological theory that explained many of the key experimental facts to exploring the use of matrix methods to study the behavior of a small number of electrons outside the Fermi surface. His approach included, of course, developing a better understanding of the role of phonons in bringing out an effective electron interaction that might be responsible for superconductivity, a problem on which we subsequently collaborated.

The polaron problem involved extending to realistic coupling constants (that were of order 3 to 6), the perturbation-theoretic calculations of the Frohlich group that were only valid for coupling constants less than 1. Soon after I started work on it, I encountered T. D. Lee in a hallway of the Physics Building. We started comparing notes on what we were working on that summer and realized, in the course of the next ten minutes, that our research could be connected, T. D., who was spending the summer in Urbana, was interested in applications in particle physics of the intermediate coupling approximation that Tomonaga had developed for meson-nucleon interactions, and we realized that it might be possible to adapt the Tomonaga approach to the strong electron-phonon interactions governing polaron motion. In the next few days we carried out the adaptation and used Tomonaga's approach to calculate the changes in the electron ground state energy and effective mass brought about by that coupling.

Later that summer, as I was describing what T. D. and I had done[4] to Francis Low (who like me had arrived in Urbana that summer) Francis argued that there had to be a simpler way to obtain our result, and he soon proved right. Thanks to Francis we found a simple and elegant expression for the polaron wave function and described it in our subsequent paper[5] (LLP) which showed that our intermediate coupling solution was equivalent to arriving at a coherent state in which the phonons in the cloud of strongly coupled phonons around the electron were emitted successively into the same momentum state.

At first sight, this work seemed to offer little insight into how one might develop a microscopic theory of superconductivity, but matters proved otherwise. Thus Bob Schrieffer has cited that LLP coherent state as the inspiration for his January 1957, invention of the "Schrieffer" wave function for the ground state of a superconductor that was the start of BCS. To see the connection, note that the LLP wave function took the form:

$$\Psi_{\rm LLP} \sim \Pi_k \exp[\Sigma_k f(k)(a_k^* + a_k)]\Psi_0$$

where Ψ_0 was the ground state wavefunction and $f(k)$ a form factor for the phonon. Schrieffer's brilliant insight was to try a ground state wavefunction for the superconductor in which the LLP phonon field was replaced by the coherent pair field of the condensate, $b_k^* = c_{k\uparrow}^* c_{-k\downarrow}^*$; his wavefunction thus took the form:

$$\Psi \sim \Pi_k \exp[\Sigma_k b_k^* f(k)]\Psi_0$$

which is easily shown to reduce to the BCS wavefunction,

$$\Psi_{\rm BCS} \sim \Pi_k[1 + b_k^* f(k)]\Psi_0$$

In the fall of 1952, I returned to my (by now long-distance) collaboration with David Bohm on developing a collective description of the quantum plasma. There appeared to be two stumbling blocks to developing a Hamiltonian approach that was isomorphic to that introduced for the magnetic interactions in a quantum plasma: clarifying the role of the subsidiary conditions that accompanied our introduction of supplementary fields to describe the quantized plasma oscillations, and justifying our use of the RPA in dealing with the long range part of the Coulomb interaction. We gradually overcame both of these in the course of a correspondence that lasted some seven months, ending in mid-April 1953. It involved the exchange of well over forty handwritten letters that came to a total of several hundred onion-skin pages, and not a little frustration as each of us began to worry that the collaboration was taking too much time away from other projects we had undertaken. Thus I found in my copies of our correspondence the following words from David Bohm to me: "I am beginning to wonder if we will ever finish the paper. It had better be soon, as I won't have time to work on it any more, as I have many other duties. Now for a few general remarks". I wrote to Bohm at about this time — April, 1953 — "I've put in about as much time as I possibly can on the writing of this paper — so while I welcome any further suggestions you might make, you can rest assured that they will probably be followed in toto. Thus I don't envisage any further tightening of this Mss on my part, and I'd be willing to send it to the *Phys. Rev.* as it is!!"

In the course of our work and correspondence we decided to divide the planned paper into two parts; Part III of the series would be written jointly and describe the collective Hamiltonian approach to the quantum plasma and overall physical picture, while I would write Part IV which dealt with the application of our results to electrons in metals.

Both papers were finally typed up in Urbana and sent off to the *Physical Review* in mid-May. In Paper III[6] we introduced the supplementary longitudinal fields that were coupled to the electrons and would, following a series of canonical transformations, describe the quantized plasma oscillations that were the major consequence of the long range part of the Coulomb interaction. We showed in an explicit calculation that if we did this only for wavevectors smaller than $k_c \sim \omega_p/v_f$ where v_f is the Fermi velocity, the corrections to the RPA description of the long-range part of the Coulomb interactions were small. We also presented arguments that the accompanying subsidiary conditions would have no influence on the ground state energy or wave function, and gave estimates of their role in general that suggested these would play a negligible role. We thus arrived at the following basic model Hamiltonian for jellium,

$$H = \Sigma_i p_i^2/2m^* + H_{\mathrm{pl}} + H_{\mathrm{sr}}$$

where $m^*(\sim m)$ was an effective mass that took into account the change in electronic mass produced by their coupling to plasmons, H_{pl} described the plasma waves brought about by the Coulomb interaction and H_{sr} described the residual short-range interaction between the electrons.

In IV[7] I showed that this Hamiltonian provided for the first time a microscopic justification of the nearly free electron model of metals. I showed it was possible to obtain a self-consistent account of the influence of electron interaction on metallic properties, and calculate, at the same level of approximation, both the correlation energy and the specific heat, with results in good agreement with those found experimentally for simple metals. The key physical argument presented there was that electrons in metals are never seen in isolation, but always as quasi-electrons, a bare electron plus its accompanying screening cloud (a region in space in which there is an absence of other electrons). It is these "quasi-electrons" that interact via a short-range screened Coulomb interaction. For many metals, the behavior of these quasielectrons is moreover protected by the adiabatic continuity that enables them to behave like the quasiparticles in the Fermi liquid theory developed subsequently by Lev Landau.

Once papers III and IV were off to the *Physical Review*, I began work on generalizing Bohm–Pines theory to study the influence of electron-electron interactions on electron-phonon interaction in metals, as a step toward developing a microscopic theory of superconductivity in which phonons could play the key role the isotope effect experiments had established. Frohlich[8] had proposed that since the phonon-induced modification in the electron self-energy could not explain superconductivity, perhaps the phonon-induced interaction between electrons might provide the answer. He found, to lowest order in the electron-phonon coupling constant, that this interaction would be attractive for electrons lying within a characteristic phonon energy of the Fermi surface. But while this seemed an interesting possibility for bringing about superconductivity, it was difficult to see how such a comparatively weak interaction could prevail over the much larger Coulomb interaction he had neglected. What I was after was a Hamiltonian description of the full problem — electrons interacting with each other and with phonons — that would tell us about the new features arising from their coupling to phonons and the role played by electron screening and repulsion in determining the overall effective electron interaction.

Before he left Princeton, David Bohm had started work there on the same problem with Tor Staver, a young Norwegian graduate student. They used the RPA to study collective motion in a coupled electron-ion plasma and were able to derive a simple expression for the longitudinal phonon dispersion relation for jellium. They then sought to develop a physical picture for the phonon-induced interaction along lines similar to those Bohm had used with Gross for classical plasmas, studying the response of an electron to the phonon wake produced by another. They were in the midst of doing this, when Staver tragically died in a ski accident before their work could be completed and published.

In developing a collective Hamiltonian approach to the full problem, I was at first successful, in that I could easily obtain the Bohm–Staver dispersion relation. However, I became stuck on how best to develop a fully self-consistent treatment of the electron-electron interaction. One morning in late 1953, in the course of discussing my lack of progress with John, he suggested that perhaps what was missing was an explicit addition of ionic motion to the canonical transformation that took one from an arbitrary supplementary field to the collective coordinates that described the collective modes of the coupled electron-electron-phonon system. We tried putting that in, and everything worked.

In the paper describing our results[9] we showed how once this was done, a straightforward generalization of the canonical transformations Bohm and I had utilized earlier led to a self-consistent account, within the RPA, of the way electron interactions modified their coupling to ions and to one another and how the combined ionic and electronic Coulomb interactions gave rise to both sound waves and plasma oscillations. We found that for electrons lying on the Fermi surface, their phonon-induced interaction was cancelled by their direct screened interaction, but that for electrons lying within a characteristic phonon energy of the Fermi surface, the phonon-induced interaction could win out, producing a net attraction. We closed our paper with the prophetic line, "The equations we have presented here should provide a good basis for development of an adequate theory."

As Nozieres subsequently pointed out, our results for the net effective frequency-dependent electron interaction at low frequencies, V_{eff}, could be put in especially simple form for jellium,

$$V_{\text{eff}}(q,\omega) = [4\pi e^2/q^2\varepsilon(q,0)][1 + \omega_q^2/(\omega^2 - \omega_q^2)]$$

where $\varepsilon(q,0)$ is the static dielectric constant. The screened Coulomb interaction sets the overall scale for the strength of the interaction, and it is evident that the net interaction will always be attractive for those electrons of momentum $\mathbf{p}$ and $\mathbf{p}+\mathbf{q}$ near the Fermi surface whose energy difference, $\omega = \varepsilon_{p+q} - \varepsilon_p$, is below the natural resonance frequency, ω_q, of the phonon that is being virtually exchanged.

6. The 1954 Solvay Congress

Thanks to Fred Seitz and John, I was invited to describe the above results in the opening talk of the 1954 Solvay Congress on "Electrons in Metals". I was more than a little apprehensive about how my talk would be received, because there in the front row was Wolfgang Pauli, who had a justly-deserved reputation for his ferocious put-downs of new theoretical results. (I was far less worried about the presence in my audience of his fellow Nobel Laureates Lawrence Bragg and Louis Neel, or that of those distinguished colleagues there who went on to win the Nobel Prize — Neville Mott, Lars Onsager, John Van Vleck, Clifford Shull and Ilya Prigogine). I began with a summary of our work that clearly demonstrated that Heisenberg's attempts to find a microscopic theory of superconductivity by a careful treatment of the Coulomb interaction were doomed to failure, and I started to relax a bit when Pauli started shaking his head vigorously to indicate his agreement with my findings. When I finished my report,[10] his was the first comment, which I quote: "I always told that fool Heisenberg that his theory of superconductivity was wrong."

7. Deciphering, Teaching, and Applying BCS

In early 1955 I left Urbana to return to Princeton to a promised tenure-track position that subsequently disappeared, and it was there, some two years later, that I received a brief letter from John reporting that he thought superconductivity had been solved. He enclosed a dittoed copy of their not-yet-published PRL. I shared the news with Elihu Abrahams, who was newly arrived at Rutgers, and my prize graduate student, Philippe Nozieres, who had come from Paris to work with me. Filled with excitement, we decided to see if we could flesh out the details of what BCS had done. After three intensive days in the living room of our house on Clover Lane we succeeded sufficiently well that I was able to teach it to my class later that spring.

In the course of these lectures, I did some simple calculations showing how the effective interaction that John and I had derived, and that formed the starting point for BCS, led in a natural way to the famous Matthias rules for the occurrence of superconductivity. I showed these to John when he came to Princeton to give what may have been his first colloquium on BCS and he encouraged me to publish them. A footnote: when I sent the paper[11] describing these results to *Phys. Rev.* in late May, its Editors were uneasy about accepting it, since John, Bob and Leon had not yet completed their full account of their theory, but John reassured them that I was in no way trying to scoop BCS, and that what I had done complemented, not competed with, their work in progress. In fact, it could easily have been written two years earlier, since it dealt with the extent to which the above effective interaction enabled one to decide what elements would, or would not, be superconducting, not the ensuing microscopic theory.

What I did in this paper was first to see how well one could do with a "minimalist" approach to calculating the average effective interaction V, which had to be attractive to bring about superconductivity, and the product $N(0)V$ that determined the superconducting transition temperature, T_c, in BCS theory. I took the effective interaction to be that John and I had derived in the RPA, with the repulsive part a screened Coulomb interaction and

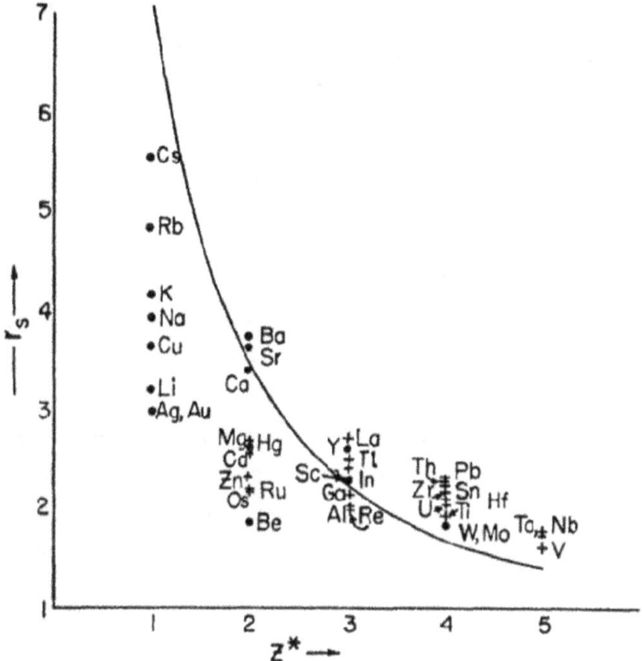

Fig. 1. The critical r_s for superconductivity as a function of Z^*. The superconducting elements are denoted by +; the nonsuperconducting elements by •.

considered separately the contributions to the attractive part coming from normal processes (in which the interacting electron momentum differences were less than a typical phonon wave vector, k_D) and the Umklapprocesses (U processes) in which their momentum difference was $q + Q > k_D$, where Q is a reciprocal lattice vector. For the latter, I further assumed that the momentum of the phonon being exchanged was always k_D, which underestimated the contribution of U processes to V. I was thus able to obtain a simple analytic expression for $N(0)V$ that depended only on the effective ion charge, Z^*, and r_s, the dimensionless interelectron spacing.

Within this simple model, the net attraction coming from N processes alone was not strong enough to overcome the screened Coulomb repulsion, so that, for example, jellium would never superconduct. When U processes were included, as Fig. 1 shows, the model turned out to do quite well at distinguishing between normal and superconducting elements in the periodic table and even had predictive power, in that it predicted the superconductivity of Mo, W, Y, Sc and Pa. I then examined more closely the influence of periodicity on the matrix elements for electron-phonon interactions, and concluded that when these are taken into account, the requirement that V be attractive and the BCS expression for T_C could together explain the empirical rules developed by Matthias.

8. BCS and Nuclear Superconductivity in Copenhagen

In the early summer of 1957 I went to Copenhagen to spend a summer at Niels Bohr's Institute for Theoretical Physics en route to Paris for a year at the Ecole Normale Superieure. My colleagues at the ITP were eager to hear about BCS, and I agreed to give two lectures describing their theory, for which a preprint was not yet available. Before I gave the first, Aage Bohr and Ben Mottelson independently gave me a strong heads-up — that I should be prepared for the possibility that Niels Bohr would interrupt early and often, as he had done a few weeks earlier during a seminar on superconductivity by John Blatt, who was, in fact, not able to finish his planned talk. So I planned accordingly. I began with a brief summary of what BCS had achieved, emphasizing that their approach was something quite new and different and that the resulting energy gap in the electronic spectrum was just what was needed to explain experiment. I also noted that BCS might be of more general applicability, since it depended solely on a net attractive interaction between fermions, and that one should therefore look into the possible superfluidity of nuclear matter and finite nuclei.

Niels Bohr listened attentively, had only a few questions as I went along, and then after my talk invited me to come to his study for a further discussion. There a few days later, in between innumerable efforts to light and relight his pipe, he explained his own unsuccessful attempts to find a way to explain the existence of a persistent current and questioned me closely as to how BCS had managed this. I left with the impression that he did not feel that either BCS or I had explained to his satisfaction the way in which the persistent current is carried by the condensate, but that overall he appeared inclined to accept its major results.

During the course of that summer, Aage, Ben and I had a number of productive discussions about applying BCS to nuclear superconductivity, and early that fall I gave a brief talk on BCS and its likely application to nuclei at an international Nuclear Conference in Rehovoth, Israel that was followed by remarks by both Aage and Ben.[12] During the following year, we wrote up our arguments concerning similarities between nuclear and superconducting behavior and suggested that the energy gap found in nuclei originated in superfluid correlations brought about by the net attractive nuclear force.[13]

Aage and Ben encouraged Spartak Beliaev to expand upon our approach, which he did, reporting on his considerable progress in explaining nuclear energy gaps and reduced inertial moments at Les Houches in the summer of 1958. The Proceedings of that seminal summer school organized by Philippe Nozieres,[14] are a gold mine for historians of science, since they also contain Bob Schrieffer's lecture notes on superconductivity and contributions from Brueckner, Bohm, me, and many others.

Our work on nuclear superfluidity stimulated Arkady Migdal to propose that neutron stars, if they existed, would be mainly made of superfluid hadrons, and Vitale Ginzburg and David Kirshnitz to note that since stars rotate, the rotational velocity of that superfluid would likely be carried by vortices. The subsequent discovery of pulsars and their identification as rotating neutron stars turned out to provide an opportunity to verify the presence of hadron superfluids in these stars and to observe directly the pinned crustal neutron superfluid that is responsible for pulsar glitches and postglitch behavior (Ref. 15 and G. Baym, this volume).

9. Why John Bardeen and Why BCS?

In the years that have passed since 1957, I have had ample opportunity to reflect on why it was that John Bardeen succeeded in his efforts to develop a microscopic theory of superconductivity, while theorists such as Landau and Feynman had failed. The answer lies, I think, in part in his approach to major problems in physics:

* Focus first on the experimental results via reading and personal contact.
* Develop a phenomenological description that ties different experimental results together.
* Explore alternative physical pictures and mathematical descriptions without becoming wedded to any particular one.
* Thermodynamic and other macroscopic arguments have precedence over microscopic calculations.
* Focus on physical understanding, not mathematical elegance, and use the simplest possible mathematical description.
* Keep up with new developments in theoretical techniques — for one of these may prove useful.
* Decide on a model Hamiltonian or wave-function as the penultimate, not the first, step toward a solution.
* Choose the right collaborators.
* DON'T GIVE UP: Stay with the problem until it is solved.

But would BCS have happened if John had stayed at Bell? I doubt it, because he did not have the resources or support there to attract postdocs and graduate students and so build a team. Would BCS have happened if he had moved to any university but UIUC? This is a good question for the historians. My own feeling is possibly not, since in Urbana, John had the support and the nurturing atmosphere that played such an important role in the development of the theory, one that enabled him to attract and inspire his youthful collaborators, Leon Cooper and Bob Schrieffer, who played such a key role in developing BCS.

Acknowledgments

In addition to the paragraphs reproduced here from the Bardeen memoir for the American Philosophical Society, in writing about my work in Urbana I have drawn upon some material that will also appear in my contribution to an online text, *Physics for the Twenty-first Century*, that is to published by Annenberg Media in early 2011.

References

1. D. Bohm and D. Pines, A collective description of electron interaction: I. Magnetic interactions, *Phys. Rev.* **82**, 625 (1951).
2. D. Pines and D. Bohm, A collective description of electron interaction: II. Collective vs. individual particle aspects of the interactions, *Phys. Rev.* **85**, 338 (1952).
3. D. Pines, *Proc. American Philosophical Society* **153**, 288 (2009).
4. T. D. Lee and D. Pines, Interaction of a non-relativistic particle with a scalar field with application to slow electrons in polar crystals, *Phys. Rev.* **92**, 883 (1953).
5. T. D. Lee, F. E. Low and D. Pines, The motion of slow electrons in polar crystals, *Phys Rev.* **90**, 297 (1955).
6. D. Bohm and D. Pines, A collective description of electron interactions: III. Coulomb interactions in a degenerate electron gas, *Phys. Rev.* **92**, 609 (1953).
7. D. Pines, A collective description of electron interactions: IV. Electron interaction in metals, *Phys. Rev.* **92**, 626 (1953).
8. H. Fröhlich, *Proc. Roy. Soc. (London) A* **215**, 291 (1952).
9. J. Bardeen and D. Pines, Electron-phonon interaction in metals, *Phys. Rev.* **99**, 1140 (1955).
10. D. Pines, The collective description of electron interaction in metals, in *Les Electrons dans les Metaux, Tenth Solvay Congress* (Institut de Physique Solvay, Bruxelles, 1955), pp. 9–57.
11. D. Pines, Superconductivity in the periodic system, *Phys. Rev.* **109**, 280 (1958).

12. D. Pines, Nuclear superconductivity, in *Proc. Rehovoth Conf. Nuclear Structure* (Interscience Press, 1957), pp. 26–27.
13. A. Bohr, B. Mottelson and D. Pines, Possible analogy between the excitation spectra of nuclei and those of the superconducting metallic state, *Phys. Rev.* **110**, 936 (1958).
14. P. Nozières (ed.), *The Many-Body Problem* (J. Dunod, Paris, 1959).
15. D. Pines and M. A. Alpar, *Nature* **316**, 27 (1985).

DEVELOPING BCS IDEAS IN THE FORMER SOVIET UNION

Lev P. Gor'kov

National High Magnetic Field Laboratory, Florida State University, 1800 East Paul Dirac Dr., Tallahassee, FL 32310, USA
gorkov@magnet.fsu.edu

The essay below is an attempt to re-create the wonderful scientific atmosphere that emerged after the basic BCS ideas first arrived in Russia in 1957. It summarizes the most significant contributions to the microscopic theory of superconductivity by Russian physicists during the next few years that gave the theory its modern form.

1. Introduction

These brief historical notes are a revised version of a lecture delivered by the author during the jubilee Symposium "BCS@50" held at University of Illinois at Urbana-Champaign (UIUC) in October, 2007.

I confine myself to a comparatively short period between the end of 1957 and the early 1960s. It was the time when the underlying ideas and methods of the microscopic theory of superconductivity itself were under careful examination in parallel with experimental efforts to test the BCS theory's predictions.[1–3]

The most significant theoretical accomplishments during that early period were the canonical transformation version of BCS (Bogolyubov), and the elaboration of superconductivity theory by using Quantum Field Theory (QFT) methods, as well as the identification of the Cooper pair wave function as the symmetry order parameter in the superconducting phase (Gor'kov). The history of these events in Russia, during the days of the former Soviet Union (USSR), is not known well in the West.

2. Science in the USSR

After the end of the Second World War exact sciences in the USSR remained practically isolated from the West. By 1957 direct contacts between scientists were cut to minimum on both sides. Soviet authorities strictly limited the Institutes of the Russian Academy's subscriptions to Western journals, partially in an attempt to save on foreign currency. New issues of *Physical Review* used to arrive after a delay. Acquiring any modern Western experimental equipment was difficult due to restrictions imposed by the US on the trade with the USSR. While these restrictions had their anticipated negative impact by introducing an unnecessary parallelism in research, ironically, it also strengthened the traditional originality of Russian Science. Russian Science continued to maintain a leading position in many areas of physics, theoretical physics and mathematics.

Participation in atomic projects had boosted the prestige of science and secured a comparative independence for physicists in Soviet society. Science in the USSR was organized quite differently than in the Western university system. Research, at least most of non-classified research, was not concentrated in universities as in the West, but mainly at the Institutes of the Academy of Sciences of the USSR. Soviet Physics and Mathematics also remained sound and healthy due to concerted governmental effort to attract and educate a whole new generation of young scientists. The pre-war generation of physicists remained active and at the height of their power; those included such figures as Petr (Pyotr) Kapitza, Lev Landau, and Igor Tamm, to name just a few.

Of especial significance for the exact sciences in the USSR was the tradition of Scientific Schools. In connection with the theory of superconductivity I mention only two of them: the Landau School and the group headed by N. N. Bogolyubov.

Among the physicists that formed the Landau School there were many internationally respected scientists, the heads of theoretical groups at the Institutes of the Academy of the USSR. Affiliation with the Landau School had never born an official character. Most common to the Landau School was a broad area of interests. The School united theorists who shared a no-nonsense attitude toward physics, and a high level of the professionalism, including, in particular, an easy use of the mathematical apparatus. The carefully elaborated educational process played an important role in achieving all of this. To be affiliated with the Landau School, one had to pass the exams of the famous Landau "Theoretical Minimum" that comprised a considerable part of all ten volumes of the current Landau and Lifshitz *Course of Theoretical Physics.* As of today, many theorists got through the school of the "Theoretical Minimum"; the beginning of Landau's efforts to educate his disciples ran back to his stay in Kharkov in the 1930s.

The only official position Landau had held since 1939 was that of the head of the small theory department at the Kapitza Institute for Physical Problems in Moscow. For a long time his group consisted of two people: E. M. Lifshitz, I. M. Khalatnikov, with A. A. Abrikosov admitted around 1950. Three younger staffers, L. P. Gor'kov, I. E. Dzyaloshinskii and L. P. Pitaevskii, joined the Landau group in the mid-1950s.

The personal style of N. N. Bogolyubov and his interests in Statistical Physics were of a more mathematical character. In the mid-1950s he was the head of the Theoretical Department at the Steklov Mathematical Institute of the Soviet Academy of Sciences in Moscow, while still preserving posts in Kiev, Ukraine. In 1956 the Soviet Government organized the Joint Institute for Nuclear Research (JINR) in Dubna, a city not far from Moscow, and Bogolyubov was one of its co-founders. JINR was designed as an international organization open to scientists from countries belonging to the Soviet Block. The participation of European countries with somewhat more liberal traditions, such as Poland, Checkoslovakia and Hungary, made it possible for scientists at JINR to enjoy a relative amount of freedom in their contacts with the West. Unlike Landau himself and members of his group, scientists from JINR traveled abroad.

In studies of the basics of superfluidity and superconductivity Soviet physicists were at the forefront at that time. The Institute for Physical Problems (The Kapitza Institute) maintained a leadership position in the area of low temperature physics. Superfluidity of He II had been discovered in 1938 by Kapitza and defined as the capability of the liquid to flow along narrow capillaries without viscosity below the *lambda*-point, $T_\lambda = 2.19\,\mathrm{K}$. The notion that superfluidity and superconductivity were two tightly related phenomena had become common soon after Landau developed the theory of He II (1940–41).

2.1. *Superfluidity and superconductivity*

In papers on helium Landau introduced one of the most important paradigms of modern Statistical Physics.[4] The basic concept was that low temperature properties of any macroscopic system could be described in terms of a gas of excitations called quasiparticles (*qps*). These *qps* are brought about as the result of interactions and, generally speaking, do not have much in common with the properties of free particles of which the system is formed. The helium *qps* must obey Bose statistics. Electrons in metals obey Fermi statistics.

Low energy excitations in He II are quantized sound waves, phonons, with a linear dispersion: $E(p) = cp$. Landau proceeded, however, with a general form of the energy spectrum, $E(p)$, which at a higher momentum may have an arbitrary shape (to account for experimental data Landau later assumed for liquid He II the spectrum with the so-called "roton" minimum).[4]

Consider $T = 0$. There are no excitations in the liquid as long as the helium is at rest. Let then helium move along the capillary with a velocity $\mathbf{V}$. Forming an excitation inside the moving liquid, $E(p)$, with the momentum $\mathbf{p}$, its energy, $E'(p)$, in the reference system where the capillary is at rest, becomes, in accordance with the Galileo's law:

$$E'(p) = E(p) + (\mathbf{p} \cdot \mathbf{V}) \tag{1}$$

In other words, forming an excitation in a moving liquid may cost less in energy (let $(\mathbf{p} \cdot \mathbf{V}) = -pV < 0$). The slope of the straight line, pV, drawn in the (E, p)-plane, increases with the velocity increase, V, and the line will finally touch the spectrum curve, $E(p)$, at some p so that $E(p) - pV = 0$. The condition determines the very moment when creation of excitations in helium becomes energetically possible for the first time, and at higher V the system would start heating itself (to "dissipate") by producing the *qps* excitations. Hence, the critical velocity, $V_{\rm cr}$, is defined as:

$$V_{\rm cr} = \min(E(p)/p) \tag{2}$$

Although the initial slope, $E(p) \cong cp$, could guarantee superfluidity, in the real He II, $V_{\rm cr}$ is less than the speed of sound, c, and is determined by the "roton" part of the spectrum.[4]

Bogolyubov contributed to the theory of superfluidity in his famous 1947 paper[5] on the energy spectrum of the non-ideal Bose gas. He showed that in the presence of a weak repulsive interaction between the Bose particles, the parabolic energy spectrum of the ideal gas, $E(p) = p^2/2M$, undergoes a fundamental change. Namely, at low momenta, due to the finite compressibility, the energy spectrum becomes linear in p, corresponding to the emergence of the "phonon" (sound) mode. Therefore, in accordance with Landau's arguments above, the non-ideal Bose gas possesses superfluidity. In his 1947 paper[5] Bogolyubov successfully exploited the very concept of the Bose-condensate phenomenon, to wit, the existence of the coherent quantum state with zero momentum occupied by the *macroscopic* number of particles. Therefore the creation and the annihilation operators, $\mathbf{a}_0^+$ and $\mathbf{a}_0$, for particles with zero momentum can be handled as c-numbers:

$$\langle N_0 + 1|a_0^+|N_0\rangle \approx \langle N_0 - 1|a_0|N_0\rangle \propto (N_0)^{1/2} \tag{3}$$

(N_0 is the number of particles in the condensate).

Returning now to superconductivity in metals, the Landau mechanism could then explain the superconductivity phenomenon as the superfluidity of the electronic liquid (abstracting from the magnetic fields produced by charge currents) if the Fermi spectrum, $E(p) = v_F(p - p_F)$, of metallic electrons in the normal phase below T_c were modified by the emergence of a small energy gap at the Fermi level. Experimental indications in favor of such gap[6] had indeed started to accumulate by 1956.

3. BCS Receives Immediate Recognition in the USSR

Before proceeding further, it should be noted that, unlike in the West, the results and ideas of BCS theory were recognized at once in Russia, at least by theorists. This fact deserves a few comments.

First, BCS arrived at the gapped spectrum at $T = 0$[3], as it was expected for the Landau mechanism to work. The retarded character of attraction between electrons *via* the virtual phonon exchange[7] seemed to be helpful for a reduction of the screened Coulomb interaction.[8]

Secondly, the proof by Cooper of the absolute instability of the Fermi sea in the presence of an arbitrary *weak* electron–electron attraction[1] was perceived as a *qualitative* idea, capable of explaining why superconductivity was wide-spread along the Mendeleev Chart, although the temperatures of transition, $T_c \sim 1 - 10$ K, for the superconductors known at that time were so low compared to the typical energy scales in metals.

There were no "great expectations" as to obtaining an "ideal" agreement between theory and experiment because of the well-known strong anisotropy of the Fermi surfaces in metals (actually, as we know now, the agreement turned out to be remarkably good for the isotropic model leading, for instance, to the non-trivial explanation of such a tiny feature as the Hebel–Slichter peak[9] in 1957).

The Russian community became familiar with the main ideas of BCS in the much simpler language of the Bogolyubov canonical transformation or the formulation of the new theory of superconductivity in terms of Quantum Fields Theory methods developed by Gor'kov. The Soviet theorists, for instance, never had any difficulties with the gauge invariance of the theory.

Finally, the transparency of the theory and its beauty was a very strong argument in its favor, at least in the eyes of Landau and his group.

4. Bogolyubov Canonical Transformation

The first of the Bogolyubov papers[10] was submitted to the Russian editors on October 10, 1957. Bogolyubov had formulated his method at $T = 0$ by departing from the phonon Fröhlich Hamiltonian:

$$
\begin{aligned}
H_{Fr} &= \sum_{k,s} E(k) a^{+}_{ks} a_{ks} + \sum_{q} \omega(q) b^{+}_{q} b_{q} + H' \\
H' &= \sum_{k,q=k'-k,s} g \left\{ \frac{\omega(q)}{2V} \right\}^{1/2} a^{+}_{ks} a_{k's} b^{+}_{q} + \text{h.c.}
\end{aligned}
\tag{4}
$$

(We use units with Planck's constant $\hbar = 1$.) The phonons are then integrated out by going to the second approximation in g, so that, actually, Bogolyubov solved exactly the same Hamiltonian as the BCS paper.[3]

In Ref. 10 was suggested that the operators a^{+}_{ks} and a_{ks} in the superconducting state transform into a set of operators for a new *qps*:

$$
\begin{aligned}
a_{k,1/2} &= u_k \alpha_{k0} + v_k \alpha^{+}_{k1} \\
a_{-k,-1/2} &= u_k \alpha_{k1} - v_k \alpha^{+}_{k0} \\
u_k^2 + v_k^2 &= 1
\end{aligned}
\tag{5}
$$

The substitution of (5) into the initial Hamiltonian leads to non-diagonal terms, $(\alpha_{k,1}\alpha_{k,0} + \text{h.c.})$.

To find the coefficients u_k, v_k, Bogolyubov resorted to what he called the "principle of compensation of dangerous diagram". Here is how the "principle" was formulated in his paper.[10] Suppose one considered the perturbation corrections to the new vacuum. Taking separately, non-diagonal terms in the transformed Hamiltonian would produce in the intermediate states the denominators of the form: $2E(k')$. Quoting the Russian text:[10] "the energy denominator $1/2E(k')$ becomes dangerous for integrating"... Then again:[10] "Thus, in the choice of the canonical transformation, it must be kept in mind that it is necessary to guarantee the mutual compensation of the diagrams which lead to virtual creation from the vacuum of pairs of particles with opposite momenta and spins."

It is difficult to understand Bogolyubov's reasons for using this rather obscure wording. By "dangerous" terms with $2E(k')$ in the denominators,

Bogolyubov in all probability was alluding to the logarithmic divergences in the Cooper channel. Bogolyubov[10] does not reference the Cooper article,[1] though he does reference the BCS letter.[2]

It was soon realized that, rather than utilizing Bogolyubov's reasoning in terms of "dangerous" diagrams, one should merely argue that in the new (superconducting) state at $T = 0$ all terms in the Hamiltonian containing products of operators $\alpha_{k,1}\alpha_{k,0}$ must give zero when applied to the wave function of the new vacuum.

Calculating various superconducting properties in terms of Bogolyubov *qps* turns out to be much simpler, hence its popularity (the method was soon generalized by many authors, including Bogolyubov himself, to the case of finite temperatures). In Russian literature the theory of superconductivity was often referred to as the "Bardeen, Cooper and Schrieffer and Bogolyubov theory".

The main triumph of BCS, from a fundamental point of view, was its derivation of the gapped energy spectrum from whence superconductivity follows in accordance with the Landau criterion. Note in passing that the BCS theory meant clean superconductors. Neither BCS, nor the Bogolyubov formulation provides the definition of the order parameter and its symmetry in the superconducting state. The theory had yet to be generalized to spatially non-homogeneous problems, especially for alloys. Besides, it was not clear how to go beyond the weak coupling approximation of BCS. This was all achieved within the framework of the Quantum Field Theoretical (QFT) method.[11]

5. Quantum Field Theory Methods ($T = 0$)

Quantum Electrodynamics was still a busy field, even in the early 1950s. It would be untimely to discuss this activity here. For our purposes, suffice to say that many from those days knew the Feynman diagrammatic methods perfectly well. For instance, my PhD thesis was on the "Quantum Electrodynamics of charged particles with zero-spin" in 1956.

Applying the diagrammatic quantum field approach to the needs of condensed matter physics (at $T = 0$) began in Russia around 1956–57. Indeed, the generalization of the methods of Quantum Field Theory to Fermi systems looked rather straightforward with the Fermi sea playing at $T = 0$ the role of the vacuum. A systematic discussion of the diagrammatic rules and some applications has been published in Ref. 12, but the technique actually was in use even before. Thus, the electron–phonons interactions in metals had been studied by Migdal[13] in 1957. Landau had applied the method of microscopic derivation to the theory of the Fermi Liquid[14] in 1958. It is also worth adding that the atmosphere at the Landau School was very friendly, and after talks at the Landau seminar the results were discussed broadly before publication.

6. Bogolyubov Talk at the Landau Seminar

By fall 1957 I was the junior scientist in the Landau group and had published papers on Quantum Electrodynamics, Hydrodynamics and helium.

Sometime in October 1957 it got abroad that N. N. Bogolyubov had finished the paper on the theory of superconductivity. He was invited to give a talk at the Landau Seminar in the Kapitza Institute.

The seminar started with somewhat heated debates. Bogolyubov focused on the formal part, i.e. on the details of his method of the canonical transformation, Landau, as usual, preferred to first hear the physics behind it. It was difficult for him to get through the formal Bogolyubov's 'principle of compensation of "most dangerous" diagrams'. Indeed, as we have seen it above, the "principle" itself was not very transparent, to say the least! Landau wanted to know the nature of the new vacuum. Here I need to explain that neither the Cooper paper[1] published in 1956, nor the short BCS letter[2] had attracted the attention of anyone in the Landau group. After the seminar break, N. N. Bogolyubov finally resorted to mentioning Cooper's result. He repeated the calculations by Cooper on the blackboard. Its transparent physics had the immediate effect of pacifying Landau.

As I was listening, it crossed my mind that the instability of the Fermi sea in the presence of a weak attraction between electrons that results in the spontaneous formation of pairs, also involves the emergence of a bosonic degree of freedom, and I decided to play with the idea.

7. Developing the QFT Approach for Theory of Superconductivity

7.1. $T = O$ *and beyond*

I cannot help but show the title and the abstract of my first paper[11] on superconductivity, which as I understand it, did not attract much attention from main players in the West. The paper came out in the Russian

Journal ZhETF in March of 1958. By that time references to BCS could be added.

SOVIET PHYSICS JETP VOLUME 34(7), NUMBER 3 SEPTEMBER, 1958

ON THE ENERGY SPECTRUM OF SUPERCONDUCTORS

L. P. GOR'KOV

Institute for Physical Problems, Academy of Sciences, U.S.S.R.

Submitted to JETP editor November 18, 1957

J. Exptl. Theoret. Phys. (U.S.S.R.) 34, 735-739 (March, 1958)

A method is proposed, based on the mathematical apparatus of quantum field theory, for the calculation of the properties of a system of Fermi particles with attractive interaction.

I had been working with the same four-fermion interaction Hamiltonian as BCS[3] and Bogolyubov:

$$H_{\text{int}} = (1/2) \sum_{k,k',\sigma,\sigma'} V_{k,k'}\, \hat{c}^{+}_{k,\sigma}\hat{c}^{+}_{k',\sigma'}\hat{c}_{k',\sigma'}\hat{c}_{k,\sigma} \tag{6}$$

with $V_{k,k'} = g < 0$ negative and nonzero only for the energies of electrons inside an interval $\varepsilon(k)$, $\varepsilon(k') < \hbar\omega_D$, a typical phonon frequency. As far as the energies of all the electrons involved would remain well below $\hbar\omega_D$, ($T_c \ll \hbar\omega_D$), the interaction in the Hamiltonian (6) could be considered as the local one in space:

$$H = \int \left\{ -\left(\hat{\psi}^{+} \frac{\Delta}{2m} \hat{\psi} \right) + \frac{g}{2}(\hat{\psi}^{+}(\hat{\psi}^{+}\hat{\psi})\hat{\psi}) \right\} d^3r \tag{7}$$

In the spatial representation the field operators, $\hat{\psi}_a(r)$, $\hat{\psi}^{+}_{\theta}(r)$, are:

$$\begin{aligned}\hat{\psi}_{\alpha}(r) &= V^{-1/2}\sum_{k,\sigma} \hat{c}_{k\sigma}\, s_{\sigma\alpha} \exp(ikr)\\ \hat{\psi}^{+}_{\beta}(r) &= V^{-1/2}\sum_{k,\sigma} \hat{c}^{+}_{k\sigma} s^{*}_{\sigma\beta} \exp(ikr)\end{aligned} \tag{8}$$

It was then straightforward to write down the equations for the field operators in the Heisenberg representation:

$$\begin{aligned}\{i\partial/\partial t + \Delta/2m\}\hat{\psi}(x) - g(\hat{\psi}^{+}(x)\hat{\psi}(x))\hat{\psi}(x) &= 0\\ \{i\partial/\partial t - \Delta/2m\}\hat{\psi}^{+}(x) + g\hat{\psi}^{+}(x)(\hat{\psi}^{+}(x)\hat{\psi}(x)) &= 0\end{aligned} \tag{9}$$

(Note the notations $x = (\mathbf{r}, t)$.)

The Green function is defined as usual:

$$G_{\alpha\beta}(x-x') = -i\langle T(\widehat{\psi}_\alpha(x)\widehat{\psi}^+_\beta(x'))\rangle \tag{10}$$

When the first of Eqs. (9) is applied to the operator $\widehat{\psi}_\alpha(x)$ in the definition (10) of $G_{\alpha\beta}(x-x')$, one obtains the equation that contains the average of a block of four of the field operators. The main physical idea then was that new terms must appear in this product, those of a Bose condensate type, to account for the bosonic degree of freedom that emerges due to the presence of Cooper pairs at $T=0$. Correspondingly, the aforementioned block of the four field operators had been decoupled as:

$$\begin{aligned}
&\langle T(\widehat{\psi}_\alpha(x_1)\widehat{\psi}_\beta(x_2)\widehat{\psi}^+_\gamma(x_3)\widehat{\psi}^+_\delta(x_4))\rangle \\
&\quad = -\langle T(\widehat{\psi}_\alpha(x_1)\widehat{\psi}^+_\gamma(x_3))\rangle\langle T(\widehat{\psi}_\beta(x_2)\widehat{\psi}^+_\delta(x_4))\rangle \\
&\quad + \langle T(\widehat{\psi}_\alpha(x_1)\widehat{\psi}^+_\delta(x_4))\rangle\langle T(\widehat{\psi}_\beta(x_2)\widehat{\psi}^+_\gamma(x_4))\rangle \\
&\quad + \langle N|T(\widehat{\psi}_\alpha(x_1)\widehat{\psi}_\beta(x_2))|N+2\rangle\langle N+2|T(\widehat{\psi}^+_\gamma(x_3)\widehat{\psi}^+_\delta(x_4))|N\rangle
\end{aligned} \tag{11}$$

Contributions from the first two terms into the equation of the Green function could be omitted assuming the weak coupling limit, but the last two averages in Eq. (11) introduced into the theory the two *anomalous* functions that are nonzero *only* in the superconducting state:

$$\begin{aligned}
\langle N|T(\widehat{\psi}_\alpha(x)\widehat{\psi}_\beta(x'))|N+2\rangle &= \exp(-2i\mu t)F_{\alpha\beta}(x-x') \\
\langle N+2|T(\widehat{\psi}^+_\alpha(x)\widehat{\psi}^+_\beta(x'))|N\rangle &= \exp(2i\mu t)F^+_{\alpha\beta}(x-x')
\end{aligned} \tag{12}$$

The two functions are akin to the *c*-numbers, Eq. (3), introduced by Bogolyubov for the Bose gas in his 1947 paper and bear the coherent macroscopic origin. The product of the two last terms in (11) would be proportional to the condensed Cooper pairs' density.

Exponential factors containing the Josephson time dependence in Eq. (12) can be removed by using the chemical potential instead of the total number of particles as the thermodynamic variable: $H \Rightarrow H-\mu N$. In the new variables the system of coupled equations has the following form:

$$\begin{aligned}
&\{i\partial/\partial t + \Delta/2m + \mu\}\widehat{G}(x-x') - ig\widehat{F}(0+)\widehat{F}^+(x-x') = \delta(x-x') \\
&\{i\partial/\partial t - \Delta/2m - \mu\}\widehat{F}^+(x-x') + ig\widehat{F}^+(0+)\widehat{G}(x-x') = 0
\end{aligned} \tag{13}$$

In Eq. (13) we denoted, for instance:

$$\widehat{F}_{\alpha\beta}(0+) = \langle \widehat{\psi}_\alpha(x)\widehat{\psi}_\beta(x)\rangle \qquad (14)$$

It is easy to see that

$$\begin{aligned} \widehat{F}_{\alpha\beta}(0+) &= i(\sigma_y)_{\alpha\beta}F \\ \widehat{F}^+_{\alpha\beta}(0+) &= -i(\sigma_y)_{\alpha\beta}F^* \end{aligned} \qquad (15)$$

The antisymmetric spinor structure of the pair wave function corresponds to the *singlet* pairing. Introduce the notation:

$$\Delta = gF \qquad (16)$$

By substituting $i\partial/\partial t \Rightarrow E$ in Eqs. (13) re-written in the momentum representation, the BCS gapped energy spectrum for a homogeneous superconductor will immediately follow as the eigenvaules of the L.H.S. operator for the system (13):

$$E(p) = \pm\sqrt{[v_F^2(p-p_F)^2 + |\Delta|^2]} \qquad (17)$$

One may now summarize some main results obtained in the paper[11] by the above formalism. To begin with, while $|\Delta|$ is the magnitude of the energy gap, *the true order parameter* in the superconducting state is Δ, (or Δ^+), i.e. *the wave function of the Cooper pair* or, more broadly, the anomalous functions themselves. The symmetry broken at the transition is the gauge symmetry, $U(1)$. Equations (13) obviously possess the gradient-invariant form when a magnetic field is introduced by the usual substitution: $-i\partial \Rightarrow (-i\partial - (e/c)A(r))$.

Interactions with the magnetic field and any other perturbations if added into the Hamiltonian (7) can be studied at $T = 0$ with the routine diagrammatic technique for the matrix composed of the Green function, G, and the anomalous functions F and F^+.

Expressions for the Free Energy and thermodynamics have been obtained in few lines.[11]

Note that Eq. (16) defines the gap, Δ, self-consistently through the F-function at equal arguments. In turn, the expression for the latter in the momentum space can be obtained by solving the system of Eqs. (13). There is no need in the variational procedure or in the Bogolyubov "principle of compensation of "dangerous" diagrams".

The Green functions at *finite* temperatures can be unambiguously found from Eqs. (13) by making use of the relations between the advanced and retarded (casual) Green functions. For normal metals it was derived by Landau in the form:[15]

$$\operatorname{Re} G(\omega) = -\frac{1}{\pi}\int_{-\infty}^{+\infty} \coth\frac{x}{2T}\frac{\operatorname{Im} G(x)}{\omega - x}dx \tag{18}$$

The provisions imposed by Eq. (18) may complicate calculations at finite temperatures since the automatism of the $T = 0$ diagrammatic technique is now lost. I do not stay on this any longer, because in such cases it is preferable to apply the thermodynamic technique that was soon elaborated for the needs of the superconductivity theory, as described below.

7.2. *Superconducting alloys*

In 1958 A. A. Abrikosov joined me. Together we started the application of the above diagram method to superconducting alloys. We published two papers on Electrodynamics and Thermodynamics of alloys.[16,17]

At calculating the transport and other kinetic properties in normal metals in the presence of defects one routinely uses the well-known Boltzmann Equation. Now, to account for the role of defects in the superconducting state, we faced the necessity of somehow including scattering on defects into the general diagrammatic approach. Without entering into details, first we had to develop[16] the so-called "cross-technique," giving the means to treat scattering of the electrons on defects diagrammatically. Our then newly created[17] "Matsubara" technique[18] (see in Sec. 7.3) was applied for the first time to alloys at nonzero temperatures.

I will skip the results that account for the role of defects in the electromagnetic properties of a superconductor (i.e. the Meissner effect). Our calculations showed, in particular, that all Green functions, including the *anomalous* function, F and F^{+}, after being averaged over the impurities' positions, acquire in the coordinate representation the exponentially decaying factors describing the loss of the coherence due to the scattering:

$$F(t-t', R) \Rightarrow F(t-t', R)\exp(-R/l) \tag{19}$$

(And similarly for the rest of the Green functions; l is the mean free path.) Hence, with the gap being defined, according to Eq. (16), at $R = 0$, the thermodynamics of a superconductor does not change in the isotropic model. This result, first obtained by Abrikosov and myself,[16] is known in the West as "the Anderson Theorem."[19]

7.3. *Developing thermodynamic QFT technique for statistical physics*

Although the transition temperature, T_c, of most early superconductors was rather low, the temperature dependence of different superconducting characteristics had been extensively studied experimentally. The needs of the theory of superconductivity had propelled in 1958 the broad implementation of diagrammatic methods for the general Statistical Physics.[18]

At first, in 1955 Matsubara[20] showed that there is a formal analogy between the so-called $\widehat{S}$-matrix in the Quantum Field Theory (at $T = 0$) and the expression of the statistical matrix for the Gibbs distribution. The latter can be written in the form:

$$\exp\left[\frac{\mu\widehat{N} - \widehat{H}}{T}\right] = \exp\left[\frac{\mu\widehat{N} - \widehat{H}_0}{T}\right] \times \widehat{S}(1/T) \tag{20}$$

The "S-matrix" here is:

$$\widehat{S}(1/T) = \widehat{T}_\tau \exp\left(-\int_0^{1/T} \widehat{H}_{\text{int}}\,(\tau)d\tau\right) \tag{21}$$

The inverse temperature, $1/T$, may be taken formally as an imaginary time, τ. The diagrammatic expansion can now be developed for the new "Green functions" defined as:

$$\bar{G}(1,2) = -\frac{\langle\langle \widehat{T}(\widehat{\bar{\psi}}\,(1)\widehat{\bar{\psi}}^{+}(2)\widehat{S})\rangle\rangle}{\langle\langle \widehat{S}\rangle\rangle} \tag{22}$$

In Eq. (22) the double brackets $\langle\langle\ldots\rangle\rangle$ mean the *grand canonical ensemble* average; the field operators now depend on the space coordinates and on the *imaginary* "time."[20]

The difficulty is that in Eq. (21) the "imaginary time" τ varies only inside the finite interval $(0 < \tau < 1/T)$ and the representation in the form of the Fourier integrals for the "time variables" is not possible. Abrikosov *et al.*[18] suggested instead to expand the temperature Green function (22) into the *Fourier Series*:

$$\bar{G}(\tau - \tau') = T\sum_n \bar{G}(\omega_n)\exp(-i\omega_n(\tau - \tau')) \tag{23}$$

where $\omega_n = \pi T n$ with n-even for Bose system, n-odd for the Fermi case.

The diagrammatic rules for diagrams with the new Green functions, $\bar{G}(\omega_n)$, turn out to be basically the same as at $T = 0$, except differences in the numeric coefficients and the fact that in all matrix elements for diagrams, the *summations* now substitute for the integrations over "frequency" variables. The analytical continuation in the complex frequency plane, $z = \omega$, allows us to connect the thermodynamic Green functions (23) calculated *at the points*, $z = i\omega_n$, along the "*imaginary*" (or the Matsubara) axis, with the casual Green functions on the real axis.[18]

Gor'kov equations (13) in the coordinate space can now be written in the thermodynamic notation:[21]

$$\begin{aligned}
&\left\{i\omega_n + \frac{1}{2m}\left(\frac{\partial}{\partial r} - i\frac{e}{c}A(r)\right)^2 - \mu\right\}\bar{G}(\omega_n; r, r') + \Delta(r)\bar{F}^{+}(\omega_n; r, r') \\
&\quad = \delta(r - r') \\
&\left\{- i\omega_n + \frac{1}{2m}\left(\frac{\partial}{\partial r} + i\frac{e}{c}A(r)\right)^2 - \mu\right\}\bar{F}^{+}(\omega_n; r, r') - \Delta^*(r)\bar{G}(\omega_n; r, r') \\
&\quad = 0
\end{aligned} \tag{24}$$

Here the "gaps" are determined through the anomalous functions with the coinciding arguments as, for instance:

$$\Delta(r) = gT\sum_n \bar{F}(\omega_n; r, r) \tag{25}$$

With the gauge transformation $A(r) \Rightarrow A(r) + \nabla\varphi(r)$ the "gap" transforms correspondingly: $\Delta(r) \Rightarrow \Delta(r)\exp(i(\frac{2e}{c})\phi(r))$.

We have mentioned that the use of the Bogolyubov *qps* significantly simplifies derivation of the Free Energy in the superconducting phase compared to Ref. 3. However, it is also not so easy to find the *qps* energy spectrum in the field's presence. Another example is given by alloys where defects are randomly distributed along the superconducting sample. In these instances the following expression for the Gibbs' potential is extremely helpful:[21]

$$\Omega_S - \Omega_N = -\int d^3r \int_0^g \frac{\delta g}{g^2}|\Delta(g; r)|^2 \tag{26}$$

7.4. *Ginsburg–Landau equations from microscopic theory*

The Ginsburg–Landau theory[22] was one of the most important phenomenological breakthroughs prior to the creation of the microscopic theory of superconductivity. The theory[22] made it possible to account for countless data on non-linear magnetic properties of superconductors in good agreement with the main experimental findings. It was therefore important to find out whether it can be substantiated on the microscopic level. This had been done early in 1959 in two of my papers,[21,23] for clean superconductors and for superconducting alloys, correspondingly. Without entering into the actual calculations, I only list below the main results.

Close to the temperature of transition, $|T - T_c| \ll 1$, Eq. (25) can be simplified. Indeed, near T_c $\Delta(r)$ is small and Eqs. (24) for $\bar{G}$ and $\bar{F}$, $\bar{F}^+$ can be solved perturbatively. When calculating $\bar{F}(\omega_n; r, r)$ in Eq. (25), one can restrict oneself by a few nonzero terms in $\Delta(r)$, $\Delta^*(r)$ and $A(r)$.[21] For clean superconductors such expansion results in the equation of the form:

$$\left\{\frac{1}{2m}\left(\partial - i\frac{e^*}{c}A(r)\right)^2 + \frac{1}{\lambda}\left[\frac{T_c - T}{T_c} - \frac{2}{N}|\Psi(r)|^2\right]\right\}\Psi(r) = 0 \qquad (27)$$

Equation (27) has the familiar form of the GL-equation[22] for the "gap" parameter:

$$\Psi(r) = \Delta(r)\sqrt{7\varsigma(3)N}/4\pi T_c\,.$$

Note $e^* = 2e$ that stands for the charge of the Cooper pair! (N is the density of electrons, $\lambda = 7\varsigma(3)E_F/12(\pi T_c)^2$.)

The expression for the electrical current is derived in the same way:

$$j(r) = -\frac{ie^*}{2m}\left(\Psi^*\frac{\partial\Psi}{\partial r} - \Psi\frac{\partial\Psi^*}{\partial r}\right) - \frac{e^{*2}}{mc}A|\Psi|^2 \qquad (28)$$

The GL-theory, as it is known,[22] contains one important dimensionless parameter, κ. In terms of the observable characteristics of a material, Ginsburg and Landau[22] expressed it in the following way:

$$\kappa = (\sqrt{2}e^*/\hbar c)H_{cr}\delta_L^2 \qquad (29)$$

(Here H_{cr} is the thermodynamic critical field and δ_L is the penetration depth.) Now κ can be written through the microscopic parameters of a metal:[21]

$$\kappa = 3T_c(\pi/v_F)^{3/2}(c/ep_F)(2/7\varsigma(3))^{1/2} \qquad (30)$$

The parameter κ determines the behavior of a superconductor in strong magnetic fields. Abrikosov showed[24] that depending on whether κ is small or large, a superconductor would belong to the one of two classes: Type I or Type II, correspondingly. From Eq. (30) one may conclude that the microscopic theory imposes no limitations on the κ-value, so that even a pure metal can be a Type II superconductor. I predicted[21] that this could be the case for the *elemental* niobium, Nb.

Type II superconductivity is more common in alloys. The GL equations for an alloy, derived in Ref. 23, look quite similar to (27, 28) but with the coefficients now depending on the transport meantime, τ_{tr}:

$$\left\{\frac{1}{2m}\left(\partial - i\frac{e^*}{c}A(r)\right)^2 + \frac{1}{\lambda_\tau}\left[\frac{T_c - T}{T_c} - \frac{2}{N\chi(\rho)}|\Psi(r)|^2\right]\right\}\Psi(r) = 0$$
$$j(r) = -\frac{ie^*}{2m}\left(\Psi^*\frac{\partial\Psi}{\partial r} - \Psi\frac{\partial\Psi^*}{\partial r}\right) - \frac{e^{*2}}{mc}A|\Psi|^2 \tag{31}$$

The GL wave function is now connected with the "gap" as:

$$\Psi(r) = [\chi(\rho)7\varsigma(3)N/16\pi^2T_c^2]^{1/2}\Delta(r) \tag{32}$$

and:

$$\chi(\rho) = \frac{8}{7\varsigma(3)\rho}\left[\frac{\pi^2}{8} + \frac{1}{2\rho}\left\{\psi\left(\frac{1}{2}\right) - \psi\left(\frac{1}{2} + \rho\right)\right\}\right] \tag{33}$$

In Eq. (33) $\rho = \frac{1}{2\pi T_c\tau_{tr}}$, and $\psi(z)$ is the logarithmic derivative of the Γ-function. (In Eq. (31) $\lambda_\tau = \lambda\chi(\rho)$).

The penetration depth and the value of the GL parameter κ also suffers the change:

$$\delta_L = \delta_{L0}/\sqrt{\chi(\rho)}; \quad \kappa = \kappa_0/\chi(\rho) \tag{34}$$

In the short mean free path limit (so-called "dirty" alloys) κ is expressible in terms of the observable quantities for the *normal* phase only:

$$\kappa = 0.065ec\gamma^{1/2}/\sigma k_B \tag{35}$$

(Here γ and σ stand for the coefficient in the linear electronic specific heat and for conductivity, respectively.)

Together with the GL and Abrikosov phenomenology[24] the above results are known as the GLAG-theory (after Ginsburg, Landau, Abrikosov and Gor'kov).

7.5. *Paramagnetic impurities*

Alloying by ordinary impurities or defects does not change either the energy gap, or T_c, as we have shown in Sec. 7.2. In 1960 Abrikosov and Gor'kov[25] studied alloys containing paramagnetic centers, i.e. atoms or ions with the nonzero spin. It turned out that the presence of such centers is detrimental to superconductivity. The transition temperature, $T_c(x)$ decreases with increase of the concentration x of the paramagnetic impurities and vanishes at some critical concentration, x_{cr}. This behavior might have been expected, for the potential for scattering on such a defect does now include, in addition, the interaction with the external spin on a center, s,

$V(p-p') \Rightarrow V(p-p') + \tilde{V}(p-p')(\boldsymbol{\sigma}\cdot\mathbf{s})$, that tends to misalign spins of the Cooper pair.

The dependence of $T_c(x)$ on the concentration is given[25] by the equation:

$$\ln(T_{c0}/T) = \psi\left(\frac{1}{2} + \frac{1}{T_c\tau_s}\right) - \psi\left(\frac{1}{2}\right) \tag{36}$$

where $1/\tau_s$ is the inverse mean time for scattering with the overturn of the electronic spin and, as in Sec. 8.4, $\psi(x) = \Gamma'(x)/\Gamma(x)$. The value of the critical concentration is given by:

$$\frac{1}{\tau_{c,\text{cr}}} = \frac{\pi T_{c0}}{2\gamma} \equiv \frac{\Delta_0}{2} \tag{37}$$

The gap in density of states (DOS) also decreases with the x increase. It was, however, somewhat unexpected to find that the energy gap in DOS closes *before* the superconductivity is fully destroyed at x_{cr} given by Eq. (37): there exists a narrow range of concentrations below x_{cr} in which superconductivity persists in spite of the zero energy gap in DOS. The result is another proof that it is the nonzero superconductivity *order parameter*, Δ, that controls the capability of a superconductor to carry persistent currents, not the finite gap value. The Landau criterion of Eq. (2) is, hence, of the lesser generality.

8. Applications of Quantum Field Theory Methods

While for many properties (kinetic properties, such as the sound absorption or thermal conductivity, or the effects, like tunneling, and others) working in terms of the Bogolyubov *qps* was more straightforward, it is not so as far as the electrodynamics of superconductors is concerned, because the new *qps* does not possess the fixed electric charge. There were other fundamental questions for which finding the solution without the diagrammatic methods would be next to impossible.

Thus, the extension of BCS theory to the case of strong electron–phonon coupling had been done by Eliashberg,[26] who in 1960 extended Migdal's paper[13] on electron–phonon interactions for normal metal by introducing, instead of the mere gap parameter, Δ, Δ^+, in Eqs. (24) for the "local" Hamiltonian (7), the new self energy parts build up on the anomalous functions F, F^+:

$$\Delta(\omega_n; p) = T\sum_m \int D(\omega_n - \omega_m; p-p')F(\omega_m; p')\frac{d^3p'}{(2\pi)^3} \tag{38}$$

In Eq. (38) $D(\omega_n; k)$ stands for the Green function of phonons multiplied by the square of the e-ph coupling constant:

$$D(i\omega_m; p - p') = g^2 \frac{\omega_0^2(p - p')}{(i\omega_m)^2 - \omega_0^2(p - p')} \,. \tag{39}$$

As was first demonstrated in Ref. 13, the adiabatic approximation, $\omega_0 \ll E_F$, makes it possible to neglect the so-called "vertex" corrections both in the normal and the anomalous self-energy parts. In the new equations, replacing Eqs. (24), the "gap" functions $\Delta(\omega_n; p)$ and $\Delta^*(\omega_n; p)$ appear together with the self-consistency condition of Eq. (38). The Bogolyubov principle of "dangerous diagrams" could not be relevant in the first place, because there are no divergences in (38) due to the frequency dependence of the phonon Green function.

It is a point to emphasize here that the Quantum Field Theory formulation also makes it easy to generalize to the case of multi-band superconductors, or to take into account the anisotropy of the electronic spectrum.[27] The latter allows the rigorous microscopic derivation of the anisotropic GL equations.[28]

The Gor'kov method, as shown by Eilenberger[29] in 1968, can be advanced even further. These considerable simplifications arise with the explicit use of the so-called "quasi-classic" approximation: T_c, $\omega_0 \ll E_F$.[29] Under this provision that obviously has a quite general character, one may re-write the Gor'kov equations into the master equations for the Green functions integrated over the energy variable:

$$\int \hat{\hat{G}}(\omega_n; \mathbf{p}, \xi) \frac{d\xi}{\pi i} = \begin{pmatrix} g(\omega_n; \mathbf{p}) & f(\omega_n; \mathbf{p}) \\ f^+(\omega_n; \mathbf{p}) & \bar{g}(-\omega_n; \mathbf{p}) \end{pmatrix} \tag{40}$$

(Here momentum $\mathbf{p}$ may run on the Fermi surface.)

The emerging set of non-linear equations is local in space and is easier for the numerical analysis and for the treatment of the inhomogeneous problem. We shall not dwell here upon these results.

9. Conclusion

The purpose of the sketchy review above was to trace how the development of powerful new theoretical methods made it possible to extend BCS to a much broader class of phenomena. Many new notions and results of a fundamental character came to light since the early 1960s, both experimentally and theoretically. The BCS model grew into the mighty theory of superconductivity, one of the most accomplished theories in condensed matter physics. I would like to stress again that, to a considerable extent, the main advances came about through the formulation of the theory in terms of Green functions.

Acknowledgments

The work was supported by NHMFL through the NSF Cooperative agreement No. DMR-0654118 and the State of Florida.

References

1. L. N. Cooper, *Phys. Rev.* **104**, 1189 (1956).
2. J. Bardeen, L. N. Cooper and J. R. Schrieffer, *Phys. Rev.* **106**, 162 (1957).
3. J. Bardeen, L. N. Cooper and J. R. Schrieffer, *Phys. Rev.* **108**, 1175 (1957).
4. L. D. Landau and E. M. Lifshitz, *Course of Theoretical Physics*: Vol. 9, *Statistical Physics*, part 2 (1998); *ibid*, Vol. 6, *Fluid Mechanics* (1999) (Butterworth-Heinemann, Oxford).
5. N. N. Bogolyubov, *J. Phys. USSR* **11**, 23 (1947).
6. W. S. Corak, B. B. Goodman, C. B. Satterthwaite and A. Wexler, *Phys. Rev.* **102**, 656 (1956); W. S. Corak and C. B. Satterthwaite, *ibid* 662 (1956); R. E. Glover and M. Tinkham, *ibid* **104**, 844 (1956); M. Tinkham, *ibid* 845 (1956).
7. H. Fröhlich, *Proc. Roy. Soc. (London)* **A215**, 291 (1952).
8. J. Bardeen and D. Pines, *Phys. Rev.* **99**, 1140 (1955).
9. L. C. Hebel and C. P. Slichter, *Phys. Rev.* **107**, 901 (1957).
10. N. N. Bogolyubov, *Sov. Phys. JETP* **7**, 41, 51 (1958).
11. L. P. Gor'kov, *Sov. Phys. JETP* **7**, 505 (1958).
12. V. M. Galitski and A. B. Migdal, *JETP* **7**, 95 (1958).
13. A. B. Migdal, *Sov. Phys. JETP* **7**, 996 (1958).
14. L. D. Landau, *Sov. Phys. JETP* **8**, 70 (1959).
15. L. D. Landau, *Sov. Phys. JETP* **7**, 182 (1958).
16. A. A. Abrikosov and L. P. Gor'kov, *Sov. Phys. JETP* **8**, 1090 (1959).
17. A. A. Abrikosov and L. P. Gor'kov, *Sov. Phys. JETP* **9**, 220 (1959).
18. A. A. Abrikosov, L. P. Gor'kov and I. E. Dzyaloshinskii, *Sov. Phys. JETP* **9**, 636 (1959); E. S. Fradkin, *ibid* 912 (1959).
19. P. W. Anderson, *J. Phys. Chem. Solids* **11**, 26 (1959).
20. T. Matsubara, *Progr. Theor. Phys.* **14**, 351 (1955).
21. L. P. Gor'kov, *Sov. Phys. JETP* **9**, 1364 (1959).
22. V. L. Ginsburg and L. D. Landau, *Zh. Eksp. Teor. Fiz.* **20**, 1064 (1950) (in Russian).
23. L. P. Gor'kov, *Sov. Phys. JETP* **10**, 998 (1960).
24. A. A. Abrikosov, *Sov. Phys. JETP* **5**, 1174 (1957).
25. A. A. Abrikosov and L. P. Gor'kov, *Sov. Phys. JETP* **12**, 1243 (1961).
26. G. M. Eliashberg, *Sov. Phys. JETP* **11**, 696 (1960); *ibid* **12**, 1000 (1961).

27. V. L. Pokrovskii, *Sov. Phys. JETP* **13**, 628 (1961); V. L. Pokrovskii and N. S. Ryvkin, *ibid* **16**, 67 (1963).
28. L. P. Gor'kov and T. K. Melik-Barkhudarov, *Sov. Phys. JETP* **18**, 1031 (1964).
29. G. Eilenberger, *Z. Phys.* **214**, 195 (1968).

BCS: THE SCIENTIFIC "LOVE OF MY LIFE"

Philip W. Anderson
Department of Physics, Princeton University,
Jadwin Hall, Princeton, NJ 08544, USA

After short comments on my early addenda to BCS — gauge invariance and the Anderson–Higgs mechanism, the dirty superconductor "theorem," and the spinor representation — I focus on the interaction mechanisms which cause electron–electron pairing. These bifurcate into two almost non-overlapping classes. In order to cause electrons to pair in spite of the strong, repulsive, instantaneous Coulomb vertex, the electrons can evade each others' propinquity on the same site at the same time either dynamically, by retaining Γ^0 (s-wave) relative symmetry, but avoiding each other in time — called "dynamic screening" — or by assuming a non-symmetric relative wave function, avoiding each other in space. All simple metals and alloys, including all the (so far) technically useful superconductors, follow the former scheme. But starting with the first discovery of "heavy-electron" superconductors in 1979, and continuing with the "organics" and the magnetic transition metal compounds such as the cuprates and the iron pnictides, it appears that the second class may turn out to be numerically superior and theoretically more fascinating. The basic interaction in many of these cases appears to be the "kinetic exchange" or superexchange characteristic of magnetic insulators.

Ever since the day in the spring of 1957 when I fell in love at first sight with the BCS theory, a very large fraction of my scientific activity has been related to that theory and its implications. Thus when asked to contribute to this volume I had to choose among 50 years of work and of memories.

Earliest, as I have described elsewhere,[1] was my strong feeling of defensiveness in the face of the attacks on the theory by Wentzel, Kohn and others on the basis of gauge invariance. After an initial rather awkward — but essentially correct — explication[2] of the physics of gauge invariance, more or less in parallel with papers of Nambu and of Bogoliubov–Shirkov,[3] this led to the complete (within RPA) theory of the "Anderson-Higgs" mechanism[4] which allows superconductivity with a true gap not sullied by "Goldstone modes". (As was clearly indicated by experiment already at that time.) In the course of this work I developed the spinor representation of BCS pairs, which Nambu elaborated into the scheme that Schrieffer used in his indispensable book. For me, it has always been a perspicuous way of understanding the coherence properties of the BCS theory. Personally, it came in very handy on my one contact with the formidable Lev Landau; intrigued that there was yet another way to describe the BCS theory, he let me run on for an hour or more in his dreaded seminar.

A second early interest was in the problem of impurity scattering. HBG Casimir's famous remark about "A Mile of Dirty Lead Wire" intrigued me, as did the contrast between the effects of magnetic impurities and of alloying or deformation that my friend Bernd Matthias kept pointing out to me. As I remember, it was during a stroll on the Berkeley campus with Jim Phillips (I was a summer visitor in 1958, he an assistant professor) that the solution came to me. I called it the "n-representation": in the presence of static disorder, let us consider the actual "scattered" — but extended — eigenfunctions φ_n of the one-electron potential, with energies E_n. If the one-electron potential is static and time-reversal invariant — which is as true of spin-orbit interactions as of ordinary potentials — every one of these must have a time-reversed partner φ_n^* which is distinct from it, by Kramers' theorem. (In the absence of spin-orbit effects, to get the partner one may just reverse the spin of a real function.) Then one makes up one's Cooper pairs out of these two functions, and proceeds with BCS as before. Of course, there are two conditions for this representation to give nearly the same result as the pure perfect crystal. One is that the density of states of the E_n be nearly the same as that of the pure crystal. For relatively weak scattering this will almost always be the case unless the pure crystal has deep structure near the Fermi surface — which is indeed the case for Bi, and Bi is indeed very sensitive to purity — a kind of "exception proves the rule" case. The second is that the interaction not be enormously anisotropic. The physical reason for this not being the case will become clear as we discuss the interactions; at that early date we were all still using BCS's single parameter $N(0)V$.

The n-representation turned out to be useful in the hands of de Gennes;[5] I myself found it valuable in explaining the finite Knight shifts in small particles.[6]

But let me finally get to the topic I was invited to discuss here: "electron–electron interactions in superconductors." I will follow a line here that I adumbrated in 2002 in a general talk[7] which started out from the incontrovertible fact that electrons, if brought close enough to each other, experience the strong Coulomb repulsion e^2/r, so that if one is to bind them together in pairs one must somehow evade this hard core repulsion.

In 1959 David Pines decided to move from his assistant professorship at Princeton to the University of Illinois, where his field of many-body physics was more appreciated, and he, John Bardeen and Gordon Baym formed a very strong nucleus in that field. At Princeton he left behind the second of his brilliant French graduate students, Pierre Morel (the first was Philippe Nozieres). Pierre was employed by the French Consulate in New York and could not follow him, so David bequeathed him to me — perhaps the greatest of the many great favors David has done for me in the course of a long association. Socially Pierre was a little rich for my blood (one of my most vivid memories of him and his very beautiful wife was their elegantly seeing us off on our cut rate Icelandair flight to Cambridge in Sept 1961 with an enormous bouquet of flowers), but physics wise he was just what I needed, extremely able mathematically yet not too enamored of his own ideas.

I had a problem all ready for him, the possibility of exotic BCS-like states with anisotropic gaps, which I had scribbled down in a notebook a few months previously. I had in mind a mere theoretical exercise, useful for the unusual properties it might exhibit in principle, but shortly it was to become more physical, as we will see presently. But let me take the story out of chronological order and write first about the other problem we worked on, the interactions responsible for the then-known, true BCS superconductors. In fact, Pierre had already on his CV a published paper on this subject,[8] done under David's direction, and it was a bit of an imposition for me to make him revisit the subject — but with supreme self-confidence, or blatant cheek, I did.

Not that the previous work was wrong — it makes the perfectly valid point that polyelectronic metals tend to have higher transition temperatures because of the Umklapp processes allowed by the large Fermi surfaces. But actually the arguments for this effect were not very clearly presented. (It may be characterized as a "crystal field" effect that enhances the electron–phonon coupling somewhat over that of a pure "jellium" model.) There was a great deal of physics yet to be done.

I had two physical insights that I wanted to explicate. The first was the idea that the long-wavelength phonons produced only a tiny portion of the interaction. Presented rather clumsily in our eventual paper,[9] we appealed to the understanding that at long wavelengths the electron wave functions move with the lattice as a neutral whole, so that the Coulomb energy to couple the electrons and phonons is decreased by a factor q which is proportional to ω, which makes the coupling, second-order in this energy, down by ω^2. Indeed, later, when we were later able to measure the spectrum of the interactions the low-frequency part seemed always to be vanishingly small. (This contradicts recent claims[10] that a supposed peak in forward scattering could cause d-wave superconductivity in the cuprates.)

The second insight was that the phonon interaction is essentially local. We put it that the phonons could be better approximated (taking into account that the long-wavelength spectrum is ineffective) by an Einstein model (somewhat broadened) than by a Debye one. (In the end, the successful approximate model for lead was two Einstein modes rather than one — one each for longitudinal and transverse modes.) That is, there is no very relevant structure in momentum space, especially after averaging over the whole Fermi surface, so that the entire problem is reduced to a local dynamics with the only relevant variation taking place in frequency space. This gives one a very attractive visualization of the interaction process as the first electron whizzing past an atom site and giving its neighbors an attractive impulse, and then the second electron coming by a little later, encountering the attractive local change in ion charge density after the original electron has long since gone but before the displacement of the ions has died out. One may, that is, to a good first approximation, neglect the k-dependence of the interaction in solving the gap equation.

A third insight was not mine alone but had already been noticed by Bogoliubov in his early papers,[11] in the course of solving the BCS gap equations. This is that the repulsive screened Coulomb interaction could be replaced by a weaker pseudopotential insofar as its effect on low-energy processes is concerned. In fact, we argued that measures of the average strengths of the Coulomb repulsion and the phonon-mediated attraction were necessarily about equal, and the predominance of the latter stemmed only from the pseudopotential weakening of the repulsive term.

The screened Coulomb potential is almost instantaneous on the scale of phonon frequencies — its energy scale is the Fermi energy — and essentially local — basically a constant. If the real action takes place at phonon frequencies, we would wish to eliminate repeated scatterings up to much higher energies. Since the early days of many-body theory we have known to take this into account by replacing the "hard core" like potential U by a scattering matrix K obeying the Dyson equation

$$K(E, E') = U + U \sum_{E'' > \omega_D} G(E, E'')K(E'', E')$$

since most E'''s are $\gg \Delta$, G is just the pair propagator $-1/(2E)$, and this may be solved as

$$K = \frac{U}{1 + N(0)U \ln[E_F/\omega_D]} \,. \tag{1}$$

From these insights and with the help of two other contributions from Russian workers came the now accepted theory of conventional, "BCS", superconductors, appropriately called the "dynamic screening" method. These two other contributions were the time-dependent version of the energy gap equations worked out by Eliashberg,[12] and Migdal's observation that higher-order terms in the phonon self-energy were smaller in the ratio $[m(\text{electron})/M(\text{phonon})]^{1/2}$. Pierre's attention was called to the Russian work by Schrieffer during a visit to Illinois. (We had been using a cruder version omitting the imaginary parts.) In essence, in this scheme, even though the parameters λ and μ quantifying the attraction and the repulsion (see below) are roughly equal, the wave function for the pair manages to avoid the repulsive region in time — not in space — and feel a net attraction. In order to do so it requires the large logarithm in Ref. 1, i.e. a very broad band.

In the original reference,[10] we normalized such quantities as $N(0)V$ and $N(0)U$ to dimensionless coupling constants appropriate for an averaged, jellium-like metal — it makes no sense to separate out an $N(0)$ since the coupling constants do not depend any more simply on density of states than does, for instance, the phonon frequency. The dimensionless quantities

$$\lambda = \frac{1}{2}\left[\frac{k_s^2}{k_s^2 + \frac{3}{5}q_D^2}\right]^2 \quad \text{for the attractive phonon interaction,}$$

$$\text{and } \mu = \frac{k_s^2}{8k_F^2} \ln\left[\frac{k_s^2 + 4k_F^2}{k_s^2}\right] \quad \text{for the repulsion} \tag{2}$$

were relatively crude but physically sound estimates based on Fermi–Thomas screening theory. Of course, details of the phonon spectrum can give some variations on this very coarse-grained picture of the conventional polyelectronic metals, but we at the time were interested primarily in achieving understanding of the broad picture. Here k_s is the screening wave number, q_D the root mean square phonon momentum, and k_F the Fermi momentum.

Inserting these parameters, and solving the gap equation numerically, we obtained a now familiar equation for the energy gap,

$$\ln\left(\frac{2\omega_D}{\Delta}\right) = \left[\lambda - \frac{\mu}{1 + \mu \ln(\varepsilon_F/\omega_D)}\right]^{-1} = [\lambda - \mu^*]^{-1}, \tag{3}$$

ω_D here is not the conventional upper limit of the phonon spectrum but the hypothetical single Einstein phonon by which we are approximating it.

The simple estimates in (2) for the parameters give λ and μ very similar values, each of order a bit less than 1/2, and in general with μ slightly larger. λ after all simply represents the effect of phonon screening in reducing the Coulomb repulsion described by μ. For elements, as we showed, this gave a reasonably good account of the general run of T_c's; and, particularly, of measured isotope effect coefficients which are all less than the BCS 1/2, and often much less. A few elements, and even more compounds, show somewhat enhanced values of λ, which I attribute to local field effects. μ^* for good metals seems to retain a low value in the range 0.10–0.14. (We also remarked that eventually almost all simple metals would be found to be superconducting, a guess which has come true in the long run.)

During the winter of 1961–62, while in Cambridge, I gave a number of seminars at English universities on this work. Rudolf Peierls attended the one in Birmingham and asked whether I had thought about experimental ways of seeing whether there was phonon-like structure in the gap, as the above implies. After some thought I replied that I supposed it would show up in tunneling spectroscopy — and it was only two weeks later I heard from John Rowell at Bell Labs that he had seen just that, features in the tunneling spectrum of Pb which seemed to correlate with (energy gap) + prominent peaks in the phonon spectrum (known from inelastic neutron scattering). There is actually a prior paper by Ivar Giaevar showing such a feature,[13] which Schrieffer had identified correctly; John Rowell's contribution was primarily to turn vague "features" into a spectroscopic tool using differentiation techniques.

I had also seen a preprint from Bob Schrieffer in which he was using the then completely new technique of online computation to solve the Eliashberg equations, on which Pierre and I had to use a complicated iteration procedure. It seemed to me therefore, once John and I got the data sorted out, that the wise way to proceed was to hand over the data to Bob's group, together with our suggested phonon spectrum (which, as mentioned above, consisted of two broadened Einstein modes, one representing longitudinal phonons and the other transverse) and to ask them to try to fit the rather detailed series of peaks which John saw. The results were beyond expectation, and came out in two simultaneously submitted letters, one by John, myself and his technical assistant;[14] and one by Bob and two students, Doug Scalapino and John Wilkins.[15] These seemed to the group of us to be clinching proof that the dynamic screening mechanism was correct for the classic superconductors. I called it, in a historical talk in 1987, the moment when "the fat lady sang."

Bill McMillan took over working with Rowell on this problem at Bell Labs with great competence, particularly finding a computer method for inverting the Eliashberg equation in order to deduce the frequency spectrum of the interaction $\alpha^2F(\omega)$, as we named it, in detail from given tunneling data, thus dispensing with our rough "Einstein Mode" approximation (but still neglecting propagation). He and Rowell successfully analyzed some half-dozen further simple superconductors,[16] and the program was further refined and utilized by a number of later authors. McMillan,[17] in a very insightful paper, expanded on the type of general estimation which Morel and I had attempted, and created a formula very like our (3) above, with a few frills and attempting to also estimate the Debye frequency parameter ω_D in a similar way.

In that period there were a number of suggestions extant as to how to achieve the "Holy Grail" of appreciably higher superconducting transition temperatures. One that was unaccountably popular was W. A. Little's "excitonic" mechanism[18] where he envisaged a metallic polymer chain with the electron–electron interactions screened by electronic excitations in side-groups attached to the chain. These excitations could be assumed to have an arbitrary resonant frequency, and, taking the BCS gap equation literally, T_c would scale with this frequency and be as large as one likes. This kind of idea was taken up by others to justify other types of hypothetical inhomogeneous systems.[19]

A glance at the formula (3) shows that the Debye frequency appears in it in two places, as the scale factor outside and as the large logarithm in the exponent. These two have opposite effects on T_c, and for given λ and μ there is actually an optimum ω_D, which tends to be only slightly above the physical value. (In the case of MgB_2, I think the optimum is very close to being achieved.) Marvin Cohen and I wrote a very short, simple paper[20] pointing out these rather obvious conclusions and remarking that there seemed therefore to be an upper limit to T_c achievable by the dynamic screening mechanism, in the neighborhood of 40° K. So far this prediction has proved to be rather precisely correct. The iron pnictides and the cuprates, which strongly violate this rule, clearly use quite different mechanisms for pairing. The "Bucky-ball" M_3C_{60} superconductors do not violate it literally but do not satisfy the conditions for dynamic screening,[21] and may be presumed to require a special mechanism involving their unusual band structure[22] and strong correlation effects.

Our paper aroused considerable opposition; it seemed that there was a strong emotional commitment on the part of Bardeen's group, as well as some of the Russian school, to the dream that someday there might appear a room-temperature superconductor; in particular, our claim that values of λ much greater than μ would lead to dynamic instability was attacked strongly as a rigorous statement rather than the practical rule of thumb that it actually is. Fortunately, we now know that dynamic screening is not the only mechanism, and await hopefully some further still undiscovered mechanism that may yet achieve the Holy Grail.

I promised to come back to Pierre's first problem. As I said on entering the subject of electron–electron interactions, the crucial problem is to somehow evade the hard core screened Coulomb repulsion that a pair of electrons inevitably incurs. In the dynamic screening mechanism, the evasion is in time — the electrons are not on the same atom at the same time. There is only one other possibility — to ensure that the electrons avoid each other in space, i.e. that the pair wave function, the solution of the BCS gap equation, is zero at the origin. For two particles in empty space, interacting via a symmetric potential, it can be shown that the lowest energy bound state necessarily is the lowest s-wave, which cannot vanish at the origin; but the case of a BCS pair is quite different because the kinetic energy necessary to create the node at the origin can be provided by the Fermi zero-point energy. For this reason it is quite reasonable that the best solution may be asymmetric around the Fermi surface and belong to a degenerate representation of the crystal point group (or, in an approximately isotropic case, to a nonzero angular momentum).

This idea occurred independently to Lev Pitaevskii[23] and to me in 1959, and we both published in 1960; I had been influenced by a remark of John Fisher and some words in the book by David Thouless.[24] I had suggested that Pierre look into it, and I also described it to Keith Brueckner during his visit to Bell in the summer of 1959. Keith, who also knew of the work at Los Alamos on the Fermi liquid properties of liquid He_3, immediately pricked up his ears and suggested that He_3 was a good candidate for such a state, resulting from the interactions of not electrons but He_3 atoms, which are also Fermions and below 1–2° Kelvin behave as an isotropic Fermi sea. (The atoms have an even harder repulsive core than electrons.) I agreed and relayed the suggestion to Pierre, who set out to work out the most likely angular momentum given the He–He interaction, and to find the properties of such a state.

Within a month or two I was a little shocked to receive a preprint from Keith along with a student, Soda, in which he estimated transition temperatures for $L = 1$ and $L = 2$ states for He_3 as well as several possible higher L states for nuclear matter. I quite frankly did not have any confidence in such estimates, using the bare interatomic potential, but, feeling very defensive of Pierre's rights in the problem, I argued that we deserved co-authorship for thinking of the problem, and Keith was very accommodating. It turned out he was right to jump in, there were at least two other sets of authors with similar ideas and we only barely achieved priority for the idea of anisotropic superfluid He_3, though our choice of state and estimate of T_c were both very wrong (as were the competitors'). The world had to wait 12 years before the real thing came along.

What I thought was worthwhile in what we were doing was to establish some matters of principle. Were these anisotropic states real superconductors? (In spite of the fact that they violated Landau's criterion for superfluidity, that excitations must have a minimum velocity, and that most of them were gapless in that the gap had zeroes.) They were indeed "super," and this was very enlightening as to the true nature of superfluidity. I still find Landau's criterion quoted in the textbooks, but it became meaningless in 1959. Also, they turned out to exhibit a number of different states for any given L and S, all with the same T_c because the equation for T_c is linear, but with different free energies; how did they choose among them? What was the meaning of the apparent possibility of orbital ferromagnetism, and how would it manifest itself? All of these questions really had to wait for final solution until we had the real stuff at hand, but we could begin to ask sensible questions. I gave a preliminary talk on these kinds of matters at Utrecht in summer 1960; and Pierre wrote a long and detailed paper in '61 on a possible $L = 2$ state for He_3, which fortunately mentioned in passing the other possibility of $L = 1$, $S = 1$.

I must stop indulging myself in He_3 nostalgia and get back to the subject of electron–electron interactions — though actually the issues involved are not all that dissimilar. But there is one vital aspect to consider — the response of the superconducting state to disorder and scattering. It is very hard to make He_3 impure, nothing dissolves in it — even He_4 — and the only "disordered" He_3 is that contained in porous media, so in spite of its minute T_c of 2.6mk He_3's condensation was finally observed in 1972 and confirmed to be in an $L = 1$, triplet pair state. But standard metals and alloys all tend to exhibit the "dirty superconductor" immunity to scattering, which, it was easy to see, was not at all characteristic of anisotropic pair states, since for these ordinary scattering around the Fermi surface was "pair-breaking", equivalent to magnetic scattering of a BCS superconductor. So even when Balian and Werthamer[25] pointed out an error in our treatment of spin in the triplet case, which allowed there to be a fully gapped state which, however, was not immune to pair-breaking by ordinary scattering, it was not considered plausible that any of the conventional superconductors were other than BCS, dynamic-screening states.

There the matter stood until 1979–80, when two unexpected, but utterly different, types of materials were discovered to be superconducting, neither of which were likely to be caused by dynamic screening. The first discovery,[26] in 1979, was a so-called "heavy electron" mixed valence metal, $CeCu_2Si_2$. These are compounds of certain rare earth metals, Ce, Yb and the actinide U, in which the f-shell metal is a magnetic ion at high temperature, but its electrons form very narrow metallic f-bands at low T. The electrons that become superconducting are dominated by the high state density in the F band; although the transition temperatures are low, they are not much smaller than the effective band width so that the entropy gained at T_c, and hence the specific heat, is much larger than for conventional superconductors. Three more such compounds, UPt_3, UBe_{13}, and URu_2Si_2, were discovered in 1983 by the Los Alamos group around Z. Fisk and H. R. Ott, and the community ceased to think of Steglich's compound as a rare anomaly. In fact, in recent years a dozen or so more such compounds have been found, many due to the efforts of Lonzarich's group at Cambridge.[27] The fact that the f-shell ion is magnetic at high temperatures proves that the intra-atomic Coulomb repulsion (which is often designated by U) dominates the Fermi energy in the f-shell bands and that there can be no pseudopotential renormalization to μ^* as in the elemental metals. Prima facie one must assume that the pairing is unconventional, involving an asymmetric pair function of some sort. This conclusion is supported by the fact that T_c is very structure-sensitive in most of these compounds.

It is somewhat inexplicable why this simple conclusion was resisted for so long by the community consensus. Somehow, the quite subtle, complicated dynamic screening mechanism became the old reliable accepted standard, while the simpler mechanism, in principle, of building a pair function with a node at the origin became thought of as strange and exotic. For decades the burden of proof remained on anyone who claimed that the new classes of superconductors were not phonon-motivated and involved an unsymmetric gap function.

In 1980 this new class of superconductors was joined by another, with the discovery of an "organic superconductor,"[28] which also turned out to be the first of a large and growing family of compounds unlike either of the first two groups. Following work in the 70's which led to the "organic metal" TTF-TCNQ,[29] Bechgaard's salts were "charge-compensated" stacks of moderately large aromatic molecules like TTF. (See figure for a typical such molecule):

$$(\mathrm{TMTSF})_2\mathrm{X},\quad \mathrm{X} = \mathrm{PF}_6^-,\ \mathrm{ClO}_4^-,\ldots$$

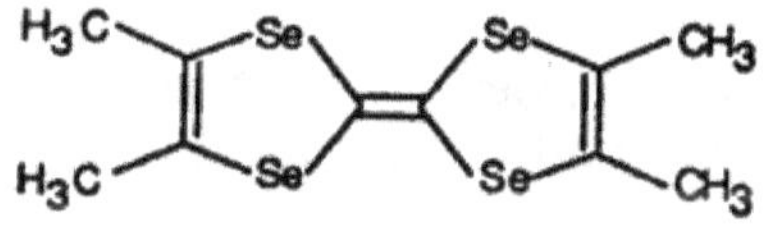

The chemical structure and electronic properties of such a stack structure are quite different from the covalently-bonded polymers like polyacetylene envisaged in Ref. 19. The electrons shared along the stack, which make the substance a metal, are from the highest occupied π molecular orbital (the HOMO) of the unsaturated part of the molecule. At ambient pressure the substance is often an antiferromagnetic Mott insulator, and is induced to become metallic and superconducting only by applying pressure, in many cases. The first discoveries were of approximately one-dimensional systems, the Bechgaard salts, but these were followed by systems which formed approximately two-dimensional nets,[30] and the number of such materials continues to grow, as does T_c, within limits.

Even more than the Steglich heavy-electron group, these materials cried out for treatment as a "strongly-interacting" system where U is greater than the bandwidth. In fact, at first sight they look like simple one-band Hubbard models with anisotropic hopping energies t_{ij}. But in actuality the approximately one-dimensional cases exhibited very complex behavior, especially at high magnetic fields, much of which seems to be associated with Landau levels and with nesting. The only very noteworthy fact about the superconductivity, which occurs below a large but reasonable Hc_2, seems to be that it is triplet in at least one case.[31]

The strongly two-dimensional materials have certain resemblances to the cuprates except that they cannot be doped and often are just on the metallic side of a Mott transition. Hesitantly, we classify them as being what Bernevig *et al.*[32] called "gossamer" superconductors, that is superconductors just on the smaller U side of a Mott transition.

The next class of superconductors to be discovered were the notorious "high T_c" cuprate superconductors, in 1987.[33] These are by far the most exciting, and ostensibly the most controversial, superconductors. Nonetheless I argue here that they are, in terms of electron–electron interactions, by far the simplest and best understood case. Most of the physics excitement (which still fully occupies those of us still in this field) has to do not with controversy about the causative interactions but with the many fascinating and unusual phenomena that Nature has chosen to tease out of what is in essence a rather simple system, so simple that we can use it as a guide to the others I have just been describing.

I was the first to point out[34] that the crystal and electronic structure of the cuprates was conducive to being described by the "Hubbard Model" Hamiltonian in an extremely simple case. This Hamiltonian (which was used by Hubbard but probably not invented by him) is simply

$$H = \sum_{i,j,\sigma} t_{ij} c^*_{i,\sigma} c_{j,\sigma} + U \sum_i n_i^{\uparrow} n_i^{\downarrow} \tag{4}$$

where the sites i lie on some simple lattice (in the cuprate case, a layer structure of simple square planar Cu's.) When U is small, (4) describes a simple metal, but when U is large and there is one electron per atom, we have a "Mott insulator". As I first pointed out,[35] the Mott insulator is usually antiferromagnetic because of the "superexchange" effect (I used the preferable designation "kinetic exchange"), which is the second-order perturbative effect of the hopping integrals t; the exchange constant J is proportional to t^2/U. But the cuprate has the almost unique ability to become a **doped Mott insulator**, that is to embody the Hamiltonian (4) in the large U case without having exactly one electron per site.

The superexchange effect can be made explicit by means of a canonical transformation due to Rice and Ueda.[36] This transformation $U_R = e^{iS}$ may be derived by gradually turning on the first term of (4) while requiring that matrix elements between different eigenvalues of the second term remain zero. The result, in the lowest energy subspace, is the "t-J Hamiltonian"

$$H_{t-J} = U_R H U_R^{-1} = P \left(\sum_{i,j,\sigma} t_{ij} c^*_{i,\sigma} c_{j,\sigma} \right) P + J \sum_{i,j} S_i \cdot S_j; \quad J \sim t^2/U$$

$$P = \Pi_i (1 - n_i^{\uparrow} n_i^{\downarrow}) \tag{5}$$

here c^* and c are electron destruction and creation operators on sites i and j, S is the spin on the designated site, and P is the "Gutzwiller" projection operator which eliminates all doubly-occupied sites.

(Parenthetically, (5) can be re-expressed in terms of the "hat" or projected operators $\hat{c}^*_{i,\sigma} = (1-n_{i,-\sigma})c^*_{i,\sigma}$ and $\hat{c}_{i,\sigma} = (1-n_{i,-\sigma})c_{i,\sigma}$ that eliminate the need for explicit P operators surrounding the kinetic energy. This is the device used in Anderson's "Hidden Fermi Liquid" theory of the overdoped cuprates.)

The exchange (J) term in (5) is a product of four fermion operators and thus can be looked at as an interaction vertex in the particle–particle channel. Therefore it can be the source of electron pairing, as may first have been pointed out by Scalapino and Hirsch[37] in the context of heavy-electron superconductors; if antiferromagnetic, it will tend to favor singlet pairs, and if one is to assume that a node at the origin is required, that suggests d-wave pairing, which they indeed mentioned. Rather confusingly, these and other authors who followed them described this as "pairing caused by spin fluctuations" which implies to the incautious that the pairing is caused by propagating or diffusing low-frequency modes in a comparable mechanism to the BCS phonons; but the derivation of J above shows that the J interaction is an essentially instantaneous vertex (on T_c scales) in the case of the doped Mott insulators, i.e. the cuprates. The spins that interact are created by a process that resembles the spin fluctuation hole-particle resummation of Berk and Schrieffer,[38] so there is a sense in which the original claim is true, but misleading rather than wrong. But common sense[39] tells us that the large interaction J must be the key to the pairing, and successful calculations[40] deriving the observed $d(x^2 - y^2)$ pairing using (5) alone have made that certain, in my opinion. Nonetheless controversy on the dynamic nature of the interaction continues, much of it fueled by appeals to the irrelevant Eliashberg formalism.[41] In my experience common sense usually trumps formalism.

The convoluted history of cuprate theory will be treated in more detail by other authors in this book, so I will not go further than the above expression of my conviction that the simple Hubbard model can account for almost all of the fascinating phenomenology which is exhibited by these marvelous compounds.

Yet another whole class of magnetic-type superconductors has recently come to light, the iron pnictides such as FeAsLaO, doped by substitution of F for O.[42] The T_c's are not up to the cuprates — maximum about 55° K — but the sheer number of quite different chemical structures staggers the mind.

All, however, have as the working portions a square planar Fe layer bound by out-of plane pnictide atoms in a very characteristic structure. The undoped system is usually magnetic but not a true insulator, more like a semimetal. What is in common with cuprates is this square planar structure and that it is both dopable and magnetic; but there is no escaping that it cannot be a one-band Mott system, and in fact multiple Fermi surfaces are observed. Moderate success has been achieved with rather complex renormalization schemes based on the assumption that the interaction is purely electronic as in the cuprates.

Without going in detail into the enormous literature on these four large classes of unconventional superconductors, I will repeat and expand the remark I made previously[7]: It seems very likely that they are all based on the second method of avoiding the hard Coulomb core of the electron–electron interaction, and therefore are all cases of unconventional, non-BCS pairing (although of course they are all, at a deeper level, based on the BCS coherent state wave-function.) Therefore an unbiased count of superconductors might well find that this is the majority case. In most of these there is good reason for ignoring the phonon coupling; and the prevalence of nearby magnetic states strongly suggests that the magnetic interaction exemplified by the t-J renormalized Hamiltonian is the major culprit in the pairing.

In summary: The electron–electron interaction which is responsible for turning most metals into superconductors at low enough temperature seems to belong to one or the other of two almost non-overlapping classes. For the simple metals and alloys, the dynamic screening mechanism elaborated from the original BCS phonon scheme is one of the best-attested truths of quantum materials theory. This is responsible for almost all technically useful superconductors, so far. But there is a numerically far larger, if technologically unimportant, category of materials that depend on exotic Pitaevskii-Thouless pairs with complex internal wave functions, vanishing at the origin in order to evade the Coulomb self-interaction U.

References

1. P. W. Anderson, presented at "BCS50", 2008.
2. P. W. Anderson, *Phys. Rev.* **110**, 827 (1958).
3. N. N. Bogoliubov, V. Tolmachev and D. V. Shirkov, *New Method in the Theory of Superconductivity*, Academy of Sciences, Moscow (1958), Translation, Consultants' Bureau, NY, 1959.
4. P. W. Anderson, *Phys. Rev.* **112**, 1900 (1958).
5. P. G. De Gennes, *Superconductivity of Metals and Alloys* (Benjamin, NY, 1966).

6. P. W. Anderson, *Phys. Rev. Lett.* **3**, 325 (1959).
7. P. W. Anderson, *Annales Henri Poincare* **4**, 1 (2003).
8. P. Morel, *J. Phys. Chem. Solids* **10**, 277 (1959).
9. P. Morel and P. W. Anderson, *Phys. Rev.* **125**, 1263 (1962).
10. A. A. Abrikosov, *Phys. Rev. B* **55**, 11955 (1995).
11. N. N. Bogoliubov, *Uzpekhi Fiz Nauk* **67**, 549 (1959).
12. G. M. Eliashberg, *Soviet Physics — JETP* **11**, 696 (1960).
13. I. Giaevar, H. R. Hart and K. Megerle, *Phys. Rev.* **126**, 941 (1962).
14. J. M. Rowell, P. W. Anderson and D. E. Thomas, *Phys. Rev. Lett.* **10**, 334 (1963).
15. J. R. Schrieffer, D. J. Scalapino and J. W. Wilkins, *Phys. Rev. Lett.* **10**, 336 (1963).
16. W. L. McMillan and J. M. Rowell, in *Superconductivity*, ed. R. D. Parks, p. 561 (Dekker, NY, 1969), for instance.
17. W. L. McMillan, *Phys. Rev.* **167**, 331 (1968).
18. W. A. Little, *Phys. Rev.* **134**, A1416 (1964). This paper is an example of highly fertile scientific fields being stimulated by questionable ideas. There is an enormous literature on organic superconductors and another on polymer metals, both fruitful systems but structurally neither of which much resemble the Little scheme. Nonetheless they can be truthfully said to have been stimulated by Little.
19. D. Allender, J. Bray and J. Bardeen, *Phys. Rev. B* **7**, 1020 (1973).
20. M. L. Cohen and P. W. Anderson, *AIP Conf. Proc.* **4**, 17 (1972).
21. P. W. Anderson, *Theories of Fullerene T_c's Which Will not Work*, unpublished preprint 1992 (submitted to *Nature*).
22. M. Capone, M. Fabrizio, C. Castellani and E. Tosatti, *Revs. Mod. Phys.* **81**, 943 (2009).
23. L. P. Pitaevskii, *Sov. Phys. JETP* **10**, 1267 (1960).
24. D. Thouless, *The Quantum Mechanics of Many-Body Systems* (Academic Press, NY 1961). David soon provided a fuller account: *Ann. Phys.* **10**, 553 (1960) but did not go very far into the properties of such a state.
25. R. Balian and N. R. Werthamer, *Phys. Rev.* **131**, 1553 (1963).
26. F. Steglich, J. Aarts *et al.*, *Phys. Rev. Lett.* **43**, 1892 (1979).
27. G. G. Lonzarich, in *Electron*, ed. M. Springford, Ch. 6, p. 109 (1997).
28. D. Jerome, A. Mazaud, M. Ribault and K. Bechgaard, *J. Phys. (Paris) Lett.* **41**, L95 (1980).
29. A. N. Bloch *et al.*, *Phys. Rev. Lett.* **34**, 1561 (1975).
30. G. Saitoh *et al.*, *Sol. St. Comm.* **42**, 557 (1982).
31. M. Y. Choi, P. M. Chaikin and R. L. Greene, *Phys. Rev. B* **34**, 7727 (1983) and subsequent work.
32. R. B. Laughlin, A. Bernevig *et al.*, *Phys. Rev. Lett.* **91**, 147003 (2003).
33. J. G. Bednorz and K. A. Muller, *Z. Phys. B* **64**, 69 (1986); M. K. Wu *et al.*, *Phys. Rev. Lett.* **52**, 408 (1987).
34. P. W. Anderson, *Science* **235**, 1196 (1987).
35. P. W. Anderson, *Phys. Rev.* **115**, 2 (1959).

36. T. M. Rice and K. Ueda, *Phys. Rev. B* **34**, 3420 (1986); further described in C. Gros, R. Joynt and T. M. Rice, *Phys. Rev. B* **36**, 387 (1987).
37. J. E. Hirsch and D. J. Scalapino, *Phys. Rev. Lett.* **53**, 706 (1984) et seq.
38. N. F. Berk and J. R. Schrieffer, *Phys. Rev. Lett.* **17**, 433 (1966).
39. PWA, *Science* **316**, 1705 (2007).
40. G. Kotliar and J. Liu, *Phys. Rev. B* **38**, 5142 (1988); F. C. Zhang *et al.*, *Supercond. Sci. & Tech.* **1**, 36 (1988).
41. T. A. Maier, D. Poilblanc and D. J. Scalapino, *Phys. Rev. Lett.* **100**, 237001 (2008); J. Hwang, T. Timusk *et al.*, *Phys. Rev. B* **75**, 155708 (2007).
42. Y. Kamihara, H. Hosono *et al.*, *J. Am. Chem. Soc.* **128**, 10012 (2006).

II. Fluctuations, Tunneling and Disorder

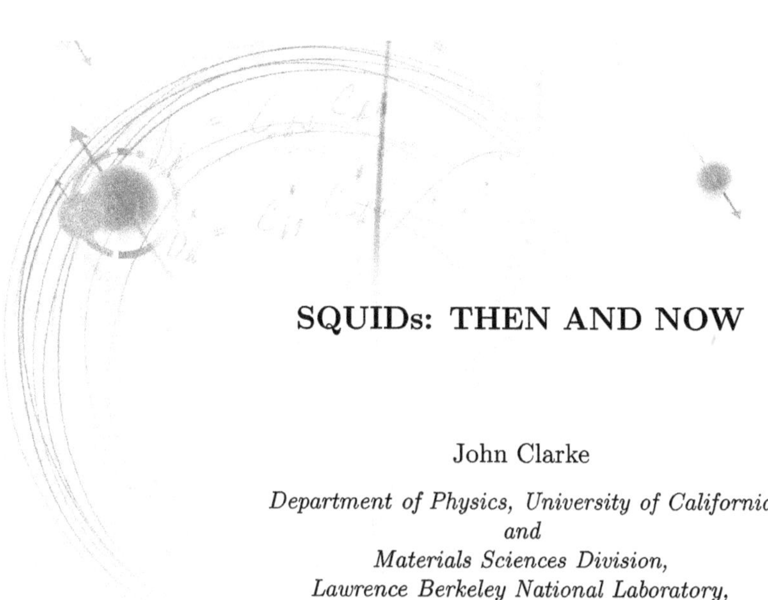

SQUIDs: THEN AND NOW

John Clarke

Department of Physics, University of California
and
Materials Sciences Division,
Lawrence Berkeley National Laboratory,
Berkeley, CA 94720, USA
jclarke@berkeley.edu

This chapter is dedicated to the memory of Brian Pippard.

Following Brian Josephson's prediction in 1962, Anderson and Rowell observed Josephson tunneling in 1963. The following year, Jaklevic, Lambe, Silver and Mercereau demonstrated quantum interference in a superconducting ring containing two Josephson tunnel junctions. Subsequently, the first practical devices emerged, including the point-contact dc and rf SQUIDs (Superconducting QUantum Interference Devices) of Zimmerman and Silver and Clarke's SLUG (Superconducting Low-inductance Undulatory Galvanometer) — a blob of solder frozen around a length of niobium wire. The return to the tunnel junction as the Josephson element was heralded by the cylindrical SQUID in 1976. The square washer dc SQUID developed by Ketchen and Jaycox in 1982 remains the workhorse design for most applications. Theories for the dc and rf SQUIDs were worked out in the 1970s. Today, SQUIDs (mostly dc) are used in a variety of configurations — for example, as magnetometers, gradiometers, cryogenic current comporators, low-frequency and microwave amplifiers, and susceptometers — in applications including magnetoencephalography, magnetocardiography, geophysics, nondestructive evaluation, precision gyroscopes, standards, cosmology, nuclear magnetic resonance, reading out superconducting quantum bits, and a myriad of one-of-a-kind experiments in basic science. Experiments are described to search for galaxy clusters, hunt for the axion, and perform magnetic resonance imaging in microtesla fields.

1. Introduction

The pairing of electrons, predicted by Leon Cooper[1] in 1956, is the cornerstone of the BCS (Bardeen, Cooper, Schrieffer) theory of superconductivity[2] that appeared in 1957. Among the myriad consequences of Cooper pairs are two that underlie the physics of the SQUID (Superconducting QUantum Interference Device): flux quantization and Josephson tunneling. The first implies that the magnetic flux embraced by a superconducting loop is quantized in units of the flux quantum $\Phi_0 \equiv h/2e \approx 2.07 \times 10^{-15}$ Tm^2; here, $h \equiv 2\pi\hbar$ is the Planck constant and e is the electronic charge. The concept of the flux quantum was in fact introduced much earlier by Fritz London,[3] who, obviously not knowing about Cooper pairs, predicted that its value should be h/e. The experimental observation of the quantization of flux in 1961 by Deaver and Fairbank[4] and Doll and Näbauer[5] — needless to say, in units of $h/2e$ — was a striking confirmation of pairing and of the existence of a *macroscopic* wave function in the superconducting state. The wave function

$$\Psi(\mathbf{r},t) = |\Psi(\mathbf{r},t)| \exp[i\phi(\mathbf{r},t)]\,, \tag{1}$$

which depends on position $\mathbf{r}$ and time t, describes the entire condensate of Cooper pairs in a given superconductor; $\phi(\mathbf{r},t)$ is a phase factor. Flux quantization arises from the fact that $\Psi(\mathbf{r},t)$ must be single-valued in going once around a superconducting loop. In the absence of any applied currents or magnetic fields, the phase $\phi(\mathbf{r},t)$ has the same value throughout the superconductor. When the loop is threaded by a magnetic flux, however, the phase around the loop changes by $2\pi n$, where the integer n is the number of enclosed flux quanta.

In the year preceding the observation of flux quantization, Ivar Giaever[6] demonstrated the tunneling of single electrons between a superconductor (S) and a normal metal (N) separated by a thin insulating barrier (I). Subsequently, he observed single-electron tunneling through SIS junctions.[7] Morel Cohen, Leo Falicov and Jim Phillips[8] explained these effects in terms of a tunneling Hamiltonian. Taking this calculation to higher order, in 1962 Brian Josephson predicted the tunneling of Cooper pairs through a barrier between two superconductors.[9] Brian showed that the phenomenon of Josephson tunneling can be summarized by two succinct equations. First, the supercurrent I flowing through the junction can be expressed as

$$I = I_0 \sin\delta\,. \tag{2}$$

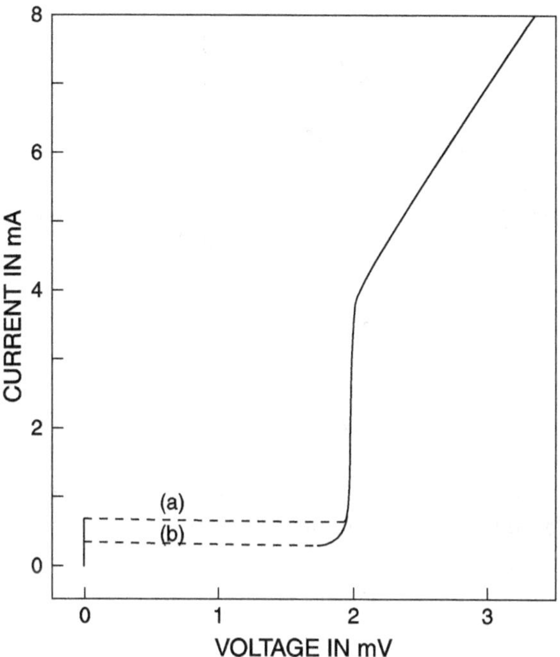

Fig. 1. Current-voltage characteristic for a tin-tin oxide-lead tunnel junction at about 1.5 K. (Reproduced with permission from Ref. 10.)

Here, $\delta(t) = \phi_1(t) - \phi_2(t)$ is the difference between the phases ϕ_1 and ϕ_2 of the condensates in the superconducting electrodes 1 and 2, and I_0 is the critical current — the maximum supercurrent the junction can sustain. Second, in the presence of a voltage V across the junction, $\delta(t)$ evolves with time as

$$d\delta/dt = 2eV/\hbar = 2\pi V/\Phi_0 \,. \tag{3}$$

Equations (2) and (3) describe the "dc" and "ac" properties of a Josephson junction. In the presence of an external current through the junction, the flow of Cooper pairs constitutes a supercurrent and, in the absence of fluctuations, the voltage across the junction remains zero until the current exceeds I_0. At higher currents, there is a voltage across the junction and $\delta(t)$ evolves according to Eq. (3). Subsequently, it became clear that this behavior is generally valid for all weakly-coupled superconductors, although there may be significant departures from the sinsusoidal current-phase relation.

One year later, Phil Anderson and John Rowell[10] observed the dc Josephson effect in a thin-film, Sn-SnOx-Pb junction cooled to 1.5 K. The I–V characteristic is reproduced in Fig. 1. Subsequently, Rowell[11] showed that

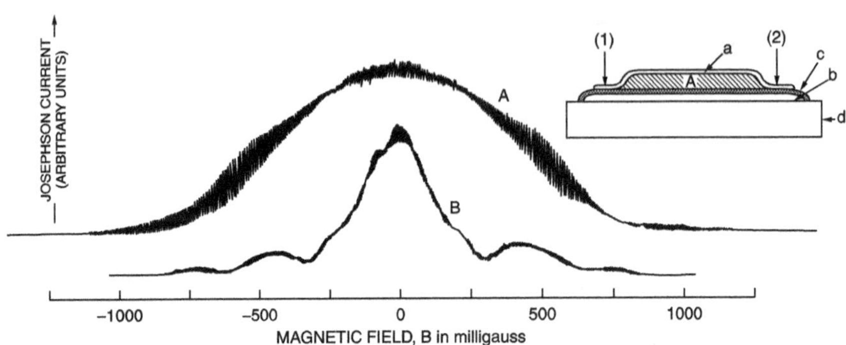

Fig. 2. The first dc SQUID. Josephson critical current versus magnetic field for two Sn-SnOx-Sn junctions (1) and (2) in parallel (inset) showing interference effects. Magnetic field applied normal to the area between junctions. Curve (A) shows interference maxima spaced at $\Delta B = 8.7 \times 10^{-3}$ G, curve (B) spacing $\Delta B = 4.8 \times 10^{-3}$ G. Maximum Josephson current is approximately 1 mA. (Reproduced with permission from Ref. 12.)

a magnetic field B, applied in the plane of the films, caused the critical current modulation

$$I_0(\Phi) = I_0(0)|\sin(\pi\Phi/\Phi_0)/(\pi\Phi/\Phi_0)|\,. \tag{4}$$

Here, $\Phi = Bw(d + \lambda_1 + \lambda_2)$; w is the width of the junction perpendicular to B, d is the barrier thickness and λ_1 and λ_2 are the penetration depths of the two superconductors. The observation of this Fraunhofer-like result — analogous to the diffraction of coherent, monochromatic light by a slit — is a dramatic demonstration of the validity of the current-phase relation.

A few months later, Bob Jaklevic, John Lambe, Arnold Silver and Jim Mercereau[12] at the Scientific Laboratory of Ford Motor Company, Dearborn, Michigan demonstrated quantum interference between two thin-film Josephson junctions connected in parallel on a superconducting loop. The dependence of the critical current on the applied magnetic field is shown in Fig. 2. The slowly varying modulation arises from the Fraunhofer-like diffraction of the two junctions [Eq. (4)]. The rapid oscillations are due to quantum interference between the two junctions, and their period is given by the field required to generate one flux quantum in the loop: thus, critical current maxima occur at $\Phi/\Phi_0 = 0, \pm 1, \pm 2, \ldots$. The observation of these oscillations — analogous to two-slit interference in optics — was the birth of the dc SQUID.

2. Early dc SQUIDs

The lead and tin thin-film Josephson junctions used in the pioneering experiments proved to be fragile and subject to failure after even a single thermal cycling. Later in 1964, Jim Zimmerman and Silver[13] came up with a simple alternative technology consisting of a narrow sheet of niobium with a thin niobium wire bent over the top of it to form two junctions. The nature of the junctions was not entirely clear. Either the native oxide on the surface of the niobium formed a tunnel barrier or there was a very narrow metallic constriction between the pieces of niobium. Nonetheless, the device behaved as a SQUID.

This was the scene when I became a research student in the Royal Society Mond Laboratory — part of the Cavendish Laboratory — at the University of Cambridge in October 1964. The Mond was presided over — with great kindness — by David Shoenberg. My thesis supervisor was Brian Pippard. He gave me a project to investigate the electrical resistance of the superconductor-normal metal (SN) interface — a topic that later became of considerable interest.[14] Brian pointed out that the way to measure this resistance was to make SNS sandwiches in which the normal metal was too thick to sustain a supercurrent — and that one needed to measure tiny voltages, on the order of 1 pV. Brian asked me to look into improving the performance of the cryogenic galvanometer he and George Pullan had previously developed.[15] This device involved a mirror and magnet suspended by a quartz fiber inside a vacuum can, surrounded by liquid helium, at the center of a Helmholtz coil through which one passed the current to be measured.

My thinking about ways to upgrade the galvanometer was brought to an abrupt end, however, by a Mond seminar given by Brian Josephson. In his talk, Brian described the experiments demonstrating flux quantization,[4,5] his own theory of pair tunneling,[9] and the observation of quantum interference[12] earlier that year. To me, as a new research student, it was utterly fascinating, and I remember thinking how I would like to work on these new ideas. My opportunity came much sooner than I could possibly have expected! The very next day, Brian Pippard bounced into my laboratory with a big smile on his face. "John, how would you like a voltmeter with a resolution of 2×10^{-15} volt in 1 second?" he asked. Not quite knowing how to answer, I somewhat cautiously admitted that this did seem like an interesting possibility. Brian — with a good deal of excitement — explained that he had dreamed up his new concept of a voltmeter the previous evening after hearing Brian Josephson's seminar. He sketched on the blackboard a voltage source $\mathcal{V}$ in series with a resistance R and an inductance L that was perfectly coupled to a SQUID, also with inductance L. The current through the resistor and inductor required to generate the flux quantum in the SQUID was simply Φ_0/L, and the corresponding static voltage $\mathcal{V}$ was thus $(\Phi_0/L)R$. Setting the time constant $\tau = L/R = 1$ s, Brian pointed out that the resolution was 2×10^{-15} V. It is interesting to note that Brian's initial idea was to make a *digital* voltmeter, each digit being a flux quantum.

Needless to say, I was thrilled — not least because my enthusiasm for dangling a tiny magnet and mirror from a quartz fiber inside a vacuum can had already begun to wane. I talked to John Adkins, who had previously made thin-film Josephson junctions in the Mond. He was not encouraging about the longevity of such devices, and I looked around for better alternatives. I came across the Zimmerman-Silver[13] article, which seemed like a good place to start. I made a similar device with the additional twist that I could adjust the vertical force on the bent wire from the top of the cryostat. I remember rolling out a piece of niobium wire to make the narrow piece of sheet. In those days, one took one's glass cryostat to the helium liquifier for Paul Booth to fill it — portable storage dewars for liquid helium seemed not to exist in 1964. After returning my dewar safely to my lab, I connected up the simple current-voltage measurement system I had built — and my SQUID worked! I spent much of that day adjusting the force on the wire, and discovered that I could vary the critical current quite readily. Sometimes I observed oscillations in critical current when I varied the applied fields and sometimes I did not. I made several variants of this design, and they all more or less worked, but I was not entirely convinced that the device was sufficiently stable to use routinely as a voltmeter.

An important part of one's life in the Cavendish was — and still is — coffee at 11 am and tea at 4 pm. The Mond students and staff always sat at the same table. One tea-time, I discussed how I was looking for a Josephson junction technology that did not require thin films yet was mechanically stable. Paul Wraight — with whom I shared my lab — suddenly looked at me and said something like "How about a blob of solder on a piece of niobium wire? Solder is a superconductor and you keep telling me that niobium has a surface oxide layer." We rushed back to our lab, where fortunately I still had some liquid helium left in my dewar from the experiment I had been running earlier in the day. I made two devices consisting of a blob of lead-tin solder melted onto a short length of niobium wire, attached some leads and lowered the devices into the helium bath. They both worked! The critical currents were roughly 1 mA. Paul and I were thrilled. The next morning,

Brian Pippard wandered into our lab to see how things were going, and Paul and I proudly showed him one of our new gadgets. Brian contemplated it thoughtfully for a while, and then — with a smile — said, "It looks as though a slug crawled through the window overnight and expired on your desk!"

I tried hard to make a SQUID by freezing a solder blob on a piece of niobium wire doubled back on itself so that there were two junctions. I applied a magnetic field to the loop of wire sticking out of the solder blob. This never worked, in retrospect because the inductance of the loop was too high. One day, I decided instead to pass a current along the wire, and immediately saw oscillations in the critical current when I changed the current. It did not take me very long to discover that the loop was irrelevant — I needed to pass the wire through the solder only once, and apply a current to the wire [Fig. 3(a)]. How did it work? Apparently, there were typically just two or three dominant junctions between the wire and the solder. The current in the wire generated a magnetic field in the penetration depths of the wire and solder, so that the area of the "SQUID" was given by the sum of the penetration depths times the separation of the junctions. The periodicity in current ranged randomly from about 0.2 to 1 mA. However unlikely, the majority of these devices showed interference between two or three junctions, and they generally survived at least scores of thermal cyclings. It is interesting to note that, as I increased the current through the niobium wire, I could generally observe thousands of oscillations in critical current with no evident diminution in amplitude. This lack of a discernible Fraunhofer pattern suggested that the junctions were essentially points.

The name for this new device had already been unwittingly provided by Brian Pippard and we simply had to work out what it stood for. And so the "Superconducting Low-inductance Undulatory Galvanometer" (SLUG) was born.

Brian's original concept was a digital voltmeter, but I soon realized that one could readily measure changes in flux that were much less than one flux quantum. I applied a static current and a sinusoidal current through the SLUG [leads "I" in Fig. 3(a)] so that voltage pulses appeared across the voltage leads [leads "V" in Fig. 3(a)]. The area under these pulses depended on the critical current of the SLUG, enabling me to determine the critical current, and hence the current in the niobium wire [leads "I_B" in Fig. 3(a)] to about ± 1 μA, using simple electronics. Once I had this worked out, I immediately put the SLUG together as a voltmeter. A postdoctoral fellow in the Mond, Steve Lipson, was interested in measuring the resistance of a single crystal of copper (to be used for measurements of magnetoresistance)

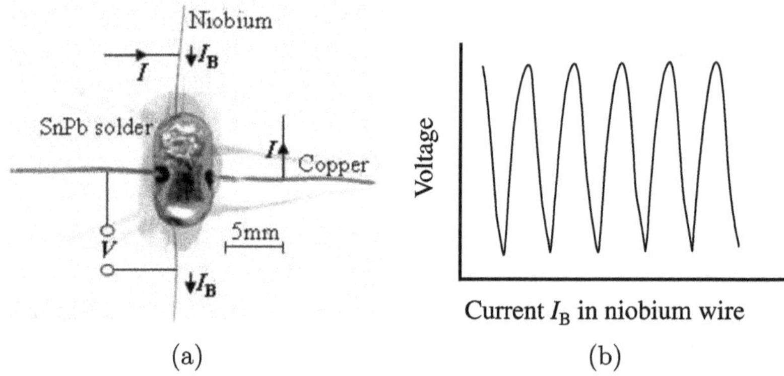

Fig. 3. The SLUG. (a) Photograph showing attachment of current (I), voltage (V) and flux bias (I_B) leads. (b) Voltage V versus I_B.

that he estimated to be about 10^{-8} Ω. We made a voltmeter by connecting the Cu block in series with a manganin wire, with a measured resistance of about 10^{-5} Ω, and the niobium wire through the SLUG. The idea was to make a potentiometer, adjusting the currents in the two resistors to produce zero current in the SLUG. After Steve and I cooled down our potentiometer, I was chastened to observe that the output from the SLUG was horribly noisy. Further inspection showed that the "noise" was in fact due to currents induced in the loop by 50-Hz magnetic fields. For the next run, Steve soldered up some lead foil to make a box that enclosed the circuit, and our pickup problem was solved. This was an invaluable lesson for me: if you have a sensitive magnetometer or voltmeter, you have to protect it from the real world, which is a very noisy place.

I used the SLUG as a voltmeter for the rest of my graduate career, mostly to investigate SNS Josephson junctions.[17] With a typical circuit resistance of 10^{-8} Ω, I could measure a voltage of about 10^{-14} V in a second. Other members of the Mond subsequently used the SLUG, notably Eric Rumbo[18] for measuring thermoelectric voltages and Pippard, Shepherd and Tindall[14] for measuring the resistance of the SN interference — my original thesis topic.

I moved to the Physics Department at the University of California, Berkeley in January 1968 as a postdoctoral scholar, and used SLUGs in various measurements. I devised an experiment to test the accuracy of the Josephson voltage-frequency relation in junctions made from different superconductors. I irradiated two SNS junctions with the same radiofrequency magnetic field, and measured the voltage difference between steps of a given order induced on the two I-V characteristics. I found that the voltages were the same to

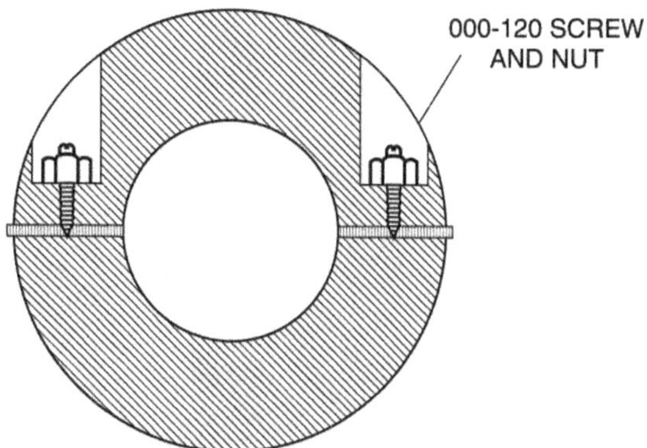

Fig. 4. Point contact dc SQUID. Reproduced with permission from Ref. 25.)

an accuracy better than 1 part in 10^8, corresponding to a voltage difference of less than 2×10^{-17} V, regardless of the nature of the superconductor.[19] This result helped to establish the Josephson junction as a means of maintaining the standard volt in terms of a frequency.[20] Much later, Shen Tsai and co-workers[21] repeated the experiment to the remarkable accuracy of two parts in 10^{16}.

Subsequently, I used a SLUG voltmeter to measure the pair-quasiparticle potential difference in nonequilibrium superconductors.[22] Single electrons tunneling from a normal electrode into a superconducting film generate an imbalance between the electron-like and hole-like quasiparticles. This "charge imbalance" is detected by a normal electrode in tunneling contact with the reverse side of the superconductor: a voltage is induced on the normal electrode relative to the chemical potential of the Cooper pairs measured some distance away from the nonequilibrium region of the superconductor. Shortly after I had completed this experiment, in early 1972 I returned to the Mond Laboratory on sabbatical leave. Mike Tinkham was also there on leave, writing his celebrated book on superconductivity, and he worked out the theory of charge imbalance.[23] Subsequently, charge imbalance was investigated in a variety of other nonequilibrium superconducting configurations.[24]

During the time I was working on the SLUG, Zimmerman and Silver[25] were busily developing the "point contact" dc SQUID, shown in Fig. 4. The body was machined from niobium, and the two halves were clamped together with a thin insulator separating them. The junctions were formed by two sharpened niobium screws that could be adjusted from outside the cryostat to provide optimal current-voltage characteristics. This simple means

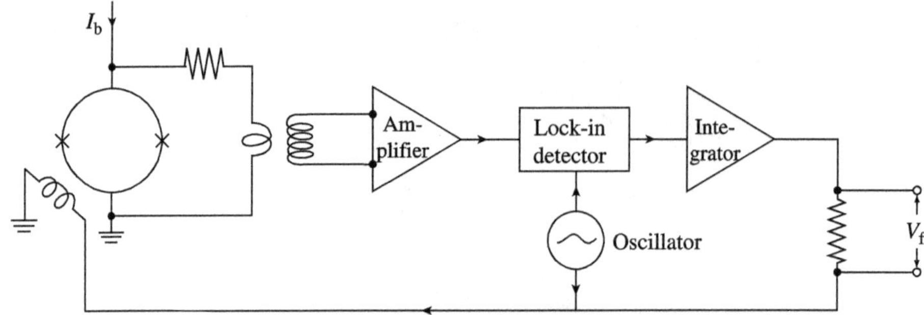

Fig. 5. dc SQUID in flux-locked loop.

of making a Josephson junction or SQUID was emulated in many laboratories. Single-junction versions were used by undergraduates[26] to demonstrate constant-current steps induced by microwaves on the current-voltage characteristics at voltages $nhf/2e$ (n is an integer, f is the microwave frequency), and thus verify the Josephson ac relation to an accuracy of 1% or better.

The observation of oscillations in voltage as the flux was varied was a beautiful piece of physics, but for practical applications, one needs a linear response. I built and operated a version of the flux-locked loop,[27] but a better one — that is still used today — was developed by Robert Forgacs and Alan Warnick[28] at the Ford Laboratory. Feedback serves several purposes: it makes the response linear in the applied flux, it allows one to track changes in flux corresponding to many flux quanta, and it enables one to detect flux changes of a tiny fraction of a flux quantum. As illustrated in Fig. 5, a sinusoidal magnetic flux with a peak-to-peak amplitude of $\Phi_0/2$ and a frequency f_m of 100 kHz is applied to the SQUID. When the static flux in the SQUID is $n\Phi_0$, the resulting oscillating voltage across it is a "rectified" sine wave, with no component at f_m. When this signal is coupled to a lock-in detector referenced to the same oscillator, the output is zero. When, on the other hand, the static flux in the SQUID is increased from $n\Phi_0$ to $(n + 1/4)\Omega_0$, there is a steady increase in the amplitude of the component at f_m, and an increasing (say) voltage out of the lock-in. If instead, we decrease the static flux from $n\Omega_0$, the voltage out of the lock-in is negative. The output from the lock-in detector is smoothed, and fed back via a resistor into a coil inductively coupled to the SQUID. The action of the flux-locked loop is to maintain the sum of the applied and fed-back fluxes in the SQUID at a constant value. A flux $\delta\Phi_a$ applied to the SQUID is canceled by an opposing flux $-\delta\Phi_a$ from the flux-locked loop. The output voltage V_f across the feedback resistor is proportional to $\delta\Phi_a$. In the early days, a typical flux noise was of the order of 10^{-3} Φ_0 $\mathrm{Hz}^{-1/2}$ ("per root Hz") — meaning an integration time of 0.25 s — at frequencies where the noise is white. At lower frequencies, f, the noise spectral density scales approximately as $1/f$. This ubiquitous "$1/f$ noise" or "flicker noise" is still actively investigated today. Virtually all SQUIDs — except those used at high frequencies — are operated in a flux-locked loop. There are several other schemes, at least one of which does not involve flux modulation.[29]

3. Early rf SQUIDs

Shortly afterwards, in 1967, the rf SQUID[30] appeared. This device involved only a single Josephson junction, and the first version was based on the machined niobium dc SQUID (Fig. 4), with one junction shorted. The rf SQUID was coupled to the inductor of a resonant circuit [Fig. 6(a)]. When the tank circuit was driven with a radio-frequency current at the resonant frequency, typically 30 MHz, the behavior shown in Fig. 6(b) was observed.[31] Here, the voltage across the tank circuit has been amplified, rectified with a diode and smoothed, so that one observes the amplitude of the rf signals. When the flux in the SQUID is $n\Phi_0$, one sees a series of "steps and risers" as the rf drive is increased. At $(n + 1/2)\Phi_0$, the steps split into a half-step above and below the steps. Thus, when the SQUID is biased at the midpoint of one of the steps the voltage is periodic in the applied flux, as shown in Fig. 6(c); needless to say, the period is Φ_0. From this point of view, the rf SQUID behaves operationally as a dc SQUID. Particularly in those early days, when making a Josephson junction was a challenge, the fact that the

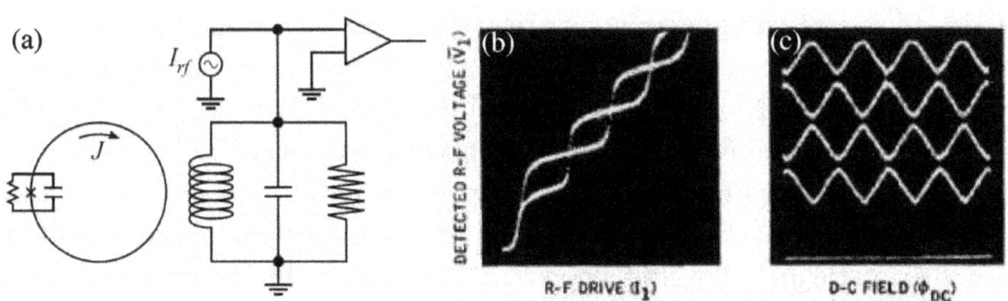

Fig. 6. The rf SQUID. (a) Schematic of rf SQUID and its readout circuit. (b) Steps and risers: Amplitude of rf signal across the tank circuit versus rf drive current for two values of applied flux. (c) Amplitude of rf signal across the tank circuit versus applied flux for four values of drive current. [(b) and (c) reproduced with permission from Ref. 31.]

rf SQUID required only a single junction gave it an edge over the dc SQUID. The rf SQUID is actually misnamed, however, as no interference takes place!

4. Tunnel Junctions Revisited

During the early 1970s, there was a general shift towards using the rf SQUID. SQUIDs made of thin films deposited on cylindrical substrates,[32,33] with a narrow microbridge as the single junction, achieved a white flux noise of about 2×10^{-4} Φ_0 $\text{Hz}^{-1/2}$. The strong dependence of the critical current of the microbridge on temperature, however, limited the operating temperature range. Such devices, as well as machined niobium devices, became commercially available.

While on sabbatical leave in the Mond in 1972, I wrote a review article on SQUIDs.[34] At the time, there was no theory for flux noise in either the dc or rf SQUID, but I made a simple estimate for the dc SQUID assuming that the noise arose from Nyquist noise in the junction resistances. This led me to conclude that the flux noise should be a few $\mu\Phi_0$ $\text{Hz}^{-1/2}$, far lower than anything that had been achieved. I concluded that the flux noise was dominated by voltage noise in the preamplifier connected to the SQUID, and that an appropriate matching network might significantly improve the performance.

After I returned to Berkeley, a new postdoctoral scholar, Wolf Goubau, and a new graduate student, Mark Ketchen, and I set to work. We borrowed three ideas from the rf SQUID: a cylindrical geometry (which gives a large area for a low inductance), a tank circuit readout and thin films. I had wanted to move away from microbridge junctions because of their narrow temperature range of operation. Fortunately, Paul Hansma, at the University of California, Santa Barbara, had achieved good reproducibility and longevity with Nb-NbOx-Pb tunnel junctions.[35] We decided to copy this technique, and acquired a sputtering system to deposit niobium. Figure 7(a) shows the geometry of our SQUID, grown on a 3-mm-diameter quartz tube. We used shadow masks to pattern the various films, which had a minimum linewidth of 75 μm. We first deposited the PbIn cylinder, followed by the gold film that formed the resistive shunt for each junction (to eliminate hysteresis on the current-voltage characteristic). We next sputtered the Nb film, which we oxidized in air in a closed oven (12 min at 130 °C was the recipe). Immediately afterwards, we deposited the PbIn "T" that completed the junctions. We scribed a slit in the PbIn cylinder with a razor blade. Subsequently, we submerged the SQUID in a solution of Duco cement

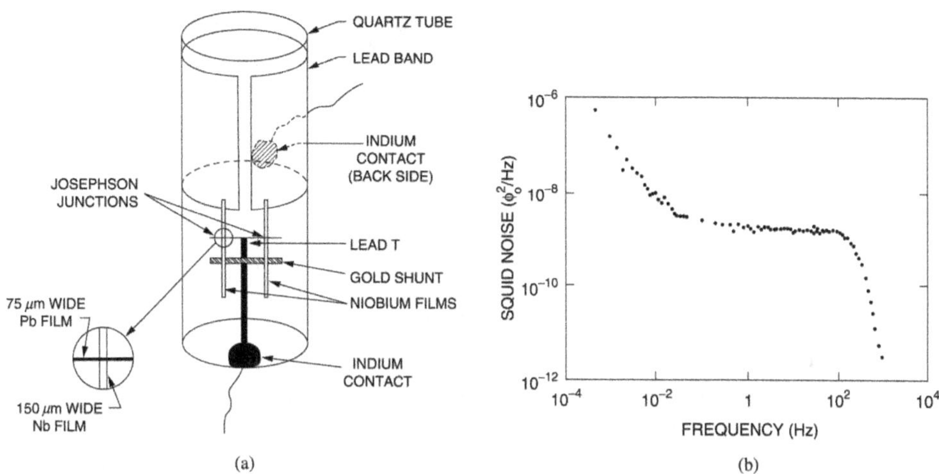

Fig. 7. Cylindrical dc SQUID. (a) Schematic. (b) Representative power spectrum of flux noise in a flux-locked loop.

dissolved in acetone — this was our favorite insulator, and it proved to be very tough. The last step was to deposit a PbIn film over the slit and the various metal strips to reduce their inductances. We attached leads with pressed indium beads. We wound the input coil from Nb wire on a Teflon rod, potted it in Duco cement, and removed the coil by dipping the whole thing in liquid nitrogen and slipping it off. The potted input coil would slide neatly over the cylindrical SQUID, giving a high coupling efficiency. We positioned the Nb modulation-feedback coil, wound on an insulating rod, inside the SQUID. This entire process sounds extraordinarily primitive by today's standards, but we could cool these devices to liquid helium temperature many times with no degradation.

To test the SQUIDs, we built a flux-locked loop with flux modulation. We boosted the resistance of the SQUID ($\sim 0.5\ \Omega$) to the optimum noise impedance of our preamplifier (a few kΩ) with a cold resonant circuit: the SQUID was shunted with an inductor and capacitor in series, and the preamplifier was connected across the capacitor. The result was a substantially enhanced signal into the preamplifier, and a correspondingly reduced flux noise. A representative power spectrum,[36] $S_\Phi(f)$ versus f, is shown in Fig. 7(b). The white noise, $3.5 \times 10^{-5}\ \Phi_0\ \mathrm{Hz}^{-1/2}$, was a good deal lower than that achieved previously, and the large loop area yielded a magnetic field noise of about 10 $\mathrm{fTHz}^{-1/2}$. It is also noteworthy that the $1/f$ noise "knee" — about 2×10^{-2} Hz — is substantially lower than that achieved in today's SQUID magnetometers.

We used our cylindrical SQUIDs in a variety of experiments. For example, Wolf Goubau, Tom Gamble and I used two, three-axis SQUID magnetometers for magnetotellurics. In this technique, one measures the impedance of the Earth at depths down to 50 km using low-frequency electromagnetic waves (10^{-4}–100 Hz) propagating to the Earth's surfaces from the magnetosphere and ionosphere. The low magnetic field noise of our SQUIDs ultimately led us to invent "remote reference magnetotellurics," a technique that eliminates noise induced by vibrational motion of magnetometers in the Earth's field by cross-correlating the signals from two distant ($\sim$5 km) magnetometers.[37,38]

I think the cylindrical SQUID played a role in turning the attention of the community back to the dc SQUID. In fact, to my knowledge, no other group adopted this design. One reason may have been the realization that higher performance required tunnel junctions with smaller areas. Nobody was excited about using photolithography — just being imported from the semiconductor industry — on a cylindrical surface!

Mark Ketchen moved to IBM, Yorktown Heights to work in the Josephson computer project. He and Jeffrey Jaycox developed the square washer SQUID,[39] shown schematically in Fig. 8(a), which is the basis of virtually all SQUIDs used today. Mark described this device as "The result of putting your thumb on one end of the cylindrical SQUID and squashing it flat, transforming it into a washer with a spiral input coil." The SQUID itself is a thin-film square washer, typically 1 mm across, with a hole in the middle and a slit that runs to the outer edge where two tunnel junctions are grown. The junctions are completed with an upper film that connects them, thus

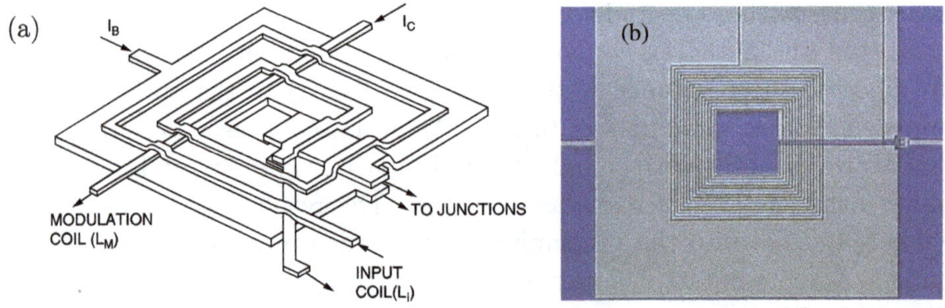

Fig. 8. Square washer dc SQUIDs. (a) Schematic of original design. (reproduced with permission from Ref. 39.) (b) Typical practical device. The two resistively shunted Josephson tunnel junctions are at the right-hand edge, one on each side of the slit. The Nb washer (light blue) is 1 mm across.

closing the SQUID loop. An insulating layer is deposited over the square washer, followed by a thin-film, spiral coil. A current passed through the coil generates a magnetic flux that the washer focuses into the hole, giving efficient coupling between the coil and the washer.

These SQUIDs were fabricated on silicon wafers using the photolithographic patterning techniques developed by the semiconductor industry. Although the stage was set for today's SQUIDs, one obstacle remained. The Ketchen-Jaycox SQUIDs were made from the PbInAu alloy used in the IBM computer project, and not as robust as one would have liked. Fortunately, in the early 1980s John Rowell and coworkers[40] and subsequently Gurvitch *et al.*[41] developed the niobium-based junction technology that is universally used today. One first deposits a Nb film on a silicon wafer, followed immediately by an Al film that is a few nanometers thick. The Al film is subsequently oxidized in an O_2 (or sometimes an ArO_2 mixture) atmosphere with a controlled pressure for a prescribed time. The gas is pumped out and the upper Nb electrode is deposited. Subsequently, the "trilayer" is patterned to form Josephson junctions. The thickness of the Al film is chosen so that some metal remains after the upper surface has been oxidized; this oxidation process is highly controllable.

The technology for the Nb-based, square washer dc SQUID — the workhorse of today's SQUID applications — was essentially in place by the mid '80s. Today, these SQUIDs are typically made in batches of several hundred on silicon wafers that are diced to produce individual devices. These devices are virtually indestructible, and can be cycled between room and cryogenic temperatures indefinitely. At 4.2 K, the flux noise is typically 1–2 $\mu\Phi_0$ $\text{Hz}^{-1/2}$ at frequencies above the $1/f$ knee of roughly 1 Hz. The corresponding (white) noise energy, $S_\Phi(f)/2L$, is $\sim 10^{-32}$ JHz^{-1} ($\sim 100\hbar$). As we shall see, much lower noise is achieved at millikelvin temperatures.

5. The Flux Transformer

Our story of the development of SQUIDs would not be complete without mentioning the superconducting flux transformer, illustrated in Fig. 9. Flux transformers can increase the magnetic field sensitivity of SQUIDs substantially and enable one to make magnetic field gradiometers. The magnetometer [Fig. 9(a)] consists of a superconducting pickup loop coupled to the input coil of a SQUID — for example, the square washer SQUID (Fig. 8) — to form a closed superconducting circuit. A magnetic field applied to the pickup loop induces a supercurrent in the transformer — because of flux

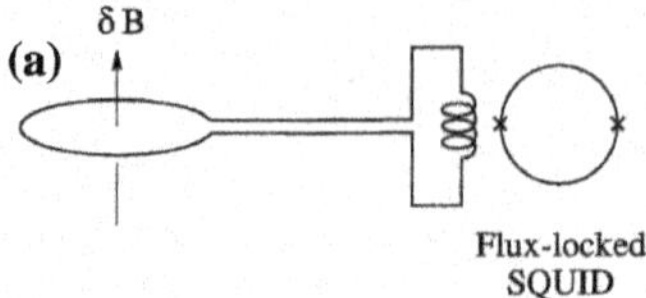

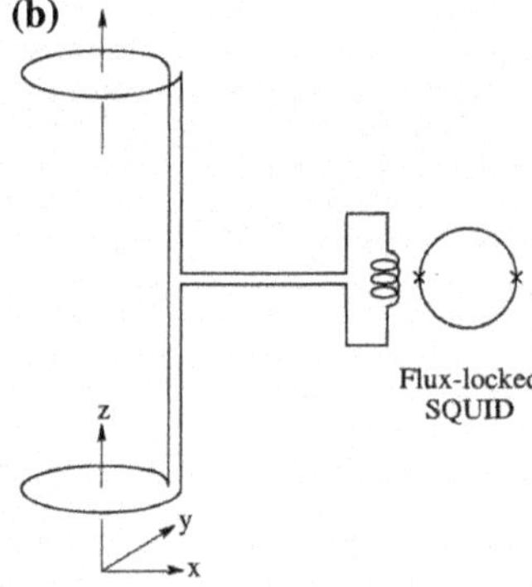

Fig. 9. Flux transformers. (a) Magnetometer and (b) first-derivative axial gradiometer, wound from niobium wire. Typical diameter of pickup loops is 50 mm.

quantization — and thus a flux in the SQUID. The transformer operates at frequencies down to zero. The transformer is optimized when the inductances of the pickup and input coils are equal. For a pickup loop diameter of 50 mm, a magnetic field noise of 10^{-15} $\text{THz}^{-1/2}$ is typical. Figure 9(b) shows a first-derivative, axial gradiometer in which the two pickup loops, nominally equal in area, are wound in opposite senses. Application of a uniform magnetic field along the axis of the loops produces no net flux in the transformer and no output from the SQUID. An applied gradient $\partial B_z/\partial z$, on the other hand, produces a current in the flux transformer and an output from the SQUID. Second-derivative axial gradiometers measuring $\partial^2 B_z/\partial z^2$ and off-diagonal gradiometers sensitive to, for example, $\partial B_z/\partial x$ are also commonly used. Gradiometers are mostly used to discriminate against distant noise sources in favor of nearby signal sources: noise from a dipole falls off with distance r as $1/r^4$ and $1/r^5$ for first- and second-derivative gradiometers. Gradiometers are commonly used to measure signals from the brain (Sec. 8).

6. A Little Theory

The early development of both dc and rf SQUIDs was largely a "seat of the pants" affair, without any detailed theory to guide their optimization.

Serious theory emerged in the '70s, and refinements continue today. I cannot possibly do justice to all the work that has been done, but will summarize the key results.

The dynamics of the Josephson junction are well explained by the McCumber[42]–Stewart[43] model, in which the Josephson element is in parallel with a resistance R (which may be an external shunt) and a capacitance C. The I–V characteristic is hysteretic when

$$\beta_c \equiv 2\pi I_0 R^2 C/\Phi_0 \equiv \omega_j RC > 1\,, \tag{5}$$

and nonhysteretic for $\beta_c < 1$. Here, $\omega_j/2\pi \equiv I_0 R/\Phi_0$ is the Josephson frequency at voltage $I_0 R$. In the absence of noise and for $\beta_c \ll 1$, the I–V characteristic reduces to $V = R(I^2 - I_0^2)^{1/2}$, which tends asymptotically to $V = IR$ for $I \gg I_0$. An external shunt resistance, which is linear, has an associated Nyquist noise current with a spectral density $S_I(f) = 4k_BT/R$ in the classical limit; k_B is the Boltzmann constant. Noise has two effects. First, it rounds the I–V characteristic at low voltages, reducing the apparent critical current.[44] To maintain a useful degree of Josephson coupling, one requires the coupling energy $I_0\Phi_0/2\pi \gg k_BT$; at 4.2 K, $I_0 \gg 0.2$ μA. The second effect of the noise current is to induce a voltage noise across the junction at nonzero voltages.[45] All computer and analytical calculations of noise in SQUIDs are based on this model.

6.1. *dc SQUID*

Claudia Tesche and I worked out the theory for the dc SQUID using this model.[46,47] That is to say, after we had agreed on the equations of motion, Claudia spent roughly a year carrying out the computer simulations including the effects of noise. The inclusion of noise requires a great deal of averaging to obtain an accurate result, and in those days, computers were of course orders of magnitude slower than they are today. To speed up the calculation, we neglected the junction capacitance, and, for given values of I_0 and junction area, chose the value of R that set $\beta_c = 1$ so that the I–V characteristic is just nonhysteretic. There is a second limitation imposed by Nyquist noise in the shunt resistors, namely $\Phi_0^2/2L \gg 2\pi k_BT$. At 4.2 K, this implies $L \ll 6$ nH.

For a typical SQUID operating at 4.2 K, we can summarize our results as follows. First, the noise energy is optimized when

$$\beta_L \equiv 2LI_0/\Phi_0 = 1\,. \tag{6}$$

The maximum response to a small change in flux $\delta\Phi_a \ll \Phi_0$ occurs when $\Phi_a \approx (2n+1)\Omega_0/4$, where the flux-to-voltage transfer coefficient

$$V_\Phi \equiv |\partial V/\partial \Phi_a|_I \approx R/L \approx 1/(\pi LC)^{1/2}\,; \tag{7}$$

we have assumed $\beta_L = \beta_C = 1$. Nyquist noise in the shunt resistors introduces a white voltage noise across the SQUID with spectral density $S_V(f)$, and a current noise around the SQUID loop with spectral density $S_J(f)$. Both spectral densities depend on the current and flux biases. At frequencies much less than the Josephson frequency and under optimum conditions, the voltage noise leads to a flux noise spectral density

$$S_\Phi(f) = S_V(f)/V_\Phi^2 \approx 16k_BTL^2/R\,. \tag{8}$$

The optimized current noise spectral density is

$$S_J(f) \approx 11k_BT/R\,. \tag{9}$$

Furthermore, the voltage and current noises are partially correlated, with a cross-spectral density

$$S_{VJ}(f) \approx 12k_BT\,. \tag{10}$$

The noise energy is

$$\varepsilon(f) = S_\Phi(f)/2L \approx 9k_BTL/R \approx 16k_BT(LC)^{1/2}\,. \tag{11}$$

We note that $\varepsilon(f)$ is not a complete characterization of the noise in the dc SQUID since it does not take into account the noise current in the loop, which induces a noise voltage into an input circuit coupled to the SQUID. This theory does not, of course, account for $1/f$ noise, which arises from fluctuations in the junction critical currents and from "flux noise," a subject that is only now beginning to be understood. We shall not delve into these subjects.

For a representative SQUID with L = 200 pH, R = 6 Ω, $\beta_L = 1$ and $\beta_C = 1$, we estimate $S_\Phi^{1/2}(f) \approx 10^{-6}\,\Phi_0$ Hz$^{-1/2}$ and $\varepsilon(f) \approx 10^{-32}$ JHz$^{-1} \approx 100\hbar$, in good agreement with measured values. Equation (11) makes it clear that "smaller is better" and that $\varepsilon(f)$ decreases linearly with temperature, in good agreement with experiment.

6.2. *rf SQUID*

The theory of the rf SQUID begins by imposing fluxoid quantization on a superconducting loop containing a single Josephson junction [Fig. 6(a)]:

$$\delta + 2\pi\Phi_T/\Phi_0 = 2\pi n\,. \tag{12}$$

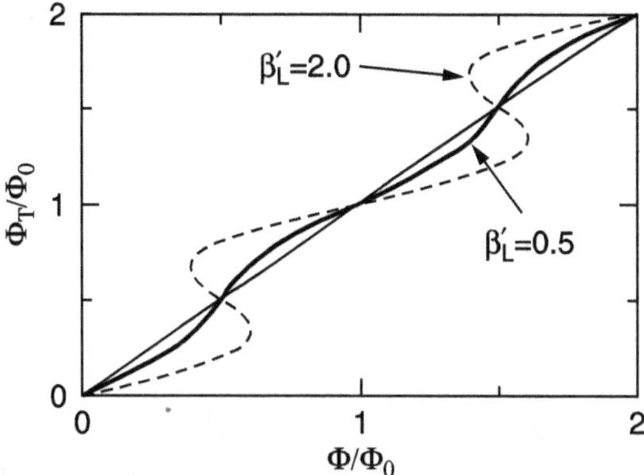

Fig. 10. Total flux Φ_T in the rf SQUID loop versus applied flux Φ_a for the dispersive ($\beta_{\rm rf} = 0.5$) and hysteretic ($\beta_{\rm rf} = 2$) regimes.

Here, Φ_T is the sum of the applied flux Φ and the flux generated by the current $J = -I_0 \sin(2\pi\Phi_T/\Phi_0)$ flowing around the loop. Thus,

$$\Phi_T = \Phi_a - LI_0 \sin(2\pi\Phi_T/\Phi_0)\,. \tag{13}$$

Inspection of Eq. (13) reveals two distinct kinds of behavior, plotted in Fig. 10. For $\beta_{\rm rf} \equiv 2\pi LI_0/\Phi_0 < 1$, the slope $d\Phi_T/d\Phi_a = [1 + \beta_{\rm rf}\cos(2\pi\Phi_T/\Phi_0)]^{-1}$ is always positive, and the Φ_T versus Φ_a curve is non-hysteretic. For $\beta_{\rm rf} > 1$, on the other hand, there are regions in which $d\Phi_T/d\Phi_a$ is positive, negative or divergent, and the Φ_T versus Φ_a plot becomes hysteretic.

In the hysteretic mode, the applied rf flux [Fig. 6(a)] causes the SQUID to make transitions between quantum states and to dissipate energy at a rate that is periodic in Φ_a. This periodic dissipation in turn modulates the Q of the tank circuit, so that when it is driven on resonance with a current of constant amplitude, the rf voltage is periodic in Φ_a. This is the origin of the behavior first observed by Zimmerman and Silver [Fig. 6(b)]. A detailed analysis[48–54] shows that the SQUID is optimized when $k^2Q \approx 1$, where k is the coupling coefficient between the SQUID loop and the inductor of the resonant circuit. Under this condition, at fixed rf current $I_{\rm rf}$ the transfer coefficient becomes

$$|\partial V_T/\partial\Phi_a|_{I_{\rm rf}} \approx \omega_{\rm rf}(L_T/L)^{1/2}/k\,. \tag{14}$$

The intrinsic noise energy is given by

$$\varepsilon(f) \approx LI_0^2(2\pi k_BT/I_0\Phi_0)^{4/3}/2\omega_{\rm rf}\,. \tag{15}$$

For SQUIDs at 4.2 K coupled to a room-temperature preamplifier, however, extrinsic noise contributions — notably preamplifier noise and loss in the line coupling the SQUID and preamplifier — may far exceed the intrinsic noise. For this reason, dissipative rf SQUIDs at 4.2 K are rarely used today. On the other hand, increasing the temperature of the SQUID from 4.2 K to 77 K has little impact on the system noise energy, whereas the noise energy of the dc SQUID scales with T [Eq. (11)]. Thus, the noise advantage of the dc SQUID over the hysteretic rf SQUID is diminished at 77 K.

In the nonhysteretic regime $\beta_{\rm rf} < 1$, the rf SQUID behaves as a flux-sensitive inductor. The underlying physics is that the inductance of a Josephson junction in the zero voltage regime is given by $L_J = \Phi_0/2\pi(I_0^2 - I^2)^{1/2}$ for $I < I_0$. Thus, an applied flux produces a circulating current and hence a change in L_J. When the tank circuit is driven off-resonance at constant amplitude and frequency, a flux change in the SQUID changes the resonant frequency, so that the rf voltage is periodic in the applied flux. In the limit $I_0\Phi_0/2\pi \gg k_BT$, the intrinsic noise energy is[51,53]

$$\varepsilon(f) \approx 3k_BTL/\beta_{\rm rf}^2R\,. \tag{16}$$

A noise energy of $23 \pm 7\hbar$ was reported for an rf SQUID operating in this regime at microwave frequencies.[55]

7. High-T_c SQUIDs

The advent of high-T_c superconductors in 1986 generated a furor of activity in the SQUID community. A successful thin-film technology for $YBa_2Cu_3O_{7-x}$ was developed that included both single-layer SQUIDs and SQUIDs with integrated coils, very much like the low-T_c device of Fig. 8(b), and planar magnetometers and gradiometers. For reasons of space, I have decided not to discuss high-T_c SQUIDs; a broad overview and copious references appear in Koelle *et al.*[56]

8. Applications of SQUIDs: Overview

The major reason I still work on SQUIDs after 46 years in the field is the sheer breadth of their applications. This diversity is driven by the extraordinary low noise energy of the SQUID, which has opened up a myriad of applications that would otherwise not exist. Many of these are one-of-a-kind experiments in basic science, and I cannot possibly hope to review them. In this section, I touch on some of the areas in which SQUIDs are used, and

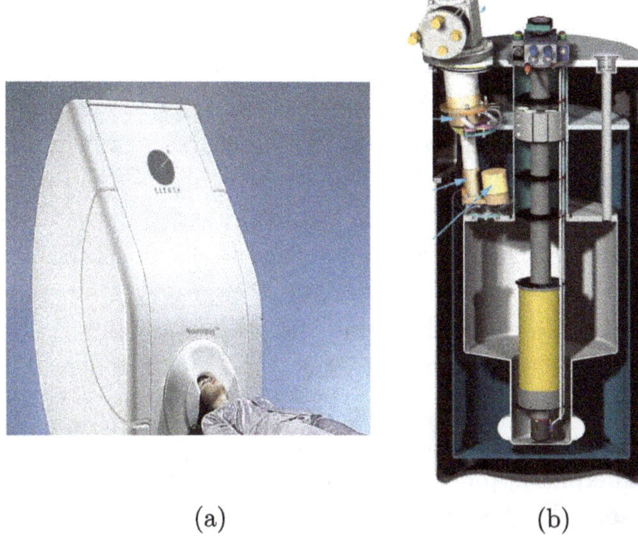

(a) (b)

Fig. 11. Commercial SQUID systems. (a) System for magnetoencephalography (courtesy Elekta). (b) Cross-section of system for measuring magnetic properties (courtesy Quantum Design).

in Secs. 9 and 10, I describe in more detail selected applications in which I have been involved personally.

Systems for magnetoencephalography (MEG)[57] involve about 300 dc SQUIDs, coupled to gradiometers, arranged in a helmet into which the subject inserts his or her head. The system — a typical version is shown in Fig. 11(a) — is generally installed in a magnetically-shielded room to attenuate external magnetic disturbances. The SQUIDs detect tiny magnetic signals produced by neuronal electrical currents in the brain. Magnetoencephalography is used in neurological and psychological research, and in clinical applications such as pre-surgical mapping of brain tumors. A second medical application is magnetocardiography (MCG),[57] for example, in the noninvasive diagnosis of heart disease such as ischemia. Other biomagnetic applications[57] include liver susceptometry to monitor the accumulation of excess iron (thalassemia) and immunoassay — the detection and identification of antibodies or antigens in blood specimens.

Almost certainly, the most commercially successful SQUID system [Fig. 11(b)] is used to measure magnetic properties such as magnetization and susceptibility at temperatures ranging from a few kelvin to above room temperature in magnetic fields up to several tesla.[58] This system is available with a cryocooler, obviating the need to supply liquid helium. A related system,[59] also highly successful, is a "rock magnetometer," with an open horizontal access to characterize core samples taken from the Earth. The "scanning SQUID microscope" enables one to move samples at room temperature in two-dimensions close to a SQUID to map their magnetic properties.[59,60] In some approaches, one measures intrinsic magnetization, while in others, one applies a low-frequency magnetic field and detects the eddy-current response. Applications include detecting flaws in aircraft wheel hubs and under the rivets of aircraft fuselages, finding tantalum inclusions in niobium sheets destined for superconducting cavities for particle accelerators, tracking corrosion by detecting the magnetic fields generated by the ion currents responsible for the oxidation, and locating defects in computer multichip modules and other semiconductor circuits. A SQUID microscope was used to map the swimming of a magnetotactic bacterium.

A quite different application is geophysical exploration[59] — in which interest has been rekindled using liquid-nitrogen-cooled SQUID magnetometers. Of particular interest is "transient electromagnetics" (TEM), in which one applies a large current pulse to the ground and measures the ensuing magnetic field. In one version of TEM, a SQUID package is towed behind an aircraft.

In the examples I have given above, the primary goal is to detect magnetic field. There is another class of applications in which the aim is to measure other physical quantities. One example, is the SLUG voltmeter. Another is in transducers for gravity wave detectors and gravity gradiometers.[61] Precision gyroscopes with SQUID readout were the basis of Gravity Probe B, intended to measure the geodetic effect (curved space-time due to the presence of the Earth) and Lense-Thirring effect (dragging of the local space-time frame due to rotation) predicted by general relativity. SQUIDs find important applications in standards and metrology,[62] for example, in verifying the universality of the Josephson voltage-frequency relation, in linking the volt to mechanical SI units, and in the cryogenic current comparator. A fascinating application of SQUIDs that came to prominence in the last decade is to superconducting qubits.[63] SQUIDs are used to detect the quantum state of flux and phase qubits, and — through the flux dependence of their inductance — to tune thin-film, microwave resonators.

In the next two sections, I briefly describe three relatively new applications of SQUIDs that illustrate their diversity. The first two are in cosmology[64] — searching for galaxy clusters and for cold dark matter (Sec. 9) — and the third (Sec. 10) is the use of SQUID gradiometers in low-frequency nuclear magnetic resonance (NMR) and magnetic resonance imaging (MRI).[64]

9. Cosmology

Measurements of the cosmic microwave background (CMB) have taught us almost everything we know about the origin of the universe — so much so that the CMB[65] is sometimes referred to as "the Cosmic Rosetta Stone". The CMB, which has a Planck distribution with a characteristic temperature of 2.726 K, originated 379,000 years after the Big Bang and has been traveling for 13.7 billion years since then. We have learned that the Universe is a rather bizarre place, consisting of 0.6% neutrinos, 4.6% baryons, 73% dark energy (DE) and 22% cold dark matter (CDM). Thus, some 95% of the universe is of unknown origin. SQUIDs are being used to investigate both DE and CDM, as I briefly describe in the next two sections.

9.1. *Dark energy: Searching for galaxy clusters*

Galaxy clusters contain typically several hundred galaxies and, with a mass of 10^{14} to 10^{15} solar masses, are the largest objects in the universe.[65] Measuring the density of galaxy clusters as a function of red-shift enables one to determine the parameters of the equation of state for DE. To make a definite test, however, one needs to find and characterize thousands of clusters. The red-shift is available from optical astronomy, but the signal is weak, making it difficult to locate large numbers of clusters. A newly implemented search technique capable of much more rapid searching involves the Sunyaev–Zel'dovich Effect (SZE).[66] This technique is based on the fact that the space between galaxies in a cluster contains a hot electron gas. When a microwave background photon passes through this gas, it has a small (1–2%) chance of being scattered to a higher energy, so that a fraction of the CMB spectrum is shifted to slightly higher frequencies. Thus, as the telescope scans across the sky, one expects to find regions in which there is a shifted Planck spectrum superimposed on the unshifted spectrum — the signature of a galaxy cluster.

Since the CMB spectrum peaks at a frequency of about 150 GHz, one needs ultrasensitive detectors in the far infrared. The detector of choice is the transition-edge sensor (TES) bolometer.[67] This consists of three elements: an optical absorber, a sensitive thermometer and a thermal conductance to a heat bath. It is convenient first to describe the thermometer (TES), which, in the groups of Bill Holzapfel, Adrian Lee and Paul Richards at Berkeley, is a thin film bilayer of Al-Ti[68] with a superconducting transition temperature of about 0.5 K. The TES is connected in series with a voltage source and the input coil of a SQUID, and operated at a temperature within its

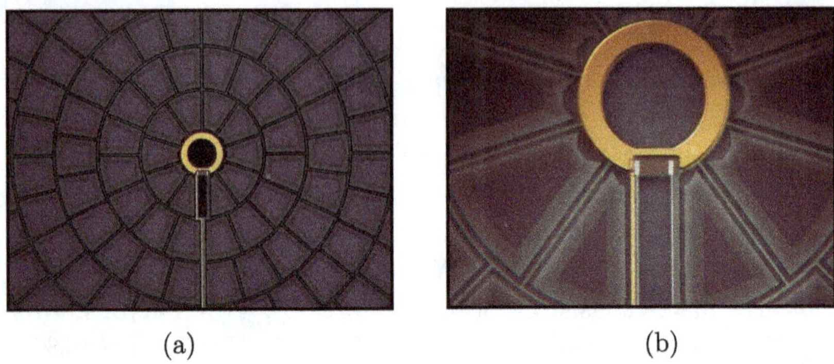

(a) (b)

Fig. 12. Far infrared bolometer. (a) Spider-web absorber, 3 mm in diameter. (b) Expanded view. The TES (lower edge of the gold ring) is electrically connected to the readout circuit by two aluminum lines (vertical in this photograph) and thermally connected to the heat bath by the gold line to the left of the left-hand aluminum line. (Reproduced with permission from Ref. 68.)

resistive transition. When a photon is absorbed, thermal energy is transferred to the TES, initially raising its temperature and increasing its resistance. As the resistance begins to increase, however, the dissipation generated by the voltage bias drops, so that the power dissipation in the TES remains constant. The change in the current flowing through the SQUID input coil is related to the absorbed energy. This electrothermal feedback results in a fast, linear response. Furthermore, the power dissipation is low, typically 1 nW. The TES is mounted at the center of the absorber, which consists of a "spiderweb" of silicon nitride coated with a thin gold film (Fig. 12). The noise in an optimized sensor is limited by fluctuations in the arrival rate of photons.

To obtain an image of a galaxy cluster, one requires hundreds or preferably thousands of such sensors mounted at the focal plane of a telescope. Needless to say, this is a challenging prospect. Although it is perfectly feasible to fabricate hundreds or thousands of SQUIDs, a serious concern is the thermal heat load generated by the leads when each TES requires its own readout SQUID. This issue caused Paul Richards to walk into my office one day in the spring of 1999 to discuss the feasibility of using a single SQUID to read out an array of TESs simultaneously with a frequency-domain multiplexer. I found this a very interesting problem, and the eventual outcome was the eight-channel multiplexer[69] shown in Fig. 13. Each TES is connected in series with a capacitor and an inductor to form a resonant circuit, each with a different frequency. An oscillator provides a comb of voltages $V_i(f_i)$ at frequencies f_i corresponding to the resonant frequencies of

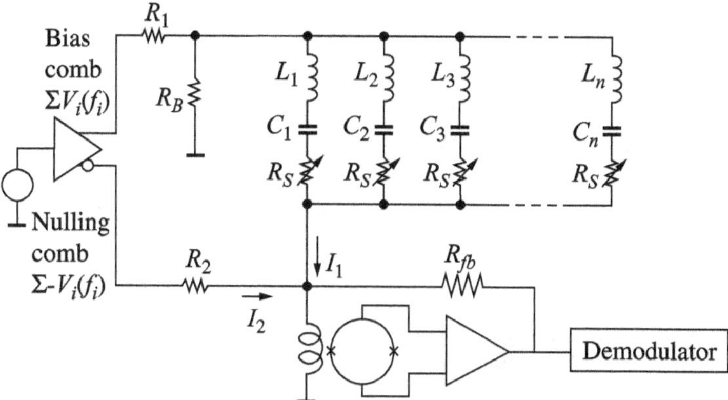

Fig. 13. Circuit for a TES frequency-domain multiplexer. A tuned circuit is connected in series with each TES. The readout SQUID consists of 100 SQUIDs in series. (Reproduced with permission from Ref. 69.)

the tuned circuit. The currents from the eight tuned circuits are summed at one end of the input coil of a SQUID (actually a 100-SQUID series array); the other end is grounded. The oscillator applies a nulling comb of voltages $-V_i(f_i)$ to the same summing point so that, in the absence of any photon flux, the total current injected into the input coil is zero. When a photon is absorbed by one of the sensors, the resulting change in current is injected into the input coil of the SQUID, which is operated in a flux-locked loop. Since the eight frequencies are distinct, all eight sensors can be read out simultaneously with the aid of a demodulator.

The Berkeley group, in collaboration with several other groups, first implemented this scheme on APEX (Atacama Pathfinder EXperiment), a 12-m telescope at about 5100 m on the Atacama plateau in Northern Chile [Fig. 14(a)]. The resolution is one minute of arc and the field of view is 22 minutes of arc. The receiver contained about 280 working channels with multiplexed readout. As a first demonstration of the system operation, Fig. 14(b) shows an SZE image of a known galaxy cluster, the "Bullet Cluster" — actually two merging clusters.[70] The signal-to-noise ratio is 20 in the central one minute of arc beam. Subsequently, the Berkeley group and others began a cluster search with SPT (South Pole Telescope), a 10-m telescope at Antarctica at about 2900 m. The resolution is one minute of arc and the field of view 60 minutes of arc; there about 660 TESs with multiplexed readout. In the next two years, SPT will survey 4000 square degrees of sky and — if the theory is correct — is expected to find thousands of galaxy clusters.

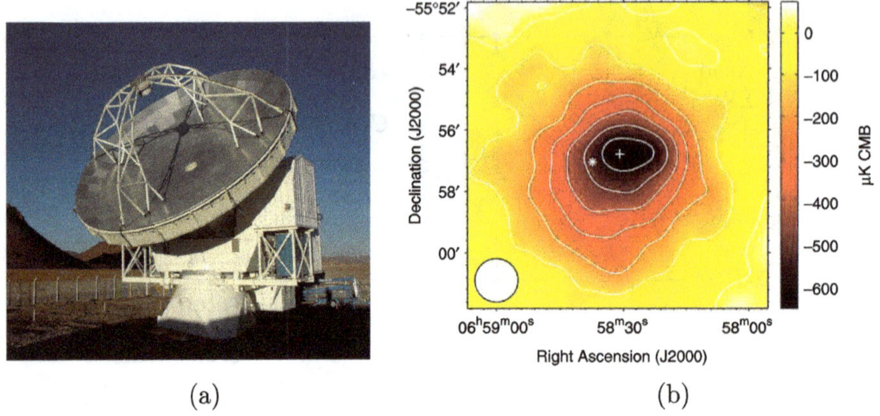

Fig. 14. APEX telescope and first results. (a) Telescope at Atacama. (b) SZE image of the "Bullet Custer"; the contour interval is 100 μK_{CMB}. Disk at lower left represents the full width at half-maximum (FWHM) resolution. (Reproduced with permission from Ref. 70.)

9.2. *Cold dark matter: The hunt for the axion*

Since we do not know what CDM is, we have to postulate some kind of particle and then search for it. One of the two leading candidates is the WIMP (Weakly Interacting Massive Particle).[71] Supersymmetry theories predict that the WIMP has a mass of 10–100 GeVc^{-2}. A WIMP search[71] currently under way involves a 250 g block of Ge. An impinging WIMP would scatter off the Ge nuclei, producing phonons, which in turn would break Cooper pairs in superconducting Al absorbers on the surface, producing quasiparticles. The absorbers are in electrical contact with 1000 parallel-connected tungsten TESs. The quasiparticles diffuse into the TESs, where they are trapped, raising the temperature. The TES array is biased with a static current, and read out with a 100-SQUID array. The second candidate is the axion, first proposed in 1978 to explain the absence of a measurable electric dipole moment on the neutron.[72–74] The axion mass m_a is predicted to be 1 μeVc^{-2} to 1 meVc^{-2} (corresponding to 0.24 to 240 GHz). Since the measured CMB density ρ_a is 0.45 GeV cm^{-3}, a mass of 1 μeV corresponds to an axion density of 4.5×10^{14} cm^{-3}.

How does one search for this light, spinless, chargeless particle? In 1983, Pierre Sikivie[75,76] suggested that it could be found via Primakoff conversion. In this scheme, in the presence of a large magnetic field B, the axion converts to a photon (together with a virtual photon) with energy given by m_a. The conversion takes place in a cooled cavity of volume V with a

Fig. 15. Axion detector installed at Lawrence Livermore National Laboratory. (Courtesy Darin Kinion.)

tunable resonance frequency. An antenna couples the signal in the cavity to a cooled, low-noise amplifier. When the cavity frequency corresponds to the frequency of photons from the putative axions, the amplifier detects a peak; off-resonance, it detects only blackbody noise. Clearly, one has to scan the cavity frequency to search for the axion. The axion-to-photon conversion power is given by[77]

$$P_a \propto B^2 V {g_\gamma}^2 m_a \rho_a \,, \tag{17}$$

where g_γ is a coupling coefficient. There are two contending theories for g_γ, the KSVZ (Kim–Shifman–Vainshtein–Zakharov) model[78,79] and the DFSZ (Dine–Fischler–Srednicki–Zhitnitsky) model,[80,81] with $g_\gamma = +0.97$ and -0.36, respectively. Evidently, the DFSZ model predicts the smaller density of photons in the cavity.

Following the first-generation detectors based on this scheme at Brookhaven National Laboratory and the University of Florida,[77] a second-generation detector was constructed at Lawrence Livermore National Laboratory (Fig. 15), and began operation in 1996. The cavity and the surrounding 7-T persistent-current magnet were cooled to 1.5 K. The cooled

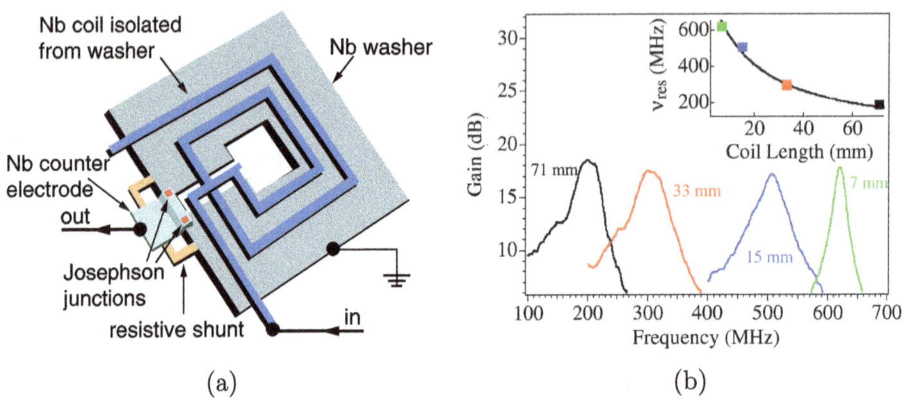

Fig. 16. Microstrip SQUID amplifier. (a) Configuration (courtesy Darin Kinion). (b) Gain versus frequency for four lengths of input coil.

HEMT (high electron mobility transistor) amplifier had a noise temperature $T_N = 1.7$ K, giving a system noise temperature $T_S = T + T_N = 3.2$ K. For the parameters of the experiment and assuming $g_\gamma = -0.36$, the time to scan from $f_1 = 0.24$ GHz to $f_2 = 0.48$ GHz is

$$\tau(f_1, f_2) \approx 4 \times 10^{17}(T_S/1K)^2(1/f_1 - 1/f_2) \text{ sec} \approx 270 \text{ years}. \tag{18}$$

Evidently, this experiment was not likely to find an axion very quickly!

In 1994, Leslie Rosenberg and Karl van Bibber approached me to ask whether it would be possible to make a SQUID amplifier with a lower noise temperature than a HEMT at frequencies around 1 GHz. At the time, this was not so simple: the essential problem with the usual SQUID washer SQUID (Fig. 8) is that the parasitic capacitance between the input coil and the washer rolls off the gain at frequencies above typically 100 MHz. Needless to say, this was a challenging problem, and, Michael Mück, Marc-Olivier André, Jost Gail, the late Christoph Heiden and I set about finding a solution.

Michael made the crucial breakthrough: instead of applying the rf signal between the two ends of the input coil, he moved one wire and applied it between one end of the coil and the SQUID washer. In this configuration, the washer serves as a groundplane for the coil, forming a microstrip or transmission line — thus making a virtue of the coil-washer capacitance [Fig. 16(a)].[82] When the length of the microstrip corresponds to a half-wavelength of the signal, the resonance produces a substantial amount of gain — typically 20 dB — as shown in Fig. 16(b). As the coil is progressively shortened, the resonance moves to higher frequency. In fact, the dependence of the resonance frequency on the coil length is quite complicated because the microstrip inductance is dominated by the inductance coupled in from the SQUID loop.[83] Suffice to say, the microstrip SQUID amplifier (MSA) achieves useful levels of gain over the frequency range[84] 0.2–2 GHz.

An important requirement for the axion detector is the ability to tune the resonant frequency of the cavity and hence of the amplifier. The resonant frequency of the MSA can be readily tuned by connecting a varactor diode between the previously open end of the coil and the SQUID washer.[85] The capacitance of the varactor — and hence the phase change of the reflected signal — is a strong function of the reverse voltage bias. One can change the resonant frequency by a factor of nearly two by adjusting the reverse bias voltage.

The next step was to measure the noise temperature of the MSA, ultimately in a dilution refrigerator since, from Eq. (11), we expect T_N to scale with T in the classical regime. In these experiments, we used a second MSA to amplify signals from the first. We found[86] that T_N was indeed proportional to T at temperatures down to about 100 mK, below which it flattened out. Separate experiments showed that this flattening was due to hot electrons[87] in the resistors shunting the SQUID junctions. At the lowest temperatures and $f = 519$ MHz, the optimized noise temperature was 47 ± 10 mK, about a factor of two above the quantum limited value $T_Q = hf/k_B \approx 25$ mK. Subsequently, using a single MSA with shunts equipped with cooling fins, Darin Kinion and I measured[88] $T_N = 48 \pm 5$ mK at 612 MHz, with a bath temperature of about 50 mK. At these frequencies, $T_Q \approx 29$ mK. This noise temperature is about 40 times lower than that of a state-of-the-art HEMT.

The potential impact on the scan rate of the axion detector is dramatic. As a scenario, suppose that the cavity is cooled to $T = 50$ mK with a dilution refrigerator, and that the readout amplifier has a noise temperature $T_N = 50$ mK. Thus, the system noise temperature $T_S = 100$ mK. Since the scan rate scales as T_S^2, the scan time in Eq. (18) becomes $270(0.1/3.2)^2$ years ≈ 100 days!

The very low noise temperature of the MSA spurred a first-step upgrade of the axion detector in which the HEMT was replaced with an MSA while the temperature was maintained at about 2 K. Since blackbody noise from the cavity is not reduced, the decrease in scan time is modest. Rather, the object of the upgrade was to demonstrate that the MSA could indeed operate as expected on the axion detector; a major challenge was effective screening of the 7-T field. In fact, the system worked extremely well at a frequency of about 842 MHz. As of writing, 88,732, 80 sec data sets have been acquired, corresponding to a net 82 days of data.[89]

Given the success of these experiments, one can be optimistic that the full upgrade that includes a dilution refrigerator to cool the cavity to (say) 50 mK will be funded. The combination of the MSA and a millikelvin-temperature cavity turns a rather impractical axion search into an extremely viable one.

10. SQUID-Detected Microtesla NMR and MRI

Magnetic resonance imaging is wonderfully successful, with perhaps 30,000 machines deployed worldwide. These machines can image most parts of the human body with a spatial resolution of about 1 mm. Probably few patients are aware of the fact that they are surrounded by roughly 100 km of NbTi wire wound in a persistent-current solenoid and immersed in about 1000 liters of liquid helium! Most machines operate at a magnetic field of 1.5 T but there is a trend towards 3-T machines. These machines are expensive, however, roughly $2M for a 1.5-T system, have a large footprint, and are heavy: it is not unknown for the cost of reinforcing the floor to exceed the cost of the machine.

MRI[90] is based on NMR,[91,92] almost always of protons, which have spin $1/2$ and magnetic moment μ_p. In the presence of an applied magnetic field B_0, the protons align either parallel or antiparallel to B_0, the energy level splitting being given (in angular frequency units) by the Larmor frequency $\omega_0 = \gamma B_0$; γ is the gyromagnetic ratio. The NMR frequency is 42.58 MHzT^{-1}. The magnetic moment of N protons in the limit $\mu_p B_0 \ll k_B T$ is $M_0 = N\mu_p^2 B_0/k_B T$ (Curie's Law). At room temperature in achievable magnetic fields, the magnetic moment is very small. For example, for $T = 300$ K and $B_0 = 1$ T, $M_0/N\mu_p = \mu_p B_0/k_B T \approx 3.6 \times 10^{-6}$.

In NMR, M_0 is initially in thermal equilibrium aligned with B_0 along the z-axis. One applies a "$\pi/2$ pulse" along the x-axis at the Larmor frequency that tips M_0 into the x-y plane in which it precesses about the z-axis at frequency $\omega_0/2\pi$. As it precesses, M_0 undergoes two relaxation processes. First, it relaxes its direction towards the z-axis, where it eventually regains thermal equilibrium in a characteristic longitudinal (spin-lattice) relaxation time T_1. Second, individual spins dephase via local field fluctuations produced at each site by neighboring spins in the transverse (spin-spin) relaxation time T_2. In water at room temperature, T_1 and T_2 are about 1 s; $T_1 \geq T_2$ always. The FWHM linewidth of the NMR line is $\Delta f = 1/\pi T_2$ ("homogeneous broadening"). Inhomogeneities in the magnetic field, however, can substantially broaden the linewidth ("inhomogeneous broadening"), and reduce T_2 to a value T_2^* given by $1/T_2^* = 1/T_2 + 1/T_2'$; T_2' is the inhomogeneous lifetime. The very important spin echo technique invented by Erwin Hahn[93] eliminates inhomogeneous broadening but not homogeneous broadening.

In MRI, one uses NMR to determine spatial structure by means of three orthogonal magnetic field gradients that define a "voxel". These gradients establish B_0 in a small volume that produces a specific NMR frequency. Gradient switching translates the voxel through the patient to construct the magnetic resonance image.

Given the success of high-field MRI, why would one consider low-field imaging and SQUID detection? To address this issue, we note first that in conventional NMR the oscillating magnetic signal induces a voltage across an inductor shunted with a capacitor to form a resonant circuit. The voltage across the tank circuit, by Faraday's Law, scales as $\omega_0 M_0$, that is, as B_0^2. Thus, at first sight, reducing B_0 would seem to be exactly the wrong thing to do. Two factors counter this thinking. First, consider replacing the tank circuit with an (untuned) flux transformer coupled to a SQUID. Since this detector responds to flux, rather than rate of change of flux, one factor of B_0 is eliminated and the output voltage scales as B_0. Second, one can prepolarize[94] the spins in a magnetic field B_p much greater than B_0. After B_p has been turned off, the spins retain a corresponding magnetic moment $M_p \gg M_0$ that decays in a time T_1, so that, although the spins precess at frequency $\gamma B_0/2\pi$, they produce a signal amplitude proportional to B_p. Thus, the amplitude of the SQUID-detected NMR signal becomes independent of B_0; one can choose B_0 at will provided that $B_0 \ll B_p$.

There is an immediate advantage of low-field NMR and MRI. For an inhomogeneity ΔB_0 in B_0, the inhomogeneous linewidth $\Delta f'$ scales as $(\Delta B_0/B_0)B_0$, that is, as B_0 for a fixed *relative* inhomogeneity $\Delta B_0/B_0$. Thus, narrow linewidths — and high spatial resolution — can be achieved in relatively inhomogeneous fields. For example, to obtain a 1 Hz linewidth in a 900 MHz NMR system, it is necessary to shim the magnetic field homogeneity to about one part in 10^9 over the volume of the sample. Although achievable, this is extremely challenging. Furthermore, spatial variations in magnetic susceptibility across a sample produce a linewidth broadening — and a loss of spatial resolution in MRI — that cannot be compensated. In contrast, at an NMR frequency of 2 kHz (for protons, corresponding roughly to the Earth's field), the field homogeneity required for a 1 Hz linewidth is only one part in 2000, which is easily obtainable; in addition, the effects of susceptibility variation are negligible.

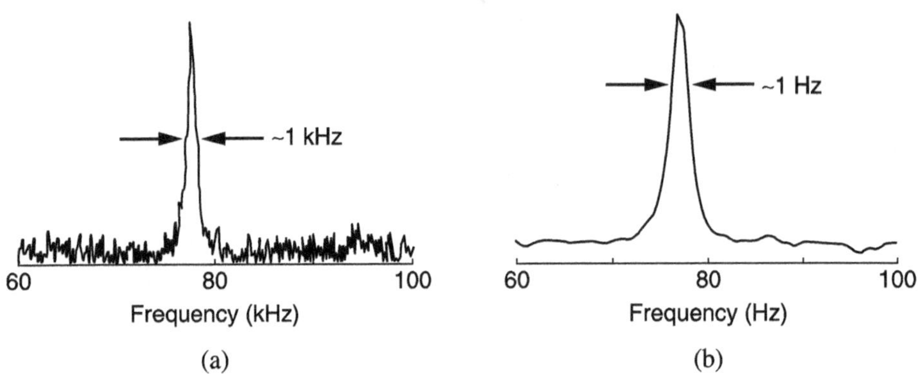

Fig. 17. NMR spectra of mineral oil. (a) $B_0 = 1.8$ mT, 10,000 acquisitions. (b) $B_p = 1.8$ mT, $B_0 = 1.8$ μT, 100 acquisitions.

Following these ideas, Robert McDermott, Andreas Trabesinger, Michael Mück, Erwin Hahn, Alex Pines and I studied NMR in protons in mineral oil.[95] The sample, maintained near room temperature, was placed in one loop of a first-derivative gradiometer coupled to a flux-locked SQUID. Separate coils supplied B_p, an oscillating field to tip the spins through $\pi/2$, and B_0 — which was deliberately made relatively inhomogeneous. Figure 17(a) shows the results of a conventional spin echo measurement with $B_0 = 1.8$ mT. The signal-to-noise ratio is poor, despite the 10,000 averages, and the linewidth is broad, about 1 kHz. In contrast, the spectrum in Fig. 17(b) was obtained at 1.8 μT following prepolarization of the spins at 1.8 mT. As expected, the linewidth was reduced by a factor of 1000, and, even with only 100 averages, the signal-to-noise ratio of the spectrum was greatly increased.

We exploited the high spectral resolution achievable in microtesla NMR to detect scalar coupling (J-coupling) in heteronuclear spin systems, for example, trimethyl phosphate, which contains a single ^{31}P atom and nine equivalent protons.[95] The ^{31}P nuclear spin interacts with the proton spins via the chemical bonds, producing a 10.4 ± 0.6 Hz splitting of the proton levels. This splitting is easily resolved in fields of a few microtesla. Since such splittings are unique to a particular chemical bond, this technique could be used as a "bond detector".

Subsequently, we constructed an MRI system,[96–98] made from wood with copper wire coils (Fig. 18). The system is generally operated at 132 μT, corresponding to a proton Larmor frequency of 5.6 kHz. Magnetic field gradients are typically 200 μT m^{-1}. The NMR signal is detected by a second-derivative gradiometer in a low-noise fiberglass dewar. A novel feature is the

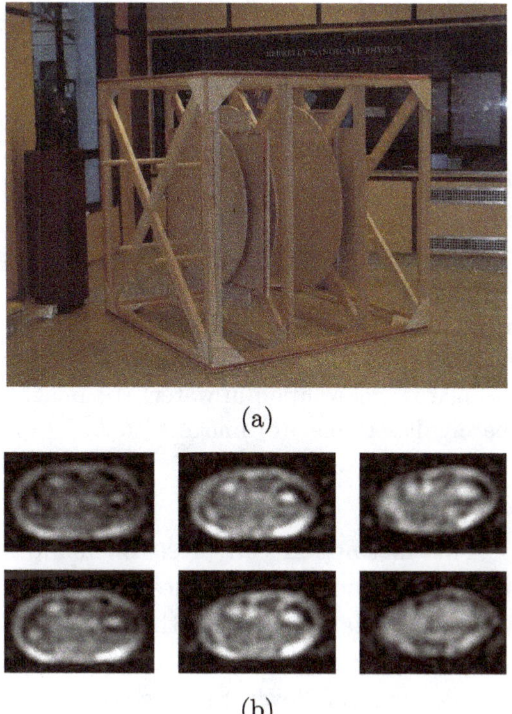

Fig. 18. MRI in microtesla fields. (a) Wooden coil frame. The "cube" carries coils to cancel the Earth's magnetic field, and supports coils to provide the Larmor field, three sets of field gradients, the polarizing field and the pulsed field. It also supports the helium dewar containing the SQUID. (b) MRI slices of a forearm obtained with $B_p = 40$ mT, $B_0 = 132$ μT and gradients of 150 μT/m. The slice thickness is 10 mm and the in-plane resolution about 2 mm.

array of 25 unshunted Josephson junctions in series with the gradiometer coils and the input coil of the SQUID. When the polarizing coil or gradient coils are being switched, the current in the superconducting circuit causes the junctions to switch to the normal state, restricting the current to a low value. Once the fields stabilize, the junctions rapidly revert to the superconducting state, enabling us to acquire the data. The magnetic field noise referred to the lowest loop of the gradiometer is typically 0.3–0.5 fT $\mathrm{Hz}^{-1/2}$. Experiments on "phantoms" with $B_p = 85$ mT, $B_0 = 132$ μT and gradients of 150 $\mu\mathrm{T\ m}^{-1}$ demonstrated an inplane resolution of 0.7 mm. Figure 18 shows six MRI slices of a forearm acquired with $B_p = 40$ mT. The inplane resolution is about 2 mm.

An important issue in MRI is distinguishing different tissue types — for example, healthy and cancerous tissues — even though the proton densities

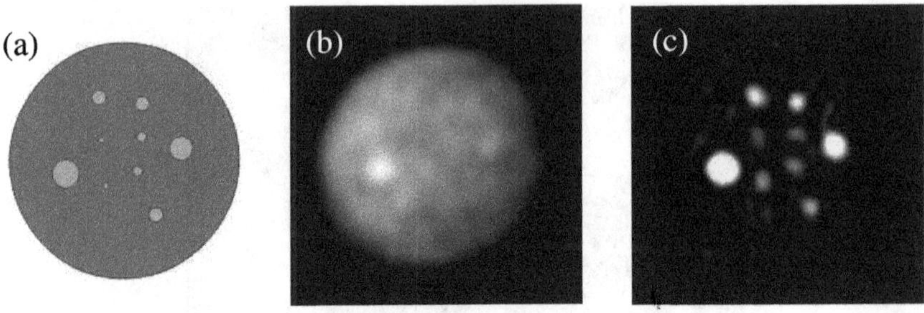

Fig. 19. T_1-weighted contrast imaging. (a) Phantom consisting of nine water-filled drinking straws, with diameters ranging from 1 to 6 mm, standing vertically in a tube containing 0.5% agarose gel solution in water. (b) Image with $B_0 = 100$ mT; the 6-mm straw is barely discernible. (c) Image with $B_0 = 132$ μT; all straws are highly visible.

may be identical. This distinction can often be made using T_1-weighted contrast imaging if the T_1-values in the tissues are different. Low-field MRI is readily adapted to this modality, and has the important virtue of offering much higher contrast than high-field MRI in many instances.

Our low-field technique is as follows. Suppose two regions, A and B, have relaxation times T_{1A} and T_{1B}, with $T_{1A} > T_{1B}$. After polarizing the proton spins, we turn off B_p, and, following a delay of a few tens of milliseconds during which the spins partially relax, we apply the imaging sequence. The image of region A will be brighter than that of region B since the magnetization at the beginning of the imaging sequence is higher. To illustrate this principle, we studied samples with 0.25% and 0.5% concentrations of agarose gel in water.[99] At fields below 500 μT, the T_1 value for the 0.25% concentration is a factor of two longer than for the 0.50% concentration, whereas above about 10 mT there is essentially no contrast. The impact on imaging capability is illustrated in Fig. 19, which shows images of a phantom consisting of plastic straws containing water standing vertically in a 0.5% agarose gel solution. At 100 mT, the image shows virtually no contrast, whereas at 132 μT, the contrast is stark. The last two figures provide a vivid demonstration of the enhanced T_1-contrast at very low fields.

Encouraged by these results, Sarah Busch, Michael Hatridge, Michael Mössle, Jeff Simko (a pathologist at the University of California, San Francisco) and I embarked on a study of *ex vivo* prostate tissue. After a cancerous prostate was removed surgically, it was taken immediately to the pathology laboratory where two small specimens were excised. One was judged (by eye) to be cancerous and the other normal. The specimens were enclosed in a biohazard bag, placed on ice, and rushed to our MRI machine where their T_1-values were measured simultaneously in a gradient field (which separates the NMR frequencies of the two specimens). Subsequently, the specimens were returned to UCSF, where Jeff characterized the nature of the tissues. In preliminary measurements on specimens from eight patients, we found that T_1 was significantly higher for healthy tissue than for cancerous tissue: the average values were 77 ± 2 ms and 47 ± 5 ms respectively. Further studies are in progress.

These *ex vivo* results suggest that microtesla MRI with T_1-contrast may be a viable technique for *in vivo* imaging of prostate cancer. Only *in vivo* studies, however, can determine the sensitivity and specificity. If indeed the method proves successful, what might its role be? It is unlikely to be adopted as a screening technique, since prostate cancer is generally diagnosed by means of a PSA (prostate specific antigen) test. Managing prostate cancer, however, poses a dilemma because of the wide range of malignancy; treatments include "watchful waiting", radiation therapy and surgery. Thus, reliable, cost-effective imaging would greatly benefit staging of prostate cancer. In fact, there are MRI-based techniques that can image prostate cancer,[100] but these techniques require a 3 T system, and their complexity unfortunately results in a cost that is too high for routine clinical use.

Thus, if microtesla MRI were able to image prostate cancer, it might well play a key role in its diagnosis and treatment. In addition to assessing the severity of the disease, microtesla MRI could provide a reference image to guide subsequent biopsy. Because of its potentially low cost per image, it could be used routinely to monitor the progression of prostate cancer during watchful waiting or radiation therapy. Radiation therapy involves the accurate placement of metallic radioactive seeds. Microtesla MRI — with its insensitivity to the presence of metallic objects — could be used to monitor the insertion of the seeds with the aid of a previously acquired reference image. A broader issue concerns the applicability of microtesla MRI to imaging other cancers, for example, breast, brain or abdominal.

In the meantime, other groups have begun research on microtesla NMR and MRI. A proton linewidth of 0.034 Hz was measured for benzene in the Earth's field using a high-T_c SQUID,[101] and a linewidth of 0.34 Hz in a J-coupling spectrum.[102] Volegov and co-workers[103] made simultaneous measurements of a somatosensory response and microtesla NMR from the brain (no claim is made that the two are correlated). Zotev *et al.*[104] used a seven-channel SQUID system to acquire three-dimensional images of the brain at 46 μT using a polarizing field of 30 mT. A European-Union-funded consortium led by Aalto University is focused on combining microtesla MRI with a 300-channel system for MEG. Such a system would allow one to perform MRI and MEG with the same system, eliminating co-registration problems and potentially reducing costs. The use of 300-SQUID gradiometers to acquire the MRI signal could greatly increase the signal-to-noise ratio.

11. Epilogue

SQUIDs have come a long way since those blobs of solder and machined pieces of niobium. In those early days, none of us could have imagined that SQUIDs would have such diversity of applications — in physics, chemistry, materials science, biology, medicine, geophysics, cosmology, quantum computing — and that in each case they are the best device for the job. SQUIDs are remarkably broadband — ranging from 10^{-4} Hz in geophysical surveying to 10^9 Hz in the axion detector. SQUID amplifiers approach the quantum limit. In the three applications I have described in some detail — searching for galaxy clusters, the hunt for the axion and microtesla MRI — the SQUID plays an essential role. In each case, the application would simply not exist without the extraordinarily low noise energy of the SQUID.

Acknowledgments

I thank the members of my research group and numerous other collaborators for their countless contributions to the research described in this chapter. I am grateful to Mark Ketchen, Adrian Lee, Bernard Sadoulet and Paul Richards for insightful discussions during the preparation of this manuscript. The writing of this article was supported by the Director, Office of Science, Office of Basic Energy Sciences, Materials Sciences and Engineering Division, of the U.S. Department of Energy under Contract No. DE-AC02-05CH11231.

References

1. L. N. Cooper, *Phys. Rev.* **104**, 1189 (1956).
2. J. Bardeen, L. N. Cooper and J. R. Schrieffer, *Phys. Rev.* **108**, 1175 (1957).
3. F. London, *Superfluids* (Wiley, New York, 1950).
4. B. S. Deaver and W. M. Fairbank, *Phys. Rev. Lett.* **7** (1961).
5. R. Doll and M. Näbauer, *Phys. Rev. Lett.* **7**, 51 (1961).
6. I. Giaever, *Phys. Rev. Lett.* **5**, 147 (1960).
7. I. Giaever, *Phys. Rev. Lett.* **5**, 464 (1960).
8. M. H. Cohen, L. M. Falicov and J. C. Phillips, *Phys. Rev. Lett.* **8**, 316 (1962).
9. B. D. Josephson, *Phys. Lett.* **1**, 251 (1962).

10. P. W. Anderson and J. M. Rowell, *Phys. Rev. Lett.* **10**, 230 (1963).
11. J. M. Rowell, *Phys. Rev. Lett.* **11**, 200 (1963).
12. R. C. Jaklevic, J. Lambe, A. H. Silver and J. E. Mercereau, *Phys. Rev. Lett.* **12**, 159 (1964).
13. J. E. Zimmerman and A. H. Silver, *Phys. Lett.* **10**, 47 (1964).
14. A. B. Pippard, J. G. Shepherd and D. A. Tindall, *Proc. Roy, Soc. (London) A* **324**, 17 (1971).
15. A. B. Pippard and G. T. Pullan, *Proc. Camb. Phil. Soc.* **48**, 188 (1952).
16. J. Clarke, *Phil. Mag.* **13**, 115 (1966).
17. J. Clarke, *Proc. Roy. Soc. A* **308**, 447 (1969).
18. E. Rumbo, *Phil. Mag.* **13**, 689 (1969).
19. J. Clarke, *Phys. Rev. Lett.* **21**, 1566 (1968).
20. B. N. Taylor, W. H. Parker, D. N. Langenberg and A. Denenstein, *Metrologia* **3**, 89 (1967).
21. J. S. Tsai, A. K. Jain and J. E. Lukens, *Phys. Rev. Lett.* **51**, 316 (1983).
22. J. Clarke, *Phys. Rev. Lett.* **28**, 1363 (1972).
23. M. Tinkham and J. Clarke, *Phys. Rev. Lett.* **28**, 1366 (1972).
24. J. Clarke, in *Nonequilibrium Superconductivity*, eds. D. N. Langenberg and A. I. Larkin (Elsevier Science Pub. B.V., Amsterdam, 1986), p. 1.
25. J. E. Zimmerman and A. H. Silver, *Phys. Rev.* **141**, 367 (1966).
26. P. L. Richards, S. Shapiro and C. C. Grimes, *Amer. J. Phys.* **36**, 690 (1968).
27. J. Clarke, Ph.D. thesis, University of Cambridge (1968).
28. R. L. Forgacs and A. Warnick, *Rev. Sci. Instrum.* **38**, 214 (1967).
29. D. Drung, R. Cantor, M. Peters, H. J. Scheer and H. Koch, *Appl. Phys. Lett.* **57**, 406 (1990).
30. A. H. Silver and J. E. Zimmerman, *Phys. Rev.* **157**, 317 (1967).
31. J. E. Zimmerman, P. Thiene and J. T. Harding, *J. Appl. Phys.* **41**, 1572 (1970).
32. J. E. Mercereau, *Phys. Rev. Appl.* **5**, 13 (1970).
33. M. Nisenoff, *Phys. Rev. Appl.* **5**, 21 (1970).
34. J. Clarke, *Proc. IEEE* **61**, 8 (1973).
35. C. M. Falco, W. H. Parker, S. E. Trullinger and P. K. Hansma, *Phys. Rev. B* **10**, 1865 (1974).
36. J. Clarke, W. M. Goubau and M. B. Ketchen, *J. Low Temp. Phys.* **25**, 99 (1976).
37. T. D. Gamble, W. M. Goubau and J. Clarke, *Geophys.* **44**, 53 (1979).
38. T. D. Gamble, W. M. Goubau and J. Clarke, *Geophys.* **44**, 959 (1979).
39. M. B. Ketchen and J. M. Jaycox, *Appl. Phys. Lett.* **40**, 736 (1982).
40. J. M. Rowell, M. Gurvitch and J. Geerk, *Phys. Rev. B* **24**, 2278 (1981).
41. M. Gurvitch, M. A. Washington and H. A. Huggins, *Appl. Phys. Lett.* **42**, 472 (1983).
42. D. E. McCumber, *J. Appl. Phys.* **39**, 3113 (1968).
43. W. C. Stewart, *Appl. Phys. Lett.* **12**, 277 (1968).
44. V. Ambegaokar and B. I. Halperin, *Phys. Rev. Lett.* **22**, 1364 (1969).
45. K. K. Likharev and V. K. Semenov, *JETP Lett.* **15**, 442 (1972).
46. C. D. Tesche and J. Clarke, *J. Low Temp. Phys.* **29**, 301 (1977).

47. C. D. Tesche and J. Clarke, *J. Low Temp. Phys.* **37**, 397 (1979).
48. J. Kurkijarvi, *J. Appl. Phys.* **44**, 3729 (1973).
49. L. D. Jackel and R. A. Buhrman, *J. Low Temp. Phys.* **19**, 201 (1975).
50. G. J. Ehnhohn, *J. Low Temp. Phys.* **29**, 1 (1977).
51. K. K. Likharev, *Dynamics of Josephson Junctions and Circuits* (Gordon Breach, New York, 1986).
52. T. Ryhänen, H. Seppa, R. Ilmoniemi and J. Knuutila, *J. Low Temp. Phys.* **76**, 287 (1989).
53. B. Chesca, *J. Low Temp. Phys.* **110**, 963 (1998).
54. P. K. Hansma, *J. Appl. Phys.* **44**, 4191 (1973).
55. L. S. Kuzmin, K. K. Likharev, V. V. Migulin, E. A. Polunin and N. A. Simonov, in *SQUID'85, Superconducting Quantum Interference Devices and their Applications*, eds. H. D. Hahlbohm and H. Lübbig (Walter de Gruyter, Berlin, 1985), p. 1029.
56. D. Koelle, R. Kleiner, F. Ludwig, E. Dantsker and J. Clarke, *Rev. Mod. Phys.* **71**, 631 (1999).
57. J. Vrba, J. Nenonen and L. Trahms, in *The SQUID Handbook. Vol. II: Applications of SQUIDs and SQUID Systems*, eds. J. Clarke and A. I. Braginski (Wiley-VCH GmbH & Co., Weinheim, 2006), Ch. 11.
58. R. C. Black and F. C. Wellstood, in *The SQUID Handbook. Vol. II: Applications of SQUIDs and SQUID Systems*, eds. J. Clarke and A. I. Braginski (Wiley-VCH GmbH & Co., Weinheim, 2006), Ch. 12.
59. T. R. Clem, C. P. Foley and M. N. Keene, in *The SQUID Handbook. Vol. II: Applications of SQUIDs and SQUID Systems*, eds. J. Clarke and A. I. Braginski (Wiley-VCH GmbH & Co., Weinheim, 2006), Ch. 14.
60. H.-J. Krause and G. Donaldson, in *The SQUID Handbook. Vol. II: Applications of SQUIDs and SQUID Systems*, eds. J. Clarke and A. I. Braginski (Wiley-VCH GmbH & Co., Weinheim, 2006), Ch. 13.
61. H. J. Paik, in *The SQUID Handbook. Vol. II: Applications of SQUIDs and SQUID Systems*, eds. J. Clarke and A. I. Braginski (Wiley-VCH GmbH & Co., Weinheim, 2006), Ch. 15.
62. J. Gallop and F. Piquermal, in *The SQUID Handbook. Vol. II: Applications of SQUIDs and SQUID Systems*, eds. J. Clarke and A. I. Braginski (Wiley-VCH GmbH & Co., Weinheim, 2006), Ch. 9.
63. J. Clarke and F. K. Wilhelm, *Nature* **453**, 1031 (2008).
64. J. Clarke, A. T. Lee, M. Mück and P. L. Richards, in *The SQUID Handbook. Vol. II: Applications of SQUIDs and SQUID Systems*, eds. J. Clarke and A. I. Braginski (Wiley-VCH GmbH & Co., Weinheim, 2006), Ch. 8.
65. P. James, E. Peebles, L. A. Page Jr. and R. B. Partridge, *Finding the Big Bang* (Cambridge University Press, Cambridge, England, 2009).
66. R. A. Sunyaev and Y. B. Zel'Dovich, *Comments on Astrophysics* **4**, 173 (1972).
67. K. D. Irwin, *Appl. Phys. Lett.* **66**, 1998 (1995).
68. D. Schwan, P. A. R. Ade, K. Basu, A. N. Bender, B. A. Benson, F. Bertoldi, H.-M. Cho, G. Chon, J. Clarke, M. Dobbs, D. Ferrusca, R. Güsten, N. W. Halverson, W. L. Holzapfel, B. Johnson, A. Kovács, J. Kennedy, Z. Kermish, R. Kneissl, T. Lanting, A. T. Lee, M. Lueker, J. Mehl, K. M. Menten,

D. Muders, M. Nord, F. Pacaud, T. Plagge, C. L. Reichardt, P. L. Richards, R. Schaaf, P. Schilke, F. Schuller, H. Spieler, C. Tucker, A. Weiss, B. Westbrook, O. Zahn, *Rev. Sci. Instrum.*, submitted.
69. T. M. Lanting, H.-M. Cho, J. Clarke, W. L. Holzapfel, A. T. Lee, M. Lueker, P. L. Richards, M. A. Dobbs, H. Spieler and A. Smith, *Appl. Phys. Lett.* **86**, 112511 (2005).
70. N. W. Halverson, T. Lanting, P. A. R. Ade, K. Basu, A. N. Bender, B. A. Benson, F. Bertoldi, H.-M. Cho, G. Chon, J. Clarke, M. Dobbs, D. Ferrusca, R. Güsten, W. L. Holzapfel, A. Kovács, J. Kennedy, Z. Kermish, R. Kneissl, A. T. Lee, M. Lueker, J. Mehl, K. M. Menten, D. Muders, M. Nord, F. Pacaud, T. Plagge, C. Reichardt, P. L. Richards, R. Schaaf, P. Schilke, F. Schuller, D. Schwan, H. Spieler, C. Tucker, A. Weiss and O. Zahn, *Astrophys. J.* **701**, 42 (2009).
71. B. Sadoulet, *Science* **315**, 61 (2007).
72. R. Peccei and H. Quinn, *Phys. Rev. Lett.* **38**, 1440 (1977).
73. S. Weinberg, *Phys. Rev. Lett.* **40**, 223 (1978).
74. F. Wilczek, *Phys. Rev. Lett.* **40**, 279 (1978).
75. P. Sikivie, *Phys. Rev. Lett.* **51**, 1415 (1983).
76. P. Sikivie, *Phys. Rev. D* **32**, 2988 (1985).
77. R. Bradley, J. Clarke, D. Kinion, L. J. Rosenberg, K. van Bibber, S. Matsuki, M. Mück and P. Sikivie, *Rev. Mod. Phys.* **75**, 777 (2003).
78. J. E. Kim, *Phys. Rev. Lett.* **43**, 103 (1979).
79. M. A. Shifman, A. I. Vainshtein and V. I. Zakharov, *Nucl. Phys. B* **166**, 493 (1980).
80. M. Dine, W. Fischler and M. Srednicki, *Phys. Lett. B* **104**, 199 (1981).
81. A. P. Zhitnitsky, *Yad. Fiz.* **31**, 497 (1980) [*Sov. J. Nucl. Phys.* **31**, 260].
82. M. Mück, M.-O. André, J. Clarke, J. Gail and Ch. Heiden, *Appl. Phys. Lett.* **72**, 2885 (1998).
83. M. Mück and J. Clarke, *J. Appl. Phys.* **88**, 6910 (2000).
84. M. Mück, C. Welzel and J. Clarke, *Appl. Phys. Lett.* **82**, 3266 (2003).
85. M. Mück, M.-O. André, J. Clarke, J. Gail and Ch. Heiden, *Appl. Phys. Lett.* **75**, 3545 (1999).
86. M. Mück, J. B. Kycia and J. Clarke, *Appl. Phys. Lett.* **78**, 967 (2001).
87. F. C. Wellstood, C. Urbina and J. Clarke, *Phys. Rev. B* **49**, 5942 (1994).
88. D. Kinion and J. Clarke, *Appl. Phys. Lett.*, submitted.
89. S. J. Asztalos, G. Carosi, C. Hagmann, D. Kinion, K. van Bibber, M. Hotz, L. Rosenberg, G. Rybka, J. Hoskins, J. Hwang, P. Sikivie, D. B. Tanner, R. Bradley and J. Clarke, *Phys. Rev. Lett.* **104**, 041301 (2010).
90. E. M. Haacke, R. W. Brown, M. R. Thompson and R. Venkatesan, *Magnetic Resonance Imaging: Physical Principles and Sequence Design* (Wiley & Sons, New York, 1999).
91. A. Abragam, *The Principles of Nuclear Magnetism* (Clarendon Press, Oxford, 1961).
92. C. P. Slichter, *Principles of Magnetic Resonance*, 3rd edition (Springer, New York, 1989).
93. E. L. Hahn, *Phys. Rev.* **80**, 580 (1950).

94. M. Packard and R. Varian, *Phys. Rev.* **93**, 941 (1954).
95. R. McDermott, A. H. Trabesinger, M. Mück, E. L. Hahn. A. Pines and J. Clarke, *Science* **295**, 2247 (2002).
96. R. McDermott, S.-K. Lee, B. ten Haken, A. H. Trabesinger, A. Pines and J. Clarke, *Proc. Natl. Acad. Sci.* **101**, 7857 (2004).
97. R. McDermott, N. Kelso, S.-K. Lee, M. Mössle, M. Mück, W. Myers, B. ten Haken, H. C. Seton, A. H. Trabesinger, A. Pines and J. Clarke, *J. Low Temp. Phys.* **135**, 793 (2004).
98. J. Clarke, M. Hatridge and M. Mössle, *Annu. Rev. Biomed. Eng.* **9**, 389 (2007).
99. S.-K. Lee, M. Mössle, W. Myers, N. Kelso, A. H. Trabesinger, A. Pines and J. Clarke, *Magn. Reson. Med.* **53**, 9 (2005).
100. J. Kurhanewicz, D. Vigneron, P. Carroll and F. Coakley, *Curr. Opin. Urol.* **18**, 71 (2008).
101. L. Qiu, Y. Zhang, H.-J. Krause, A. I. Braginski, M. Burghoff and L. Trahms, *Appl. Phys. Lett.* **91**, 072505 (2007).
102. J. Bernarding, G. Buntkowsky, S. Macholl, S. Hartwig, M. Burghoff and L. Trahms, *J. Am. Chem. Soc.* **128**, 714 (2006).
103. P. Volegov, A. N. Matlachov, M. A. Espy, J. S. George and R. H. Krause, Jr., *Magn. Reson. Med.* **52**, 467 (2004).
104. V. S. Zotev, A. N. Matlashov, P. L. Volegov, I. M. Savukov, M. A. Espy, J. C. Mosher, J. J. Gomez and R. H. Kraus, Jr., *J. Magn. Reson.* **194**, 115 (2008).

RESISTANCE IN SUPERCONDUCTORS

Bertrand I. Halperin*, Gil Refael† and Eugene Demler*
*Physics Department, Harvard University, Cambridge, MA 02138, USA
†Department of Physics, California Institute of Technology, Pasadena, CA 91125, USA

In this pedagogical review, we discuss how electrical resistance can arise in superconductors. Starting with the idea of the superconducting order parameter as a condensate wave function, we introduce vortices as topological excitations with quantized phase winding, and we show how phase slips occur when vortices cross the sample. Superconductors exhibit non-zero electrical resistance under circumstances where phase slips occur at a finite rate. For one-dimensional superconductors or Josephson junctions, phase slips can occur at isolated points in space-time. Phase slip rates may be controlled by thermal activation over a free-energy barrier, or in some circumstances, at low temperatures, by quantum tunneling through a barrier. We present an overview of several phenomena involving vortices that have direct implications for the electrical resistance of superconductors, including the Berezinskii–Kosterlitz–Thouless transition for vortex-proliferation in thin films, and the effects of vortex pinning in bulk type II superconductors on the nonlinear resistivity of these materials in an applied magnetic field. We discuss how quantum fluctuations can cause phase slips and review the non-trivial role of dissipation on such fluctuations. We present a basic picture of the superconductor-to-insulator quantum phase transitions in films, wires, and Josephson junctions. We point out related problems in superfluid helium films and systems of ultra-cold trapped atoms. While our emphasis is on theoretical concepts, we also briefly describe experimental results, and we underline some of the open questions.

1. Introduction

The ability of a wire to carry an electrical current with no apparent dissipation is surely the most dramatic property of the superconducting state. Under favorable conditions, the electrical resistance of a superconducting wire can be very low indeed. Mathematical models predict lifetimes that far exceed the age of the universe for sufficiently thick wires under appropriate conditions. In one experiment, a superconducting ring was observed to carry a persistent current for more than a year without measurable decay, with an upper bound for the decay rate of a part in 10^5 in the course of a year.[1] However, in other circumstances, as for sufficiently thin wires or films, or in the presence of penetrating strong magnetic fields, non-zero resistances are observed. Over the past fifty years, a great deal of theoretical and experimental effort has been devoted to obtaining better qualitative and quantitative understanding of how this resistivity arises, and how superconductivity breaks down, in the variety of possible situations.

Let us first recall how superconductors can exhibit negligible resistance in favorable situations. In a paper that appeared seven years before BCS, Ginzburg and Landau (GL) proposed that a superconductor should be characterized by a complex-valued function of position, $\Psi(\mathbf{r})$, referred to as the *superconducting order parameter.* They proposed, further, that one could define a free-energy functional F, which depends on $\Psi(\mathbf{r})$, and which would be minimized, in the case of a homogeneous superconductor with no external magnetic field, at temperatures below the critical temperature T_c, by an order parameter $\Psi(\mathbf{r})$ whose magnitude was a constant $\Psi_0 > 0$, determined by the temperature T. In order to minimize the free energy, the phase of the complex order parameter should be independent of position, but the value of the free energy would be independent of the choice of this constant's overall phase.

Following BCS, we now understand that the microscopic origin of the GL order parameter is the condensate wave function for Cooper pairs. Except for an arbitrary normalization constant, this is equal to the anomalous expectation value $\langle\psi_\uparrow(\mathbf{r})\psi_\downarrow(\mathbf{r})\rangle$, where $\psi_\uparrow(\mathbf{r})$ and $\psi_\downarrow(\mathbf{r})$ are the annihilation operators for an electron at position $\mathbf{r}$, with spin up and spin down, respectively. The existence of this non-zero expectation value signifies that the superconducting state has a broken symmetry, namely the $U(1)$ symmetry, commonly referred to as gauge symmetry, associated with the conservation of charge or electron number.

The GL assumption enables us to understand the phenomenon of persistent currents. Consider a superconducting ring in a situation where the magnitude of the order parameter is a constant, but where the phase varies around the ring. Since the order parameter Ψ must be single-valued at any given position, the net phase change around the ring must be an integral multiple, n, of 2π. A state with non-zero winding number n will have an *excess* free energy, proportional to n^2 for moderate values of n. As we shall see below, it will also carry an electrical current proportional to n. One finds that in order for the initial state to decay to a state with a smaller winding number, and hence, a smaller value of the free energy and a smaller value of the current, it is necessary for the system to pass through an intermediate configuration of $\Psi(\mathbf{r})$ where the free energy is larger than the free energy of the initial current carrying state, by an amount ΔF. Because $\Psi(\mathbf{r})$ is the wave function for a large number of Cooper pairs, the free energy barrier ΔF can be very large compared to the thermal energy $k_{\mathrm{B}}T$, and the probability to be carried over the barrier by a thermal fluctuation can be exceedingly small. On the other hand, by adjusting parameters, including the temperature and the dimensions of the system, one may reduce ΔF to the point where transitions occur at a small but measurable rate.

Quite generally, if we know the minimum value of the free energy barrier ΔF separating a state with winding number n from a state with $n-1$, we may estimate a thermally-activated rate of transitions between the two states as

$$\eta = \Omega\, e^{-\Delta F/T}\,, \tag{1}$$

where the prefactor Ω depends on details of the system, but is generally a less-rapidly varying function of the parameters than the exponential factor, in the region of interest. (We measure temperature in energy units, so that $k_{\mathrm{B}} = 1$.) The most important step, therefore, in estimating the decay rate of persistent current in a superconducting loop is to identify the process with smallest barrier and to calculate ΔF as accurately as possible. After that, one must estimate the associated prefactor Ω.

The most favorable path for a change in winding number, and the corresponding value of ΔF, can be quite different in different geometries, as we shall discuss below. However, there are some general considerations. The winding number n may be defined as $(2\pi)^{-1}$ times the integral of the phase gradient $\nabla\phi$ along a closed path around the ring, embedded in the superconductor. If we assume that $\Psi(\mathbf{r})$ varies continuously as a function of position and time, however, the only way that the accumulated phase can jump discontinuously at some time t, is if Ψ vanishes at some point on the path. Since Ψ has both a real and imaginary part, the locus of points $(\mathbf{r}, t)$ with $\Psi = 0$ will generically form a set of co-dimension 2. In the case of a two-dimensional superconductor, at a fixed time t, zeroes of Ψ should occur at isolated points, known as *vortex points*, which are described as positive or negative, depending on whether the phase changes by $\pm 2\pi$ as one winds

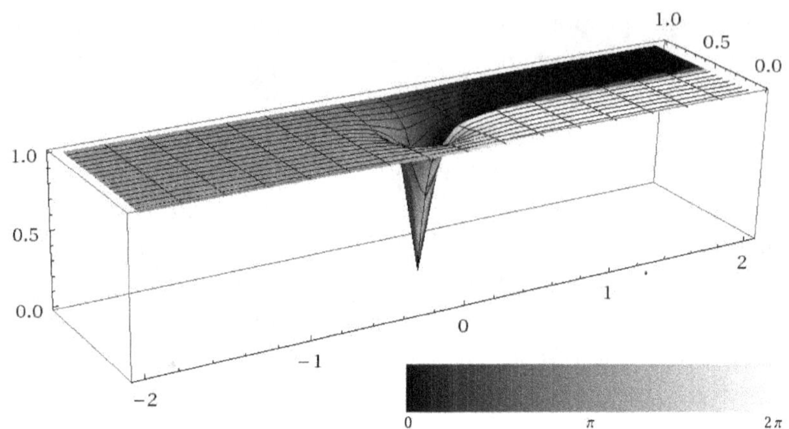

Fig. 1. An illustration of a vortex in a superconducting strip. The height of the function corresponds to the magnitude of the order parameter $\Psi(\mathbf{r})$, and the grayscale to its phase $\phi(\mathbf{r})$, mod 2π. The grayscale discontinuity along the line $y = 0.5$, for $x > 0$, does not represent a discontinuity in Ψ, because $e^{i\phi}$ is continuous. The coherence length used here is $\xi = 0.1$.

around the vortex in a counterclockwise sense. In three dimensions, the vortex core, where $\Psi = 0$, becomes a line segment ending at the boundaries of the superconductor, or possibly a closed loop, embedded within the material. For a given location of the vortex core, the state of minimum free energy will generally occur when the magnitude of Ψ returns to its equilibrium value over a healing distance comparable to the BCS coherence length $\xi(T)$. The structure of a vortex is depicted in Fig. 1.

Consider a superconducting annulus with a given initial phase-winding number $n > 0$. One way to change n with the smallest barrier ΔF would be to introduce a single positive vortex at the outer edge of the annulus, and move it across the ring until it disappears at the inner edge. Alternatively, we could move a negative vortex from the inner edge to the outer. In either case, any closed curve around the ring would have been cut once by the vortex, and the winding number would have changed by -1. The free energy cost for introducing a vortex into the film will depend on the details of the material, but will clearly increase with increasing thickness of the film. The free energy barrier can therefore be made arbitrarily large by going to a wire of sufficient thickness.

In order to go further, it is helpful to make some assumptions about the form of the free energy functional $F(\Psi)$. The specific form proposed by

Ginzburg and Landau may be written as

$$F = \int d^3\mathbf{r} \left[-\frac{\alpha}{2}|\Psi|^2 + \frac{\beta}{4}|\Psi|^4 + \frac{\gamma}{2}\left| \left(\nabla - \frac{2ie}{\hbar c}\mathbf{A} \right) \Psi \right|^2 \right], \tag{2}$$

where $\mathbf{A}(\mathbf{r})$ is the electromagnetic vector potential (we use Gaussian cgs units throughout this chapter). The parameters α, β, γ, are all positive below T_c, but α is assumed to vanish linearly as $T \to T_c$. If there is no magnetic field present, the free energy is minimized when $|\Psi| = (\alpha/\beta)^{1/2}$, and the resulting free energy reduction per unit volume is $f_0 = (\alpha^2/4\beta)$. The parameters of (2) result in a coherence length ξ and a magnetic penetration depth λ given by

$$\xi = (\gamma/\alpha)^{1/2}, \quad \lambda = \kappa\xi, \quad \kappa^2 \equiv \frac{\hbar^2 c^2 \beta}{16\pi e^2 \gamma}. \tag{3}$$

We shall be interested primarily in "type II" superconductors, where $\kappa \gg 1$. The free energy cost for creating a vortex is typically of order $f_0 \xi^2 L_v$, where L_v is the length of the vortex.

The free energy (2) implies that in a state of stable or metastable equilibrium, there will be an electrical current ("supercurrent"), whose density is given by

$$\mathbf{j}_s(\mathbf{r}) = -c\frac{\delta F}{\delta \mathbf{A}(\mathbf{r})} = \frac{2e\gamma}{\hbar}|\Psi|^2 \left(\nabla\phi - \frac{2e}{\hbar c}\mathbf{A} \right), \tag{4}$$

where $\phi(\mathbf{r})$ is the phase of $\Psi(\mathbf{r})$. Although the specific forms (2) and (4) are quantitatively correct only close to T_c, they are qualitatively correct more generally, and they will serve well for our purposes. The BCS theory gives a prediction for the Ginzburg–Landau parameters in terms of microscopic parameters of the material, such as the electron density of states, the effective mass, the mean-free-path, and the electron–phonon interaction strength.

The decay rate of a persistent current can be changed dramatically if there is an applied magnetic field exceeding the threshold $H_{c1} = \frac{hc/2e}{4\pi\lambda^2} \ln \kappa$ for penetration into the superconductor. In this case one finds that there are vortices present even in thermal equilibrium, with a vortex density n_v proportional to the magnetic field:

$$n_v = B/\Phi_0, \tag{5}$$

where $\Phi_0 \equiv hc/2e$ is the *superconducting flux quantum.* Although the vortices will ideally form a triangular Abrikosov lattice, in practice the lattice is usually distorted and the vortices tend to be pinned by various types of inhomogeneities in the material. In this case, since vortices are already present, the free energy barrier for relaxation of a supercurrent at low temperatures will be the activation energy necessary for vortices to become unpinned so they can move across the current-carrying sample.

An analysis of the decay rate for persistent current in a ring can be extended directly to the onset of resistance in an open superconducting wire carrying a current between two contacts. Because $\Psi(\mathbf{r})$ represents a wave function for Cooper pairs of charge $2e$, there is a commutation relation between the phase ϕ and the total electron number N, given by

$$[\phi, N] = 2i\,. \tag{6}$$

Then, because electrical charge is conserved, it can be shown that in a state of local equilibrium, at any temperature T, the phase ϕ must evolve in time according to the Josephson relation

$$d\phi/dt = 2eV/\hbar\,, \tag{7}$$

where V is the electrochemical potential (i.e. the voltage) at the point $\mathbf{r}$. (Equivalently, (7) may be understood as a consequence of gauge invariance.) If the two ends of a superconducting wire were each in local equilibrium, with their voltages differing by an amount $\Delta V \neq 0$, and if no vortices or phase slips were allowed to cross the intervening wire, the phase difference between the two ends of the wire would increase linearly in time, and the resulting supercurrent would increase without limit. In order to reach a steady state with a constant current, there must be a net flow of vortices or phase slips across the wire, at a rate η that just relaxes the phase build up due to the voltage, which requires

$$\Delta V = \pi\eta\hbar/e\,. \tag{8}$$

If, for a given current, there is a large free energy barrier, so that the rate η given by Eq. (1) is very small, then the resulting voltage will be proportionally small.

In the presence of a non-zero electric field, there can be an additional contribution to the current from the "normal fluid", which arises from thermally excited quasiparticles. When the phase slip rate η is small, the normal current will be negligible compared to the supercurrent in a dc measurement, but it can be an important source of dissipation in ac applications, as discussed in Sec. 4.

Our general considerations may be applied to a variety of situations. In wires that are thin compared to the coherence length ξ, one may neglect the variation of Ψ across the diameter of the wire, and consider that the order parameter is only a function of x, the distance along the wire. In this case, the winding number can change if the order parameter passes through zero at some location x_0 along the wire. Such events are often referred to as *point phase slips*. Phase slips can also occur at weak links, or *Josephson junctions*, where the free energy has the form of a periodic function of phase difference across the link, as opposed to the quadratic function of the phase gradient that is most commonly applicable to a bulk superconductor or film.

The various geometries and mechanisms for resistance by thermal activation will be discussed below in Sec. 2, which will concentrate on the effects of thermally activated phase slips or vortex motion on dissipation, in single Josephson junctions, thin wires, films and bulk materials. In Sec. 2.4, we also discuss the role of vortices in the thermodynamic phase transition between the superconducting and normal states of a thin film. Dissipation arising from vortices induced by an applied magnetic field will be discussed in Sec. 2.6.

In very small Josephson junctions, very thin wires or highly disordered thin films, at sufficiently low temperatures, the mechanism for relaxation of supercurrent may be quantum tunneling of phase slips, rather than thermal activation over a barrier. Quantum phase slips will be discussed in Sec. 3.

While the focus of this chapter is on superconductors, the dissipation mechanisms discussed are common to many other types of systems. Examples can be found among neutral superfluids such as helium and ultracold atoms (see Refs. 2–6 for reviews). For example, theoretical analyses of transport of two-dimensional systems near a superconducting or superfuid transition were strongly motivated by experiments on films of ^{4}He.[7] Another interesting case, where dynamics of the condensate order parameter determines transport properties, can be found in bilayer quantum Hall systems at the filling factor $\nu = 1$.[8] In such systems one expects to find spontaneous interlayer coherence, which is analogous to exciton condensation,[9] so that both interlayer tunneling and antisymmetric longitudinal resistivity are determined by the dynamics of the condensate order parameter. Current research on systems of ultracold atoms will be briefly discussed in Sec. 5.

The emphasis in this chapter will be on the theoretical concepts necessary to understand resistance in superconductors. We include only a limited discussion of experimental results, with limited references to the corresponding literature. We cite numerous references containing detailed theoretical analyses and quantitative calculations, but here too we are far from complete.

It should emerge from the discussions below that while we believe we have a good understanding of the basic mechanisms responsible for resistance in superconductors, there remain many puzzles and unanswered questions about the interpretation of experimental data. Particularly in regimes where quantum fluctuations are important, the subject remains quite active.

2. Phase Slips Produced by Thermal Activation

2.1. *Phase slips in Josephson junctions*

We begin by examining how phase slips arise in a Josephson junction between two finite superconducting electrodes. Tunneling of Cooper pairs between the electrodes gives rise to the following junction energy:

$$U(\phi) = -E_J \cos\phi \tag{9}$$

where ϕ is the phase difference between the superconducting electrodes, and E_J is the Josephson energy (we have assumed here that any magnetic flux through the junction is negligible compared to Φ_0). A non-zero value of ϕ will lead to a supercurrent across the junction, given by

$$I_\phi = (2e/\hbar)\partial U/\partial\phi = (2e/\hbar)E_J \sin\phi\,. \tag{10}$$

For a tunnel junction between two BCS superconductors with an energy gap Δ, E_J is given by the Ambegaokar–Baratoff formula:

$$E_J = \frac{\hbar\Delta}{4e^2 R_T}\tanh\left(\frac{\Delta}{2T}\right), \tag{11}$$

where R_T is the resistance of the junction in the normal state right above T_c.

If the superconducting electrodes are thick enough, such that there is a large energy barrier for the order parameter to vanish inside them, then the time dependence of ϕ should be given by the Josephson relation Eq. (7), with V being the instantaneous voltage difference between the electrodes. This voltage may fluctuate in time because of thermal noise or because of quantum-mechanical fluctuations. Here we consider only the classical thermal contributions.

The precise dynamics of ϕ will depend on the way the junction is inserted in an external circuit. Suppose that the junction is connected to an ideal constant current source, with current I. If the phase ϕ at some instant of time is such that $I_\phi \neq I$, there will be a build up of the charge difference q between the two sides of the junction, given by $dq/dt = I - I_\phi$. This, in turn, will give rise to a voltage difference $V = q/C$, where C is the capacitance of the junction. (We neglect here the capacitance of the external circuit, which is in most cases small compared to that of the junction.) The voltage V, in turn, will give a non-zero time-derivative of ϕ. If the only current across the

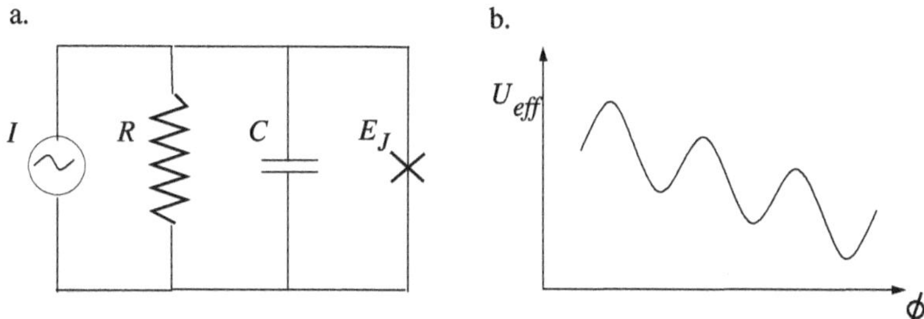

Fig. 2. Josephson junction connected to a current source with current I. Panel (a) shows the RSJ model, where the tunnel junction is shunted by a resistance R and a capacitance C. In most cases, C is determined by the internal capacitance of the junction, while R^{-1} may be the sum of an external shunt conductance and an internal conductance due to tunneling of normal quasiparticles at non-zero temperatures. Panel (b) shows the "washboard" effective potential, $U_{\text{eff}}(\phi) = -E_J \cos\phi - (\hbar/2e)I\phi$, for the phase ϕ.

junction is supercurrent I_ϕ, then we have a closed set of equations for ϕ and q which we may write in a Hamiltonian form

$$\frac{d\phi}{dt} = \frac{2e}{\hbar}\frac{\partial \mathcal{H}}{\partial q}, \qquad \frac{dq}{dt} = -\frac{2e}{\hbar}\frac{\partial \mathcal{H}}{\partial \phi}, \tag{12}$$

$$\mathcal{H} = U(\phi) - \frac{\hbar}{2e}I\phi + \frac{q^2}{2C}. \tag{13}$$

These equations suggest that we should identify $\hbar q/2e$ as a momentum canonically conjugate to ϕ, which is consistent with the commutation relation $[\phi, q] = 2ie$ that one would expect based on Eq. (6).

If I is smaller than a critical current $I_c = 2eE_J/\hbar$, the ϕ-dependent terms in H_{eff} have the form of a "washboard potential", illustrated in Fig. 2(a), with a local minimum and a local maximum in each unit cell. The local minima are stable solutions of the equation $I_\phi = I$. If the system is initially placed at such a point, and it initially has $q = 0$, then in the absence of external noise it will stay there forever, with a constant current I and $V = 0$. If the system is displaced slightly from one of the minima, it will oscillate forever about this minimum, unless some additional coupling supplies enough energy to get it over the barrier separating it from one of the neighboring minima. As long as the system is trapped near one of the minima, the time-average of $d\phi/dt$ will still be zero, so we again have a current carrying junction with $V = 0$.

When $I = 0$, the energy necessary to go over a barrier will be equal to $2E_J$, regardless of whether the system moves in the direction of increasing or decreasing ϕ. When $I \neq 0$, however, the barriers $\Delta F^{\pm}$ for increase or decrease of ϕ will differ from each other by an amount $-\pi\hbar I/e$. If $I/I_c \ll 1$, one finds simply

$$\Delta F^{\pm} \approx 2E_J \mp \frac{\pi\hbar I}{2e}\,. \tag{14}$$

The energy to get over the barrier can come from coupling to thermal fluctuations in the external circuit, or from thermally excited quasiparticles tunneling across the junction. At non-zero temperatures, in addition to the supercurrent, there will generally be a normal current across the junction, due to thermally excited quasiparticles, when the voltage difference $V(t)$ is different from zero. This can be represented, at least approximately, by a model in which there is a shunt resistor, with resistance R, across the Josephson junction, as shown in Fig. 2(b). If the external circuit is not an ideal current source, then R^{-1} should be a sum of the contribution from the quasiparticles and the differential conductance of the external circuit. The shunt resistance will lead to an added term, $-q/RC$ in the equation for dq/dt, which will lead to damping of the coupled oscillations q and ϕ. In addition, one must include a white noise term in the equation of motion, proportional to T/RC, so that for $I = 0$, the system can reach thermal equilibrium, with a probability distribution $P(\phi) \propto \exp[-U(\phi)/T]$. If one uses the same white noise source for $I \neq 0$, one finds a transition rate, in the direction of increasing or decreasing ϕ, which may be written in the form

$$\eta_{\pm} = \Omega e^{-\Delta F^{\pm}/T}\,, \tag{15}$$

where the prefactor Ω depends, in general on the parameters R and C, as well as on the Josephson energy E_J. In the limit $I/I_c \ll 1$ one can neglect the difference in the prefactors for forward and backwards transitions. Then the mean voltage across the junction, which is proportional to the net difference $\eta = \eta_+ - \eta_-$, in accord with Eq. (8) is given by

$$\langle V\rangle = (2\pi\hbar/e)\,\Omega\, e^{-2E_J/T}\sinh(\pi\hbar I/2eT)\,. \tag{16}$$

In the limit of small currents ($I \ll eT/\hbar$) one obtains an effective dc resistance for the junction:

$$R_{\text{eff}} = \langle V\rangle/I = (\pi\hbar^2/e^2T)\,\Omega e^{-2E_J/T}\,. \tag{17}$$

When the barrier is large compared to T, so that R_{eff} is small compared to the shunt resistance or the normal resistance of the junction, then essentially all of the current through the junction is the supercurrent, and the total current I is essentially the same as the mean value of I_ϕ.

It should be noted that the barrier ΔF^+ tends to zero as $I \to I_c$. Thus, even if E_J/T is quite large, so that phase slips are negligible for $I/I_c \ll 1$, there may be a regime close to I_c where thermally activated phase slips become important.

The prefactor Ω can actually be calculated for the RSJ model. In the absence of damping, a junction near a minimum of the cosine potential $U(\phi)$ will oscillate about it at the Josephson plasma frequency

$$\Omega_{JC} \equiv \frac{1}{\hbar}\left(\frac{2e^2 E_J}{C}\right)^{1/2} . \tag{18}$$

Then for an underdamped junction, where $\Omega_{JC}RC$ is larger than 1 (but not larger than E_J/T), in the limit $I \to 0$, the pre-factor, may be written[10,11] $\Omega = \Omega_{JC}/2\pi$. For an overdamped junction, with $\Omega_{JC}RC \ll 1$, one finds in the limit $I \ll I_c$, that[11]

$$\Omega = \frac{e^2 E_J RC}{\pi\hbar^2} . \tag{19}$$

2.2. *Thermal phase slips in a thin wire close to T_C*

When a superconducting wire is narrow compared to the coherence length $\xi(T)$, it is generally correct to neglect variations in the order parameter across the diameter of the wire, and to treat Ψ, at any time t, as a function of a single position variable x, the distance along the wire. A vortex, in this case, degenerates to a single point where $\Psi(x) = 0$, and a phase slip event is an isolated point in space and time, where Ψ passes through zero and the phase difference along the wire jumps by $\pm 2\pi$.

For temperatures relatively close to T_c, one may use the Ginzburg–Landau functional (2) to calculate the free energy barriers $\Delta F^{\pm}$ for creation of a phase slip in the wire. In analogy to our treatment of the Josephson junction, if the wire is connected to an external current source with a current I, we must add to the GL functional a term $-(\hbar/2e)I\Delta\phi$, where $\Delta\phi$ is the difference in the phases $\phi(x)$ measured at the two ends of the wire. (We may assume that $\Psi \neq 0$ at the two ends, so that the phases at the ends may be defined to be continuous functions of time.) Alternatively, one may calculate the free energy barrier to produce a phase slip in a closed superconducting loop that carries a current I.

The free energy activation barrier for phase slips in a wire was first worked out, using the GL theory, by Langer and Ambegaokar.[12] For a wire loop of length L, with a specified phase change $2\pi n$, the order parameter configuration that leads to a local minimum of the free energy functional has a uniform phase gradient $k = 2\pi n/L$ and a constant magnitude of Ψ, with

$$\Psi_k(x) = \sqrt{\alpha/\beta}e^{ikx}(1 - \xi^2 k^2)^{0.5}\,. \tag{20}$$

A phase slip will reduce k by $2\pi/L$. For that to occur, however, the order parameter must go through an intermediate state $\Psi_s(x)$ which is a saddle point of the G-L free energy, where at some point x_0 in the wire, the order parameter nearly vanishes. The unwinding of the phase is concentrated close to x_0, and the magnitude of the order parameter is depressed in a region of length ξ about that point. An explicit solution for $\Psi_s(x)$, for arbitrary current, was given by McCumber and Halperin.[13] Once the order-parameter configuration reaches the saddle-point configuration Ψ_s, the configuration can evolve with a continuously decreasing free energy through the actual phase slip event, where the order parameter passes though zero at some point on the line.

Roughly speaking, the free energy cost of reaching the saddle point is the cost of suppressing the order parameter in the phase slip region. More precisely, in the limit of zero current, the barrier is found to be

$$\Delta F_0 = K_0 A \xi f_0\,, \tag{21}$$

where A is the cross-sectional area of the wire, $K_0 = 8\sqrt{2}/3$, and $f_0 = \alpha^2/4\beta$ is the condensation energy per unit volume. (The quantity f_0 is related to the critical field H_c by $f_0 = H_c^2/8\pi$.) The barrier ΔF_0 decreases $\propto (T_c - T)^{3/2}$ when $T \to T_c$.

For currents that are non-zero but small compared to the critical current, $k \ll \xi^{-1}$, the barriers for forward or backward phase slips have the form $\Delta F^{\pm} = \Delta F_0 \mp \pi\hbar I/2e$, as was found for the single Josephson junction. Then, if we assume that the rate of phase slips has an activated form similar to Eq. (15), we obtain a voltage drop $\langle V \rangle$ identical to (16), with the zero-current activation energy $2E_J$ replaced by ΔF_0.

In order to calculate the pre-exponential factor $\Omega(T)$, one must make additional assumptions about the dynamic equations of motion for the order parameter. This was done by McCumber and Halperin[13] using the simplest possible model, the time-dependent Ginzburg-Landau (TDGL) theory. In this model, the order parameter obeys an equation of motion of the form

$$\frac{1}{\Gamma}\frac{\partial \Psi}{\partial t} = -\frac{\delta F}{\delta \Psi^*} + \eta(t)\,, \tag{22}$$

where $\eta(x,t)$ is a Gaussian white noise source with correlator $\langle \eta(x,t)\eta(x',t')\rangle = 2k_BT\,\Gamma^{-1}\delta(t-t')\delta(x-x')$, and

$$\frac{\delta F}{\delta \Psi^*} = -\frac{\gamma}{2}\left(\nabla - \frac{2ie}{\hbar c}A\right)^2 \Psi - \frac{\alpha}{2}\Psi + \frac{\beta}{2}\Psi\,|\Psi|^2 . \tag{23}$$

The constant Γ is temperature-independent near T_c, and is chosen so that $\Gamma\alpha = 8k_B(T_c - T)/\pi\hbar \equiv 1/\tau_{GL}$. With these equations, one finds that in the absence of a current, the function $\Psi(x)$ will have an equilibrium distribution $P\{\Psi\} \propto e^{-F/T}$, and that an initial state close to equilibrium will relax to it at a rate $1/\tau_{GL}$, which goes to zero as $T \to T_c$.

The explicit calculation of the activation rate from the TDGL equation uses not only the values of F at the minimum and the saddle point, but also the eigenvalues of the second derivative matrices at the two points. The positive eigenvalues contribute entropy corrections, which modify the numerical value of Ω, while the negative eigenvalue at the saddle point determines the overall time scale. For a translationally invariant system, such a calculation always results, up to a factor of order unity, in the product of the number of "independent" phase-slip configurations (e.g. where along the wire phase slips can occur), the inverse of the TDGL time constant, and the square root of the free-energy cost of the saddle configuration divided by the temperature.[14] In the limit $I \to 0$, the explicit result for the prefactor, obtained by McCumber and Halperin, is

$$\Omega(T) = \frac{0.156}{\tau_{GL}}\frac{L}{\xi}\left(\frac{\Delta F_0}{T}\right)^{1/2}, \tag{24}$$

where L is the length of the wire. For $T \to T_c$, the prefactor varies as $(T_c - T)^{9/4}$. The electrical resistance is given by $R = (\pi\hbar^2/e^2T)\Omega e^{-\Delta F_0/T}$, in analogy to (17), using (21) and (24).

Above T_c, the order parameter fluctuations predicted by TDGL theory give a temperature-dependent contribution to the electrical conductivity, in one, two and three dimensions, which was first calculated by Aslamazov and Larkin. An interpolation between this regime and the McCumber–Halperin formula therefore gives a prediction for the entire temperature dependence near T_c, within the TDGL theory.

The McCumber–Halperin formula for the resistance of a wire seems to work surprisingly well in fitting resistivity data on thin wires close to T_c (see, e.g. Ref. 15), and even at lower temperatures.[16] Nevertheless, the formula comes with some caveats. The TDGL theory is derived by integrating out the fermionic degrees of freedom, and assuming that they can respond rapidly to changes in the order parameter. This is probably a good assumption for calculating time-dependent fluctuations in the order-parameter at temperatures slightly above T_c. However, its use below T_c is problematic. One problem is that the energy gap of most superconductors becomes rapidly bigger than k_BT, at a small distance below T_c. Even where this has not occurred, there can be some very long relaxation times associated with small rates for inelastic scattering of quasiparticles, restoration of "branch imbalance", etc. In principle, these effects could lead to a very large reduction in the value of the prefactor, which could change considerably the interpretation of the data. Even above T_c, there are additional contributions to the electrical conductivity in addition to the Aslamazov–Larkin term which may be important in the experiments. The overall situation remains to be sorted out.

2.3. *Planar geometries*

For a thin film superconductor, whose thickness d is small compared to the penetration depth λ, we can generally neglect the effects of magnetic fields produced by currents in the film. Therefore we can take $\mathbf{A}$ to be the vector potential due only to an applied external magnetic field. In the absence of such a field, we may drop $\mathbf{A}$ from the equations. Moreover, if d is small compared to ξ, we may neglect variations in Ψ over the thickness of the film. One may then look for the planar configuration $\Psi(x,y)$ that minimizes (2) subject to the constraint that there exist a pair of vortices of opposite sign, with a given separation $2\mathbf{D}$, which we choose to be large compared to the coherence length ξ but small compared to the distance to the nearest boundary. The resulting free energy, relative to the free energy of the ground state without the vortices, is

$$\delta F = 2\pi K d[\log(D/\xi) + 2\epsilon_c]\,, \quad K = \gamma\alpha/\beta\,, \tag{25}$$

where ϵ_c is a constant of order unity. If there is a non-zero background supercurrent density $\mathbf{j}$, however, resulting from a uniform gradient of the phase ϕ, one finds an additional term in the free energy of the form

$$\delta F_j = -(\pi\hbar/e) j D_\perp d\,, \tag{26}$$

where $D_\perp$ is the component of the separation perpendicular to the current. This additional term may be derived by adding a term $-(\hbar/2e)I\Delta\phi$ to the effective Hamiltonian, in analogy to our treatment of the Josephson junction and one-dimensional wire, where $\Delta\phi$ is the phase difference between the two ends of the film and I is the total current, integrated across the width of the sample. Minimization of the free energy in the presence of a pair of vortices leads to an incremental change in $\Delta\phi$ which is equal to $2\pi D_\perp/W$, where W is the width of the film. The additional term may also be understood as arising from the magnus force on a vortex, which is proportional to the circulation of the vortex and the background current, and is perpendicular to both. Similar formulas apply to a single vortex interacting with its own image charge near an edge of the superconductor, except that in this case the right-hand sides of (25) and (26) must be divided by two, while $\mathbf{D}$ is the distance to the edge. In either case, we see that if the displacement $\mathbf{D}$ is increased in the direction of the magnus force, the free energy will eventually decrease, as the linear term will win over the logarithm. The free energy maximum occurs when $D_\perp \approx 2eK/j\hbar$. The free energy barrier for nucleating a vortex at one edge of a sample, and freeing it from its image charge, is thus found to be, for small j,

$$\Delta F \approx \pi K d \log(2eK/j\hbar\xi) . \tag{27}$$

At low temperatures, $T \ll Kd$, this results in a dissipative response which gives a voltage of the form

$$V \propto e^{-\Delta F/T} \propto j^{x(T)} , \tag{28}$$

where $x(T) \approx \pi K d/T$. A more accurate analysis, which includes the contribution to the pre-exponential factor arising from the entropy associated with the position of the vortex, predicts that[17,18]

$$x(T) \approx 1 + \pi K d/T . \tag{29}$$

Thus vortex nucleation processes produce dissipation in a superconducting strip at any finite temperature and current. However, if $x(T)$ is large, the differential resistance arising from Eq. (28) vanishes as a large power law as the supercurrent approaches zero. We shall see that the exponent is always larger than 3 in the superconducting phase.

In a film of finite width, the logarithmic increase of ΔF will be cut off when the current is so small that $D_\perp$ reaches $W/2$. Also, if one takes into account the magnetic field produced by a vortex, the increase in ΔF will be cut off if $D_\perp$ exceeds the length scale $\lambda_\perp = 2\lambda^2/d$ for magnetic screening in a film (also known as the Pearl penetration length). Thus for sufficiently small currents, one should recover a linear voltage-currrent relation, with a small value of the resistivity.

2.4. *Thermally excited vortices and the BKT transition*

At temperatures comparable to the phase stiffness Kd of a superconducting film, vortices may arise as thermal excitations. In an infinitely thin film, these vortices are described as an interacting gas of Coulomb particles, i.e. with an interaction that depends logarithmically on distance (in a film with a finite thickness, the logarithmic interaction will prevail at distances shorter than the Pearl penetration length of $\lambda_\perp = 2\lambda^2/d$). The bare fugacity of the vortices is dictated by their core energy, $E_c = 2\pi Kd\epsilon_c$, and is given by:

$$\zeta \approx \frac{1}{\xi^2} e^{-2\pi Kd\epsilon_c/T}, \tag{30}$$

which may be taken as a rough estimate for the total density of vortices in a film.

The intervortex interaction leads to a subtle vortex-pairing transition, known as the Berezinskii–Kosterlitz–Thouless transition.[19–21] At high temperatures, the vortices behave as a charged, but unbound, plasma. As the temperature drops under a critical value, T_{KT}, vortices form neutral vortex and anti-vortex bound pairs. The temperature where this transition occurs can be obtained by simple thermodynamic consideration.[22] If we consider a finite, but large, system of size L, we can compare the energy cost of adding an uncompensated vortex with its entropy gain. The energy increase due to an uncompensated vortex in a film of side L follows from Eq. (25), and is given by $\Delta U = \pi Kd[\log(L/\xi) + 2\epsilon_c]$. On the other hand, such a vortex has an entropy which is roughly $\Delta S = \ln(L^2/\xi^2) = 2\ln L/\xi$. The net increase of free energy through the nucleation of a vortex is thus:

$$\Delta F \approx (\pi Kd - 2T)\ln L/\xi \tag{31}$$

where we ignore the film-size independent core energy. We see that this free energy cost diverges in the thermodynamic limit, if $T < T_{KT}$, defined by

$$T_{KT} = \frac{\pi Kd}{2}. \tag{32}$$

Vortices are then logarithmically confined into neutral pairs, and free vortices do not exist. The system is still a superfluid, though the value of ρ_s, and hence of K will be somewhat reduced because of the polarizability of bound vortex pairs in the presence of a current. (It is this temperature-dependent renormalized value of K that should be used in (32) to determine T_{KT}.) For $T > T_{KT}$, single vortices would proliferate and form a free plasma. As free vortices should have a finite diffusion constant, they would move in response to an arbitrarily small electrical current, giving a non-zero resistivity.

Another perspective on the BKT transition is obtained by considering the spatial correlations of the superconducting order parameter in thin films. Because two dimensions is the lower critical dimension for the $U(1)$ superconducting order parameter, there is no true long range order at non-zero temperatures. Instead, the correlation function in the superfluid phase is predicted to decay slowly at large distances, according to the power law,

$$\langle e^{i(\phi(r)-\phi(0))}\rangle \sim \frac{1}{r^{T/\pi Kd}}. \tag{33}$$

(The exponent here is necessarily <1/4 in the superfluid phase.) For $T > T_{KT}$ the correlation function falls off exponentially with a decay length $\xi_+(T)$.

A proper analysis of the BKT transition makes use of a renormalization group (RG) approach, in which one repeatedly integrates out the effects of pairs separated by a distance smaller than a running cutoff of form $a_l = \xi e^l$, and keeps track of the renormalization of K and the vortex fugacity ζ. For $T > T_{KT}$ the renormalization must be stopped at a length scale where the density of vortices is comparable to a_l^{-2}, and the correlation length ξ_+ is identified with this value of a_l. The RG analysis predicts specifically that ξ_+ should diverge at T_{KT} with an essential singularity:

$$\xi_+ \sim e^{-a/\sqrt{|T-T_{KT}|}}, \tag{34}$$

where a is a constant. The confinement length of the bound vortex pairs should diverge in a similar manner for $T \to T_{KT}^-$. Because the correlation length diverges so rapidly above T_{KT}, there should be only a very weak essential singularity in the specific heat at the BKT transition point itself: all derivatives of the free energy should be continuous there.

The linear resistivity for $T > T_{KT}$, should be proportional to the density of free vortices $\sim 1/\xi_+^2$. For $T < T_{KT}$, the resistivity vanishes in the the limit $j \to 0$, but finite currents can cause disassociation of bound vortex pairs, producing a non-zero voltage. For small currents, this voltage should be described by the power law (28). Comparing (29) with (32), we see that the exponent $x(T)$ is > 3 for $T < T_{KT}$, and it approaches 3 for $T \to T_{KT}^-$. One also predicts[17,18] that for $T = T_{KT}$, the induced voltage should have the universal dependence $V \propto j^3$.

The BKT transition as presented above applies equally well to thin films of superfluid ^{4}He as to thin-film superconductors. Indeed the theory has been supported by a number of beautiful experiments in the helium case.[7] An important feature of the superconductor, however, is that for a bulk sample, the thermal variation of the superfluid density and other parameters is very well given by a mean field theory, such as the BCS or Ginzburg–Landau theories, except in an extremely narrow range near the bulk transition temperature T_{c0}. The transition temperature T_{KT} in a film is far enough below T_{c0} that the mean field theory may be used to estimate the bare parameters of the RG analysis. The parameter a in (34), as well as the pre-exponential factors, can be estimated from the mean field parameters and the measured normal conductivity σ_n.[17] Similarly, using the time-dependent Ginzburg–Landau theory, one may estimate quantitatively the enhancement of the conductivity above σ_n due to incipient superconducting fluctuations for a range of temperatures above T_{KT}.

A number of experiments on thin superconducting films have indeed observed the predicted forms of the resistivity.[23,24] However, there are also experiments, particularly involving cuprate superconductors, which have been fit to different functional forms. A detailed analysis by Strachan, Lobb, and Newrock[25] suggests that the apparent discrepancies may be due to a combination of uncertainties in the choice of the mean-field T_c and problems where the vortex separation length $D_\perp$ may be exceeding $\lambda_\perp$ or the film width. The reader is referred to that article for a detailed discussion, as well as for citations to the experimental and theoretical literature.

2.5. *dc resistance in bulk superconductors without magnetic fields*

Bulk superconductors support a more robust superconducting state, and dissipative effects in them are much more suppressed than in lower-dimensional superconductors. We first consider a wire that is thick compared to the coherence length ξ, but thin compared to the magnetic penetration depth λ, which is possible in an extreme type II material. Then the mechanism for producing dissipation in the presence of a supercurrent density j is the thermal nucleation of vortex rings, which can expand across the diameter of the wire, and change the phase by 2π. When the vortex ring is small, the line tension due to the vortex energy will try to collapse the ring, but eventually, for a large vortex, this will be overcome by the Magnus force due to the current, which will favor expansion of the ring.

The rate of vortex loop formation is thus inhibited by an energy barrier, which the thermal fluctuation may overcome.[26] Following the same logic that led to Eq. (28), we note that the energy of a vortex ring of radius R is:

$$E_{\text{ring}} \approx 2\pi R E_c - \pi R^2 j \Phi_0 \,, \tag{35}$$

with E_c being the vortex energy per unit length. Note that E_c also contains a slowly varying contribution $\sim \ln(R/\xi)$ when $\xi < R < \lambda$; this does not affect the scaling behavior we discuss. The vortex-ring energy has a maximum at $R_{\max} = E_c/j\Phi_0$ with energy $E_B \approx \pi E_c^2/j\Phi_0$. From this energy barrier we infer that the finite-current resistivity of a bulk superconductor should vary as $\rho \sim e^{-J_T/j}$ with:

$$J_T = \frac{\pi E_c^2}{T\Phi_0} . \tag{36}$$

This leads to a rapidly vanishing linear resistivity in the limit of $j \to 0$ as expected. In practice, this exponential behavior is so strong, that there would typically be a threshold, say $J_T/30$, below which no voltage drop could be measured in practice.

At sufficiently high current densities, when j exceeds the Ginzburg–Landau critical current $J_F \sim \frac{E_c}{\Phi_0 \xi}$, superconductivity would break down not due to thermal fluctuations, but due to the mean-field energy cost of the current, according to the G-L free-energy functional, Eq. (2). (J_F is also the current density where the energy-maximum radius would be comparable to the coherence length, ξ, so vortex rings stop being a useful concept.) The condition $J_T < J_F$, occurs only extremely close to T_c, in the Ginzburg fluctuation regime where mean field theory breaks down, and even the condition $J_T/30 < J_F$ for measurable dissipation due to the formation of vortex rings may be difficult to achieve.

For a superconducting wire that is thicker than the magnetic penetration depth, the situation is more complicated because the current is confined to a skin depth λ. This can make it even more difficult for a vortex ring to expand across the interior of the wire, where the current density is small. Again, if the current is sufficiently low, one can easily enter the regime where dissipation rates are unmeasurably small, in the absence of applied magnetic fields.

2.6. *Resistance in an applied magnetic field*

For a bulk type II superconductor, when the applied magnetic field exceeds the lower critical field H_{c1} of the material, the magnetic field will penetrate the superconductor, giving rise to a finite density of vortices, even in thermal equilibrium. If there is also a non-zero macroscopic electric current flowing in the superconductor, the vortices will feel a magnus force, which tries to move them in the direction perpendicular to the current flow. If the vortices can move in response to this force, there will be an induced voltage, proportional to the density of vortices and their mean velocity of motion, which will lead to dissipative resistance.

If the current is not too large, and if the temperature is low, the net rate of vortex motion can be extremely small, as the vortices will tend to be pinned by imperfections in the material. The origin of pinning can be point-like crystal defects due to interstitial atoms, impurities or vacancies, or due to extended defects such as grain boundaries, twin boundaries, or dislocations. Variations in material composition or crystal structure could also cause pinning. In a thin film, variations in the film thickness could lead to variations in the core energy which could lead to pinning. For applications of superconductors as high current transmission lines or in superconducting magnets, one wants to have pinning forces that are as strong as possible, to prevent the motion of vortex lines and to suppress dissipation. For this reason, pinning sites are often added by adding impurities or by cold working. Pinning is most effective when the pinning objects have size comparable to the coherence length ξ, which is the size of the vortex core.

In the absence of disorder, vortices in a bulk superconductor in a magnetic field will tend to form a triangular array, known as an Abrikosov vortex lattice.[27] Since the interactions between vortices try to keep them in a regular array, vortex motion in the presence of a macroscopic current is actually the result of the competition between the magnus force, the lattice stiffness, and the random pinning potential. This interplay was qualitatively discussed by Larkin and Ovchinnikov[28] [LO] under the assumption that the pinning is caused by a large density n_p of pinning objects, each having a weak effect on the vortex lattice. The energy gain for a pinning site inside a vortex core was assumed to have a value u for a site at the center of the core, falling to zero smoothly over the core radius. Then the root-mean square pinning force exerted by a pinning at a random position is given by $f_p \approx u/a$, where $a \approx (\Phi_0/B)^{1/2}$ is the distance between vortices in the Abrikosov lattice. LO assume that the lattice is fragmented into domains where the lattice order is maintained, while the lattice distorts slightly to accomodate the random pinning potential, and they estimate the domain size and shape that will optimize the free energy gain due to the pinning potentials. They find that the optimum domain volume V_c and the pinning energy δF_p for each domain are given by:[1]

$$V_c \approx \frac{C_{\text{tilt}}^2 C_{\text{shear}}^4 \xi^6}{n_p^3 f_p^6}, \quad \delta F_p \approx -\frac{n_p^2 f_p^4}{C_{\text{tilt}} C_{\text{shear}}^2 \xi^2} \tag{37}$$

where $C_{\text{tilt}} = B^2/4\pi$ is the tilt modulus of the Abrikosov lattice, and $C_{\text{shear}} \approx (1 - B/H_{c2})^2 B H_{c1}/16\pi$ is its shear modulus. LO assume that the maximum pinning force per unit volume $\approx f_p(n_p/V_c)^{1/2}$ determines the critical current J_c for the onset of vortex motion at zero temperature, which leads to the result

$$J_c B \sim \frac{n_p^2 f_p^4}{C_{\text{tilt}} C_{\text{shear}}^2 \xi^3}. \tag{38}$$

Following the same arguments for a superconducting film yields:

$$J_c B \sim \frac{n_p f_p^2}{C_{\text{shear}} \xi d}. \tag{39}$$

It is interesting to note that according to this logic any amount of pinning would render a superconductor dissipationless for sufficiently low currents. However, in the LO weak pinning limit, the stiffer the vortex lattice, the smaller is the critical current. This is different from the case of a small density of very strong pinning sites, where the lattice stiffness could enhance the effects of pinning.

At finite temperature vortex motion may arise below J_c due to thermal fluctuations. Even a qualitative understanding of these effects is difficult since it requires understanding not only the characteristic forces that vortices encounter, but the shape of the collective pinning potential as a function of current. Early collective flux pinning theories proposed by Anderson and Kim[29] suggested that the maximum potential barrier for a vortex to depin is $U(j) \propto (1 - j/J_c)$. A more modern approach[26] emphasizes the fact that the potential barrier, due to collective effects, actually diverges near $j \to 0$ as:

$$U(j) \sim U_0 \left(\frac{J_c}{j}\right)^{\mu}, \tag{40}$$

where μ is an exponent ≤ 1. The voltage drop V in a superconductor with current density j should then be proportional to the Boltzmann factor, giving

$$V(j) \sim \exp\left(-\frac{U_0}{T}\left(\frac{J_c}{j}\right)^{\mu}\right). \tag{41}$$

This modified Arrhenius law leads to a vanishing linear resistance in the limit of zero current, but any finite current at a finite temperature will experience dissipation. Combining this with the Anderson–Kim model near $j \sim J_c$, Eq. (40), leads to the prediction that a superconducting loop that initially carries a current close to J_c will experience a current decay which becomes slower and slower with time, with a form

$$j(t) \approx J_c \left(1 + \frac{\mu k_B T}{U_0} \ln(1 + t/t_0)\right)^{-1/\mu}, \tag{42}$$

where t_0 is a microscopic time scale.

If disorder is sufficiently strong, it may obliterate the Abrikosov lattice altogether. In this case it is thought that at low temperatures a *vortex glass* phase replaces the Abrikosov lattice phase.[26] The properties of this phase, which has many consequences for high T_c cuprate superconductors, lie beyond the scope of this review. An exhaustive review article on the rich topic of vortex pinning and motion has been given by Blatter *et al.*[30]

For currents larger than J_c at low temperatures, and even for weak currents close to T_c if the disorder is small, the vortex lattice may be unpinned from the defects, and may flow freely under the Magnus force of the current. We would then expect to find a vortex drift velocity v_d proportional to the current density j, with a coefficient η_v^{-1} that depends on the temperature and the material at hand, but may be relatively insensitive to the quantity of defects. (We are concerned here only with the component of motion parallel to the Magnus force, which means perpendicular to **j**.) This gives an electrical resistivity, in the flux flow regime, given by

$$\rho = \frac{\pi \hbar B}{\eta_v e \Phi_0} = \frac{B}{\eta_v c} . \tag{43}$$

A crude estimate of η_v was obtained by Bardeen and Stephen, who modeled the vortex core as a region of normal fluid, and estimated the rate of energy dissipation in the core of a moving vortex by calculating the normal current induced by an effective field proportional to the product of drift velocity and an effective magnetic field of $\Phi_0/\pi\xi^2$. This led to an estimate

$$\eta_v \approx \sigma_n \Phi_0 / \pi \xi^2 c , \tag{44}$$

where σ_n is the electrical conductivity of the normal metal. Using the relation between ξ and the upper critical field H_{c2} and ξ, we find an approximate form for the flux-flow resistivity:

$$\rho \approx \frac{B}{\sigma_n H_{c2}} . \tag{45}$$

Despite the crude approximations involved, this formula seems to work surprisingly well in many cases.

3. Quantum Fluctuations in Junctions, Wires, and Films

In mesoscopic superconducting devices, phase slips may occur due to quantum flucutations rather than thermal fluctuations. As mentioned above, the phase of the superconducting order parameter is canonically conjugate to the charge density, or Cooper pair density. Therefore charging terms in the

Hamiltonians describing such superconducting devices will produce fluctuations of the phase variable, which can lead to dissipation. In superconducting wires and higher dimensional arrays the competition between charging effects and the Josephson coupling terms may give rise to a zero-temperature quantum phase transition, and not just to finite current dissipation. In this section we will touch upon these effects.

3.1. *Quantum phase slips in Josephson junctions*

Let us return to the model of a capacitative Josephson junction connected to an ideal constant current source, considered in Sec. 2.1. For the moment we assume there is no shunt resistor or other sources of dissipation, so that the system may be described by the Hamiltonian $\mathcal{H}$ defined in (13), with the previously stated commutation rule $[\phi, q] = 2ie$. We shall now treat the problem quantum mechanically, however, rather than taking the classical limit. It is convenient to think about wave functions which depend on the variable ϕ, so that q is represented by the operator $q = -2ei\partial/\partial\phi$.

In a capacitatively shunted Josephson junction, quantum mechanics has several effects. We shall focus on the case were $E_J > E_C$ where $E_C = 2e^2/C$ is the Cooper pair Coloumb blockade energy, and we shall first consider the situation where the external current $I = 0$. Suppose that the system is initially trapped near the cosine minimum at $\phi = 0$, so that $-E_J \cos\phi \approx \frac{1}{2}E_J\phi^2 - E_J$. The approximate Hamiltonian is now that of a harmonic oscillator, with resonance frequency given by Eq. (18), $\Omega_{JC} = (E_J E_C)^{1/2}/\hbar$. Quantum mechanics predicts a series of energy levels, separated by $\hbar\Omega_{JC}$, near the bottom of each cosine well.

The second quantum effect is the possibility for ϕ to tunnel between two adjacent wells. This process is a quantum phase slip, and its amplitude can be estimated from a WKB calculation:

$$\zeta \sim \Omega_{JC}\sqrt{S}e^{-S} \tag{46}$$

where the action barrier S is given by[14]

$$S = -\int_0^{2\pi} d\phi \left(\frac{E_J(1 - \cos\phi)}{E_C} \right)^{1/2} = 4\sqrt{2}\sqrt{E_J/E_C}\,. \tag{47}$$

For a Josephson junction consisting of two superconductors separated by an insulating layer of fixed thickness, the coupling energy E_J will be proportional to the area A of the junction, while $E_C \propto A^{-1}$. Thus the frequency Ω_{JC} is independent of A, while the action S is proportional to A.

We shall be interested in small junctions, where S is larger than unity but not so large that the tunneling rate is completely negligible.

We can now ask at which temperatures does the quantum tunneling process become more pronounced than the thermal phase slip, whose rate is given by (14) and (15). We see that quantum tunneling should become more important than thermal slip processes when $e^{-2E_J/T} < e^{-S}$, which means that T should be less than the crossover temperature $T_Q = \hbar\Omega_{JC}$.

To treat the tunneling more quantitatively, we may use the analogy with a particle in a periodic potential. When $I = 0$, the eigenstates of the Hamiltonian (13) may be characterized by a "wave vector" k in the first Brillouin zone, which is equal to the charge q modulo $2e$. For energies less than E_J we find a series of narrow tight binding bands, of width 4ζ, separated from each other by energy gaps $\approx \hbar\Omega_{JC}$. For the lowest energy band we have

$$E_k \approx -E_J + \frac{1}{2}\hbar\Omega_{JC} - 2\zeta \cos(2\pi k/2e) \,. \tag{48}$$

For energies above E_J, we find free running bands, separated by narrow energy gaps at the Brillouin zone boundaries $k = \pm e$ and at the zone center, $k = 0$.

If we now connect the junction to an ideal current source with current $I \neq 0$, we must take into account the term proportional to I in (13). We thus obtain the Hamiltonian for a quantum particle in a tilted washboard potential, like that shown in Fig. 2(a).

The current term acts like a force which changes continuously the quasi-momentum k. If I is not too large, a particle initially in a low energy state with $k \approx 0$ will accelerate until it reaches the Brillouin zone edge, $k = e$ and then will be back-scattered by a reciprocal lattice vector, into $k = -e$. (This is the origin of Bloch oscillations, where pulling on an electron in a periodic potential produces an oscillatory motion back and forth, in the absence of dissipation.) Physically, when the charge on the capacitor plates reaches e, a Cooper pair is transmitted through the junction and makes the charge $-e$, and the process of charging is repeated.[31–33] During this process the phase ϕ oscillates back and forth, but there is *zero* average voltage drop, while the time-average supercurrent is equal to the input current I.

There is, however, another type of energy eigenstate, where the particle starts out with an energy above the top of the cosine potential and then accelerates to larger and larger velocities, with increasing value of ϕ. Moreover, a particle that is initially trapped in a low-energy Bloch-oscillation state will eventually tunnel out, by a series of Zener processes through higher tight binding bands into the runaway states. The time scale for this will be very long, if I is sufficiently small, but eventually it should happen, if the system is truly described by the dissipationless washboard model. In the runaway state, the voltage steadily increases as the charge q builds up continuously on the capacitor plates, while the time average supercurrent through the junction is zero. Thus the Josephson junction has become an insulator.

3.2. *Resistively shunted Josephson junction*

The case of a junction shunted only capacitatively is clearly a rather pathological limit. In fact, the situation changes radically if we add a shunt resistor in parallel to the junction. The resistor provides a damping force which, depending on its size, can either stabilize the system in the state of localized ϕ, where it carries a supercurrent with vanishing voltage drop, or can stabilize a modified form of the runaway state, where ϕ increases linearly in time, giving rise to a finite voltage drop V, but with vanishing time-average supercurrent. In this latter case, the current I is carried entirely by the shunt resistor, so that the junction itself is essentially an insulator.

A systematic way of modeling the resistively-shunted Josephson junction (RSJ) is provided by the Caldeira–Legett model, to be described below. However, we shall first try to understand qualitatively where a transition might occur between an insulating and superconducting state in the limit of low currents. Essentially, the phase of the junction is determined by the extent of uncertainty in the phase of the junction. Since we are considering the limit of low currents, let us ignore the current source altogether, and consider what happens when a Cooper pair is transferred across the junction. The charge $2e$ creates a voltage imbalance which leads, in turn, to a current through the resistor, until the imbalance is relaxed. As current flows through the resistor, there is a voltage drop $V = I(t)R = \hbar\dot{\phi}/2e$ across the system, which causes the phase ϕ to wind. The total winding amount is:

$$\Delta\phi = \int dt\dot{\phi} = \int dt 2eI(t)R/\hbar = \frac{(2e)^2}{\hbar R}\,. \tag{49}$$

If the phase winding as a result of a Cooper-pair tunneling is higher than 2π, we expect that the phase coherence cannot be maintained. Thus we may conclude that for R larger than some critical value, of order

$$R_Q = \frac{h}{4e^2} = 6.45k\Omega \tag{50}$$

the junction will be in its insulating phase in the limit of zero current. The quantity R_Q is the "quantum resistance" associated with Cooper pairs. One can construct a dual argument and calculate the amount of charge that gets transferred across the resistor in case of a phase slip. This yields by an analogous calculation $\Delta Q = 2eR_Q/R$, and therefore indicates that for $R < R_Q$ we can have a superconducting phase, since the charge fluctuations are larger than a Cooper pair. Analysis of the Caldeira–Legget model indicates that in the limit of zero temperature and current, there is indeed a sharp transition between superconducting and insulating states, and that this transition occurs precisely at $R = R_Q$. This transition was originally predicted by Schmid[34] (see also Refs. 35 and 36) and for a related system by Chakravarty.[37]

The Caldeira–Leggett model, for $E_J \gg E_C$ is described by the imaginary time action:

$$S_{RSJ} = \frac{1}{\hbar}T\sum_{\omega}\left[\frac{1}{2E_J}\frac{\hbar^2}{4e^2}|q_\omega|^2\omega^2 + \frac{1}{2}|\omega||q_\omega|^2 R\right] - \frac{1}{\hbar}\int d\tau\zeta\cos 2\pi\frac{q}{2e}, \quad (51)$$

where the sum is over Matsubara frequencies $\omega = 2\pi n\hbar/T$ and the integral is over imaginary times τ. The first term is the inductive energy in the Josephson junction due to a current. The last term describes the hopping between two tight binding states of the junction at $\phi = 2\pi n$ and $\phi = 2\pi(n \pm 1)$. The middle term is the Caldeira–Leggett term,[38] which imitates a term describing the damping force due to a resistor. Caldeira and Leggett constructed the dissipative term by considering the Junction coupled to many oscillators, with a frequency distribution chosen to give the correct damping rate, and integrating out the oscillators.

The partition function of the RSJ, calculated as a path integral over all imaginary time trajectories with a Boltzmann factor exponential in the action (51), can be understood by expansion in powers of ζ. This yields the partition function of a one-dimensional (imaginary time) gas of interacting phase slips, with "charge" ± 1 indicating the phase-slip's direction. The "interaction" between phase-slips with charges p_1 and p_2 at times τ_1 and τ_2 give a contribution to the action of

$$G(\tau_1, \tau_2) = 2p_1p_2\frac{R_Q}{R}\log(\Omega_{JC}|\tau_1 - \tau_2|)\,. \quad (52)$$

As we found previously for vortices in a film, the logarithmic interaction between phase slips induces a phase transition between a superconducting state where phase slips are bound in neutral pairs, and a resistive state with unpaired phase slips. This transition, however, is a quantum transition between two zero-temperature ground states. The depairing transition occurs when the action to add an additional uncompensated phase slip matches its "quantum entropy" in imaginary time $\tau_T = \hbar/T$:

$$\frac{R_Q}{R}\log\Omega_{\rm JC}\tau_T - \log\Omega_{\rm JC}\tau_T = 0\,. \tag{53}$$

Therefore, when the shunt has $R > R_Q$, the junction itself is insulating, and all current is forced to go through the shunt. Quite generally, the interaction strength between two phase-slips is $2R_Q$ divided by the combined shunt dissipation. This is a useful principle for a quick analysis of quantum Josephson junction systems.

A combination of a renormalization group analysis similar to that for the Kosterlitz–thouless transition and heuristic arguments provide us with the resistance of the RSJ as a function of temperature. The phase slip fugacity renormalization is:

$$\frac{d\zeta}{d\ell} = \left(1 - \frac{R_Q}{R}\right)\zeta\,, \tag{54}$$

where the upper frequency cutoff is set at $e^{-l}\Omega_{JC}$. The initial value ζ_0 (ζ at $l = 0$) is given by Eq. (46). If we carry out the RG flow until the point $l = l^*$, where $2e^{-l}\Omega_{JC} \sim T/\hbar$, we can obtain an expression for the resistance due to quantum phase slips. If $\zeta \to \Omega_{JC}$ during any stage of the RG flow, then superconductivity in the junction breaks down. Otherwise, the probability rate of a phase-slip occuring is:

$$p_{\rm ps} \sim \left(\frac{\zeta_{l^*}}{\Omega_{JC}}\right)^2 . \tag{55}$$

The rate r of occurrence of phase slips of either sign, is given by the product of this probability and the renormalized frequency scale: $r = e^{-l^*}\Omega_{JC}p_{\rm ps}$. In the presence of a non-zero current I, the potential drop V is determined by the difference of the rates for forward and backward phase slip rates, which is a product of r and the factor $\sinh(hI/2eT)$, as in Eq. (16), if we assume $\hbar I \ll eT$. These arguments lead to a linear resistance of the Josephson junction (to be understood as parallel to the shunt resistor)[39]:

$$R(T) \sim R_Q\left(\zeta_{l^*}/\Omega_{JC}\right)^2 \sim R_Q\left(\zeta_0/\Omega_{JC}\right)^2 \frac{1}{T^{2(1-R_Q/R)}}\,. \tag{56}$$

Note that the qualitative behavior of a Josephson junction in the quantum regime depends crucially on the properties of the external circuit through the shunt resistance R. In general, if the temperature is far below the energy gap of the superconductors on either side of the junction, there should be no contribution to the shunt conductance from tunneling of excited quasiparticles across the junction. This contrasts with the results in the classical regime, where the external circuit was found to influence the pre-exponential factor but not the activation energy for resistance in the junction. Experimentally, a superconductor to insulator quantum phase transition in a single Josephson junction tuned by resistance of the external circuit has been demonstrated in Ref. 40.

We have seen that even a small shunt resistance or dissipative coupling can have major effects on the dc conductance of a Josephson junction in the quantum regime. However, in high-frequency experiments, it may be possible to ignore dissipation, if the latter can be made sufficiently small. This is the driving principle in designs to use superconducting circuits as elements to construct a quantum computer.[41–46] Although the general subject is outside the scope of this review, we mention one recent experiment where, after embedding a small Josephson junction in a superconducting circuit with high kinetic inductance, it was possible to observe coherent quantum tunneling between two adjacent wells of the $\cos\phi$ potential, with Rabi oscillations at a frequency 350 MHz.[47] (This is much smaller than the classical oscillation frequency within a well, $\Omega_{JC}/2\pi \approx 13.5$ GHz.)

Before concluding this section we would like to discuss another perspective on the interplay of quantum fluctuations and dissipation in Josephson junctions. Consider first the case of an underdamped junction in the limit where $E_J \ll E_C$, which is opposite to the regime we have been considering so far. Since the shunt resistance is large compared to R_Q, the junction will be in the usual Coulomb blockade regime, where there is an energy gap $E_B \approx E_C$ for electrical transport. The vanishing of the linear conductance of the Josephson junction in this regime appears quite natural. RG analysis, however, predicts insulating behavior of Cooper pairs for underdamped junctions even in the limit $E_J \gg E_C$, when one would naively expect Coulomb blockade effects to be suppressed. The RG argument can be formulated as follows: in the underdamped regime the probability of quantum phase slips increases with lowering the temperature as $\sim T^{-2(1-R_Q/R)}$. However the prefactor in this expression involves ζ_0, the probability of QPS at the microscopic scale Ω_{JC} (see Eq. (56)). The latter is given by Eq. (47) and is exponentially small. Thus observing insulating behavior of underdamped Josephson junctions in the regime $E_J \gg E_C$ requires working at exponentially low temperatures and currents.[40,48] We remark that nonlinear transport at non-zero voltages can be quite complicated in this regime and we shall not attempt to discuss this here. Results depend on many details of the environment.[49]

In the discussion above, Ohmic dissipation was introduced in the form of a Caldeira–Leggett heat bath of harmonic oscillators. This is the simplest quantum mechanical model which produces the correct classical equations of motion. One may also consider more realistic microscopic models of dissipation, such as quasiparticle tunneling (see e.g. Ref. 50). These models are more challenging for theoretical analysis and result in a richer set of phenomena and more complicated phase diagrams (see e.g. Ref. 51).

3.3. *Quantum phase slips in wires: the quantum K-T transition*

As we saw in Sec. 2.2, thin superconducting wires, like mesoscopic Josephson junctions, will have a finite phase-slip related resistance at any non-zero temperature. One may also ask, however, about the possibility of phase-slip events caused by quantum tunneling processes, which might be important at sufficiently low temperatures. According to our current theoretical understanding, as discussed below, an infinitely long wire of superconducting material can show a phase transition at zero temperature, as a function of wire thickness, in which superconductivity is destroyed by unbinding of phase slips in the space-time plane, analogous to the finite-temperature Berezinskii–Kosterlitz–Thouless transition in a two-dimensional film, or the zero-temperature phase transition in a single junction connected to a shunt resistor.

The simplest way to understand the phase-slip proliferation transition in a wire is to think of it as a chain of superconducting grains with self-capacitance, that are connected via Josephson junctions. Each grain roughly represents a segment of length $a \sim \xi(0)$ of the wire, and the Lagrangian describing the wire is then a sum over segments i of

$$\mathcal{L}_i = \frac{1}{2}Ca\left(\frac{\hbar}{2e}\frac{\partial \phi_i}{\partial \tau}\right)^2 + \frac{J}{a}\cos\left(\phi_{i+1} - \phi_i\right), \tag{57}$$

where C is the capacitance per unit length, and J is proportional to the one-dimensional superfluid density in the wire (we have assumed here that the capacitance to ground is more important than the capacitance between grains at the wavelengths of the important fluctuations). The *effective* impedance shunting a given Josephson junction is calculated by assuming that the other junctions are perfectly superconducting, and therefore behave as inductors for small fluctuations in the current. The effective impedance of two semi-infinite telegraph lines (one on each side of the junction) with per-length capacitance C and inductance $\hbar^2/4e^2J$ is

$$Z = 2\sqrt{\hbar/2eJC}\,. \tag{58}$$

As we discuss below, the superconductor-insulator transition happens when $Z = R_Q/2$, where the extra factor of 2 is due to the entropy arising from the spatial degree of freedom of phase slips.[52]

A more quantitative analysis of the wire can continue along the lines of the single junction analysis. The partition function of a wire, like a single junction, can be written as that of a neutral gas of interacting phase slips in space-time with a logartihmic interaction.[53,54] The strength of the interaction is determined by the effective dissipation of the chain, given in Eq. (58), and for two phase slips separated by a space-time vector (x, τ), it is:

$$G(x, \tau) = p_1 p_2 \sqrt{JC/4e^2} \log\left(\Omega_{\mathrm{JC}}\left(x^2/v_{MS}^2 + \tau^2\right)^{1/2}\right), \tag{59}$$

where $v_{MS}^2 = (4e^2/\hbar^2)J/C$ is the Mooij–Schoen mode: the speed with which phase fluctuations propogate in the superconducting wires. By thinking of the imaginary time direction τ as a second space direction, we see that this Lagrangian coincides with the energy density of films, Eq. (4). Phase slips are clearly the space-time analog of vortices in 2D films. Since we now have the restriction $|\tau| < T/\hbar$, a quantum chain at finite T corresponds to the classical behavior of a film of finite width.

If we use $y = v_{MS}\tau$, the dimensionless stiffness K of the film is replaced by

$$K_Q = \sqrt{JC/4e^2}\,. \tag{60}$$

With this classical-quantum mapping, we can infer all properties of the wires. We can use an RG analysis to describe the flow of the plasmon-interaction strength K_Q, and a phase-slip fugacity ζ, as we integrate out modes of the phase ϕ with large frequencies and wave vectors, and rescale both space and time. Skipping technical details, the flow equations one obtains are

$$\begin{aligned} \frac{dK_Q}{dl} &= -\frac{\pi}{2}K_Q^2\zeta^2 \\ \frac{d\zeta}{dl} &= \zeta\left(2 - \frac{K_Q}{2}\right) \end{aligned} \tag{61}$$

If we expand about the transition point $K_Q = 4$, these equations have the same form as the Kosterlitz–Thouless flow equations. At $T = 0$, if the initial value of K_Q is sufficiently large, and ζ is small, one flows to a point on a "fixed line", with $\zeta = 0$ and $K_Q > 4$. This implies that the wire is a superconductor at $T = 0$, with a renormalized value of the superfluid density, or equivalently of J, which is related to K_Q by Eq. (60). For temperatures that are non-zero but sufficiently small, one finds a resistivity that decreases

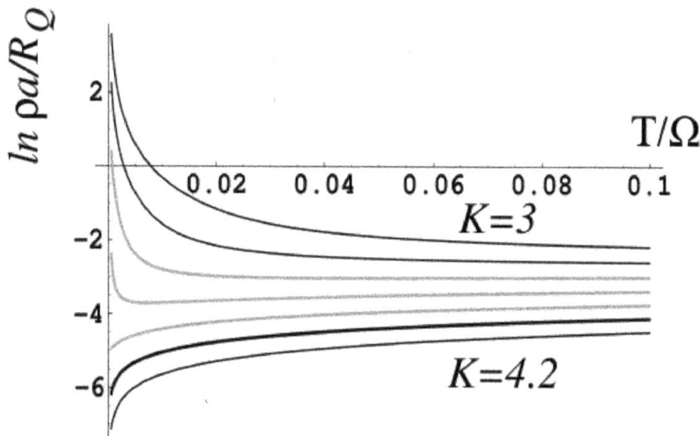

Fig. 3. Resistivity of a superconducting wire as a function of temperature in the vicinity of a Berezinskii–Kosterlitz–Thouless zero-temperature phase slip unbinding transition. The parameter K takes the values $K = 3, 3.2, 3.4, \ldots, 4.2$ with $K = 4$ being the quantum critical point. The gray lines correspond to the insulating phase of the wire, but, interestingly, they initially show a reduction in resistance as temperature decreases, and only at the lowest temperatures their resistivity curves up. In experiments, such effects will give rise to non-monotonic behavior of the resistivity.

with T according to the power law[55]

$$\rho(T) \sim T^{K_Q - 3}. \tag{62}$$

If at some point in the RG flow, the value of K_Q becomes smaller than 4, the fugacity ζ will begin to increase, and K_Q will then decrease to zero. (This can happen if the wire is too thin.) The wire will then be an insulator at $T = 0$. Mirroring the behavior of ζ, one predicts that for wires that are slightly on the insulating side of the transition, the resistivity should first decrease and then increase with decreasing temperature, eventually diverging as $T \to 0$. Figure 3 shows the traces of resistance versus temperature according to the K-T RG picture.

3.4. *Experiments on nanowires*

Superconductivity and quantum phase slips in ultra thin quantum wires were investigated in several experiments recently.[56–59] Particularly germane to the discussion above were the nanowire experiments done by the Tinkham and Bezryadin groups on MoGe amorphous nanowires (see Ref. 60 for a review). These experiments followed the resistance as a function of temperature

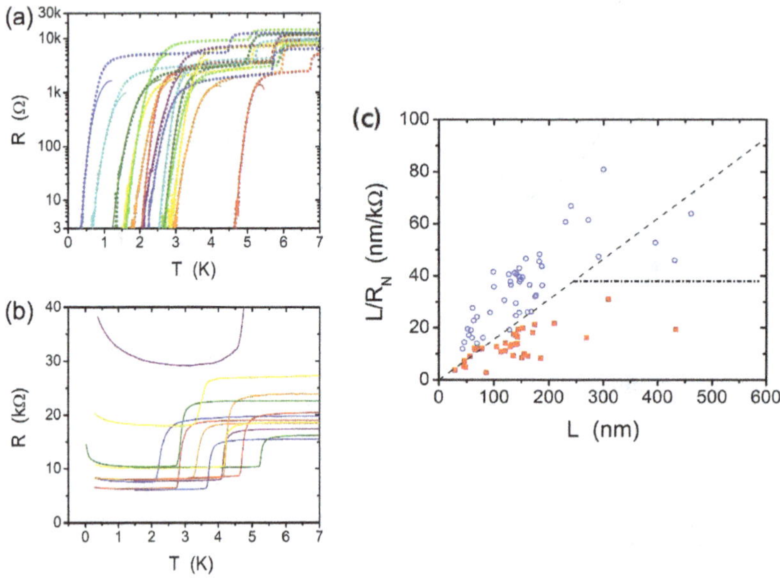

Fig. 4. MoGe nanowires experiments, taken from Ref. 59. (a) Resistance versus temperature of superconducting samples. (b) Resistance versus temperature of insulating samples. (c) Phase diagram of all wires in a and b in terms of the normal state resistance (the resistance just after the leads turn superconducting, indicated by the "knee" in the $R(T)$ curves) and the conductivity. The dashed line corresponds to $R_N = R_Q$, and the dashed dotted line is added here as the suggested long-wire critical conductance.

for wires of various lengths (100 nm to 1 μm) and widths (5 nm to 25 nm). Figure 4 summarizes some of these experiments.

The MoGe nanowire experiments clearly showed a transition between weakly insulating behavior and superconducting behvior at low temperature. Furthermore, the persistant resistance of the wires at low temperatures indicated that this transition is driven by quantum fluctuations. As Fig. 4 shows, the location of the transition is consistent with a global transition at $R_N = R_Q = 6.5k\Omega$ for short wires ($L < 200$ nm), where the wires behave like single shunted junctions, and with a local infinite-wire like transition for longer wires. This behavior is expected on the basis of phenomenological two-fluid models,[61–63] and provides support to the theory of quantum phase-slip proliferation.

Nevertheless, a point of controversy is the detailed temperature dependence of the resistance on the temperature. The measurements of Bezryadin, in particular, show a rapid decay of resistance with decreasing temperature for short superconducting nanowires,[16] contrary to the expectation of a power-law decay [Eq. (56)]. In Ref. 64 it is shown that taking into account a finite density of phase slips in short wires in a self-consistent way, by modifying the effective shunting resistance to be $R_{\text{eff}} = R_N + \alpha\zeta^2$, with ζ being the phase slip fugacity, indeed produces sharp declines of the resistance with decreasing temperatures in good agreement with the experiments.

While it is beyond the scope of this review, we would like to emphasize that several other attempts to describe the behavior of MoGe nanowires were made. In particular, Ref. 65 showed that if pair-breaking effects which give rise to dissipation in the nanowires exist, one can describe the nanowires using the Hertz–Millis field theory which gives rise to a universal conductance at the superconductor-insulator transition point. It was later shown by Vojta's group[66] that any amount of disorder would drive this field theory into an infinite-randomness phase, which implies exotic scaling properties not yet compared to experiment. Additionally, we must mention that quantum effects are also expected to affect the mean-field transition temperature (neglecting quantum phase slips) of thin superconducting wires. The theory for this suppression was developed by Finkel'stein and Oreg, and shows good agreement with experiment.[67]

3.5. *Quantum phase transitions in films*

The study of superconductivity in thin films at low temperatures is of particular contemporary interest. Quantum effects may drive thin superconducting films into a resistive and even insulating states at low temperatures. Similarly, disorder, which is always present in experimental realizations of thin films, plays a crucial role in the fate of superconductivity in films.

Roughly speaking, films from superconducting materials undergo two classes of superconducting-insulating transitions: magnetic-field induced, and disorder induced. In both cases the transition is between a superconducting phase with a vanishing resistance at $T \to 0$, and an insulating phase with a diverging resistance as $T \to 0$. Furthermore, a rough separation is made of films that undergo such a transition into two classes: granular and amorphous films. We leave the review of the observed phenomena in films to the review article by A. M. Goldman,[68] also included in this collection. Below we will briefly recount some of the guiding principles in this topic.

The disorder effects on the superconducting state in amorphous films are observed as a suppression of the critical temperature T_c as the thickness of the film is reduced.[33,69,70] By and large, this phenomenon is explained by Finkelstein's analysis of the mean-field transition in a thin diffusive metallic film.[67,71] The main idea is that Coloumb interactions suppress the transition to the superconducting state more efficiently in diffusive thin films since the time by which charge fluctuations can relax is longer. For a review see Ref. 71.

In granular films, and in films in a finite magnetic field, it is expected that Cooper pairs form, but fail to establish phase coherence due to phase fluctuations induced by disorder and Coulomb interactions.[72] These phase fluctuations would give rise to a transition between a superconducting state at low fields or when disorder is weak, and an insulating state at the opposite limits. Some examples of disorder induced transitions are given in Ref. 73. The magnetic field induced transitions occur in materials such as InO[74,75] and TiN,[76,77] and produces insulating states with a staggering resistance in excess of $R_{\square} \sim 1G\Omega$.

One illuminating, albeit only qualitative, picture for the quantum-fluctuations induced transition is given in terms of vortices. A neutral gas of vortices describes quantum fluctuations in the zero-field limit, and in a finite normal magnetic field, there must be a net density of vortices. A formal duality maps the field theory of a bosonic gas (e.g. the Cooper pairs) to a field theory of a gas of vortices, which are also considered bosonic.[78] Since the two theories are suspected to have similar universal properties with regards to the formation of a condensed state, it also suggests that at the superconducting transition the resistance per square of the film should be of order $R_Q = h/4e^2$ (assuming a small Hall angle).[79]

To roughly see how it comes about, let us discretize the film into an array of Josephson junctions. Qualitatively, the film can be described *either* in terms of the number of bosons (Cooper pairs) in each grain and their conjugate phases ($n_i^{(CP)}$, ϕ_i) *or* in terms of the number of vortices in each plaquette and their conjugate phases ($n_i^{(V)}$, θ_i). The transition between the Cooper-pair superfluid and insulator is also a transition between a localized-vortex phase and a vortex superfluid. At the transition, both bosonic gases are diffusive, and their diffusion times should also be similar. If we consider the resistance of the film, we can also concentrate on a single representative bond (the 2D geometry guarantees that this would also be the resistance per square). The current across such a junction is $I = 2e/\tau_{CP}$ with τ_{CP} the time for a Cooper pair to cross the junction. Alternatively, in terms of vortex motion, the voltage across the junction is $V = \frac{\hbar}{2e}\frac{d\Delta\phi}{dt} \approx \frac{\hbar}{2e}\frac{2\pi}{\tau_V}$, with τ_V the time constant for vortex motion across the junction. If vortices and

Cooper pairs are close to being self dual at the transition, then we expect $\tau_V \sim \tau_{CP}$, and therefore:

$$R_c = \frac{V}{I} = \frac{\hbar}{2e}\frac{2\pi}{\tau_V}\frac{\tau_{CP}}{2e} \approx \frac{h}{4e^2}\,. \tag{63}$$

Let us write the Cooper-pair Hamiltonian, and its dual, the vortex Hamiltonian, explicitly. For simplicity, we will assume there is no disorder in the discretized model, and that the Cooper-pairs only experience short range repulsion. The Cooper-pair degrees of freedom in a granular array have the Hamiltonian:

$$\mathcal{H} = -\sum_{\langle ij\rangle} J\cos(\phi_i - \phi_j + \Phi_{ij}) + U(n_i)^2 \tag{64}$$

J is the (nearest neighbor) Josephson coupling, Φ_{ij} accounts for the (physical) vector potential between the grains, U is the charging energy of each grain. This Hamiltonian can be recast in terms of the vortex density, $n^{(V)} = \frac{1}{2\pi}\nabla \times \nabla\phi$, and an angle θ_i which is conjugate to the vortex density. Note that in the vortex context the index i refers to plaquettes bounded by Josephson junctions. The Hamiltonian is found to be

$$\mathcal{H}_V \approx -\sum_{\langle ij\rangle} t^{(V)}\cos(\theta_i - \theta_j) + U^{(V)}\delta n_i^{(V)}\delta n_j^{(V)}\ln(r_i - r_j) \tag{65}$$

where $\delta n_i^{(V)} \equiv n_i^{(V)} - B/\Phi_0$. The first term in (65) describes the hopping of vortices between adjacent plaquettes, with hopping strength $t^{(V)}$; the stronger the charging interactions (and, in principle, disorder) in the sample are, the larger is t_V. (An estimate of t_V for a pure BCS two-dimensional superconductor is described in Ref. 80.) The vortex-interaction parameter is roughly $U^{(V)} = \pi J$.

The vortex Hamiltonian (65) and the Cooper-pair Hamiltonian (64) are both bosonic, but their details differ. Thus the Cooper-pair-vortex duality is not an exact self-duality. Nevertheless, it is thought that the two actions are sufficiently similar that the resistance of films at the superconductor insulator transition should be close to the value R_Q that an exact self-duality would indicate.

The superfluid-insulator transition in 2D systems is still not fully understood, and the vortex-Cooper-pair duality is so far a guiding principle more than a theory. The experimental situation is made even more complicated due to some samples exhibiting resistance saturation in parameter regimes between the superconducting and insulating regimes, as the temperature sinks below $T \sim 100$ mK,[75,81,81,82] which by some is thought to be an intervening exotic metalic phase[83,84] whose origins are unknown. While our discussion above was referring to superconducting films, it mostly applies to superfluid Helium films as well. Like superconducting films, the appearance of superfludity in Helium on strongly disordered substrate such as vycor is still not fully clarified. (See e.g. Refs. 85 and 86).

4. ac Conductivity

As was mentioned in the introduction, at non-zero temperatures, in the presence of a non-zero potential gradient, there will be a contribution to the electrical current from the motion of thermally excited quasiparticles. This normal fluid contribution will be negligible compared to the supercurrent in a dc measurement, if the phase slip rate η is sufficiently small, since the potential gradient itself will be vanishingly small in this circumstance. In an ac experiment, however, the supercurrent will be accompanied by a non-zero reactive voltage even in the absence of vortex motion, and this voltage will lead to a non-zero normal current with associated dissipation.

In the absence of vortex motion, we can write the current $\mathbf{j}$ induced by a weak electric field $\mathbf{E}$ at frequency ω as $\mathbf{j} = \sigma(\omega)\mathbf{E}$, where

$$\sigma(\omega) = \frac{i\rho_s}{m\omega} + \tilde{\sigma}_n(\omega) \tag{66}$$

where ρ_s is the superfluid density, m is the electron mass, and $\tilde{\sigma}_n$ is the conductivity of the normal fluid, which approaches a non-zero real constant for frequencies below the scattering frequency of the quasiparticle excitations. The superfluid density is related to the previously defined quantity K by $\rho_s/m \equiv (2e)^2 K$. The first term in (66) defines a "kinetic inductance" for the superconductor.

The combination of the Cooper-pair inductance, and the normal electrons' dissipation has important consequences for electronic applications such as resonators and microwave cavities. The dissipated power per unit volume in a superconducting material with an ac current density j is given by ρj^2, where the ac resistivity ρ is defined by $\rho = \mathrm{Re}[1/\sigma(\omega)] \approx \omega^2 \tilde{\sigma}_n m^2/\rho_s^2$. Thus, in microwave cavities made of a superconducting material, the normal fluid will be responsible for losses on its surface. The skin depth δ of the radiation $\approx (mc^2/4\pi\rho_s)^{1/2}$ which is independent of ω, and since the total ac current per unit area of the surface for a given intensity of the incident microwave power is also independent of ω, the ratio between the power absorbed in the surface and the incident power is proportional to $\omega^2\tilde{\sigma}_n$.

In standard BCS superconductors, the minimum energy to create a quasiparticle is given by the energy gap Δ. The value of $\tilde{\sigma}_n$ is proportional to the number of excited quasiparticles, which goes to zero as $e^{-\Delta/T}$ at low temperatures. However, one can also have superconductivity without an energy gap, as in the presence of magnetic impurities. In such cases, the normal fluid conductance will not vanish exponentially at low temperatures, and a more complicated analysis is necessary. See e.g. Refs. 87 and 88. Of course for frequencies larger than $2\Delta/\hbar$ there would be power absorption even at $T = 0$.

5. Systems of Ultra-cold Atoms

Recent experiments with ultra-cold alkali atoms have opened a new chapter in the study of nonequilibrium dynamics of superfluids. Two features of such systems make them particularly suitable for studying dynamical phenomena: complete isolation from the environment and characteristic frequencies of the order of kHz, which are readily accessible to experimental analysis. Dynamics of atoms in optical lattices[89] has particularly close connection to the issues that we discussed earlier in the context of superconductors. Here we restrict our discussion to bosonic atoms although interesting experiments have also been done with fermions (see e.g. Ref. 90).

Optical lattices are created using standing waves of laser beams that provide an artificial periodic potential for the atoms. The strength of the optical potential can be controlled and atomic systems can be tuned between the regimes of a weak lattice, where kinetic energy dominates, and strong lattice, where repulsive interactions between atoms play the dominant role.[89,91] In the former case, one finds BEC of weakly interacting atoms and a macroscopic occupation of the state with quasimomentum $k = 0$. In the latter case, the system is in the Mott insulating state which has no long range phase coherence and strongly reduced number fluctuations. The transition between the two phases is an example of the quantum fluctuations driven phase transition which we discussed in Secs. 3.3 and 3.5. Such a transition was first observed by Greiner *et al.*[92] by measuring momentum occupation numbers in the so-called time-of-flight experiments.

Experimentally, one can also prepare cold atoms systems moving with respect to the optical lattice and study the decay of the current. Such experiments are very similar in spirit to the critical current measurements in the case of superconductors. In the weakly interacting regime, one expects the critical current to be determined by the inflection point of the single particle dispersion.[93–95] Beyond the inflection point, the effective mass becomes negative, which is equivalent to a change of interaction from repulsive to attractive, so small density fluctuations become amplified making the system unstable to fragmentation. Close to the superfluid-to-Mott transition we expect the critical current to go to zero continuously. So the question is how to connect the two regimes. This was analyzed both theoretically[96] and experimentally,[97–99] and direct signatures of both thermal and quantum phase slips have been observed.

6. Conclusion

We have seen that the mechanisms for production of resistance in superconducting materials are understood in broad outline. However, there remain many open questions, particularly at very low temperatures, when quantum fluctuations are important. In such situations, the dynamics of phase slips or vortex motion will be sensitive to couplings to sources of dissipation, internal or external, possibly at multiple frequencies, and we have only a limited understanding of how this occurs in actual experiments. Open questions exist even for presumably classical problems, such as the resistance of a superconducting wire close to near T_c, where the simple time-dependent Ginzburg–Landu model seems to work much better than it should.

There are many open questions regarding the role of disorder in the classical as well as quantum regimes. We understand only partially the collective pinning that results from the interplay of disorder and vortex-vortex repulsion for a type-II superconductor in a strong magnetic field. Issues of how to increase pinning and decrease flux creep are of great importance for practical applications of superconductors in the areas of power transmission and high field magnets.

Review articles and books, as well as articles in the original literature, point to open issues in the field. Among the helpful examples are Refs. 49, 54, 60, 100–103, 104.

Acknowledgments

The authors have benefited from discussions with many people on topics of this review, including, in recent years, M. Tinkham, Y. Oreg, A. Bezryadin, I. Aleiner, D. S. Fisher, M. P. A. Fisher, V. Galitskii, and C. L. Lobb. They acknowledge support from the Packard Foundation, a Cottrell Fellowship from the Research Corporation, and NSF grants DMR-0906475 and DMR-0705472.

References

1. M. Tinkham, *Introduction to Superconductivity* (Dover, New York, 2004).
2. R. E. Packard, *Rev. Mod. Phys.* **70**(2), 641 (1998).
3. A. Griffin, D. Snoke and S. Stringari (eds.), *Bose-Einstein Condensation* (Oxford University Press, 1995).
4. F. Dalfovo, S. Giorgini, L. P. Pitaevskii and S. Stringari, *Rev. Mod. Phys.* **71**(3), 463 (1999).
5. A. J. Leggett, *Rev. Mod. Phys.* **73**(2), 307 (2001).
6. I. Bloch, J. Dalibard and W. Zwerger, *Rev. Mod. Phys.* **80**(3), 885 (2008).
7. G. Agnolet, D. F. McQueeney and J. D. Reppy, *Phys. Rev. B* **39**, 8934 (1989).
8. S. D. Sarma and A. Pinczuk (eds.), *Perspectives in Quantum Hall Effects* (Wiley, New York, 1997).
9. J. P. Eisenstein and A. MacDonald, *Nature* **432**, 691 (2004).
10. Y. M. Ivanchenko and I. A. Zil'berman, *JETP Lett.* **8**, 113 (1968).
11. V. Ambegaokar and B. I. Halperin, *Phys. Rev. Lett.* **22**(25), 1364 (1969).
12. J. S. Langer and V. Ambegaokar, *Phys. Rev.* **164**(2), 498 (1967).
13. D. E. McCumber and B. I. Halperin, *Phys. Rev. B* **1**(3), 1054 (1970).
14. D. S. Golubev and A. D. Zaikin, *Phys. Rev. B* **64**, 014504 (2001).
15. R. S. Newbower, M. Tinkham and M. R. Beasley, *Phys. Rev. B* **5**, 864 (1972).
16. A. Rogachev and A. Bezryadin, *Appl. Phys. Lett.* **83**, 512 (2003).
17. B. I. Halperin and D. R. Nelson, *J. Low Temp. Phys.* **36**, 599 (1979).
18. V. Ambegaokar, B. I. Halperin, D. R. Nelson and E. D. Siggia, *Phys. Rev. B* **21**(5), 1806 (1980).
19. J. M. Kosterlitz and D. J. Thouless, *J. Phys. C: Solid State Phys.* **6**(7), 1181 (1973).
20. V. L. Berezinskii, *JETP* **34**, 610 (1972).
21. J. V. José, L. P. Kadanoff, S. Kirkpatrick and D. R. Nelson, *Phys. Rev. B* **16**(3), 1217 (1977). doi: 10.1103/PhysRevB.16.1217.
22. J. M. Kosterlitz, *J. Phys. C* **7**, 1046 (1974).
23. S. L. Chu, A. T. Bollinger and A. Bezryadin, *Phys. Rev. B* **70**(21), 214506 (2004).
24. A. T. Fiory, A. F. Hebard and W. I. Glaberson, *Phys. Rev. B* **28**(9), 5075 (1983).
25. D. R. Strachan, C. J. Lobb and R. S. Newrock, *Phys. Rev. B* **67**(17), 174517 (2003).
26. D. S. Fisher, M. P. A. Fisher and D. A. Huse, *Phys. Rev. B* **43**(1), 130 (1991).
27. A. A. Abrikosov, *JETP* **5**, 1174 (1957).
28. A. I. Larkin and Y. N. Ovchinikov, *J. Low Temp. Phys.* **34**, 409 (1979).
29. P. W. Anderson and Y. B. Kim, *Rev. Mod. Phys.* **36**(1), 39 (1964).
30. G. Blatter, M. V. Feigel'man, V. B. Geshkenbein, A. I. Larkin and V. M. Vinokur, *Rev. Mod. Phys.* **66**(4), 1125 (1994).
31. K. K. Likharev and A. B. Zorin, *Jpn. J. Appl. Phys.* **26** (1987).
32. D. Averin and K. Likharev, *J. Low Temp. Phys.* **62**, 345 (1986).
33. D. B. Haviland, Y. Liu and A. M. Goldman, *Phys. Rev. Lett.* **62**(18), 2180 (1989).

34. A. Schmid, *Phys. Rev. Lett.* **51**(17), 1506 (1983).
35. S. Bulgadaev, *JETP Lett.* **39**, 314 (1984).
36. S. Korshunov, *Sov. Phys. JETP* **65**, 1025 (1987).
37. S. Chakravarty, *Phys. Rev. Lett.* **49**(9), 681 (1982).
38. A. O. Caldeira and A. J. Leggett, *Ann. Phys.* **149**, 374 (1983).
39. S. Korshunov, *Sov. Phys. JETP* **66**, 872 (1987).
40. J. S. Penttilä, U. Parts, P. J. Hakonen, M. A. Paalanen and E. B. Sonin, *Phys. Rev. Lett.* **82**(5), 1004 (1999).
41. Y. Nakamura, Y. Pashkin and J. Tsai, *Nature* **398**, 768 (1999).
42. D. Vion, *Science* **296**, 886 (2002).
43. J. Martinis, S. Nam, J. Aumentado and C. Urbina, *Phys. Rev. Lett.* **89**, 117901 (2002).
44. I. Chioresc u, Y. Nakamura, C. Harmans and J. Mooij, *Science* **299**, 1869 (2003).
45. J. A. Schreier, A. A. Houck, J. Koch, D. I. Schuster, B. R. Johnson, J. M. Chow, J. M. Gambetta, J. Majer, L. Frunzio, M. H. Devoret, S. M. Girvin and R. J. Schoelkopf, *Phys. Rev. B* **77**, 180502 (2008).
46. E. Hoskinson, F. Lecocq, N. Didier, A. Fay, F. W. J. Hekking, W. Guichard, O. Buisson, R. Dolata, B. Mackrodt and A. B. Zorin, *Phys. Rev. Lett.* **102**, 097004 (2009).
47. V. E. Manucharyan, J. Koch, M. Brink, L. I. Glazman and M. H. Devoret, preprint, arXiv:0910.3039 (2010).
48. K. Likharev and A. Zorin, *J. Low Temp. Phys.* **59**, 347 (1984).
49. G. L. Ingold and Y. V. Nazarov, *Single Charge Tunneling*, Vol. 294, p. 21107. NATO ASI Series B (Plenum, 1992).
50. V. Ambegaokar, U. Eckern and G. Schön, *Phys. Rev. Lett.* **48**, 1745 (1982).
51. F. Guinea and G. Schön, *J. Low Temp. Phys.* **69**, 219 (2004).
52. D. Haviland, *Physica C: Superconductivity*, **352**(1–4), 55 (2001). ISSN 0921-4534.
53. R. M. Bradley and S. Doniach, *Phys. Rev. B* **30**(3), 1138 (1984).
54. R. Fazio and H. van der Zant, *Phys. Rep.* **355**, 235 (2001).
55. T. Giamarchi, *Phys. Rev. B* **46**(1), 342 (1992).
56. F. Altomare, A. M. Chang, M. R. Melloch, Y. Hong and C. W. Tu, *Phys. Rev. Lett.* **97**(1), 017001 (2006).
57. C. N. Lau, N. Markovic, M. Bockrath, A. Bezryadin and M. Tinkham, *Phys. Rev. Lett.* **87**, 217003 (2001).
58. A. Bezryadin and M. T. C. N. Lau, *Nature (London)* **404**, 971 (2000).
59. A. T. Bollinger, A. Rogachev, M. Remeika and A. Bezryadin, *Phys. Rev. B* **69**(18), 180503 (2004).
60. A. Bezryadin, *J. Phys: Condens. Matter* **20**, 43202 (2008).
61. G. Refael, E. Demler, Y. Oreg and D. S. Fisher, *Phys. Rev. B* **75**(1), 014522 (2007).
62. G. Refael, E. Demler and Y. Oreg, *Phys. Rev. B* **79**(9), 094524 (2009).
63. H. P. Büchler, V. B. Geshkenbein and G. Blatter, *Phys. Rev. Lett.* **92**(6), 067007 (2004).
64. D. Meidan, Y. Oreg and G. Refael, *Phys. Rev. Lett.* **98**(18), 187001 (2007).

65. S. Sachdev, P. Werner and M. Troyer, *Phys. Rev. Lett.* **92**(23), 237003 (2004); A. Del Maestro, B. Rosenow, N. Shah and S. Sachdev, *Phys. Rev. B* **77**, 180501 (2008).
66. T. Vojta, C. Kotabage and J. A. Hoyos, *Phys. Rev. B* **79**(2), 024401 (2009).
67. Y. Oreg and A. M. Finkel'stein, *Phys. Rev. Lett.* **83**(1), 191 (1999).
68. A. M. Goldman, Review in present volume.
69. J. M. Graybeal and M. R. Beasley, *Phys. Rev. B* **29**(7), 4167 (1984).
70. J. M. Valles, R. C. Dynes and J. P. Garno, *Phys. Rev. B* **40**(10), 6680 (1989).
71. A. M. Finkel'stein, *Physica (Amsterdam) B* **197**, 636 (1994).
72. M. A. Steiner, N. P. Breznay and A. Kapitulnik, *Phys. Rev. B* **77**, 212501 (2008).
73. A. Frydman, O. Naaman and R. C. Dynes, *Phys. Rev. B* **66**(5), 052509 (2002).
74. G. Sambandamurthy, L. W. Engel, A. Johansson and D. Shahar, *Phys. Rev. Lett.* **92**(10), 107005 (2004).
75. M. A. Steiner and A. Kapitunlnik, *Physica C* **422**, 16 (2005).
76. T. I. Baturina, D. R. Islamov, J. Bentner, C. Strunk, M. R. Baklanov and A. Satta, *JETP Lett.* **79**, 337 (2004).
77. B. Sacepe, C. Chapelier, T. I. Baturina, V. M. Vinokur, M. R. Baklanov and M. Sanquer. arXiv:0805.1356 (2008).
78. M. P. A. Fisher and D. H. Lee, *Phys. Rev. B* **39**(4), 2756 (1989).
79. M. P. A. Fisher, *Phys. Rev. Lett.* **65**(7), 923 (1990).
80. A. Auerbach, D. P. Arovas and S. Ghosh, *Phys. Rev. B* **74**, 064511 (2006).
81. N. Mason and A. Kapitulnik, *Phys. Rev. Lett.* **82**(26), 5341 (1999).
82. Y. Qin, C. L. Vicente and J. Yoon, *Phys. Rev. B* **73**(10), 100505(R) (2006).
83. J. Wu and P. Phillips, *Phys. Rev. B* **73**, 214507 (2006).
84. V. M. Galitski, G. Refael, M. P. A. Fisher and T. Senthil, *Phys. Rev. Lett.* **95**(7), 077002 (2005).
85. D. Finotello, K. A. Gillis, A. Wong and M. H. W. Chan, *Phys. Rev. Lett.* **61**(17), 1954 (1988).
86. P. A. Crowell, J. D. Reppy, S. Mukherjee, J. Ma, M. H. W. Chan and D. W. Schaefer, *Phys. Rev. B* **51**(18), 12721 (1995).
87. K. Maki, *Superconductivity* (Marcel Dekker, Inc., 1969).
88. P. G. de Gennes, *Superconductivity of Metals and Alloys* (Benjamin, New York, 1966).
89. D. Jaksch, C. Bruder, J. I. Cirac, C. W. Gardiner and P. Zoller, *Phys. Rev. Lett.* **81**(15), 3108 (1998).
90. D. E. Miller, J. K. Chin, C. A. Stan, Y. Liu, W. Setiawan, C. Sanner and W. Ketterle, *Phys. Rev. Lett.* **99**(7), 070402 (2007).
91. M. P. A. Fisher, P. B. Weichman, G. Grinstein and D. S. Fisher, *Phys. Rev. B* **40**(1), 546 (1989).
92. M. Greiner, O. Mandel, T. Hansch and I. Bloch, *Nature* p. 419 (2002).
93. B. Wu and Q. Niu, *Phys. Rev. A* **64**(6), 061603 (2001).
94. A. Smerzi, A. Trombettoni, P. G. Kevrekidis and A. R. Bishop, *Phys. Rev. Lett.* **89**(17), 170402 (2002).

95. L. Fallani, L. De Sarlo, J. E. Lye, M. Modugno, R. Saers, C. Fort and M. Inguscio, *Phys. Rev. Lett.* **93**(14), 140406 (2004).
96. A. Polkovnikov, E. Altman, E. Demler, B. I. Halperin and M. D. Lukin, *Phys. Rev. A* **71**(6), 063613 (2005).
97. C. D. Fertig, K. M. O'Hara, J. H. Huckans, S. L. Rolston, W. D. Phillips and J. V. Porto, *Phys. Rev. Lett.* **94**(12), 120403 (2005).
98. J. Mun, P. Medley, G. K. Campbell, L. G. Marcassa, D. E. Pritchard and W. Ketterle, *Phys. Rev. Lett.* **99**(15), 150604 (2007).
99. D. McKay, M. White, M. Pasienski and B. DeMarco, *Nature* **453**, 76 (2008).
100. G. Schön and A. D. Zaikin, *Phys. Rep.* **198**, 238 (1990).
101. K. K. Likharev, *Dynamics of Josephson Junctions and Circuits* (Gordon and Breach Publishers, 1986).
102. K. Arutyunov, D. Golubev and A. Zaikin, *Phys. Rep.* **464**, 1 (2008).
103. E. Simanek, *Inhomogeneous Superconductors* (Oxford University Press, Oxford, 1994).
104. T. Giamarchi, *Quantum Physics in One Dimension* (Oxford University Press, 2004).

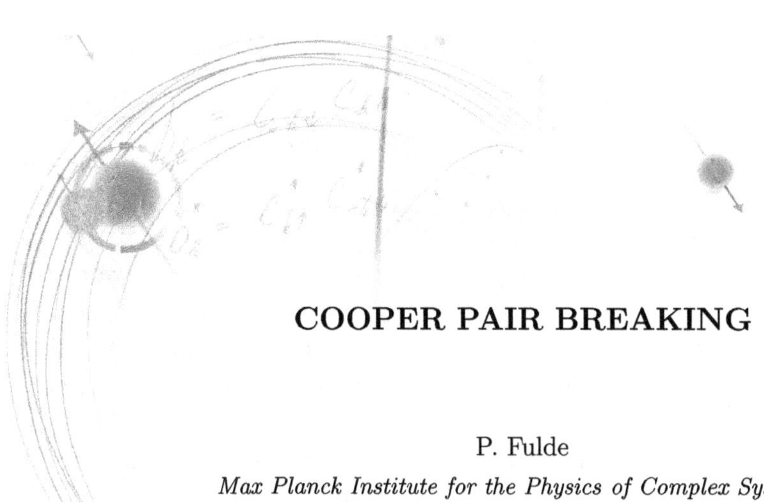

COOPER PAIR BREAKING

P. Fulde
Max Planck Institute for the Physics of Complex Systems,
Nöthnitzer Str. 38, 01187 Dresden, Germany
and
Asia Pacific Center for Theoretical Physics,
POSTECH, Pohang, Korea
fulde@pks.mpg.de

An overview is given of a number of pair-breaking interactions in superconductors. They have in common that they violate a symmetry of the pair state. In most cases pairs are formed from time reversed single-particle states, a noticeable exception being antiferromagnetic superconductors. When time reversibility is broken by an interaction acting on the electrons, the time evolution of the time-reversal operator plays an important role. Depending on whether it is nonergodic or ergodic, we deal with pair weakening or pair breaking. Numerous different interactions are analyzed and discussed. Unifying features of different pair-breaking cases are pointed out. Special attention is paid to the Zeeman effect and to scattering centers with low-energy excitations. The Kondo effect and crystalline field split rare-earth ions belong in that category. Modifications caused by strongly anisotropic pair states are pointed out. There is strong evidence that in some cases intra-atomic excitations lead to pair formation rather than pair breaking for which an explanation is provided.

1. Introduction

During the second part of the '50s several groups in Göttingen and at Bell Laboratories realized that small amounts of paramagnetic impurities, i.e. transition-metal or rare-earth ions have profound effects on the transition temperature T_c of superconductors.[49,50,58,64] This temperature was found to decrease rather rapidly with increasing impurity concentration n_I. For example, 0.6% of Gd are sufficient to suppress superconductivity in $LaAl_2$ ($T_c = 3.2$ K) completely. This stimulated F. Reif and the student M. Woolf to start work on tunneling on quenched In and Pb films in order to study the behavior of the density of states as function of impurity concentration.[59] The results were truly surprising. Not only was a rapid decrease of the superconductivity energy gap observed as T_c decreased with increasing n_I of Gd and Fe, but it was also found that for a range of impurity concentrations the gap disappeared completely despite the fact that the probes were still superconducting. The experiments revealed clearly that a gap in the excitation spectrum is not a prerequisite for the existence of a superconducting state. Rather, it is the order parameter which distinguishes it from the normal state. This is by now common knowledge. But the phenomenon of gapless superconductivity was in 1962 rather unexpected. Not long thereafter it was realized that there already existed a theory by Abrikosov and Gorkov[2] which predicted what was found experimentally. Since at that time communication between physicists in the Soviet Union and their western colleagues were rather limited, that theory had remained rather unnoticed. Another reason might have been that the theory was formulated in an elegant Green's function formalism in which only a very small number of condensed matter theorists in western countries were familiar with at that time. This changed very fast after the potential of that approach was realized (see, e.g. Refs. 4, 39, 40, 68). Nevertheless, it was of help to many working in the field that an alternative approach to gapless superconductivity based on the time evolution of the time-reversal operator was presented by de Gennes.[17] He drew attention to the fact that the exchange interaction between conduction electrons and magnetic ions changes sign when a time-reversal transformation is applied to the conduction electron system. The theory is very physical and simple but not as easily applicable to a wide range of problems as is the Green's function approach.

It was very soon realized that pair breaking and gapless superconductivity are rather general phenomena. Any external perturbation acting on Cooper pairs will lead to pair breaking when it is not time-reversal invariant. In the case of paramagnetic impurities it is the exchange interaction between the conduction electrons and the impurity which violates time-reversal symmetry. The reason is apparent. Since Cooper pairs are usually formed from time-reversed one-particle states,[5,20] any perturbation acting differently on the partners of a pair will destabilize the paired state and break it. (Note that pairing does not always take place in time reversed states. In an antiferromagnetic superconductor electrons are paired in states which are time reversed followed by a lattice translation connecting the two sublattices.[6]) Another example of time symmetry breaking is an external magnetic field applied either perpendicular to the surface of a type II superconductor[15,18,41] so that Abrikosov vortices form or applied parallel to a thin film.[39,41,75] Details of pair breaking vary considerably here, depending on mean free path and coherence length of the system.[17] Since a magnetic field acts on the orbit of an electron as well as on its spin (Zeeman effect), the relative size of the two effects plays an important role (Maki parameter).[43,76] Also the interface between a superconducting and a ferromagnetic film acts as a pair breaker and so does a supercurrent under certain conditions.

Numerous experiments have been done in which several pair-breaking perturbations act at the same time on a superconductor. An example are paramagnetic impurities and a magnetic field.[14,30] The magnetic field can spin-polarize the impurities and spin-polarized impurities act differently on Cooper pairs than unpolarized ones. Nevertheless, there are many situations where we may simply add the characteristic pair-breaking parameters of the different perturbations in order to describe their combined effect.[25]

Let us return to the starting point, i.e. to the pair-breaking properties of magnetic impurities. When the magnetic ions are subject to the Kondo effect they form singlet states with the conduction electrons. In the presence of Cooper pairing the singlet formation is affected. This may result in reentrant superconductivity, an effect pointed out by Müller-Hartmann and Zittartz[57] and measured by Maple *et al.*[33,47] Thereby a superconducting state appears at a transition temperature T_{c1} but disappears again at a lower temperature $T_{c2} < T_{c1}$. Eventually it can reappear again at an even lower temperature $T_{c3} < T_{c2}$.

When we deal with paramagnetic rare-earth ions, e.g. Pr^{3+} ions their incomplete $4f$ shell is in a Hund's rule ground-state multiplet, in the present case with total angular momentum $J = 4$. Under the influence of the crystalline electric field (CEF) of the neighboring ions this manifold is split into different CEF levels, with level spacings of order few to tens of meV. When those ions are in a superconductor, T_c and the CEF splittings are of similar magnitude and influence each other. Thereby the form of the dominant interaction between the conduction electrons and the CEF split energy levels plays a decisive role. When the exchange interaction is most important we expect Cooper pair breaking since the interaction is not time-reversal invariant with respect to the conduction electrons. However, when aspherical Coulomb charge scattering is more important we expect support of Cooper pair formation, because the interaction is time-reversal invariant. In that case the inelastic intra-atomic scattering processes have a similar effect on pair formation as Einstein oscillators, i.e. localized phonons.[23]

A number of interesting effects result from the Zeeman term in a magnetic field.[10,13,46,61] If it is small it causes a Zeeman splitting of the BCS density of states, an effect pioneered by Meservey and Tedrow.[52,72] With increasing field strength eventually Cooper pairs are broken. But before the system goes into the normal state it may go over into a gapless *inhomogeneous* state.[22,37] We face the problem of how to pair electrons when the constituents have a different chemical potential (FFLO state). In order for this state to be formed, requirements on the mean-free path and on the Maki parameter have to be met. Since that topic is covered in another contribution to this volume, we omit it here. Related to the Zeeman-interaction term is also the Jaccarino-Peter effect.[34] It is based on a partial compensation of the pair-breaking exchange interaction of polarized magnetic impurities and the Zeeman term due to an applied field. Therefore it may happen that superconductivity is no longer completely suppressed by a given concentration of magnetic ions and will reappear when the field is sufficiently high. At even higher fields, superconductivity is eventually destroyed due to the pair-breaking effect of the field. This reentrance of superconductivity was indeed observed by Meul *et al.*[53] and Hiraki *et al.*[31]

In particular in the Ginzburg-Landau regime but not only there, numerous physical effects caused by pair-breaking external perturbations have been calculated and tested by experiments. Examples are ultrasonic attenuation, Knight shift, electrical and thermal conductivity, and nuclear-spin relaxations to name a few. What is astonishing is the high degree of accuracy with which experiments could be explained and partially predicted. It is rarely matched by other theories in condensed matter physics, quantum Hall physics excluded. The robustness and accuracy of the BCS theory is the basis of these findings.

As pointed out above, the interactions between CEF-split magnetic ions and conduction electrons contain parts, e.g. a spherical charge scattering, which are time-reversal invariant and therefore favor pair formation instead of pair destruction. However, they are very often dominated by the time-reversal symmetry violating exchange interaction. But there is one clear-cut exception: the filled skutterudite $PrOs_4Sb_{12}$ has a T_c of 1.85 K, while the corresponding compound $LaOs_4Sb_{12}$ has a T_c of 0.74 K only. The only difference between the two systems are the $4f^2$ electrons of Pr^{3+} which are absent in La. Therefore the $4f$ electrons must provide the additional glue for Cooper pairs.[11] An even more surprising case is UPd_2Al_3, where there is clear evidence that non-time-reversal invariant magnetic scattering of itinerant electrons on localized ones results in superconductivity.[51] However, that implies a non-s wave pairing with an order parameter which changes sign on translation through $Q_z = \pi/c$. Here c is the lattice constant perpendicular to the layers of U. An order parameter with a nodeline was indeed confirmed experimentally.[32,36,48]

A review of pair-breaking in superconductivity which summarizes the situation of the theory in the late '60s was given by Maki,[44] a pioneer in the field. Many experimental results are also found in a review by Alloul *et al.*[3] In the following we want to discuss some of the most important concepts.

2. Time-reversal Symmetry Breaking

The Hamiltonian of an interacting electron system is time-reversal invariant, i.e.

$$[H, T_R]_- = 0\,. \tag{1}$$

Here $T_R = -i\sigma_y C$ is the time-reversal operator with C changing a function into its complex conjugate and σ_y denoting a Pauli matrix. This implies that for each one-particle eigenstate $H\psi_n = \epsilon_n \psi_n$ there exists a time reversed one $\psi_{\overline{n}} = T_R \psi_n$ with the same energy (Kramers' theorem), i.e. $H\psi_{\overline{n}} = \epsilon_n \psi_{\overline{n}}$. These two states can be paired to a ground state

$$|\Phi_{\mathrm{BCS}}\rangle = \prod_n \left(u_n + v_n c_n^+ c_{\overline{n}}^+\right)|0\rangle\,, \tag{2}$$

where c_n^+ creates an electron with wavefunction ψ_n. When momentum and spin are good quantum numbers it is $n = (\mathbf{k}, \sigma)$ and $\overline{n} = (-\mathbf{k}, -\sigma)$.

Let us assume that we add a perturbation to the system which violates time-reversal symmetry. An example is

$$H_{\mathrm{int}} = \frac{e}{2m}\sum_i (\mathbf{p}_i \mathbf{A} + \mathbf{A}\mathbf{p}_i) - \mu_B \sum_i \boldsymbol{\sigma}_i \mathbf{H} \tag{3}$$

where $\mathbf{p}_i, \boldsymbol{\sigma}_i$ are momentum and spin of the i-th electron and $\mathbf{A}$ and $\mathbf{H}$ are the vector potential and magnetic field, respectively. Another one is

$$H_{\mathrm{int}} = -(g-1)J_{\mathrm{ex}} \sum_{\mathbf{k},\mathbf{k}';\alpha\beta} c_{\mathbf{k}'\alpha}^+ \boldsymbol{\sigma}_{\alpha\beta} c_{\mathbf{k}\beta} \mathbf{J} \tag{4}$$

where g is the Landé factor and $\mathbf{J}$ is the total angular momentum of a magnetic impurity. In those cases the time evolution of $T_R(t)$ does not vanish any more and instead follows from

$$\frac{dT_R}{dt} = i\,[H_{\mathrm{int}}, T_R]_-\,. \tag{5}$$

As shown by de Gennes[17] the linearized Ginzburg-Landau equation from which the transition temperature T_c is determined can be written in the weak coupling limit in the form

$$1 = N(0)V \int d\epsilon d\epsilon' \frac{1 - f(\epsilon) - f(\epsilon')}{\epsilon + \epsilon'} \, g(\epsilon' - \epsilon) \,. \tag{6}$$

Here $N(0)V$ is the BCS parameter $f(\epsilon)$ is the Fermi function and most importantly

$$g(\epsilon) = \int \frac{dt}{2\pi} \left\langle T_R^+(0) T_R(t) \right\rangle e^{-i\epsilon t} \tag{7}$$

is the Fourier transformed correlation function of the time-reversal operator. One notices that when $H_{\text{int}} = 0$ or when H_{int} is time-reversal invariant, we obtain $g(\epsilon - \epsilon') = \delta(\epsilon - \epsilon')$ and (6) reduces to the well known BCS expression. Here we are interested in $[H_{\text{int}}, T_R] \neq 0$. In this case we must distinguish between two different long-time behaviors of the correlation function $\langle T_R^+(0) T_R(t) \rangle$. They are

$$(a) \quad \lim_{t \to \infty} \left\langle T_R^+(0) T_R(t) \right\rangle = \eta \,, \qquad 0 < \eta < 1$$

$$(b) \quad \lim_{t \to \infty} \left\langle T_R^+(0) T_R(t) \right\rangle = e^{-2t/\tau_R} \,. \tag{8}$$

In the first case one speaks of nonergodic processes while the second one implies ergodic behavior. The effect on T_c and on Cooper pairing is quite different in the two cases. Nonergodic behavior leads to *pair weakening*, while ergodic behavior results in *pair breaking*.

When pair weakening takes place the transition temperature is reduced to

$$k_B T_c = 1.13 \, \omega_D \exp\left[-\frac{1}{\eta N(0) V} \right] , \tag{9}$$

where ω_D denotes a cut-off in the frequency of the bosons which provide for the electron attractive interactions. The effect of the non-time-reversal invariant interaction is just a reduction of the effective electron–electron interaction. If, however ergodic behavior prevails we may set

$$g(z) = \frac{1}{2\pi} \frac{\tau_R}{1 + z^2 \tau_R^2 / 4} \tag{10}$$

and after integrating (6) the equation for T_c becomes

$$1 = N(0)V \sum_{n \simeq 0} \frac{1}{n + 1/2 + 1/(2\pi T_c \tau_R)} \,. \tag{11}$$

Note that here a cut off at ω_D is still missing so that the sum is divergent. After the cut off is introduced (11) becomes

$$\ln(T_c/T_{c0}) + \psi\left(\frac{1}{2} + \frac{1}{2\pi T_c \tau_R}\right) - \psi(1/2) = 0 \tag{12}$$

where $\psi(x)$ is the digamma function.[1] The transition temperature T_{c0} is obtained when $H_{\text{int}} = 0$. The last relation is the same which Abrikosov and Gorkov obtained in their original work. One finds from (12) that T_c/T_{c0} drops continuously with increasing pair-breaking parameter $1/\tau_R$ and vanishes at a critical value of

$$\left(\frac{1}{\tau_R}\right)_{\text{crit}} = \frac{\pi T_{c0}}{2\gamma}; \qquad \gamma = 1.78\,. \tag{13}$$

3. Nonergodic versus Ergodic Behavior

As pointed out above, nonergodic behavior of time-reversal symmetry breaking leads to pair weakening instead of pair breaking. Two examples are given for illustration. Consider a thin superconducting film of thickness d with a rough surface but no scattering centers inside the film, when a magnetic field is applied parallel to the film.[18] In that case

$$\begin{aligned} \frac{dT_R}{dt} &= \frac{ie}{m}(\mathbf{pA} + \mathbf{Ap})\,T_R \\ &= -i\frac{d\phi}{dt}T_R \end{aligned} \tag{14}$$

where $\phi(t)$ is the phase of the propagating electron. When the electron moves straight from one surface of the film to the other, its phase does not change (the vector potential has only a component perpendicular to the film which is symmetric with respect to the film center). Therefore the only contribution to dT_R/dt comes from the parts of the path before it hits the surface for the first time and after the last hit. Thus $\langle T_R^+(0)T_R(t)\rangle$ remains finite even in the limit $t \to \infty$. This results in pair weakening.

A second example concerns a staggered field imposed onto the conduction electrons, i.e.

$$H_{\text{int}} = -I\sum_{\mathbf{k},\mathbf{Q}} h_{\mathbf{Q}}\left(c^+_{\mathbf{k}\uparrow}c_{\mathbf{k}+\mathbf{Q}\uparrow} - c^+_{\mathbf{k}\downarrow}c_{\mathbf{k}+\mathbf{Q}\downarrow}\right)\,. \tag{15}$$

The **Q**'s are reciprocal lattice vectors of the magnetic lattice and are restricted to the first Brillouin zone. It is

$$\frac{dT_R}{dt} = 2iH_{\text{int}}T_R\,. \tag{16}$$

While $[H_{\text{int}}, T_R] \neq 0$ we find that the operator $Y = T_R R$ commutes with H_{int}, where R shifts the system by a vector connecting the two sublattices. Thus $[H_{\text{int}}, Y]_- = 0$. This is obvious: after application of T_R the spins change sign which increases their energy in the staggered field. By shifting the electrons from one sublattice to the other the spin direction is again in line with the staggered field. This implies that when $\psi_{\mathbf{k}\sigma}(\mathbf{r})$ is an eigenfunction in the staggered field so is $Y\psi_{\mathbf{k}\sigma}(\mathbf{r})$ with the same eigenvalue. Therefore we may pair $\psi_{\mathbf{k}\sigma}(\mathbf{r})$ with $e^{i\varphi}Y\psi_{\mathbf{k}\sigma}(\mathbf{r})$ where the phase φ is chosen by convenience. Being able to pair electrons properly even when $[H_{\text{int}}, T_R] \neq 0$ shows that in the presence of the interaction (15) Cooper pairs are possibly weakened but not broken.

Next we turn towards ergodic behavior of perturbations. Thereby superconductivity is discussed within the weak coupling limit. Systems with pair-breaking interactions can be divided into two groups. Group 1 includes all those cases for which a standard theory can be worked out for all temperatures. Group 2 comprises those situations for which a general pair-breaking theory can be worked out only for the Ginzburg-Landau (GL) regime where the order parameter is small. The GL equation is of the form

$$\left[\ln\frac{T}{T_{c0}} + \psi\left(\frac{1}{2}+\rho\right) - \psi\left(\frac{1}{2}\right)\right]\left\langle|\Delta(\mathbf{r})|^2\right\rangle + \frac{1}{2(2\pi T)^2} f_1(\rho)\left\langle|\Delta(\mathbf{r})|^4\right\rangle = 0 \tag{17}$$

where $f_1(\rho)$ varies from case to case. The parameter $\rho = (2\pi T\tau_R)^{-1}$ is a measure of the strength of pair breaking. The coefficient of $\langle|\Delta(\mathbf{r})|^2\rangle$ is the same as met before (see (12)) and is generic for all ergodic systems. It determines T_c. At lower temperatures the spatial variation of the order parameter requires an individual treatment which is the reason for different forms of $f_1(\rho)$.

We start with perturbations belonging to group 1. As pointed out before, for that group a theory is available in closed form for all temperatures.[2] Different pair-breaking processes lead to equivalent forms of the single-particle Green's function where they enter in form of a pair-breaking parameter $1/\tau_R$. As a consequence, the thermodynamic properties expressed in terms of this pair-breaking parameter are the same in all cases. For transport coefficients this holds true too, but with one exception. We may obtain different expressions when the relevant correlation function contains an s-wave spin triplet vertex and when in addition the momentum transfer is less than $(\ell\xi_0)^{-1/2}$ where ξ_0 is the coherence length.[45] This is the case, e.g. for the spin susceptibility $\chi_s(\mathbf{q}, \omega)$ when q is small. Otherwise different pair-breaking mechanisms are completely equivalent.[45] The equivalence holds true in particular for the density of states. In the BCS theory it is given by

$$N_S(E) = N(0)\ \mathrm{Im}\frac{E/\Delta}{\sqrt{1-(E/\Delta)^2}} \tag{18}$$

where $N(0)$ is the density of states per spin direction in the normal state. In the presence of a pair-breaking parameter $1/\tau_R$ this expression is modified to

$$N_S(E) = N(0)\ \mathrm{Im}\frac{u}{\sqrt{1-u^2}} \tag{19}$$

where $u(E)$ is a solution of the equation

$$\frac{E}{\Delta} = u\left(1 - \frac{1}{\tau_R\Delta}\,\frac{1}{\sqrt{1-u^2}}\right). \tag{20}$$

The latter has to be solved numerically and a plot is shown in Fig. 1. For parameter values $0.91 < [(\tau_R)_{\mathrm{crit}}/\tau_R] < 1$ where $(\tau_R)_{\mathrm{crit}}$ denotes the critical pair-breaking parameter at which T_c vanishes, one finds no gap in the excitation spectrum, i.e. the superconductor is gapless.

Gaplessness can again be rederived from the behavior of the time-reversal operator T_R.[17] Let us see under which circumstances the excitation energy E_n can be expanded for small order parameter Δ in the form

$$E_n = |\epsilon_n| + \left\langle |\Delta(\mathbf{r})|^2 \right\rangle \sum_m \frac{|\langle n\,|T_R|\,m\rangle|^2}{\epsilon_n - \epsilon_m} \tag{21}$$

where $|m\rangle, |n\rangle$ are single-electron states. When $[H_{\mathrm{int}}, T_R]_- = 0$ the matrix elements $\langle n|T_R|m\rangle \neq 0$ only when $|m\rangle$ and $|n\rangle$ are time reversed. Then

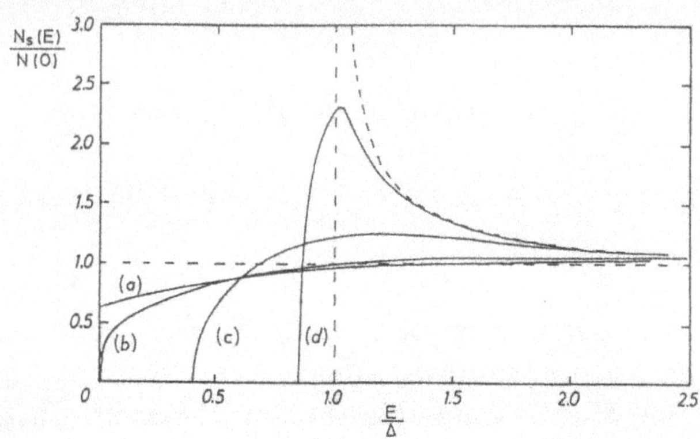

Fig. 1. Tunneling density of states (per spin) as a function of energy E/Δ for various values of τ_R^{-1}. Curves (a)–(d) correspond to $(\tau_R\Delta)^{-1} = 1.33;\ 1.0;\ 0.5;\ 0.05$, respectively. (From Ref. 68)

$\epsilon_n = \epsilon_m$ and the expression diverges. An expansion of that form is therefore not allowed. If however $[H_{\text{int}}, T_R]_- \neq 0$ than $\langle n|T_R|m\rangle \neq 0$ for a range of $|\epsilon_m - \epsilon_n| \simeq \tau_R^{-1}$. In this case we can write (21) in analogy to (6) as

$$E(\mathbf{p}) = |\epsilon(\mathbf{p})| + \left\langle |\Delta|^2 \right\rangle P \int \frac{d\epsilon'}{\epsilon(\mathbf{p}) + \epsilon'}\, g(\epsilon' - \epsilon(\mathbf{p}))$$

$$= |\epsilon(\mathbf{p})| + \frac{1}{2} \frac{\left\langle |\Delta(\mathbf{r})|^2 \right\rangle |\epsilon(\mathbf{p})|}{\epsilon(\mathbf{p})^2 + (\tau_R)^{-2}} \,. \tag{22}$$

An expansion of the form (21) is therefore possible. Note that there is no gap in $E(\mathbf{p})$ in the limit of small Δ. From the last relation we obtain

$$N_S(E) = N(0) \left[1 + \frac{\Delta^2}{2} \frac{E^2 - (1/\tau_R)^2}{\left(E^2 + (1/\tau_R)^2\right)^2} \right] \tag{23}$$

which agrees with the corresponding expression obtained from (19) and (20).

Perturbations belonging to **group 1** include:

Paramagnetic impurities. As pointed out in the introduction, superconductors with paramagnetic impurities were the first in which gapless superconductivity was observed. The Hamiltonian is the one given by (4). The spins of the magnetic impurities are treated classically. The scattering of conduction electron by the impurities is treated in Born approximation. This excludes a possible Kondo effect which is discussed later. The pair-breaking parameter is found to be

$$\frac{1}{\tau_R} = 2\pi n_I N(0)(g-1)^2 J_{\text{ex}}^2 J(J+1) \,. \tag{24}$$

The dependence of $T_c(n_I)$ follows from (12) and a plot is found below in Fig. 8.

Thin film in a parallel magnetic field. Here we have to distinguish between local and nonlocal electrodynamics. In the first case the mean free path ℓ must satisfy $\ell \ll d$ where d is the film thickness. In addition the conditions $d \ll \lambda, (\ell\xi_0)^{1/2}$ have to be fulfilled, where λ is the penetration depth. The following pair-breaking parameter is obtained[39,41]

$$\frac{1}{\tau_R} = \frac{\tau_{tr}}{18} (v_F e d H)^2 \,. \tag{25}$$

Note that τ_{tr} is the transport mean free time while v_F is the Fermi velocity. In case of nonlocal electrodynamics $d \lesssim \ell$ the expression (25) is generalized to

$$\frac{1}{\tau_R} = \frac{\tau_{tr}}{18}(v_F edH)^2 g\left(\frac{\pi\ell}{d}\right), \tag{26}$$

where $g(x) = (3/2x^3)[(1+x^2)\arctan x - x]$.[75] The limit $\ell \ll d$ reproduces (25).

Supercurrent. In case of a uniform current time-reversal symmetry is broken in the center of mass system because of scattering processes. When $\ell \ll \xi_0$ we find

$$\frac{1}{\tau_R} = \frac{2}{3}\tau_{tr} v_F^2 q^2 \tag{27}$$

where q denotes the shift of the electron distribution in momentum space in the presence of a current.[40]

In the Ginzburg-Landau regime (see (17)) the function $f_1(\rho)$ applies to all perturbations belonging to group 1

$$f_1(\rho) = -\frac{1}{2}\psi^{(2)}\left(\frac{1}{2}+\rho\right) - \frac{1}{6}\psi^{(3)}\left(\frac{1}{2}+\rho\right) \tag{28}$$

where $\psi^{(2)}(x)$ and $\psi^{(3)}(x)$ are higher derivatives of the digamma function.

Next we turn towards perturbations belonging to **group 2** according to the above classification scheme. The pair-breaking parameter τ_R^{-1} applies here to the Ginzburg-Landau regime (17) only. Two examples are discussed.

Type II superconductors. By restricting oneself to the effect of a magnetic field on the electron orbits, i.e. by neglecting the Zeeman term we obtain in the dirty limit ($\ell/\xi_0 \ll 1$),[15,18,42]

$$\frac{1}{\tau_R} = \frac{\tau_{tr} v_F^2 eH}{3}. \tag{29}$$

When a magnetic field is applied, nucleation starts at the surface. In the surface regime[60]

$$\frac{1}{\tau_R} = 0.59\,\frac{\tau_{tr} v_F^2 eH}{3}. \tag{30}$$

In the bulk as well as in the surface regime the function $f_1(\rho)$ is

$$f_1(\rho) = -\frac{1}{2}\psi^{(2)}\left(\frac{1}{2}+\rho\right). \tag{31}$$

Contact between a superconducting and a magnetic film. Boundary conditions require that, at an interface between a superconductor and a magnetic metal, the order parameter vanishes.[16] In the Ginzburg-Landau regime close to T_c the order parameter varies normal to the interface $x = 0$ like $\Delta(x) \sim \cos \frac{\pi x}{2d_s}$ where d_s is the thickness of the superconducting film which extends from $0 < x < d_s$. In the limit of $\ell \ll d_s$ and $(\ell \xi_0)^{1/2} < d_s$ the pair-breaking parameter is[24,26]

$$\frac{1}{\tau_R} = \frac{\tau_{tr} v_F^2}{6} \left(\frac{\pi}{2d_s} \right)^2 , \tag{32}$$

while $f_1(\rho)$ is given by

$$f_1(\rho) = -\frac{1}{2} \psi^{(2)} \left(\frac{1}{2} + \rho \right) + \frac{1}{18} \psi^{(3)} \left(\frac{1}{2} + \rho \right) . \tag{33}$$

By comparison with (28) one notices that the term proportional to $\psi^{(3)} \left(\frac{1}{2} + \rho \right)$ depends sensitively on the spatial variation of $\Delta(\mathbf{r})$.

4. Zeeman Splitting of Quasiparticles

When a magnetic field is applied parallel to a very thin film, e.g. of Al with 20 Å thickness, the effect of the field on the spins dominates the one on the orbits to the extent that the latter may be neglected. A similar argument holds true for type II superconductor when the κ value, i.e. the ratio of penetration depth to coherent length is very large. In both cases the transition to the normal state is of first order provided that the spin-orbital mean free path $\ell_{s0} = v_F \tau_{s0}$ is not too small. Generally the density of states splits into two parts

$$N_T(E) = \frac{N(0)}{2} \operatorname{sgn} E \, \Re e \left\{ \frac{u_+}{\sqrt{u_\pm^2 - 1}} + \frac{u_-}{\sqrt{u_\pm^2 - 1}} \right\} . \tag{34}$$

The $u_\pm(E)$ are the solutions of the following coupled equations

$$\frac{E \mp \mu_B H}{\Delta} = u_\pm + \frac{i}{3\tau_{s0}\Delta} \frac{u_\pm - u_\mp}{\sqrt{u_\mp^2 - 1}} . \tag{35}$$

The root has to be chosen so that $(u_\pm^2 - 1)^{1/2} \to u_\pm \operatorname{sgn}(Re\ u_\pm)$ when $E \to \pm\infty$. In the special case that $\tau_{s0}\Delta_0 \ll 1$ (35) reduces to (20) with the pair-breaking parameter given by

$$\frac{1}{\tau_R} = \frac{3\tau_{s0}}{2v_F} (\mu_B H)^2 , \tag{36}$$

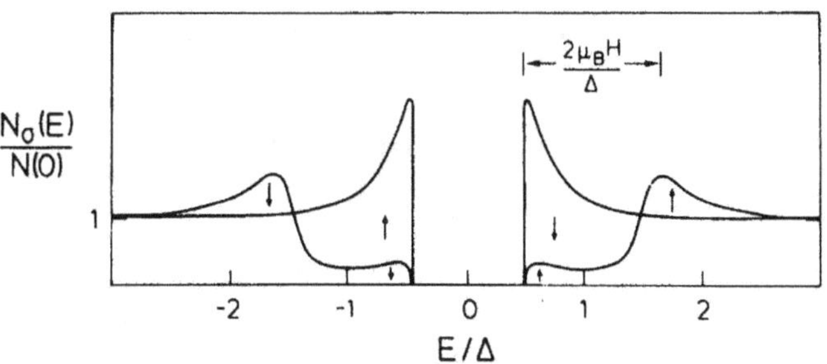

Fig. 2. Zeeman split quasiparticle density of states $N_\sigma(E)$. Due to the presence of small spin-orbit scattering the two components are slightly mixed.

i.e. we are back to the standard pair-breaking theory. The Zeeman-split density of states has been observed by Meservey and Tedrow in tunneling experiments.[52,72] Since electrons do not change their spin during the tunneling process, the method can be applied as a tool to measure magnetic properties.[73] An example of $N_T(E)$ based on (34) is shown in Fig. 2.

When the effect of the field on the orbits cannot be neglected, the prefactor α of the $\langle|\Delta(\mathbf{r})|^2\rangle$ term in the Ginzburg-Landau equation is of the form

$$\alpha = \left[\ln\frac{T}{T_{c0}} + \frac{1}{2}\left(1+\frac{b}{\sqrt{b^2-h^2}}\right)\psi\left(\frac{1}{2}+\rho_-\right)\right.$$
$$\left. + \frac{1}{2}\left(1-\frac{b}{\sqrt{b^2-h^2}}\right)\psi\left(\frac{1}{2}+\rho_+\right) - \psi\left(\frac{1}{2}\right)\right]. \tag{37}$$

Here the abbreviations $b = (3\tau_{s0}\Delta_0)^{-1}$ and $h = \mu_B H/\Delta_0$ have been introduced. Furthermore

$$\rho_\pm = \frac{\Delta_0}{2\pi T}\left\{a \pm \sqrt{b^2-h^2}\right\}. \tag{38}$$

The pair-breaking parameter a depends on the particular situation. For a type II superconductor in the dirty limit it is given by

$$a = \frac{\tau_t v_F^2 eH}{3\Delta_0} + b \tag{39}$$

while for a thin film in a parallel field it is in the same limit

$$a = \frac{\tau_t}{18\Delta_0}(v_F edH)^2 + b. \tag{40}$$

The two forms of a should be compared with (29) and (25).

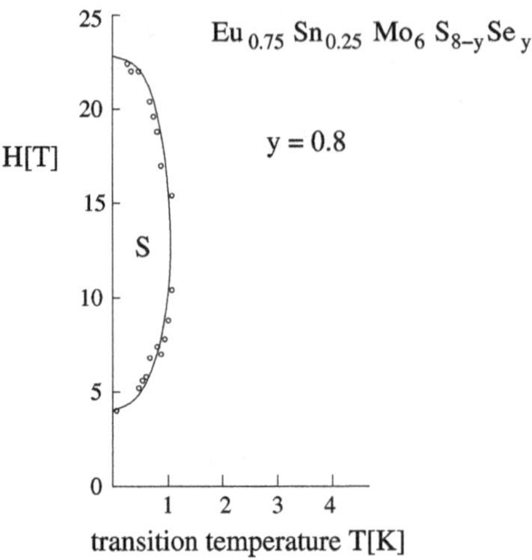

Fig. 3. Field induced superconductivity based on the Jaccarino-Peter effect. (From Ref. 53)

Equation (37) for a type II superconductor in a high field was first derived by Werthamer, Helfand and Hohenberg.[76] It had been obtained before for the case that the Zeeman term $\mu_B \boldsymbol{\sigma} \cdot \mathbf{H}$ is caused by spin-polarized magnetic impurities, i.e. by a term $-n_I(g-1)J_{\text{ex}}\sigma\langle J_z\rangle$.[25] Adding both terms, i.e. working with a total Zeeman interaction of the form

$$H_{\text{Zeem}} = -\sum_i \sigma_i \left(\mu_B H - n_I(g-1)J_{\text{ex}} \langle J_z\rangle\right) \tag{41}$$

can lead to a Jaccarino-Peter effect.[34] Here a compensation takes places between the effect of an external field and that of polarized impurities which couple via exchange to the conduction electrons. As a result, superconductivity may become magnetic field induced. This will be the case when in the absence of a field the (unpolarized) magnetic ions have destroyed superconductivity and when their effect is reduced by the aforementioned compensation effect. Clearly that requires a delicate balance between different material properties. Nevertheless, field induced superconductivity has been observed in $Eu_{0.5}Pb_{0.5}Mo_6S_8$[53] (see Fig. 3) and apparently also in λ-$(BETS)_2$ $FeCl_4$.[31]

The order of the phase transition from the superconducting to the normal state in the presence of pair-breaking perturbations is obtained from the sign of the function $f_1(\rho)$ in (17). When $f_1(\rho) = 0$ a system changes from a second- to a first-order phase transition. For perturbations belonging to group 1 the phase transition is always of second order, while for those belonging to group 2 it depends on the special case. The most important case is, of course, that of type II superconductors (Shubnikov phase) with a large κ value. Here the order of transition depends not only on the ratio T_c/T_{c0} but also on the Maki parameter and the spin-orbital scattering parameter b.

5. Scattering Centers with Low-Energy Excitations

As previously discussed, the field of Cooper pair breaking started with a study of the effects of paramagnetic impurities. Originally the scattering of conduction electrons off those ions was studied by treating the interaction (4) in Born approximation. A few important generalizations took place since then. One consists in including interactions between impurities.[29] Another is going beyond the Born approximation. This was done by treating the impurity spin classically[65] as well as quantum mechanically, i.e. by including the Kondo effect in the latter case. For a review see, e.g. Ref. 56. A third generalization takes into account that magnetic impurities, here especially rare earth impurities have often intra-atomic excitations with energies in the meV regime. Since these energies are of the same order as $k_B T_c$ one expects a strong interplay between the two.[23]

In generalizing the Born approximation, it is advantageous to work with the one-particle $T_1(\omega)$ and two-particle $T_2(\omega, \omega')$ scattering matrices.[7,8] In terms of those functions the transition temperature is determined by solving the coupled equations

$$\begin{aligned} \tilde{\omega}_n &= \omega_n - n_I \mathrm{Im}\, T_1(\tilde{\omega}_n) \\ \tilde{\Delta}_n &= \Delta + n_I \pi N(0) T_c \sum_m T_2(\tilde{\omega}_n, \tilde{\omega}_m) \frac{\tilde{\Delta}_m}{|\tilde{\omega}_m|} \end{aligned} \tag{42}$$

where $\omega_n = 2\pi T_c(n + 1/2)$ with n being an integer. The first of the two equations describes the scattering of a conduction electron in the normal state giving rise to a self-energy while the second one accounts for a modification of the pairing interactions due to inelastic impurity scattering. From the self-consistency equation (11) one can derive in the limit $n_I \to 0$ the following lowering of the transition temperature due to pair breaking

$$\begin{aligned} &-\left.\frac{dT_c}{dn_I}\right|_{T_{c0}} \\ &= -T_{c0}^2 \pi \sum_{n=-\infty}^{+\infty} \left[\frac{\mathrm{Im}\, T_1(\omega_n)}{\omega_n^2} \mathrm{sgn}\, \omega_n + \frac{\pi N(0)}{|\omega_n|} T_{c0} \sum_{m=-\infty}^{+\infty} T_2(\omega_n, \omega_m) \frac{1}{|\omega_m|} \right]. \end{aligned} \tag{43}$$

We want to include the Kondo effect, which shows up to order $J_{\rm ex}^3$ but requires going beyond perturbation expansion. For this it is advantageous to start instead of (4) from the more basis Anderson Hamiltonian. It is written down for a single ion with a ν_f degenerate f orbital of energy ϵ_f which is hybridizing with matrix element $V_{m\sigma}(\mathbf{k})$ with the conduction electrons. Electrons in the f orbital repel each other with energy U. Therefore

$$H=\sum_{\mathbf{k}\sigma}\epsilon(\mathbf{k})c^+_{\mathbf{k}\sigma}c_{\mathbf{k}\sigma}+\epsilon_f\sum_{m}^{\nu_f}n^f_m+\frac{U}{2}\sum_{m\neq m'}n^f_m n^f_{m'}+\sum_{\mathbf{k}m\sigma}\left(V_{m\sigma}(\mathbf{k})f^+_m c_{\mathbf{k}\sigma}+h.c.\right)\,. \tag{44}$$

All energies are measured from the Fermi energy E_F. As common $n^f_m = f^+_m f_m$. It is supposed that $U \to \infty$, which implies that the f-site is either occupied with one electron or empty. This corresponds to Ce^{3+}, a well studied Kondo ion. There are three different regimes between which one may distinguish. They depend on the ratio ϵ_f/Γ where $\Gamma = \pi\nu_f N(0)[V_{m\sigma}(k_F)]^2$:

(a) $\epsilon_f/\Gamma \gg 1$: the f orbital remains empty and only ordinary, nonmagnetic scattering takes place;
(b) $|\epsilon_f/\Gamma| \simeq 1$: the f-site is in the intermediate-valence or valence-fluctuating regime;
(c) $-\epsilon_f/\Gamma \gg 1$: this is the regime in which magnetic scattering and the Kondo effect take place and on which we shall concentrate.

The essential feature of the Kondo effect is the formation of a singlet state, in this case of the $4f$ electron with the conduction electrons. The $4f$ electron of Ce^{3+} with $J = 5/2$ can hop via $V_{m\nu}(\mathbf{k})$ from orbital ν into the conduction band while a conduction electron returns to the site in orbital ν'. The singlet formation energy is characterized by a temperature T_K, the Kondo temperature. It is often of order meV. For temperatures $T \gg T_K$ the Kondo ions have a Curie-Weiss susceptibility and an effective magnetic moment

$$\mu^2_{\rm eff}(T) = 3T\chi(T)\,, \tag{45}$$

while for $T \ll T_K$ the susceptibility for the singlet state is Pauli-like. The aim is to understand the pair-breaking effect of Kondo ions. The conduction electron scattering rate is

$$\operatorname{Im} T_1(\omega, T) = -\frac{1}{2}\left(\Gamma/N(0)\right)\rho_f(\omega, T) \tag{46}$$

and therefore proportional to the many-body spectral density $\rho_f(\omega, T)$ for addition or removal of an f-electron. The latter consists of a Lorentzian centered at ϵ_f (removal of an f electron) of width $\nu_f\Gamma$ and height $(\pi\nu_f^2\Gamma)^{-1}$

and a small resonance-like feature at the Fermi energy (Abrikosov-Suhl resonance). There is also a broad peak at $\epsilon_f + U$ corresponding to the addition of a second f-electron. But since U is large it is of no interest here. It is observed in inverse photoemission, i.e. Bremsstrahlung experiments.

For Ce ions $\epsilon_f \simeq -2$ eV, while the narrow resonance at E_F has a small spectral weight of $(1 - n_f) = \pi k_B T_K/(\nu_f \Gamma)$ and is due to spin fluctuations. When (46) is set into (43) the reduction of T_c is found to be

$$-\left.\frac{dT_c}{dn_I}\right|_{T_c0} = \frac{\pi^2}{4}\nu_f N(0) J_{\text{ex}}^2 \,, \tag{47}$$

where here $J_{\text{ex}} = -V_{\text{hyb}}^2/|\epsilon_f|$. This can be compared with the corresponding expression of the Abrikosov-Gorkov theory which for $\nu_f = 2$ and $g = 2$ is

$$-\left.\frac{dT_c}{dn_I}\right|_{T_c0} = \frac{3\pi^2}{8} J(J+1) N(0) J_{\text{ex}}^2 \,. \tag{48}$$

Starting from the Anderson model shows that *charge* fluctuations from f^1 to f^0 are the ones which cause the pair-breaking in the Abrikosov-Gorkov theory. The strength of the Abrikosov-Suhl resonance is a measure of the probability of a $f^0 \to f^1$ transition when an f electron is added to the ion. Remember that the ground state of a Ce^{3+} impurity is a coherent superposition of a f^0 and f^1 electron state. At low T and low ω the resonance dominates the scattering rate of the conduction electron. It depends on the scaled quantities T/T_K and ω/T_K only.

The effect of $T_1(\omega_n)$ is supplemented by the effective interaction between electrons of opposite spin $T_2(\omega_n, \omega_m)$. It results from a virtual polarization of the impurity. Here we concentrate on elastic scattering where $\omega_m = \omega_n$. It gives the dominant contribution to dT_c/dn_I and counteracts the effect of $\operatorname{Im} T_1(\omega_n, T_{c0})$ because it reflects the increasingly nonmagnetic character of a Kondo ion for large ratios of T_K/T_{c0}. When it is added in (43) to the contribution of $\operatorname{Im} T_1(\omega_n)$ we obtain the dependence of $-(dT_c/dn_I)_{T_{c0}}$ shown in Fig. 4. Pair-breaking is strongest for a ratio $T_K/T_{c0} \simeq 5$. In the limit of large T_K/T_{c0}, pair-breaking reduces to zero because of a strongly bound singlet state. The superconducting state is a pair condensate of quasiparticles formed by a coherent superposition of conduction and f electrons. The transition temperature T_c of that system will be generally lower than that of the pure one. We speak here of pair weakening because the system behaves like a BCS superconductor except for the change in T_c. This is easily understood. A large T_K implies a large $\nu_f \Gamma$ and a value of $(1-n_f)$ which is in the intermediate valence or valence-fluctuating regime. In this case Friedel's model of a nonmagnetic virtual bound state applies.[21] The impurity level

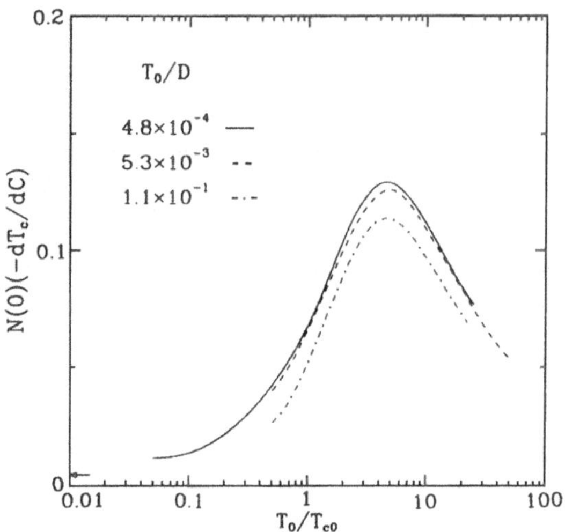

Fig. 4. Initial depression of T_c with impurity concentration n_I as function of (T_0/T_{c0}), i.e. the ratio of the Kondo temperature ($T_0 = T_K$) and the superconducting transition temperature. The pair-breaking effect of an impurity is largest for $T_0/T_{c0} \simeq 5$. (From Ref. 8)

at ϵ_f is broadened and partially occupied but U is not large enough to generate a local moment. At the impurity site two electrons of opposite spins repel each other and therefore the averaged attractive interaction of paired electrons is diminished.[8,35,66,78] Therefore we deal here with pair weakening rather than pair breaking.

The nonmonotonous behavior of dT_c/dn_I has an interesting consequence, i.e. a possible reentrant behavior of superconductivity.[57] For a range of impurity concentrations for which pair breaking changes strongly with T_K/T_c, one finds as function of temperature two and even three transition temperatures T_c. This effect has indeed been observed and an example is shown in Fig. 5.

Rare-earth ions with well localized $4f$ electrons in an incomplete f-shell have a $(2J+1)$-fold degenerate ground state where J denotes the lowest Hund's rule multiplet. The latter is split in the crystalline electric field (CEF) of the surrounding of an ion. Conduction electrons cause transitions between the CEF energy levels. Of particular interest are non-Kramers ions which have an even number of $4f$ electrons and may have a nondegenerate singlet CEF ground state. An example is Pr^{3+} which is in a $4f^2$ configuration. Two interactions with the conduction electron are of particular importance. One is the isotropic exchange interaction (4) while the second

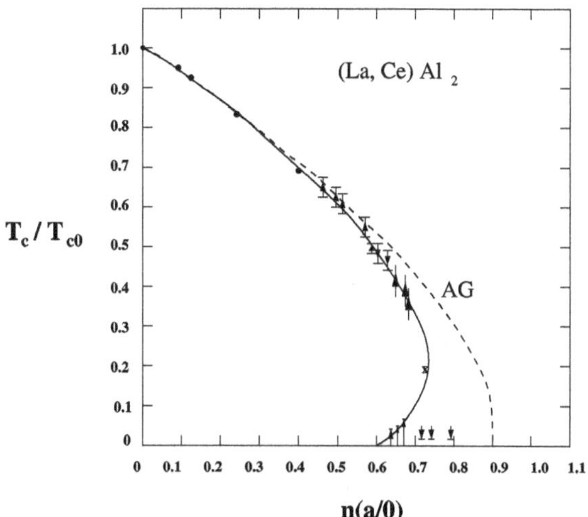

Fig. 5. Reduced transition temperature versus impurity concentration for (La, Ce) Al_2 with reentrant behavior due to the Kondo effect. AG denotes the result of the Abrikosov-Gorkov theory. (From Ref. 47)

is the aspherical Coulomb charge scattering

$$H_{\mathrm{AC}} = \left(\frac{5}{4\pi}\right)^{1/2} \sum_{kk'\sigma} \sum_{m=-2}^{+2} I_2\left(k's; kd\right) Q_2[Y_2^m(\mathbf{J}) c^+_{k's\sigma} c_{kdm\sigma} + h.c.]\,. \quad (49)$$

Here Q_2 is the quadrupole moment of, e.g. the Pr^{3+} ion and the Coulomb integrals $I_2(k's; kd)$ can be found in Ref. 23. The $c_{kdm\sigma}$ destroy a conduction electron with momentum $k = |\mathbf{k}|$, in a $\ell = 2$ state with azimuthal quantum number m and spin σ, while $c^+_{k's\sigma}$ creates an electron with momentum k' in a $\ell = 0$ state. The operators $Y_2^m(\mathbf{J})$ are spherical harmonics written in terms of J_α operators. For example, $Y_2^{\pm 1}(\mathbf{J}) = \pm\left(J_z J^\pm + J^\pm J_z\right) / \left[(2/3)^{1/2}(2J^2 - J)\right]$. The Hamiltonian transfers angular momentum $\ell = 2$ between conduction electrons and the $4f^2$ shell.

Note that the conduction-electron part of H_{AC} is time-reversal invariant. Therefore transitions between CEF levels caused by H_{AC} contribute to pair formation rather than to pair breaking. Their effect resembles that of localized phonons. As it turns out, the exchange matrix elements are usually dominant and for that reason CEF split non-Kramers ions usually break Cooper pairs. A notable exception is pointed out below. However, the depression of T_c with impurity concentration n_I deviates strongly from the Abrikosov-Gorkov theory. Consider a CEF splitting of the ions as shown in Fig. 6 and the effect of H_{int} given by (4).

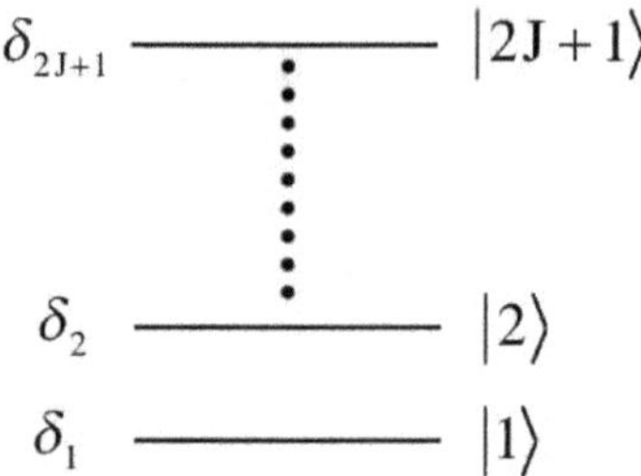

Fig. 6. Schematic view of CEF eigenstates $|i\rangle$ and energies δ_i of an ion some of which may be degenerate.

The CEF eigenstates $|i\rangle$ and energies δ_i are assumed to be known, e.g. from inelastic neutron scattering experiments. The matrices T_1 and T_2 can be expressed in terms of the magnetic susceptibility

$$\chi(\omega_n, \omega_m) = \sum_{ij} \frac{|\langle i|J|j\rangle|^2 (n_i - n_j)}{(\omega_n - \omega_m)^2 + \delta_{ij}^2} \delta_{ij} \,. \tag{50}$$

The thermal population of the different states $|i\rangle$ is denoted by n_i, and $\delta_{ij} = \delta_i - \delta_j$. In the simplest case of two singlets separated by an energy δ and all other CEF levels neglected, we obtain

$$\chi(\omega_n, \omega_m) = \frac{2\delta \tanh(\delta/2T)}{(\omega_n - \omega_m)^2 + \delta^2} |\langle 0|J|1\rangle|^2 \,. \tag{51}$$

The T matrices are shown in Fig. 7.

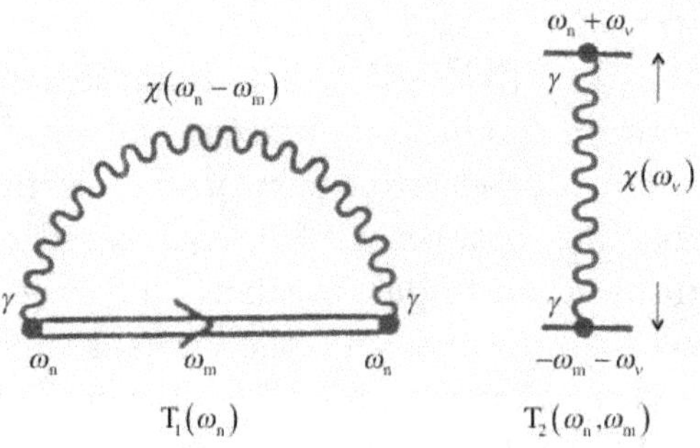

Fig. 7. Scattering matrices $T_1(\omega_n)$ and $T_2(\omega_n, \omega_m)$ due to impurities with internal degrees of freedom $\chi(\omega_\nu)$ is the spin susceptibility and $\gamma = (g-1)N(0)J_{\text{ex}}$. Solid lines refer to electronic Green's functions.

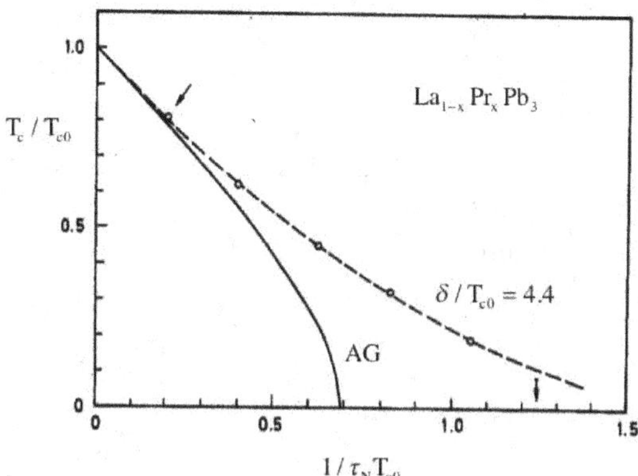

Fig. 8. T_c as function of impurity concentration for $La_{1-x}Pr_xPb_3$. The dashed line is the theoretical curve for $\delta/T_{c0} = 4.4$. The point denoted by an arrow has been fitted to give an initial slope of one. AG denotes the Abrikosov-Gorkov results. (From Ref. 9)

From (42) and (51) we obtain

$$\ln \frac{T_c}{T_{c0}} + \psi\left(\frac{1}{2} + \frac{2}{\pi^2 T_c \tau_N} \frac{y(\delta/2T_c)}{y(\delta/2T_{c0})}\right) - \psi\left(\frac{1}{2}\right) = 0 , \tag{52}$$

where τ_N is defined so that the slope $d(T_c/T_{c0})/d(\tau_N T_{c0})^{-1} = -1$. This has the advantage that we can compare the result with those of the Abrikosov-Gorkov theory. The function $y(x)$ has the value $y(0) = 2$ and is monotonously decreasing with x. For $x = 10$ it has dropped to $y(10) \simeq 0.6$. Thus with increasing δ/T_c the pair-breaking effect of a single ion decreases continuously because virtual CEF excitations are reduced. Therefore with higher concentration n_I the pair-breaking effect of a single impurity decreases (for more details see Ref. 23). With increasing ratio (δ/T_c), i.e. with decreasing T_c a virtual excitation to the upper CEF becomes less favorable. Experiments by Bucher *et al.*[9] on $La_{1-x}Pr_xPb_3$ have confirmed the above results (see Fig. 8 where the results of the Abrikosov-Gorkov theory are shown too). The differences to the standard pair-breaking theory are less pronounced for Kramers ions, which always have a degenerate ground state.

6. Breaking of Anisotropic Pair States

In the high-T_c superconducting cuprates pairing takes place in the form of a strongly anisotropic pair state. An isotropic s-wave order parameter cannot explain many experimental data. In particular the low-temperature thermodynamic properties such as the specific heat require instead strongly anisotropic pairing. Phase sensitive measurements favor d-wave pairing. Also in other superconductors with strong electron correlations, strongly anisotropic pair states prevail. Examples are the Fe-pnictides or f-electron systems with heavy quasiparticles.[12,70] A generalization of the theory of pair breaking and of pair weakening to strongly anisotropic pairing is therefore mandatory. Equation (42) is generalized for a momentum dependent order parameter[27] to

$$\tilde{\omega}_n(\hat{\mathbf{k}}) = \omega_n - n_I \mathrm{Im}\, T_1(\hat{\mathbf{k}}, \omega_n)$$

$$\tilde{\Delta}_n(\hat{\mathbf{k}}) = \Delta(\hat{\mathbf{k}}) + n_I N(0) \int d\hat{\mathbf{k}}' n(\hat{\mathbf{k}}') T_2(\hat{\mathbf{k}}, \hat{\mathbf{k}}', \omega_n) \frac{\tilde{\Delta}_n(\hat{\mathbf{k}}')}{|\tilde{\omega}_n(\hat{\mathbf{k}}')|} . \tag{53}$$

The two-dimensional unit vector $\hat{\mathbf{k}}$ runs over the Fermi surface. The function $\hat{n}(\hat{\mathbf{k}})$ is normalized to one, i.e. $\int d\hat{\mathbf{k}} n(\hat{\mathbf{k}}) = 1$. Within the Born approximation the scattering matrices of a nonmagnetic impurity with scattering potential $u(\hat{\mathbf{k}}, \hat{\mathbf{k}}')$ are

$$T_1\ (\hat{\mathbf{k}}', \omega_n) = -i\pi \mathrm{sgn}\, \omega_n N(0) \int d\hat{\mathbf{k}}' n(\hat{\mathbf{k}}') |u(\mathbf{k}, \mathbf{k}')|^2$$

$$T_2\ (\hat{\mathbf{k}}, \hat{\mathbf{k}}', \omega_n) = |u(\hat{\mathbf{k}}, \hat{\mathbf{k}}')|^2 . \tag{54}$$

Similarly (43) is generalized to

$$-\left.\frac{dT_c}{dn_I}\right|_{T_{c0}} = \pi^2 T_{c0}^2 \sum_n \frac{N(0)}{\omega_n^2} \int d\hat{\mathbf{k}}' d\hat{\mathbf{k}} n(\hat{\mathbf{k}}') n(\hat{\mathbf{k}}) \left\{ |\Delta(\hat{\mathbf{k}})|^2 |u(\mathbf{k}, \mathbf{k}')|^2 \right.$$

$$\left. - \Delta(\hat{\mathbf{k}}) |u(\hat{\mathbf{k}}, \hat{\mathbf{k}}')|^2 \Delta(\hat{\mathbf{k}}') \right\} . \tag{55}$$

Hereby we used a normalization of the order parameter of the form

$$\int d\hat{\mathbf{k}} n(\hat{\mathbf{k}}) |\Delta(\hat{\mathbf{k}})|^2 = 1 . \tag{56}$$

When the order parameter $\Delta(\hat{\mathbf{k}})$ is constant, the right-hand side of (55) vanishes. This is quite different for an order parameter with

$$\int d\hat{\mathbf{k}} n(\hat{\mathbf{k}}) \Delta(\hat{\mathbf{k}}) = 0 , \tag{57}$$

a situation encountered, e.g. in d-wave superconductors. Here we have to account for anisotropies in the scattering matrix elements. The reduction in T_c is largest for isotropic scattering, i.e. when $u(\hat{\mathbf{k}}, \hat{\mathbf{k}}') = \mathrm{const}$. The effect can be reduced to a large extent when the scattering is strongly anisotropic. An anisotropic order parameter can be nearly insensitive to impurity scattering if the latter is also strongly anisotropic. For example, it may happen that a superconducting, nearly constant order parameter does exist only on parts of the Fermi surface. The latter may consist of several disconnected pieces. If impurity scattering were to connect only parts of the Fermi surface where the order parameter is nonzero, then its effect on superconductivity would be very small.

7. Pair Formation versus Pair Breaking

After the discovery of high-T_c superconductivity it was suspected from an early stage on, that the dominant pairing interaction does not result from phonons.[54,55,62,69] The same holds true for superconductors with heavy quasiparticles[70] although here, because of the low transition temperatures, the argument is not as compelling. In the following we discuss two cases where there is strong evidence for non-phononic pairing interactions. Contrary to the above, the electronic excitations are associated with localized electrons which couple to itinerant ones.

The filled skutterudite $PrOs_4Sb_{12}$ has a transition temperature T_c ($= 1.85$ K) which is more than twice as large as the corresponding one of $LaOs_4Sb_{12}$ ($T_c = 0.74$ K). Both superconductors seem to be of conventional s-wave type.[38,71,77] Since the only difference are the $4f^2$ electrons of Pr, they must be responsible for the large increase in pairing interactions. Inelastic neutron-scattering experiments have demonstrated[28] that the CEF ground state of the $J = 4$ multiplet is a Γ_1 singlet with a low-lying excited triplet Γ_t of T_d symmetry at 8 K. All other levels are so high in energy that their influence can be neglected. The triplet is a superposition of two triplets of O_4 symmetry,[67]

$$|\Gamma_t, m\rangle = \sqrt{1-d^2}\,|\Gamma_5, m\rangle + d\,|\Gamma_4, m\rangle\ , \quad m = 1, 2, 3\,. \tag{58}$$

From experiments one can deduce $|d| = 0.26$, which implies that $|\Gamma_t, m\rangle$ has mainly $|\Gamma_5, m\rangle$ character. But transitions $|\Gamma_1\rangle \to |\Gamma_5, m\rangle$ can take place only via $H_{\rm AC}$ but not via $H_{\rm ex}$. Therefore these processes are pair forming. The boson propagator due to intra-atomic CEF excitations is

$$\chi(\mathbf{q}, \nu_n) = \sum_{\alpha\beta m} \left|\Lambda^m_{\alpha\beta}(\hat{\mathbf{q}})\right|^2 \frac{2\delta}{\nu_n^2 + \delta^2} \tag{59}$$

with

$$\Lambda^m_{\alpha\beta}(\hat{\mathbf{q}}) = \frac{\sqrt{3}g}{2}\hat{q}_\alpha \hat{q}_\beta \left\langle \Gamma_1 \left|(J_\alpha J_\beta + J_\beta J_\alpha)\right| \Gamma_t^m \right\rangle \tag{60}$$

and $\nu_n = 2\pi T n$. The matrix element results from H_{AC} which is rewritten for cubic symmetry of site i in the form

$$H_{\text{AC}}(i) = \frac{\sqrt{3}}{2} g \sum_{\mathbf{kq}\sigma} \sum_{\alpha\beta\text{cycl.}} (J_\alpha J_\beta + J_\beta J_\alpha)_i \, \hat{q}_\alpha \hat{q}_\beta \; c^+_{\mathbf{k}-\mathbf{q}\sigma} c_{\mathbf{k}\sigma} e^{i\mathbf{kR}_i} \,. \qquad (61)$$

The Eliashberg equations[19] allow for the determination of T_c when the dynamic two-particle interactions are known. The latter are mediated by phonons in the case of $\text{LaOs}_4\text{Sb}_{12}$ and by phonons supplemented by $\chi(\mathbf{q}, \nu_m)$ in the case of $\text{PrOs}_4\text{Sb}_{12}$.[11] A detailed discussion of Eliashberg's equations is found, e.g. in Ref. 63. When the equations are solved, one obtains the observed enhancement of T_c in $\text{PrOs}_4\text{Sb}_{17}$ for a coupling constant of $g =$ 0.04 eV. This is a very reasonable value. It also gives the observed effective electron mass enhancement in the normal state of the system when the self-energy due to $\chi(\mathbf{q}, \nu_n)$ is evaluated.

Another system with pairing interactions based on intra-atomic excitations is UPd_2Al_3, a superconductor with $T_c = 1.8$ K and an antiferromagnet with $T_N = 14.3$ K. Due to strong intra-atomic or Hund's rule correlations the $5f$ electrons of U ions remain localized in some of the f orbitals while they delocalize in others.[79] Another way of stating the same is by saying that intra-atomic correlations renormalize the hopping matrix-elements of some of the $5f$ orbitals to zero. This Dual Model has been successful in explaining a number of experiments. In UPd_2Al_3 two of the nearly 2.5 $5f$-electrons with $j_z = 5/2$ and $1/2$ remain localized and form a Hund's rule ground state. Inelastic neutron scattering shows dispersive CEF excitations (magnetic excitons) below the Neél temperature. They result from cooperative effects of localized electrons. Coupling of conduction electrons to these excitations explains the strongly anisotropic, heavy-quasiparticle masses in that system. The same excitations act as bosons which provide for binding of electrons to Cooper pairs. However, a special form of the order parameter is required, i.e.

$$\Delta(\mathbf{p}) = \Delta_0 \cos(cp_z) \quad or \quad \Delta(\mathbf{p}) = \Delta_0 \sin(cp_z) \qquad (62)$$

with c denoting the lattice constant perpendicular to the U planes. Eliashberg's equations are solved with a magnetic exciton propagator given by

$$\chi\left(q_z, \nu_n\right) = \frac{I^2}{2} \frac{\omega_{\text{ex}}}{\nu_n^2 + \omega_E(\mathbf{q})} \,. \qquad (63)$$

The experimentally observed exciton dispersion can be fitted with

$$\omega_E(\mathbf{q}) = \omega_{\text{ex}} \left[1 + \beta \cos(cq_z)\right] \,, \qquad (64)$$

where $\omega_{ex} = 5$ meV and $\beta = 0.8$.[74] The size of the coupling constant I follows from the heavy-quasiparticle masses in that compound. The equations have a solution for an order parameter of the form (62). The $\sin(cp_z)$ solution can be excluded, because it requires a spin triplet and gives a lower T_c than the $\cos(cp_z)$ solution. For the latter a value of $T_c \simeq 3$ K is found which is of the right order.[51] Note that there is no solution existing for an order parameter of s-wave symmetry.

Why is s-wave pairing excluded while an order parameter of the form of (62) gives a seizable T_c? Pair formation is not expected since excitations of magnetic excitons violate time-reversal symmetry. Therefore they break Cooper pairs of electrons in time reversed states. However, when the order parameter is of the form of (62) electrons are paired in states which are time reversed followed by a lattice vector translation. Expressed differently, electrons forming a pair are located preferably on different sublattices. Therefore magnetic excitons can provide a pair forming electron interaction here.

The two examples of $PrOs_4Sb_{12}$ and UAl_2Pd_3 round off our understanding of pair breaking and its relation to time-reversal symmetry.

Acknowledgments

Part of this chapter is based on Ref. 27. I would like to thank G. Zwicknagl and P. Thalmeier for numerous stimulating discussions and T. Takimoto for careful reading of the manuscript.

References

1. M. Abramowitz and I. A. Stegun, *Handbook of Mathematical Functions* (Dover Publ., New York, 1965).
2. A. A. Abrikosov and L. P. Gorkov, *Zh. Eksp. Teor. Fiz.* **39**, 1781 (1960); Engl. Transl.: *Sov. Phys.-JETP* **12**, 1243 (1961).
3. H. Alloul, M. Gabay, J. Bobroff and P. Hirschfeld, *Rev. Mod. Phys.* **81**, 45 (2009).
4. V. Ambegaokar and A. Griffin, *Phys. Rev.* **137**, A1151 (1965).
5. P. W. Anderson, *Phys. Rev. Lett.* **3**, 325 (1959).
6. W. Baltensperger and S. Straessler, *Phys. Kondens. Mater.* **1** 20 (1963).
7. N. E. Bickers, D. L. Cox and J. W. Wilkins, *Phys. Rev. B* **36**, 2036 (1987).
8. N. E. Bickers and G. Zwicknagl, *Phys. Rev. B* **36**, 6746 (1987).
9. E. Bucher, K. Andres, J. P. Maita and G. W. Hull, *Helv. Phys. Acta* **41**, 723 (1968).
10. B. S. Chandrasekhar, *Appl. Phys. Lett.* **1**, 7 (1962).

11. J. Chang, I. Eremin, P. Thalmeier and P. Fulde, *Phys. Rev, B* **76**, 220510(R) (2007).
12. A. V. Chubukov, D. F. Efremov and I. Eremin, *Phys. Rev. B* **78**, 134512 (2008).
13. A. M. Clogston, *Phys. Rev. Lett.* **9**, 226 (1962).
14. J. E. Crow, R. P. Guertin and R. D. Parks, *Phys. Rev. Lett.* **19**, 77 (1967).
15. P. G. de Gennes, *Phys. Kondens. Mater.* **3**, 79 (1964a).
16. P. G. de Gennes, *Rev. Mod. Phys.* **36**, 225 (1964b).
17. P. G. de Gennes, *Superconductivity of Metals and Alloys* (W. A. Benjamin Inc., New York, 1966).
18. P. G. de Gennes and M. Tinkham, *Phys. Kondens. Mater.* **1**, 107 (1964).
19. G. M. Eliashberg, *Zh. Eksp. Teor. Fiz.* **38**, 666 (1960); Engl. Transl.: *Soviet Phys.-JETP* **11**, 696 (1960).
20. R. A. Ferrell, *Phys. Rev. Lett.* **3**, 262 (1959).
21. J. Friedel, *Nuovo Cimento Suppl.* **12**, 1861 (1958).
22. P. Fulde and R. A. Ferrell, *Phys. Rev.* **135**, A550 (1964).
23. P. Fulde, L. L. Hirst and A. Luther, *Z. Phys.* **238**, 99 (1970).
24. P. Fulde and K. Maki, *Phys. Rev. Lett.* **15**, 675 (1965).
25. P. Fulde and K. Maki, *Phys. Rev.* **141**, 275 (1966a).
26. P. Fulde and K. Maki, *Phys. Kond. Mater.* **5**, 380 (1966b).
27. P. Fulde and G. Zwicknagl, *Superconductivity, Springer Series in Solid-State Sciences*, Vol. 90 (Springer, Heidelberg, 1990).
28. E. A. Goremychkin, R. Osborn, E. D. Bauer, M. B. Maple, N. A. Frederick, W. M. Yuhasz, F. M. Woodward and J. W. Lynn, *Phys. Rev. Lett.* **93**, 157003 (2004).
29. L. P. Gorkov and A. I. Rusinov, *Sov. Phys.-JETP* **19**, 922 (1964).
30. F. Heiniger, E. Bucher, J. P. Maitra and L. D. Longinotti, *Phys. Rev. B* **12**, 1778 (1975).
31. K. Hiraki, H. Mayaffre, M. Horavić, C. Berthier, A. Uji, T. Yamaguchi, H. Tanaka, A. Kobayashi, H. Kobayashi and T. Takahashi, *J. Phys. Soc. Jpn.* **76**, 124708 (2007).
32. M. Hiroi, M. Sera, N. Kobayashi, Y. Haga, E. Yamamoto and Y. Onuki, *J. Phys. Soc. Jpn.* **66**, 1595 (1997).
33. J. G. Huber, W. A. Fertig and M. B. Maple, *Solid State Comm.* **15**, 453 (1974).
34. V. Jaccarino and M. Peter, *Phys. Rev. Lett.* **9**, 290 (1962).
35. A. B. Kaiser, *J. Phys. C* **3**, 409 (1970).
36. Y. Kitaoka, H. Tou, K. Ishida, N. Kimura, Y. Onuki, E. Yamamoto, Y. Haga and K. Maezawa, *Physica B* **281–282**, 878 (2000).
37. A. I. Larkin and Y. N. Ovchinnikov, *Zh. Eksp. Teor. Fiz.* **47**, 1136 (1964); Engl. Transl.: *Sov. Phys.-JETP* **20**, 762 (1965).
38. D. E. MacLaughlin, J. E. Sonier, R. H. Heffner, O. O. Bernal, B. L. Young, M. S. Rose, G. D. Morris, E. D. Bauer, T. D. Do and M. B. Maple, *Phys. Rev. Lett.* **89**, 157001 (2002).
39. K. Maki, *Progr. Theor. Phys. (Kyoto)* **29**, 10 (1963a).
40. K. Maki, *Progr. Theor. Phys. (Kyoto)* **29**, 333 (1963b).
41. K. Maki, *Progr. Theor. Phys. (Kyoto)* **31**, 731 (1964a).
42. K. Maki, *Phys. Kond. Mater.* **1**, 21 (1964b).

43. K. Maki, *Phys. Rev.* **148**, 362 (1966).
44. K. Maki, *Superconductivity*, ed. R. D. Parks, Vol. 2 (Dekker, New York, 1969).
45. K. Maki and P. Fulde, *Phys. Rev.* **140**, A1586 (1965).
46. K. Maki and T. Tsuneto, *Progr. Theor. Phys. (Kyoto)* **31**, 945 (1964).
47. M. B. Maple, W. A. Fertig, A. C. Mota, L. E. DeLong, D. Wohlleben and R. Fitzgerald, *Sol. State Comm.* **11**, 829 (1972).
48. K. Matsuda, Y. Kohori and T. Kohora, *Phys. Rev. B* **55**, 15223 (1997).
49. B. T. Matthias and R. M. Bozorth, *Phys. Rev.* **109**, 604 (1958).
50. B. T. Matthias, H. Suhl and E. Corenzwit, *J. Phys. Chem. Solids* **13**, 156 (1960).
51. P. McHale, P. Thalmeier and P. Fulde, *Phys. Rev. B* **70**, 014513 (2004).
52. R. Meservey, P. M. Tedrow and P. Fulde, *Phys. Rev. Lett.* **25**, 1270 (1970).
53. H. W. Meul, C. Rossel, M. Decroux, Ø. Fischer, G. Remenyi and A. Briggs, *Phys. Rev. Lett.* **53**, 497 (1984).
54. P. Monthoux, A. V. Balatsky and D. Pines, *Phys. Rev. Lett.* **67**, 3448 (1991).
55. P. Monthoux and G. G. Lonzarich, *Phys. Rev. B* **66**, 224504 (2002).
56. E. Müller-Hartmann, *Magnetism*, Vol. 5, eds. G. T. Rado and H. Suhl (Academic Press, New York, 1973).
57. E. Müller-Hartmann and J. Zittartz, *Z. Phys.* **234**, 58 (1970).
58. W. Opitz, *Z. Phys.* **114**, 263 (1955).
59. F. Reif and M. A. Woolf, *Phys. Rev. Lett.* **9**, 315 (1962).
60. D. Saint-James and P. G. de Gennes, *Phys. Rev. Lett.* **7**, 306 (1963).
61. G. Sarma, *J. Phys. Chem. Solids* **24**, 1029 (1963).
62. D. J. Scalapino, E. Loh and J. E. Hirsch, *Phys. Rev. B* **34**, 8190 (1986).
63. J. R. Schrieffer, *Theory of Superconductivity* (W. A. Benjamin Inc., New York, 1964).
64. K. Schwidthal, *Z. Phys.* **158**, 563 (1960).
65. H. Shiba, *Progr. Theor. Phys.* **40**, 435 (1968).
66. H. Shiba, *Progr. Theor. Phys.* **50**, 50 (1973).
67. R. Shiina, *J. Phys. Soc. Jpn.* **73**, 2257 (2004).
68. S. Skalski, O. Betbeder-Matibet and P. R. Weiss, *Phys. Rev.* **136**, A1500 (1964).
69. J. Spalek, *Phys. Rev. B* **38**, 208 (1988).
70. G. R. Stewart, *Rev. Mod. Phys.* **56**, 755 (1984).
71. H. Suderow, S. Vieira, J. D. Strand, S. Bud'ko, and P. C. Cranfield, *Phys. Rev. B* **69**, 060504(R) (2004).
72. P. M. Tedrow and R. Meservey, *Phys. Rev. Lett.* **26**, 192 (1971).
73. P. M. Tedrow and R. Meservey, *Phys. Rev.* **238**, 173 (1994).
74. P. Thalmeier, *Eur. Phys. J. B* **27**, 29 (2002).
75. R. S. Thompson, *Phys. Rev. Lett.* **15**, 971 (1965).
76. N. R. Werthamer, E. Helfand and P. C. Hohenberg, *Phys. Rev.* **147**, 295 (1966).
77. M. Yogi, N. Nagai, Y. Imamura, H. Mukuda, Y. Kitaoka, D. Kikuchi, H. Sugawara, Y. Aoki, H. Sato and H. Harima, *J. Phys. Soc. Jpn.* **75**, 124702 (2006).
78. M. J. Zuckermann, *Phys. Rev.* **140**, A899 (1965).
79. G. Zwicknagl and P. Fulde, *J. Phys.: Condens. Matter* **15**, S1911 (2003).

SUPERCONDUCTOR-INSULATOR TRANSITIONS

A. M. Goldman
School of Physics and Astronomy, University of Minnesota,
116 Church St. SE, Minneapolis, MN 55455, USA
goldman@physics.umn.edu

Superconductor-insulator transitions, especially in thin films, can provide the simplest examples of the continuous quantum phase transition paradigm. Quantum phase transitions differ from thermal phase transitions in that they occur at zero temperature when the ground state of a system is changed in response to a variation of an external parameter of the Hamiltonian. In the example of the superconductor-insulator transition, this control parameter could be the parallel or perpendicular magnetic field, disorder, or charge density. Quantum phase transitions are studied through measurements at nonzero temperature of physical behaviors influenced by the quantum fluctuations associated with the transition. This review will focus on experimental aspects of superconductor-insulator transition in disordered films that are effectively two-dimensional. In particular, the evidence for quantum critical behavior in the various types of transitions will be presented. The various theoretical scenarios for the transitions will also be discussed along with the extent to which they are supported by experiment. Open questions relating to the nature of the very puzzling insulating regime and whether there are many different types of superconductor-insulator transitions will be presented. Although this research area is more than 20 years old, rather central issues are not resolved.

1. Introduction

In this short review we will focus on experimental aspects of superconductor-insulator (SI) transitions that have been studied in disordered ultrathin films that are effectively two-dimensional (2D) or quasi-2D. These provide the simplest examples of SI transitions, which are part of an important paradigm of contemporary condensed matter physics, the continuous quantum phase transition (QPT).[1] We will not discuss three-dimensional systems[2] or junction networks,[3] which have also been studied, and which in the case of junction networks can serve as model systems for films.

Continuous QPTs occur at zero temperature, when the ground state of a system is changed in response to the variation of an external parameter of the Hamiltonian. In the example of the SI transition this could be parallel or perpendicular magnetic field, disorder, or charge density. Quantum phase transitions are studied through measurements at nonzero temperature of physical behaviors influenced by quantum fluctuations. Among other systems exhibiting quantum phase transitions are ^{4}He adsorbed on random substrates, two-dimensional electron gases, and numerous complex strongly correlated electron materials including high temperature superconductors. The concept is also relevant to the physics of cold atoms, which in some instances can serve as model systems for many condensed matter phenomena.[4]

Superconductor-insulator transitions are of particular importance because of their connection with the fundamental phase-number uncertainty relation, which makes them the simplest of such transitions. Although the subject of SI transitions might be considered to be mature, there remain many open issues resulting from the study of new materials, the application of new experimental techniques, and the extension of measurements to lower temperatures and higher magnetic fields.

2. Materials and Tuning Parameters

Experiments on various materials with different tuning parameters do not clearly distinguish between competing theoretical pictures. Indeed there may be several types of SI transitions. First, the mechanism may depend upon the tuning parameter, which can be disorder, electrostatic charge, perpendicular or parallel magnetic field, surface magnetic impurities, or dissipation. Second, the mechanism and phase diagram may depend upon the type of film being investigated, i.e. whether it is an ultrathin quench condensed amorphous or granular film, an amorphous film of InO_x, MoGe, $Nb_{0.15}Si_{0.85}$, or Ta, or a polycrystalline film of TiN. Third, the character of the substrate may matter. The nature of the transition could be altered in the presence of a high dielectric constant substrate such as $SrTiO_3$, or may depend upon the nature of an underlayer. Finally another issue is the role of spin-orbit scattering. Experiments on high-Z elements (a-Bi or a-Pb) may produce different results from those carried out on low-Z elements (a-Be or a-Al), where there is minimal spin-orbit scattering.

Among some of the phenomena that have sparked renewed interest in the field are the giant magnetoresistance peak found in some nearly critical SI transition systems, leading in some instances to what has been called "superinsulating" behavior at sufficiently low temperatures.[5,6] In some systems there appears to be a "quantum metal" in the limit of very high magnetic fields.[7,8] A deep analogy may exist between the insulating regime of some of these systems and the pseudogap regime of the high temperature superconductors.[9,10] Other issues include the possibility of an intermediate metallic phase between the superconductor and insulator in some perpendicular magnetic field tuned transitions.[11]

The nature of the insulating regime, apart from the magnetoresistance peak, is an important issue, as there is new experimental evidence from a variety of sources that superconducting coherence persists over finite length scales.[12–15] Also, the Nernst effect has emerged as a means of exploring the insulating regime of the field-tuned SI transition.[16–18] The possibility of electrostatically inducing superconductivity has been demonstrated, allowing the exploration of behavior that does not depend on possible changes in morphology associated with different doping or film thickness.[19,20] Using electrostatic charging or doping it is possible to import into the study of the SI transition some of the approaches used for 2D electron gas physics.

3. Theoretical Scenarios

Interest in SI transitions arose in the context of study of the interplay between superconductivity and disorder. Anderson[21] and Abrikosov and Gor'kov[22] showed that nonmagnetic impurities do not affect the superconducting transition because Cooper pairs form from extended time-reversed eigenstates, which have disorder included. On the other hand with sufficient disorder Anderson localization occurs, and superconductivity will not occur, even in the presence of an attractive electron–electron attraction.[23] This leads to the question of how superconductivity disappears with increasing disorder.[24] At nonzero temperatures, the superconducting phase transition is of second order, so that there are fluctuations. For a transition suppressed to zero temperature, the fluctuations become quantum in character and the transition becomes a quantum phase transition.[1]

In two dimensions (2D) there is added complication as the superconducting transition is a topological, or Kosterlitz–Thouless–Berezinskii transition.[25] Initially, all 2D electronic systems were believed to be localized, even for arbitrarily weak disorder. This changed with the observation of apparent metal-insulator transitions in strongly interacting high mobility low-carrier-density MOSFETs.[26] Superconductor-insulator transitions in 2D are additional examples in which interactions play a critical role.

There have been four general approaches to explaining the demise of superconductivity in the 2D limit as a function of disorder or magnetic field. The first is based on a perturbative microscopic description of homogeneous systems and considers the interplay between the attractive and repulsive electron–electron interactions in the presence of disorder. In this microscopic description, suppression of the critical temperature follows from the renormalization of the electron–electron interaction in the Cooper channel by the long range Coulomb repulsion in the presence of disorder.[27–29] The amplitude of the superconducting order parameter is suppressed by disorder, and at a sufficient level of disorder, Cooper pairs break up into fermions, which localize in 2D. The second is a boson localization theory, which predicts a quantum critical point. Inside the critical region, Cooper pair phase fluctuations are important, and these pairs in some approximation are bosons.[30] The observation of apparently direct SI transitions in ultrathin films, in the $T \to 0$ limit[31,32] in part served to motivate this approach. This will be discussed below. The third is based on the physics of granular superconductors. It builds on dissipative models of resistively shunted Josephson junction arrays, and may have application to both granular and homogeneous 2D superconductors. The fourth involves percolation. The message conveyed by experiment regarding the applicability of various models continues to be quite mixed and indeed the explanation of some transitions may involve more than one idea.

3.1. *The Fermion localization scenario*

As long as the reduction of the transition temperature is relatively small, the perturbative approach mentioned above should provide an accurate expression for the data. In the regime where the transition would be suppressed to zero there will be quantum fluctuations, and this theory should fail, as such fluctuations are not included.[33] On the other hand, the reduction of the mean field transition temperature that is predicted by perturbative theory based on the weakening of the screening of the Coulomb interaction with increasing disorder[29] is consistent with studies of homogeneous *a*-MoGe films by Graybeal and Beasley[34] (see Fig. 1).

3.2. *The Boson localization scenario*

Boson localization theories[30,35,36] are closely related to models used to

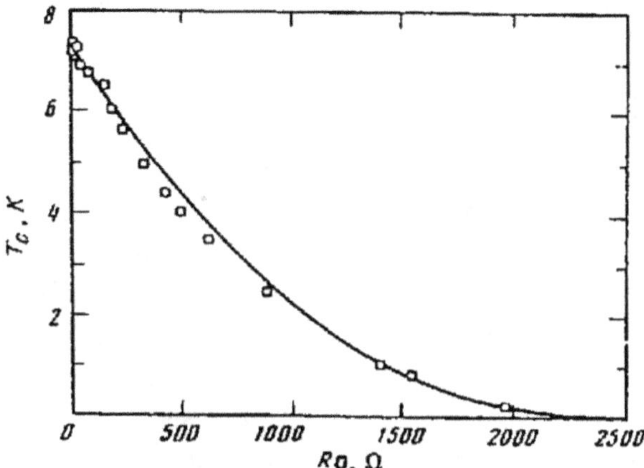

Fig. 1. Suppression of superconductivity in amorphous Mo_xGe_y with increasing disorder.[34] The solid line is a theoretical fit as discussed in Ref. 29.

describe superfluid helium and Josephson junction arrays. Doniach[37] pioneered this approach involving phase fluctuations, in the context of Josephson junction arrays. There is a quantum critical point, and the ground state can be controlled by disorder or perpendicular magnetic field. An important prediction is that at a critical level of disorder, or magnetic field, Cooper pairs move diffusively, leading to *metallic* behavior in the $T \to 0$ limit only at the quantum critical point. If one assumes complete duality between Cooper pairs and vortices[38] this resistance is universal with a value given by the quantum resistance for pairs, $h/4e^2$. The physics is related to the universal jump condition in the finite temperature Kosterlitz–Thouless–Berezinskii transition.[25]

The first real indication that disorder might drive the transition temperature down to a low enough value that quantum fluctuations could be important was the work of Haviland, Liu and Goldman[31] in which the thickness variation of resistance versus temperature $R(T)$, was studied in amorphous Bismuth (a-Bi) films. Bismuth is a semimetal in its usual form, but is metallic and superconducting under high pressure, or in thin film amorphous form, which results when it is grown on substrates held at liquid helium temperatures.[39] This investigation was carried out in an apparatus in which the films could be grown *in situ*, at low temperatures and measurements carried out in alternation with increments of growth. If such studies are carried out using substrates pre-coated with a thin layer of amorphous Ge or Sb, atomic scale rather than mesoscale disorder apparently results.

In contrast, films made using a low temperature evaporation technique but without an underlayer of amorphous Ge or Sb, are granular in that they possess mesoscale clusters. They also exhibit a separation between superconducting and insulating behavior in the limit of zero temperature, but their properties differ in detail from the nominally homogeneous amorphous films described above. In particular, the dependence of the resistance versus temperature is a nonmonotonic function of temperature.[40]

Curves of $R(T)$ for a sequentially deposited set of a-Bi films grown on a-Ge substrates are presented in Fig. 2. These resemble renormalization flows to an unstable fixed point at zero temperature, which for this particular set of films correspond to the quantum resistance for pairs, $h/4e^2$, or 6450 $\Omega/\square$. This result set the stage for consideration of the SI transition as a continuous quantum phase transition and motivated the application of Boson physics to the SI transition. Unfortunately subsequent studies of the SI transition in other materials indicated that the critical resistance was not universally $h/4e^2$. Yazdani and Kapitulnik[41] in their study of the field-driven SI transition, suggested that the nonuniversal critical point resistance could be a consequence of a parallel fermionic conductance channel at nonzero temperature, a view also supported by Gantmakher and collaborators.[42] However this approach does not describe all of the systematics of the data of Markovic *et al.*,[43,44] on the field-driven transition of quench-evaporated films of different thicknesses.

Without duality assumption, and with a realistic form for the interaction between charges, the resistance at the critical point is a nonuniversal constant. Experimentally the resistance appears to be sample-dependent but is within a factor of 2 or 1/2 of the quantum resistance for pairs, $h/4e^2 = 6450\ \Omega$,[45] although recent work[46,47] suggests that there may be regimes in which the critical resistance of the field-driven SI transition may indeed be $h/4e^2$. In elaborations of the Bose-Hubbard model, the value of the critical resistance depends on the form of interaction between charges, and the details of the calculations. (See for example Refs. 48–53.) Some of these works are concerned with including low energy quasiparticle interactions in the theory. These are claimed to be a most serious aspect of missing physics, and they introduce amplitude fluctuations into the description of the transition.

Tunneling experiments on quench evaporated homogeneous films of metals similar in preparation to those shown in Fig. 2 seem to support fermion rather than boson scenarios.[54] The mean field transition temperature and the energy gap are suppressed at a fixed ratio as sheet resistance increases,

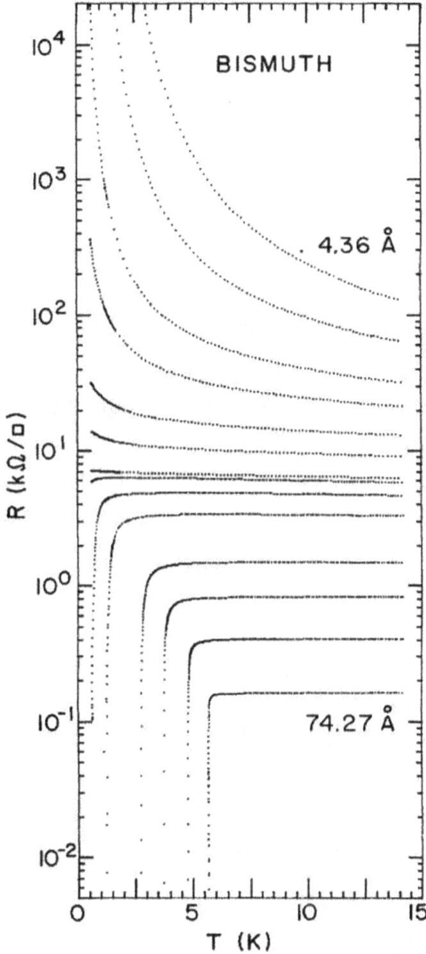

Fig. 2. Evolution of the temperature dependence of the sheet resistance $R(T)$ with thickness for an a-Bi film deposited onto Ge. Fewer than half of the traces actually acquired are shown. Film thicknesses that are shown range from 4.36 to 74.27 Å. (From Ref. 31.)

a phenomenon treated theoretically by Smith *et al.*[55] The energy gap also vanishes on the insulating side of the SI transition of homogeneously disordered films,[56] and on the insulating side of the field-tuned transition.[57] These studies suggest that there are serious amplitude fluctuations associated with the SI transition. One might also conclude from this interpretation that the vanishing of the energy gap implies that the amplitude of the order parameter vanishes in the insulating state, which would appear to falsify the

Boson-localization model. Such a conclusion should be treated with care as the vanishing of the gap feature in superconducting tunneling could result from other effects such as pair breaking by phase fluctuations.

4. Scaling Analysis of the Critical Regime of a Continuous SI Transition

Because the SI transition is so intimately connected with the idea of direct continuous quantum phase transitions, it is of interest to examine this in more detail. As mentioned, the application of the quantum phase transition (QPT) paradigm followed the publication of the data of Fig. 2. Continuous QPTs are transitions at absolute zero brought about by changing a parameter in the Hamiltonian of the system.[1,58] The quantum mechanical ground state changes when the critical point is crossed. In contrast with phase transitions at nonzero temperature, quantum effects are central to QPTs. In other phase transitions, although the order parameter itself may be quantum mechanical, classical thermal order parameter fluctuations govern its behavior at the relevant long wavelengths. In the case of QPTs, the fluctuations themselves are quantum mechanical. Near a QPT there are divergent correlation lengths, for the spatial, ξ, and temporal, ξ_τ, directions. The latter is associated with a vanishing energy scale. These lengths depend on the control parameter, $\delta = |(g - g_c)/g_c$ for the transition. Here g can be a measure of disorder, magnetic field, or charge, and g_c is the critical value of the control parameter. We then write:

$$\xi \sim |\delta|^{-\nu}, \quad \text{and} \quad \xi_\tau \sim \xi^z \tag{1}$$

This defines the correlation length exponent ν and the dynamical critical exponent z, where ξ and ξ_τ are the correlation length and dynamical correlation length, respectively. Physical quantities in the critical region close to the QPT are homogeneous functions of the independent variables in the problem. Details of issues relevant to quantum phase transitions are considered in the works cited above.

The key feature of QPTs is the interplay of dynamics and thermodynamics. As a consequence of this, a d-dimensional quantum system at finite temperature is described in the $T \to 0$ limit as a classical system of $d + z$ dimensions, with the finite extent of the system in the extra dimensions being given by $-\hbar\beta$ in units of time, where $\beta = 1/k_BT$. The extent of these extra dimensions are divergent only in $T \to 0$ limit. What is remarkable is that the universality class of the quantum transition may be one studied extensively in some classical context. This also allows for the possibility of a computational treatment of the quantum mechanical problem using simulations of the $d+z$ dimensional classical problem. Disorder on the other hand can change the universality class of the equivalent classical problem. It is also not true in general that space and time enter in the same fashion in the equivalent classical problem. For this to happen the dynamical exponent, z, must be unity, and whether or not this happens depends upon the detailed quantum dynamics of the system.

As mentioned above, the effect of considering $T \neq 0$ in the statistical mechanics is to force the "temporal" dimension of the problem to be finite. The formal model used to analyze data at nonzero temperatures is finite-size scaling. The success of finite size scaling analyses of the various superconductor-insulator transitions is part of the evidence for there being QPTs. Scaling can be used to characterize properties measured at nonzero temperatures in the regime of critical fluctuations and thus to determine the critical exponents and universality class of the transition. For the resistance near the transition, it is of the form:

$$R = R_c F(\delta/T^{1/\nu z}, \delta/E^{1/\nu(z+1)}) \tag{2}$$

Here R is the sheet resistance, R_c and F are an arbitrary constant and an arbitrary function. The energy scale for the fluctuation $\Omega \sim \xi_T^{-z}$ is cut off at nonzero temperature by $k_B T$, defining a cut off length L_T given by $k_B T \sim L_T^{-z}$. This gives rise to the first term in the argument of the arbitrary function F. The second term comes from a characteristic length associated with the electric field as compared with the correlation length. The applicability of these ideas assumes that there is a continuous and direct SI transition. Hebard and Palaanen carried out the first scaling analysis of a perpendicular field driven transition of In_2O_3 films.[32] They found that the exponent product, $\nu z \sim 1.2$ from scaling with only the first argument of Eq. (2). Later Yazdani and Kapitulnik analyzed the properties of Mo_xGe_y films, carrying out both temperature and electric field scaling.[41] They reported both νz and $\nu(z+1)$ and thus ascertain that $z = 1$ in agreement with expectations for systems with long-range Coulomb interactions.[59]

For homogeneous films of different types, including cuprates, there are numerous finite-size scaling analyses of experimental data for the perpendicular field-driven transition in addition to the work of Hebard and Paalanen and Yazdani and Kapitulnik. These include Paalanen, Hebard, and Ruel,[60] Okuma and Kokubo,[61] Inoue *et al.*,[62] Seidler, Rosenbaum and Veal,[63] Wang *et al.*,[64] Okuna, Terashima and Kokubo,[65] Gantmakher *et al.*,[42] and on the thickness and field-driven transitions together, Markovic *et al.*[43,44] These are all consistent with a quantum critical point. An important caveat is that the success of scaling by itself and the identification of the universality class of the transition from the values of the critical exponents may not elucidate the microscopic physics of the transition or the nature of the insulating state.

5. Scaling of Continuous Quantum Phase Transitions

We now consider data of $R(T, \delta)$ for a set of films different from those shown in Fig. 2. As is typical of such experiments, at some critical thickness, d_c, R becomes temperature independent, while for even thicker films it decreases rapidly with decreasing temperature, indicating the onset of superconductivity. The critical thickness is found by plotting R versus d at different temperatures (inset of Fig. 3) and identifying the crossing point for which the resistance is temperature independent, or by plotting dR/dT versus d at the lowest temperatures and finding the thickness for which $dR/dT \to 0$. In the quantum critical regime the resistance of a two-dimensional system is expected to obey the scaling law of Eq. (2), with the in-plane electric field taken to be a constant.

To analyze the data, we obtain curves of $R(d)$ at various temperatures. Having determined the critical thickness in the manner described above, we then rewrite Eq. (2) as $R(t, \delta) = R_c F(\delta t)$ where $t = T^{-1/\upsilon z}$, and treat the parameter $t(T)$ as an unknown variable which is determined at each temperature to obtain the best collapse of the data. Specifically, t is determined by performing a numerical minimization between a curve at a particular temperature and the lowest temperature curve. The exponent product υz is then found from the temperature dependence of t, which must be a power law in temperature for the procedure to make physical sense. This procedure does not require detailed knowledge of the functional form of the temperature or thickness dependence of R, or prior knowledge of the critical exponents. A different method of obtaining critical exponents was also used to check the consistency of this procedure. A log-log plot of $(\partial R/\partial d)|_{d_c}$ versus T^{-1} was constructed. Its slope is equal to $1/\nu z$ if Eq. (2) is obeyed. Exponents obtained this way were essentially identical, within the quoted experimental uncertainty, to those obtained using the first procedure.

The collapse of the data of $R(T, \delta)$ is shown in Fig. 3.[44] The exponent-product υz is found to be 1.2 ± 0.2, with the error determined from the power law fit. This agrees with theoretical predictions from which $z = 1$ would be expected for a bosonic system with long-range Coulomb interactions independent of dimensionality, and with $\nu > 1$ in 2D.

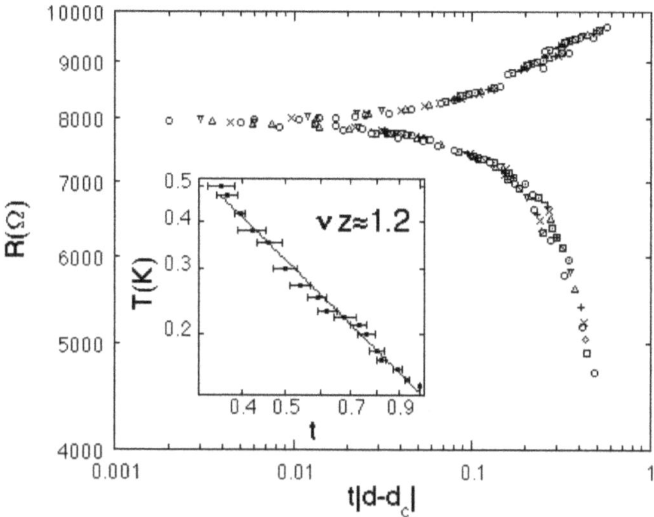

Fig. 3. Resistance per square as a function of the scaling variable, $t|d - d_c|$, for 17 different temperatures, ranging from 0.14 to 0.5 K. Different symbols represent different temperatures. Inset: temperature dependence of t. (From Ref. 44.)

When the exponent product was determined with magnetic field rather than thickness as the tuning parameter, $\nu z = 0.7 \pm 0.2$ was found. The fact that the field-tuned transition differs from the thickness-tuned transition suggests a universality class different from that of the thickness-tuned transition. If $z = 1$, this result, with $\nu \sim 0.7$, would correspond to the 3D XY model.

It is important to note several features of these two SI transitions. The transitions are direct in that there appear to be no intermediate metallic phases and there is no resistance saturation at the lowest temperatures. This is in contrast with results reported for other types of systems[11,66] or what was found in granular films.[40] It should also be noted that the films, which have been studied in the case of the perpendicular field-tuned SI transition are very close in their properties to the insulating regime. Subsequent works, in our laboratory on *a*-Bi films,[67] and by others on *a*-Pb films,[68] which were less disordered have revealed the presence of an intermediate regime that may have two phases. The precise nature of the phase diagram is not known at this time.

Parker *et al.*,[68] also explored the SI transition by decorating Pb films with magnetic impurities. They did not carry out a scaling analysis of this data.

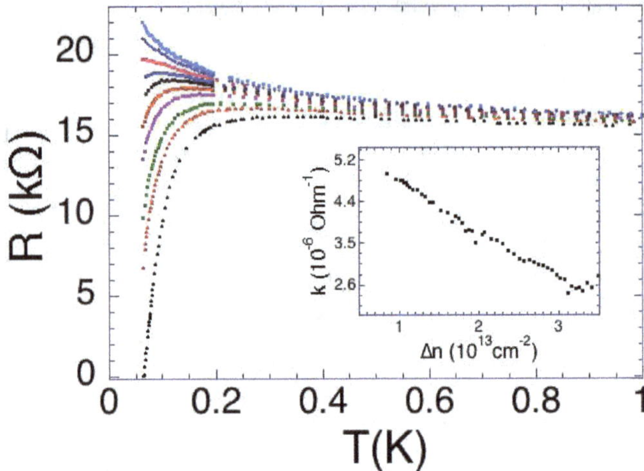

Fig. 4. Resistance versus temperature at various values of Δn for the 10.22 Å thick film with $B = 0$. Data are shown from 60 mK to 1 K. The values of Δn that are shown, from top to bottom, are 0, 0.62, 1.13, 1.43, 1.61, 1.83, 2.04, 2.37, 2.63, and 3.35_1013 cm^{-2}. 44 curves of $R(T)$ for other values of Δn are omitted from the plot for clarity. Inset: slope of ln(T) versus Δn in the normal state. (From Ref. 19.)

The superconductor-insulator transition of *a*-Bi films has been studied in parallel magnetic fields with two types of samples, those in which superconductivity was induced by electrostatic doping, which will be discussed below, and those which were intrinsic superconductors with thicknesses close to the critical thickness for the appearance of superconductivity.[69] In both instances exponent products of about 0.7 were found.

The SI transition has also been tuned electrostatically. This gets around issue of sample inhomogeneity and vortex pinning that are important in thickness and perpendicular magnetic field tuned transitions. In this approach the level of physical disorder is in principle fixed, and the outcome is not dependent upon the degree of vortex pinning.

The field effect geometry was one in which $SrTiO_3$ (STO) crystal served as both a substrate and gate insulator. A sequence of *a*-Bi films was studied.[19,69] The temperature dependence of insulating films was governed by the 2D Mott variable range hopping form.

The addition of electrons to a 10.22 Å thick film, using the field effect, induced superconductivity as shown in Fig. 4. An important finding in this work was the crossover from 2D Mott hopping in the insulating regime to a ln(T) dependence of the conductance on temperature in the normal state for films which underwent a transition to superconductivity. What was also found is that the coefficient of the $\ln(T)$ term in the conductance was a linear function of the charge transfer, as shown in the inset to Fig. 4. This coefficient is related to the screening of the electron–electron interaction.

This SI transition was successfully analyzed using finite-size scaling, employing the charge transfer Δn as the control parameter. This suggests that the electrostatically tuned transition is also a continuous quantum phase transition. In this instance, the exponent product νz was 0.7 ± 0.1, including data all the way up to 1 K if the $\ln(T)$ dependence of the conductance was first removed. This was done by assuming that there were two parallel conductance channels, one involving weak localization and electron–electron interactions, and the other the critical fluctuations.

The apparent coincidence of an insulator-metal transition with the insulator superconductor transition suggests that the charge-tuned SI transition is an electronic phenomenon and is not associated with the localization of bosons.

6. Metallic Regimes

Measurements by Mason and Kapitulnik,[70] Chervenak and Valles,[71] Qin *et al.*,[72] and Seo *et al.*,[11] support the idea of a metallic regime for certain samples. Gantmakher and collaborators[42] reported a similar behavior. The resistances of films, which have dropped significantly at high temperatures and appear to be superconducting, were found to saturate at nonzero values in the $T \to 0$ limit. This metallic regime, between the superconductor and insulator, exhibits a resistance much lower than that of the normal state. Chevernak and Valles[71] in their field-tuned studies refer to this phase as a vortex liquid. Tewari has suggested that this is a sharp crossover that appears to be first order in character from a strong superconductor to an inhomogeneous state, which is a weak superconductor.[73]

Kapitulnik *et al.* discussed the possibility that coupling to a dissipative heat bath characterized by a single parameter stabilizes the metallic phase.[74] Dissipation also plays a different role in phase-only theories of the SI transition of granular superconductors that build on the physics of resistively shunted Josephson junction arrays.[75] Dissipation can localize the phase of the system, leading to superconductivity.[76] Mason and Kapitulnik reported that a ground plane in proximity with a MoGe film seems to promote superconductivity, although it did not change the dynamical critical exponent as might be expected with enhanced screening.[66] Vishwanath *et al.*, have presented a comprehensive theoretical discussion of screening and dissipation. In their work, a ground plane introduces screening of the long-range Coulomb interaction and provides a source of dissipation due to the gapless diffusive electrons.[77] Work on hot electron effects in the parallel field transition, suggested that resistance saturation may be the result of a failure of electrons to cool.[78] This work also casts doubt on the validity of electric field scaling as discussed in Yazdani and Kapitulnik.[41]

In studies of the thickness-dependent SI transition of *granular* ultra-thin films, a metallic regime is indeed found before the transition to zero resistance.[40,79,80] In contrast with the results mentioned above, this metallic regime is found at relatively high temperatures. Das and Doniach,[81] Dalidovich and Phillips,[82] and Ng and Lee,[83] have presented theoretical approaches to intervening metallic phases. Phillips and Dalidovich reviewed the subject in a comprehensive manner.[84]

7. Insulating Regimes

There has been considerable interest in the insulating regime. Palaanen *et al.*,[60] were the first to report a peak in $R(B)$ above the critical field of a-In_xO_y films. The Hall effect also turned on near this field.[85] They suggested this was the signature of a transition or crossover between Bose and Fermi insulators and that the order parameter amplitude was non-vanishing in the insulating regime. Very recently there have been a number of papers confirming and extending this work.[42,86–89] An alternative to the Bose-to-Fermi insulator crossover at this field is that the insulating regime consists of intermixed superconducting and normal phases. The understanding of this phenomenon has attracted considerable attention as Steiner, Boebinger and Kapitulnik have noted the similarity of data on some InO_x films to observations on cuprate superconductors, but with a much larger magnetic field scale for the latter.[9] They suggested that the insulating regime of cuprates attained in high fields is a Bose insulator. The linear resistance can actually fall to zero near criticality at sufficiently low temperatures, as first reported by Sambandamurthy *et al.*,[6] for InO_x films and later for TiN by Baturina *et al.*[90] The explanation of this "superinsulating" behavior in terms of a macroscopic Coulomb blockade in large Josephson junction arrays[91–93] has generated significant controversy.[94] Recent work has shown that the apparent switching to an insulating state may be a nonlinear thermal effect.[95,96]

Other support for the existence of Cooper pairs in the insulating regime are the measurements of Markovic *et al.*,[97] of the anisotropic magnetoresistance of films on the insulating side of the disorder-tuned SI transition, which support the idea that vortices are present. Gantmakher *et al.* concluded from an analysis of magneto-transport in Cd-Sb alloys that there may be localized pairs in the insulating state of this system.[98] Recently Crane *et al.*[14] reported studies of the complex *AC* conductivity of InO_x films through the magnetic field tuned SI transition. The data revealed a nonzero frequency superfluid stiffness well into the insulating regime. Stewart *et al.*, (2007) presented results on quench-condensed *a*-Bi films patterned with a nano-honeycomb array of holes.[13] Insulating films exhibited temperature-dependent resistances and magnetoresistance oscillations dictated by the flux quantum $h/2e$, which were interpreted as of evidence of Cooper pairing in the insulating regime. Nernst effect measurements on the perpendicular field driven transition of Nb_xSi_{1-x} films also suggest the presence of superconducting coherence in the insulating regime.[16–18]

8. Quantum and Classical Percolation

Shimshoni *et al.*,[99] and Sheshadri *et al.*,[100] suggested that the SI transition is percolative. This phenomenology would become applicable if films were either physically inhomogeneous or, if actually homogenous, were rendered inhomogeneous by order parameter amplitude fluctuations.[101] Resistive transport then involves tunneling of vortices across weak-link-connected superconducting domains. This scenario blurs the distinction between granular and homogenous systems. Percolation might also explain the fact that the correlation length exponents found in finite size scaling analyses of some SI transitions are close to the percolation exponent in 2D, which is $\sim 4/3$. Considering the temperature dependencies of the length scales controlling percolation and the quantum critical point, Kapitulnik *et al.*[74] suggested that scaling carried out using high temperature data would yield percolation exponents, whereas low temperature data would be needed to find quantum critical exponents. Based on "super-universality" they also suggested that the correlation length exponent, $n = 7/3$, and the critical resistance should be $h/2e^2$, or double the quantum resistance for pairs. Dubi *et al.*,[102] discussed a model based on quantum percolation and constructed a global phase diagram.

Through a detailed study of scaling near the magnetic field-tuned superconductor-to-insulator transition in strongly disordered films, Steiner, Brezney and Kapitulnik[46] reported that results for a variety of materials could be collapsed onto a single phase diagram consisting of two clear branches, one with weak disorder and an intervening metallic phase, and

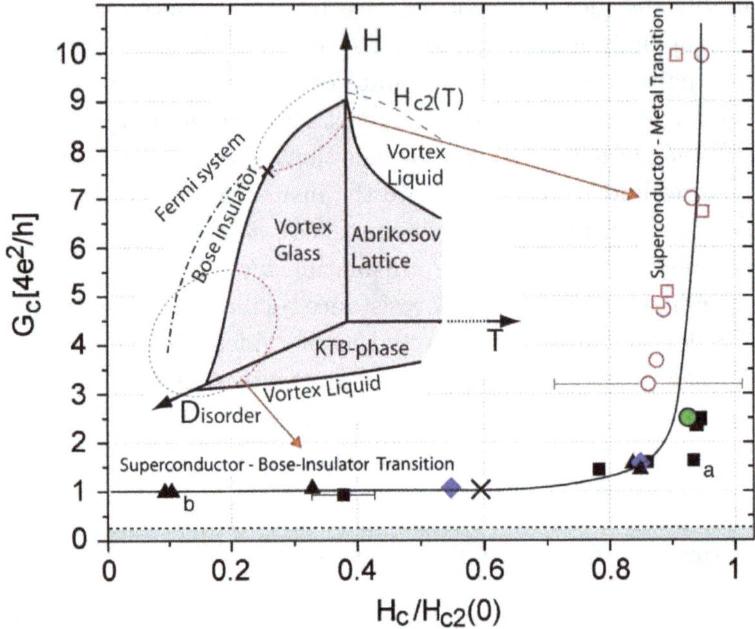

Fig. 5. Phase diagram including all samples, using reduced axes. The open circles and squares are for MoGe films. The full circle is for thin Ta films.[11] The diamonds are for InO_x (Refs. 32 and 60); full triangles and squares are for InO_x data; the solid line is a guide to the eye. The X denotes a possible critical point between a transition to a Bose insulator and a transition to a metal. The bottom-shaded area denotes the region for which the conductance is smaller than h/e^2. The inset shows the low temperature part of the generic phase diagram as proposed in Ref. 35; the crossover line between the Bose and Fermi insulators is terminated before reaching zero disorder and does not reflect the possible emergence of a quantum metallic state. (From Ref. 46.)

the other with strong disorder. Along the strongly disordered branch, they reported that the resistance at the critical point approached the quantum resistance for pairs, and the exponent product $\nu z \sim 2.3 \pm 0.2$, or 7/3 was consistent with values associated with quantum percolation. Along the weakly disordered branch there was an apparent intermediate metallic regime and the critical resistance was below the quantum resistance for pairs. Data from the strong disorder branch are all from In_xO_y films, whereas the low disorder branch consisted of Mo_xGe_y and Ta films. The proposed phase diagram is shown in Fig. 5.

This work adds a different twist to the study of the SI transition, suggesting that while a true Bose-like SI transition can be achieved in the limit of strong disorder, the other limit of weak disorder and high critical fields contains different physics in which a quantum metallic state is the central feature. The high disorder films exhibit a substantial magnetoresistance peak. It is suggested that the regime of high resistance is a Bose insulating regime, which is more extensive in magnetic field, the more disordered the film. The high field behavior is always Fermi like in character.

The exponent product for the more disordered films of 2.33, which is close to 7/3, suggests that the transition is a quantum rather than classical percolation transition and that the results are consistent with the dirty Boson picture of the SI transition, at least for the case of the strongly disordered branch of the data.

9. Discussion

Despite a long history it is clear that there are many open questions relating to SI transitions. These include the following: What different physical models govern the various SI transitions, which have different exponent products and different critical resistances? Why are the insulating states of various transitions so very different? What are the ingredients necessary for a large resistance peak? Is this regime one in which there are localized Cooper pairs in an essentially homogenous system, or in particular is disorder on a length scale greater than atomic scales necessary for such a peak? Is an intermediate metallic regime, or a two-phase regime always an intrinsic feature of some transitions or is the former an artifact of the failure of electrons to cool? Assuming that there are examples in which there is no intermediate metallic and/or two-phase regime phase, can one explore parameter space of the SI transition and determine critical exponents in the asymptotic limit so as to accurately ascertain the universality classes of these transitions? The work of Steiner, Brezney and Kapitulnik purports to do this for In_xO_y, but its results, in particular the claim that the critical resistances of highly disordered films are universal are hard to understand given the requirement of charge-flux duality.[34]

Acknowledgments

The author would like to thank numerous individuals for enlightening discussions over the years. These include, M. P. A. Fisher, Steven Girvin, Aharon Kapitulnik, Zvi Ovadyahu, Aviad Frydman, James Valles, Wenhao Wu, Boris Shklovskii, Leonid Glazman, Robert Dynes, Herve Aubin, He would also like to thank students who have worked in his laboratory on this problem. These include Bradford Orr, Heinrich Jaeger, Ying Liu, Anthony Mack, Nina Markovic, Cathryn Christiansen, Gloria Martinez-Arizala, Luis Hernandez, Kevin Parendo, Sarwa Tan, and Yen-Hsiang Lin.

References

1. S. Sachdev, *Quantum Phase Transitions* (Cambridge University Press, Cambridge, UK, 1999).
2. A. Gerber, A. Milner, G. Deutscher, M. Karpovsky and A. Gladkikh, *Phys. Rev. Lett.* **78**, 4277 (1997).
3. H. S. J. van der Zant, W. J. Elion, L. J. Geerligs and J. E. Mooij, *Phys. Rev. B* **54**, 10081 (1996).
4. C. Bruder, R. Fazio and G. Schon, *Ann. Phys. (Leipzig)* **9–10**, 566 (2005).
5. G. Sambandamurthy, L. W. Engel, A Johansson and D. Shahar, *Phys. Rev. Lett.* **94**, 017003 (2005).
6. T. I. Baturina, A. Yu. Mironov, V. M. Vinokur and M. R. Baklanov, *Phys. Rev. Lett.* **99**, 257003 (2007).
7. V. Yu. Butko and P. W. Adams, *Nature* **409**, 161 (2001).
8. T. I. Baturina, C. Strunk, M. R. Baklanov and A. Satta, *Phys. Rev. Lett.* **98**, 127003 (2007).
9. M. Steiner, G. Boebinger and A. Kapitulnik, *Phys. Rev. Lett.* **94**, 107008 (2005).
10. Y. Dubi, Y. Meir and Y. Avishai, *Nature* **449**, 876 (2007).
11. Y. Seo, Y. Qin, C. L. Vicente, K. S. Choi and J. Yoon, *Phys. Rev. Lett.* **97**, 057005 (2006).
12. C. L. Vincente, Y. Qin and J. Yoon, *Phys. Rev. B* **76**, 100507 (2006).
13. M. D. Stewart, Jr., A. Yin, J. M. Xu and J. M. Valles Jr., *Science* **318**, 1273 (2007).
14. R. Crane, N. P. Armitage, A. Johansson, G. Sambandamurthy, D. Shahar and G. Gruner, *Phys. Rev. B* **75**, 184530 (2007).
15. K. H. Sarwa, B. Tan, K. A. Parendo and A. M. Goldman, *Phys. Rev. B* **78**, 014506 (2008).
16. A. Pourett, H. Aubin, J. Lesueur, C. A. Marrache-Kikuchi, L. Berge, L. Dumoulin and K. Behnia, *Nature Physics* **2**, 683 (2006).
17. A. Pourett, H. Aubin, J. Lesueur, C. A. Marrache-Kikuchi, L. Berge, L. Dumoulin and K. Behni, *Phys. Rev. B* **76**, 214504 (2007).
18. P. H. Spathis, H. Aubin, A. Pourret and K. Behnia, arXiv:0712.2655v1.
19. K. A. Parendo, K. H. Sarwa, B. Tan, A. Bhattacharya, M. Eblen-Zayas, N. Staley and A. M. Goldman, *Phys. Rev. Lett.* **94**, 197004 (2005).
20. K. A. Parendo, K. H. Sarwa, B. Tan and A. M. Goldman, *Phys. Rev. B* **73**, 174527 (2006).
21. P. W. Anderson, *J. Phys. Chem. Solids* **11**, 26 (1959).
22. A. A. Abrikosov and L. P. Gor'kov, *Zh. Eksp. Teor. Fiz. (USSR)* **36**, 319 (1959).

23. P. A. Lee and T. V. Ramakrishnan, *Rev. Mod. Phys.* **57**, 287 (1985).
24. M. Ma and P. A. Lee, *Phys. Rev. B* **32**, 5658 (1985).
25. J. E. Mooij, in *Percolation, Localization and Superconductivity*, eds. A. M. Goldman and S. A. Wolf (Plenum Press, New York and London, 1983), p. 325.
26. E. Abrahams, S. V. Kravchenko and M. P. Sarachik, *Rev. Mod. Phys.* **73**, 251 (2001).
27. A. Maekawa and H. Fukuyama, *J. Phys. Soc. Japan* **51**, 1380 (1981).
28. H. Ebisawa, H. Fukuyama and S. Maekawa, *J. Phys. Soc. Japan* **54**, 2257 (1985).
29. A. M. Finkel'stein, *Physica B* **197**, 636 (1994).
30. M. P. A. Fisher, *Phys. Rev. Lett.* **65**, 923 (1990).
31. D. B. Haviland, Y. Liu and A. M. Goldman, *Phys. Rev. Lett.* **62**, 2180 (1989).
32. A. F. Hebard and M. A. Paalanen, *Phys. Rev. Lett.* **65**, 927 (1990).
33. A. Larkin, *Ann. Phys. (Leipzig)* **8**, 785 (1999).
34. J. M. Graybeal and M. R. Beasley, *Phys. Rev. B* **29**, 4167 (1984).
35. M. P. A. Fisher, G. Grinstein and S. M. Girvin, *Phys. Rev. Lett.* **64**, 587 (1990).
36. M.-C. Cha, M. P. A. Fisher, S. M. Girvin, M. Wallin and P. Young, *Phys. Rev. B* **44**, 6883 (1991).
37. S. Doniach, *Phys. Rev. B* **24**, 5063 (1981).
38. X. G. Wen and A. Zee, *Int. J. Mod. Phys. B* **4**, 437 (1990).
39. W. Buckel and R. Hilsch, *Z. Phys.* **138**, 109 (1959).
40. H. M. Jaeger, D. B. Haviland, B. G. Orr and A. M. Goldman, *Phys. Rev. B* **40**, 182 (1989).
41. A. Yazdani and A. Kapitulnik, *Phys. Rev. Lett.* **74**, 3037 (1995).
42. V. F. Gantmakher, M. V. Golubkov, V. T. Dolgopolov, G. E. Tsydynzhapov and A. A. Shashkin, *Physica B*, **284–288**, 649 (2000).
43. N. Markovic, C. Christiansen, and A. M. Goldman, *Phys. Rev. Lett.* **81**, 5217 (1998).
44. N. Markovic, C. Christiansen, A. M. Mack, W. H. Huber and A. M. Goldman, *Phys. Rev. B* **60**, 4320 (1999).
45. N. Markovic, A. M. Mack, G. Martinez-Arizala, C. Christiansen and A. M. Goldman, *Phys. Rev. Lett.* **81**, 701 (1998).
46. M. Steiner, N. P. Breznay and A. Kapitulnik, *Phys. Rev. B* **77**, 212501 (2008).
47. W. Wu, private communication.
48. A. Gold, *Z. Phys. B* **81**, 155 (1990); A. Gold, *Z. Phys. B* **87**, 169 (1992).
49. W. Krauth, N. Tivedi and D. Ceperley, *Phys. Rev. Lett.* **67**, 2307 (1991).
50. M. Makivic, N. Trivedi and S. Ullah, *Phys. Rev. Lett.* **71**, 2307 (1993).
51. N. Trivedi, R. T. Scalettar and M. Randeria, *Phys. Rev. B* **54**, R3756 (1996).
52. K.-H. Wagenblast, A. van Otterlo, G. Schon and G. T. Zimanyi, *Phys. Rev. Lett.* **79**, 2730 (1997).
53. P. Phillips and D. Dalidovich, *Phil Mag. B* **81**, 847 (2001).
54. J. M. Valles, Jr., R. C. Dynes and J. P. Garno, *Phys. Rev. B* **40**, 6680 (1989).
55. R. A. Smith, M. Yu. Reizer and J. W. Wilkins, *Phys. Rev. B* **51**, 6470 (1995).
56. J. M. Valles, Jr., R. C. Dynes and J. P. Garno, *Phys. Rev. Lett.* **69**, 3567 (1992).

57. S.-Y. Hsu, J. A. Chervenak and J. M. Valles, Jr., *Phys. Rev. Lett.* **75**, 132 (1995).
58. S. L. Sondhi, S. M. Girvin, J. P. Carini and D. Shahar, *Rev. Mod. Phys.* **69**, 315 (1994).
59. I. F. Herbut, *Phys. Rev. Lett.* **87**, 137004 (2001).
60. M. A. Paalanen, A. F. Hebard and R. R. Ruel, *Phys. Rev. Lett.* **69**, 1604 (1992).
61. S. Okuma and N. Kokubo, *Phys. Rev. B* **51**, 15415 (1995).
62. M. Inoue, H. Matsushita, H. Hayakawa and K. Ohbayashi, *Phys. Rev. B* **51**, 15448 (1995).
63. G. T. Seidler, T. F. Rosenbaum and B. W, Veal, *Phys. Rev. B* **45**, 10162 (1992).
64. T. Wang, K. M. Beauchamp, D. D. Berkley, B. R. Johnson, J.-X. Liu, J. Zhang and A. M. Goldman, *Phys. Rev. B* **73**, 24510 (1992).
65. S. Okuna, T. Terashima and N. Kokubo, *Phys. Rev. B* **58**, 2816 (1998).
66. N. Mason and A. Kapitulnik, *Phys. Rev. B* **65**, 220505(R) (2002).
67. Y.-H. Lin and A. M. Goldman, submitted for publication in *Phys. Rev. Lett.*
68. J. S. Parker, D. E. Read, A. Kumar and P. Xiong, *Europhys. Lett.* **75**, 950 (2006).
69. K. A. Parendo, K. H. Sarwa, B. Tan, A. Bhattachary, M. Eblen-Zayas, N. Staley and A. M. Goldman, *AIP Conf. Proc.* **850**, 949 (2006).
70. N. Mason and A. Kapitulnik, *Phys. Rev. Lett.* **82**, 5341 (1999).
71. J. A. Chervenak and J. M. Valles, Jr., *Phys. Rev. B* **61**, R9245 (2000).
72. Y. Qin, C. L. Vincente and Y. Yoon, *Phys. Rev. B* **74**, 100505 (2006).
73. S. Tewari, *Phys. Rev. B* **69**, 014512 (2004).
74. A. Kapitulnik, N. Mason, S. A. Kivelson and S. Charkravarty, *Phys. Rev. B* **63**, 125322 (2001).
75. I. V. Yurkevich and I. V. Lerner, *Phys. Rev. B* **64**, 054515 (2001).
76. A. J. Rimberg, T. R. Ho, C. Kurdak, J. Clarke, K. L. Campman and A. C. Gossard, *Phys. Rev. Lett.* **78**, 2632 (1997).
77. A. Vishwanath, J. E. Moore and T. Senthil, *Phys. Rev. B* **69**, 054507 (2004).
78. K. A. Parendo, K. H. Sarwa, B. Tan and A. M. Goldman, *Phys. Rev. B* **74**, 134517 (2006).
79. C. Christiansen, L. M. Hernandez and A. M. Goldman, *Phys. Rev. Lett.* **88**, 037004 (2002).
80. K. A. Parendo, K. H. Sarwa, B. Tan and A. M. Goldman, *Phys. Rev. B* **76**, 100508 (2007).
81. D. Das and S. Doniach, *Phys. Rev. B* **60**, 1261 (1999).
82. D. Dalidovich and P. Phillips, *Phys. Rev. Lett.* **89**, 027001 (2002).
83. T. K. Ng and D. K. K. Lee, *Phys. Rev. B* **63**, 144509 (2001).
84. P. Phillips and D. Dalidovich, *Science* **302**, 243 (2003).
85. M. P. A. Fisher, *Physica A* **177**, 553 (1991).
86. G. Sambandamurthy, L. W. Engel, A. Hohansson and D. Shahar, *Phys. Rev. Lett.* **92**, 107005 (2004).
87. M. Steiner and A. Kapitulnik, *Physica C*, **422**, 16 (2005).
88. Y. J. Lee, Y. S. Kim, E. N. Bang, H. Lim and H. K. Shin, *J. Phys.: Condens. Mater.* **13**, 8135 (2001).

89. T. I. Baturina, D. R. Islamov, Z. D. Kvon, M. R. Baklanov and A. Satta, cond-mat/0210250.
90. T. I. Baturina, A. Yu Mironov, V. M. Vinokur and M. R. Baklanov, *Phys. Rev. Lett.* **99**, 257003 (2007).
91. V. M. Vinokur, T. I. Baturina, M. V. Fistul, A. Yu. Mironov and M. R. Baklanov, *Nature* **452**, 612 (2008).
92. M. V. Fistul, V. M. Vinokur and T. I. Baturina, *Phys. Rev. Lett.* **100**, 086805 (2008).
93. M. V. Fistul, V. M. Vinokur and T. I. Baturina, arXiv:0806.4311v1.
94. K. B. Efetov, M. V. Feigel'man and P. B. Wiegmann, arXiv:0804.3775v3.
95. M. Ovadia, B. Sacepe and D. Shahar, *Phys. Rev. Lett.* **102**, 176802 (2009).
96. B. Altschuler, V. Kravtsov, I. Lerner and I. Aleiner, *Phys. Rev. Lett.* **102**, 176803 (2009).
97. N. Markovic, A. M. Mack, G. Martinez-Arizala, C. Christiansen and A. M. Goldman, *Phys. Rev. Lett.* **81**, 701 (1998).
98. V. F. Gantmaker, *Physics-Uspekhi* **41**, 211 (1998).
99. E. Shimshoni, A. Auerbach and A. Kapitulnik, *Phys. Rev. Lett.* **80**, 3352 (1998).
100. K. Sheshadri, H. R. Krishnamurthy, R. Pandit and T. V. Ramakrishnan, *Phys. Rev. Lett.* **75**, 4075 (1995).
101. A. Ghosal, M. Randeria and N. Trivedi, *Phys. Rev. B* **65**, 014501 (2001).
102. Y. Dubi, Y. Meir and Y. Avishai, *Phys. Rev. Lett.* **94**, 156406 (2005).

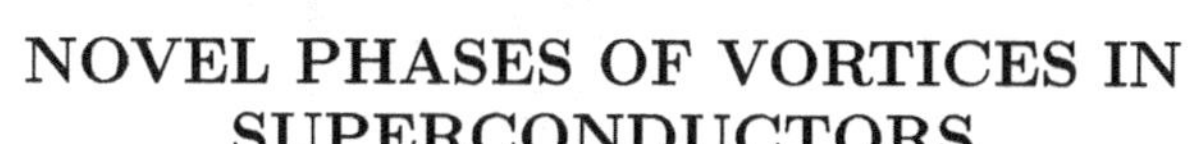

NOVEL PHASES OF VORTICES IN SUPERCONDUCTORS

Pierre Le Doussal

CNRS-Laboratoire de Physique Théorique,
Ecole Normale Supérieure, Paris 75238 Cedex 05, France
ledou@lpt.ens.fr

An overview is given of the new theories and experiments on the phase diagram of type II superconductors, which in recent years have progressed from the Abrikosov mean field theory to the "vortex matter" picture. We then detail some theoretical tools which allow to describe the melting of the vortex lattice, the collective pinning and creep theory, and the Bragg glass theory. It is followed by a short presentation of other glass phases of vortices, as well as phases of moving vortices.

1. Introduction

With the discovery of high-T_c superconductors, the physics of vortices in superconductors has experienced a revolution. Because of high temperature and the anisotropy of the new compounds, fluctuations are much enhanced and the mean field picture of a well ordered vortex lattice made of parallel straight vortices became vastly insufficient. A host of new phenomena have been discovered, most of them arising from the interplay of the enhanced fluctuations and pinning. Describing them has been a challenge to the theory which has considerably benefited from the advances in related fields such as the physics of glasses and disordered systems. As a result, new concepts in vortex physics have emerged. The development of powerful new experimental tools has allowed to image and probe the behavior of the vortices to unprecedented high precision from the microscopic to mesoscopic scale, and also to create and control new pinning environments for the vortices. This has allowed for tests of the new theories and for a fruitful interplay with experiments.

The field has grown much too vast for any exhaustive review in such short space, so we have chosen instead to focus on the sub-topic of the phase diagram of vortices. We first give an overview (Sec. 2) with a historical bent on how the phase diagram has metamorphosed from the mean field (MF) picture to the more recent (and in many respect still evolving) "vortex matter" picture. This part is more qualitative, easy to read and describes the interplay between experiments and theory. Next, we develop two specialized theoretical topics. The first is the melting of the vortex lattice (Sec. 3), with two dual approaches, one based on the elasticity theory and the second on the lowest Landau level within the Ginzburg–Landau theory. Phase coherence and decoupling phenomena are briefly evoked. The second topic is the description of the vortex lattice in the presence of pinning disorder (Sec. 4), and what is the fate of the Abrikosov lattice. It contains a short presentation of modern methods in the theory of disordered systems, which have been crucial tools to develop new concepts such as the glass phases of vortices. We then give a description of the Bragg glass phase, and, in Sec. 5, a brief description of the vortex glass, the Bose glass, as well as moving phases of vortices.

For the conventional picture we refer here to the classical articles, books and reviews[1–3] for all details and references. For the numerous developments in the last decades, we apologize in advance to many of our colleagues for omissions, and refer to the many excellent reviews on vortex physics[4–10] for a more exhaustive presentation and for different perspectives. Some of the present material already appeared in Ref. 11 but has been notably updated and expanded.

Notations: Here x is 3D coordinate (more generally, d dimensional), $x = (r, z)$ where r lives in 2D. x_1 designates one particular direction. r denotes either the vector or its norm, when no ambiguity arises. Fourier transforms are not noted differently, $q = (q_\perp, q_z)$. c is the speed of light, and c an elastic constant. T denotes kT.

2. Overview

2.1. *Conventional picture, basics*

The conventional, also called mean-field, equilibrium phase diagram of type II superconductors in an external field H is obtained by the minimization of Ginzburg–Landau (GL) energy functional. It predicts a Meissner phase for $H < H_{c1}(T)$, where the complex superconducting order parameter ψ is uniform and $B = 0$, and a mixed phase $H_{c1}(T) < H < H_{c2}(T)$ where the magnetic induction B penetrates the bulk in the form of vortex lines. Each

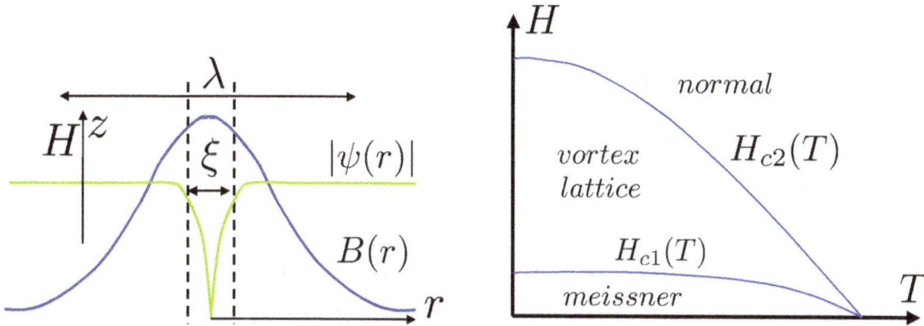

Fig. 1. Left: The vortex solution. Right: Mean-field phase diagram.

vortex line carries a quantum of magnetic flux $\Phi_0 = hc/2e$ and spans the entire system, being aligned along H. An isolated vortex consists of a normal cylindrical region of radius ξ, the core, where $\psi(r)$ is reduced (it vanishes at the center), surrounded by another cylindrical region of radius λ where the induction is screened by super-currents ($B(r) \sim e^{-r/\lambda}$ and $|\psi|^2 = cst$ away from the vortex). Its energy cost per unit length, $e_l = H_{c1}\Phi_0/4\pi = \epsilon_0 \ln(\lambda/\xi)$ is overcome by its magnetic energy gain $-H\Phi_0/4\pi$ in the mixed phase. The important energy scale (per unit length):

$$\epsilon_0 = \left(\frac{\Phi_0}{4\pi\lambda}\right)^2 \tag{1}$$

determines both the self-energy and mutual interactions between vortex lines. From minimization of the GL functional, Abrikosov predicted that these vortex lines form a regular lattice (later found to be triangular in standard systems), observed later in decoration experiments. This is natural if one thinks of vortex lines as identical mutually repelling objects. From flux quantization, the area of the unit cell is $a^2 = \Phi_0/B$ in simple relation to the averaged induction B, with a spacing $a_0 = (4/3)^{1/4}a$ for a triangular lattice. The GL theory predicts that both the coherence length ξ and the London length λ are temperature dependent with $\xi \sim (1 - T/T_c)^{-1/2}$ near T_c and $\kappa = \xi/\lambda$ being T-independent.[a] The vortex spacing decreases from $a_0 \sim \lambda$ at $B \sim H_{c1} \sim \Phi_0/\lambda^2$ to $a_0 \sim \xi$ at $H = H_{c_2} = \Phi_0/2\pi\xi^2$, leading to the mean field phase diagram pictured in Fig. 1.

In practice, superconductors are almost never in equilibrium, and one must worry about vortex dynamics, pinning and surfaces. In presence of a

[a]While GL is in principle valid near T_c, as any good theory its range of validity is considerably larger with choices of parameters from phenomenology or microscopic theory (when available).

current density j, each vortex segment aligned along unit vector n experiences a force f per unit length. In the simplest and most frequent situation (small Hall angle), it is equal to the Lorentz force $f = \frac{\Phi_0}{\mathsf{c}} j \times n$ where j is the local current density. In the absence of other forces a vortex thus moves as a result of a uniform current with velocity v given by:

$$\eta v = f \tag{2}$$

where η is the friction coefficient from scattering of normal core electrons with the substrate. The resulting time-dependent induction generates an (in-plane) electric field $E = B \times v/\mathsf{c}$, hence, from (2), a resistive response (to in-plane currents) $E \sim j$. In the vortex lattice, each vortex also feels the (repulsive) Lorentz forces from the screening currents of its neighbors. These, however, cancel for the equilibrium, or for the uniformly moving vortex lattice (VL), leading to $E = \rho_{ff} j_{\text{ext}}$ in presence of an external transport current j_{ext}, with $\rho_{ff} = \Phi_0 B/\eta \mathsf{c}^2$. Similarly at $B \approx H_{c2}$ to the resistivity of the normal metal yields the Bardeen–Stephen estimate for the friction $\eta \approx \Phi_0 H_{c2}/\rho_n \mathsf{c}^2$. Note that near a surface, the surface screening current (which pushes the vortex inside) and the image vortex (which attracts it) conspire to create, for $H_{c1} < H < H_c$ a surface barrier to vortex entry.

Thus, to achieve a true dissipationless transport current one needs that vortices be pinned. Pinning is provided by impurities in the material. Since they weaken superconductivity locally, they are energetically favorable positions for the vortex core to sit. Schematically, the conventional description of pinning replaces (2) by[b]:

$$\eta v = (f - f_c)\theta(f - f_c) \tag{3}$$

represented in Fig. 2, i.e. the vortex line remains pinned for $f < f_c$ and recovers a linear v–f characteristics for $f > f_c$. Here f_c is the critical force, qualitatively the maximum pinning force *per unit length*[c] that the impurities can exert. It corresponds to a critical current density $j_c = \mathsf{c} f_c/\Phi_0$, with $j_c < j_0 \sim \mathsf{c}\epsilon_0/\xi\Phi_0$ the depairing current. Equation (3) immediately yields to irreversible, far from equilibrium behavior, and to the Bean critical state. Consider, for example, a superconductor occupying the $x_1 > 0$ half-space and H along z. From Maxwell's equation, $j = -\frac{\mathsf{c}}{4\pi}\frac{dB}{dx_1}\hat{x}_2$. Since B (averaged over a mesoscopic region) is proportional to the vortex density, a given value

[b]Valid for a particle in a potential $V(u) = f_c|u|$. More realistic potentials with finite barriers lead to qualitatively similar $v - f$ curves, though asymptotic to $\eta v = f$ at large f, as shown in Fig. 2.

[c]Below we denote the critical force *per unit volume* $F_c = f_c/a^2$ not to be confused with an energy.

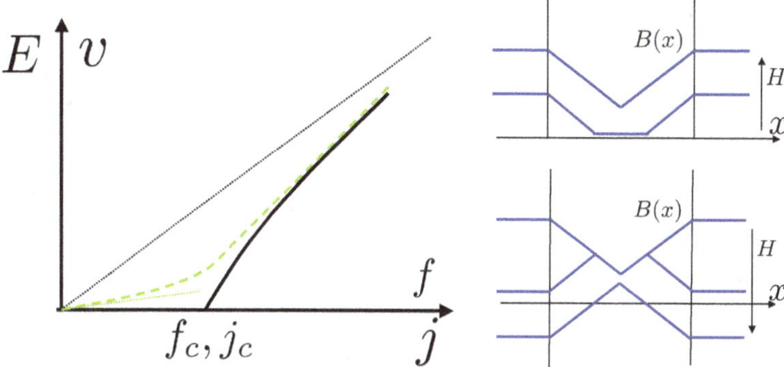

Fig. 2. Left: Conventional v–f (E–j) characteristics at $T = 0$ (solid line). Creep motion at small f is possible at $T > 0$ (dashed line) with nonzero linear resistivity (linear slope). Right: Bean profile, the slope $\left|\frac{dB}{dx}\right| \sim j_c$ is constant, upon increasing the external field (top) and upon decreasing the external field (bottom) leading to trapped flux and to hysteretic magnetization loop upon cycling the field.

of j on the curve of Fig. 2 implies a gradient in vortex density. Vortices then tend to move towards lower density regions. However, since v vanishes at j_c the system gets blocked in one of the metastable states such that everywhere $\left|\frac{dB}{dx_1}\right| = \frac{4\pi}{c} j_c$ or $B = 0$. Cycling of H under this constraint yields a sequence of Bean metastable states with irreversible magnetization $M_{\text{irr}}(H) \sim B(H) - H$. This gives rise to a hysteretic magnetization loop with area proportional to j_c times a geometric length in contrast to reversible equilibrium magnetization in the absence of pinning (and of surface and geometrical barriers).

2.2. *Limits of the conventional picture*

The above mean field picture neglects the fluctuations of the superconducting order parameter $\psi(x)$ and, in the mixed phase, also the fluctuations in the positions of the vortex lines. These fluctuations are thermal and quantum fluctuations or deformations of the vortex lines induced by pinning disorder. Amazingly, and simplifying a little, a detailed understanding of those has not proved crucial to account for the physics of conventional low T_c superconductors until the middle of the '80s.

This situation became untenable with the advent of the high T_c superconductors which led to a host of new phenomena and concepts in vortex physics and raised many questions. Some have been answered, others remain open. Many of these questions are related to the correct treatment of these fluctuations. Interestingly, this has pushed for a fruitful re-examination of some lower T_c superconductors, such as $NbSe_2$.

Some of these questions were anticipated. Ten years after Abrikosov,[3] Eilenberger[12] showed that thermal fluctuations of $|\psi|$, can become non-negligible, i.e. $\delta|\psi|^2/|\psi|^2 = O(1)$, but only in a hard to observe window (with realistic parameters at the time) near $(H_{c2} - B)/H_{c2} \approx 10^{-4}$. Interestingly, his treatment, i.e. within the lowest Landau level (LLL) led already to the idea of melting while retaining identity of vortices.

Similarly, a treatment of pinning neglecting deformations of the vortex lines is not consistent. A strictly straight vortex line would simply "average" over point impurities, leading to a vanishingly small $f_c \sim L^{-1/2}$ as its length L increases. The very nature of pinning requires vortex deformations that lower the energy. The modern theory of pinning started with the pioneering works of Larkin and Ovchinnikov.[13,14]

Below the threshold $f < f_c$, the combined effect of thermal fluctuations and pinning was known to lead to "flux creep" by relaxation over the barriers. It was described in terms of a particle (representing a blob of pinned lattice) moving by thermal activation over a 1D energy landscape with fixed barriers U_c, a model called Thermally Assisted Flux Flow (TAFF)[15] which predicts a linear response $v \sim f$ at small force, although with a reduced resistivity $\rho = \rho_{ff} e^{-U_c/T}$. The rounding of the $v - f$ curve near f_c implies thermal relaxation of the Bean profile to a uniform one. However for low T_c materials U_c/T is large and this would occur on gigantic time scales. In practice only the region $f \lesssim f_c$ relaxes. There the TAFF model predicts $v = v_0 \exp\left(\frac{U_c}{j_c T}(j - j_c)\right)$ which leads to the very slow logarithmic relaxation of what is called the *persistent current* $j(t) \approx j_c\left(1 - \frac{T}{U_c}\ln t\right)$ which was observed in low T_c materials.[16,17] For these relaxation effects, being extremely small in low T_c materials, there was no need to develop a deeper understanding of the creep process, e.g. to take into account the deformations of the vortex lattice during the motion.

2.3. *The new picture: the vortex matter*

2.3.1. *Material characteristics and early experimental facts*

The first characteristic of the new superconductors, besides high T_c, e.g. 92 K for $YBa_2Cu_3O_{7-\delta}$ (YBCO) and $T_c = 135$ K for $HgBa_2Ca_2Cu_3O_x$, is that they are layered materials, mainly *ab* superconducting planes separated along *c*-axis by insulating layers. They are insulators which become superconductors under doping. Despite the unconventional microscopic mechanism and the d wave character of the order parameter,[18] it is usually accepted that, apart from a few specific points (some mentioned below), they can be described by the *anisotropic* version of the Ginzburg–Landau model, with mainly c-axis anisotropy. One distinguishes the penetration and coherence lengths, as well as the effective masses, along ab-plane, noted here $\lambda_{ab} \equiv \lambda$, $\xi_{ab} \equiv \xi$, $m \equiv m_{ab}$ and along c-axis, ξ_c, λ_c and m_c. The anisotropy parameter $\epsilon = \frac{1}{\gamma} = \sqrt{\frac{m}{m_c}} = \frac{\lambda}{\lambda_c} = \frac{\xi_c}{\xi} \ll 1$, e.g. $\epsilon_{\mathrm{YBCO}} = 0.16$. One still denotes $\kappa = \lambda/\xi$ and uses (1). The critical fields are $H_{c2}^c = \frac{\Phi_0}{2\pi\xi^2} = \epsilon H_{c2}^{ab}$ and $H_{c1}^c = \frac{\Phi_0}{4\pi\lambda^2} \ln \kappa$.

High T_c superconductors are strongly type II i.e. $\kappa \approx 10^2$, hence $H_{c2}^c/H_{c1}^c \sim \kappa^2 \sim 10^4$ with $H_{c1} = O(100G)$ and $H_{c2} = O(100T)$. The mixed phase is thus huge and crucial for applications. The London approximation, which neglects the spatial variations of the modulus of the order parameter, works in most of the phase diagram. The vortex lattice spacing $a_0 \approx 480A/\sqrt{B(\mathrm{Tesla})}$ varies over two orders of magnitude from $a_0 \sim \lambda \sim 0.2\mu\mathrm{m}$ for $B \sim H_{c1}$ to $a_0 \sim \xi \sim 20$ Å near H_{c2} (outside the critical regions).

In the most anisotropic compounds, such as $Bi_2Sr_2CaCu_2O_{8+\delta}$ (BSCCO) with $\epsilon_{\mathrm{BSCCO}} = 10^{-2}$, the interlayer transport is essentially via Josephson effect and one must replace the anisotropic GL model by its discretized version along the c-axis, the Lawrence–Doniach model. The flux lines then become rather a superposition of so-called pancake vortices which live and can move in each ab plane. A new transition is then predicted to occur: decoupling of the layers with loss of the critical current along z.[19–21]

The importance of thermal fluctuations can be appreciated from the ratio of temperature to condensation energy in a coherence volume, the Ginzburg number $Gi = (T_c/2\sqrt{2}\epsilon\epsilon_0(0)\xi(0))^2$,[d] a material parameter which also gives the temperature window $1-t < Gi$, $t = T/T_c$ over which critical fluctuations cannot be ignored within the GL theory. In conventional superconductors $Gi \sim 10^{-8}$ and pinning is strong $j_c/j_0 \sim 10^{-2} - 10^{-1}$ while in high T_c superconductors $Gi \sim 10^{-2}$ and pinning is weaker $j_c/j_0 \sim 10^{-2} - 10^{-3}$.

A first striking feature of experiments on high T_c materials, is that one does not observe the jump in the specific heat predicted by mean field theory at the mean field $H_{c2}(T)$. Instead there is only a broad maximum around that location, which broadens as the field increases, as in Fig. 3. The maximum itself shows active degrees of freedom and suggests that there are vortices. Below we denote $H_{c2}(T)$ (and respectively $T_{c2}(H)$) the location of

[d]Here and below $\xi(0)$, etc. means $T = 0$ value.

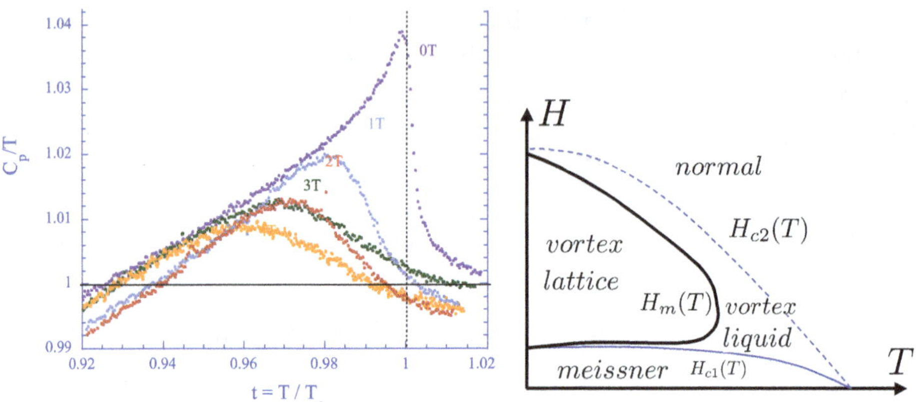

Fig. 3. Left: Specific heat of YBCO with the phonon background substracted, from.[22] Right: Sketch of the expected phase diagram in the absence of pinning disorder.

the mean field transition even though it does not correspond to a genuine phase transition.

Early experiments gave many hints that the vortex lattice could be melted in a large fraction of the phase diagram. Transport measurements[23] showed a suppression of the critical current and a long resistive tail much below the superconductor-to-normal transition in BSCCO in fields above $\sim 0.5T$. These results suggest that vortices flow freely, without a shear modulus, in response to an arbitrary small driving force over the larger part of the $B(T)$ phase diagram, down to temperatures as low as 35 K. Torque measurement on the same material found that damping of mechanical oscillations change drastically at this temperature.[24] Being dynamical in nature, these experiments were mainly an evidence for a transition from a pinned regime to an unpinned one, not a direct proof of melting. The main experimental fact was the existence[25] of an *irreversibility line* $H_{\rm irr}(T)$ far below H_{c2} separating these two regimes. At the same time, magnetic decoration on YBCO, performed at much lower fields near H_{c1}, showed that the vortex lattice, well-formed at higher fields, becomes disordered at low vortex density.[26,27] Separating dynamics from thermodynamics was thus the main challenge. The order of the transition (continuous versus first-order) was also not clear. Giant flux creep relaxation was observed[28] and another unusual feature, the fishtail or second magnetization peak was also observed at low temperature (see below).

2.3.2. *New theories*

For theorists the challenge was to develop the GL theory beyond mean field. There are two main ways to describe the fluctuations in the mixed phase, which are dual to each other. One can either use a description in terms of the complex order parameter $\psi(r)$, with its phase and modulus. This approach is more convenient near $H_{c2}(T)$, where the LLL approximation can be used. Alternatively, one can describe fluctuations in terms of the displacements of individual vortex lines $u_i(z)$, more convenient when the latter are well-separated and the London approximation can be used ($B \lesssim 0.2H_{c2}$). One must then describe the statistical mechanics of interacting line objects.

In the absence of quenched disorder, this second approach naturally led to two phases: (i) a *vortex solid* where the energy of deformations of the vortex lattice are described by standard elastic theory, adapted to a solid set of lines. The simplest description requires the in-plane bulk and shear modulus c_{11}, c_{66} and a tilt modulus c_{44} for deformations of the vortex lines away from the direction of the external field. These are both field, B, and wave-vector, q, dependent, i.e. the elasticity is non-local and the VL is softer at its smallest scales $\sim a$. They were calculated quite precisely[29–31] (ii) *a vortex liquid* phase, described as a collection of fluctuating entangled lines[32]: an analogy, most useful in the small field region $B \ll H_{c2}$, maps vortex lines in 3D onto imaginary time world lines of 2D bosons and allows one to estimate the entropy of the liquid. The two phases are separated by a *melting line* $H_m(T)$. Its position was estimated from the usual Lindemann criterion for the melting of a 3D solid, from the thermal average $\langle u^2 \rangle = c_L^2 a^2$, where the Lindemann constant $c_L = 0.1 - 0.4$ in standard solids. The simplest estimates (see Sec. 3.4) yields $T_m \sim a^3\sqrt{c_{66}c_{44}}c_L^2$, and using $c_{66} \sim \epsilon_0/a^2$ and $c_{44} \sim \epsilon^2\epsilon_0/a^2$ yields $B_m \sim \Phi_0\epsilon^2\epsilon_0^2 c_L^4/T^2$. Near T_c, and for $B_m \ll H_{c2}$ one must take into account the temperature dependence of λ, leading to $B_m \approx 5.6(c_L^4/Gi)H_{c2}(0)(1-T/T_c)^2$. More accurate predictions valid for larger B_m were made taking into account the full non-local elasticity.[29] A sketch of the melting line is shown in Fig. 3. The melting line is reentrant near $H_{c1}(T)$ since the density of vortices and the shear modulus vanish there.

Although useful, the Lindemann criterion does not provide a true theory of melting. While the superficial analogy with 3D solids suggests first-order melting, it is not so obvious for the superconductor since a detailed theory should start from the GL model and describe both the formation of the Abrikosov lattice and the fluctuations of $\psi(r)$, and, indeed, in mean-field, the transition is second order! Early attempts[33] to include fluctuations in the

LLL approach suggested that these induce a first order transition (FOT), however the approach is difficult.[34] Some success was obtained to treat fluctuations due to condensation energy and to compute smooth variations of the thermodynamics functions.[35] More recently, as experimental evidence for a FOT became overwhelming, precise calculations of the free energy of both the solid and the liquid, were performed within GL. They allow one to account for the FOT and to calculate the value of the Lindemann number.[36–40] Numerical simulations[41] have also seen a first order transition. The melting transition is developed in Sec. 3.

An order parameter for the solid is the translational order correlation, expressed in terms of vortex displacements as the translational average $\langle\cdot\cdot\rangle$ over the sample:

$$C_K(x) = \langle e^{iK\cdot u(x)} e^{-iK\cdot u(0)} \rangle \tag{4}$$

Here it is equivalent to the thermal average (denoted $\langle\cdot\cdot\rangle$). It is also equal to the Fourier transform of the vortex density correlation $S(k) = a^4 \langle \rho_v(-k)\rho_v(k)\rangle$ for $k_\perp$ near K, a 2D reciprocal lattice vector of the (e.g. triangular) vortex lattice. $S(k)$ is measured from neutron diffraction experiments, up to a form factor due to the distribution of $B(x)$ around a vortex. In the perfect solid $C_K(x)$ decays to a constant, $e^{-\frac{1}{4}K^2\langle u^2\rangle}$ the Debye–Waller factor, leading to δ-function Bragg peaks in $S(k)$ at the reciprocal lattice, equivalently $\langle e^{-iK\cdot u}\rangle \neq 0$. In the flux liquid, $S(k)$ exhibits instead broad isotropic 2D rings around $k \sim \pi/a_0$: there is no in-plane translational order, and vortex lines $u_i(z)$ wander diffusively in the z direction.[32] It can be obtained from the elastic solid by letting dislocation and disclination loops proliferate,[42] which drive the shear modulus to zero. Within the GL description, the flux liquid is described as fluctuations around $\psi(x) = 0$, i.e. it is continuously related to the normal phase, with no change in symmetry.

The outstanding question was how this picture is modified in the presence of pinning disorder. Many early experiments showed rather a continuous transition occuring at the irreversibility line.[43] Two main approaches were put forward. Both agreed on one property: the vortex lattice is changed into *a glass*. A glass induced by quenched disorder has a non-trivial disordered ground state, and the energy cost of excitations on scale L is a positive random variable which grows typically as L^θ with $\theta > 0$. The first approach, known as *vortex glass*[44,45] was inspired from spin glasses, and postulated the existence of an off-diagonal order parameter $\overline{|\langle\psi(x)\rangle|^2} \neq 0$, i.e. in each sample $\langle\psi(x)\rangle$ is non-zero but non-uniform and random. While mostly phenomenological, it could be argued assuming that the quenched disorder has destroyed the vortex lattice,[45] leaving a description only in terms of a simpler discrete XY model with a quenched random gauge field $A(x)$. Numerics[46] found that this gauge glass model could indeed exhibit a glass phase in $d = 3$, i.e. with $\theta > 0$.

The second approach retains the elastic lattice structure at small scale and describes the coupling of disorder to the vortex displacement field $u(r, z)$. The problem becomes a particular case of the more general problem of an elastic manifold in a random potential. Theoretical progress in that more general problem, described in great detail in Sec. 4, lead to the picture of collective pinning and also to a glass phase beyond the so-called Larkin length R_c. The precise nature of this phase was however unclear for a while. Larkin had found[13] that the deformations $\delta u(x) = u(x) - u(0)$ induced by disorder grow with scale x, i.e. the standard Abrikosov lattice was unstable even to the weakest disorder. A possible scenario was thus that the translational order correlation would decay fast beyond a second length scale, R_a, such that $u \sim a$, and that beyond a length $R_d \sim R_a$ dislocations in the vortex lattice could become favorable, leading to a topologically disordered state, as argued in the vortex glass picture.

It was then argued[47] that the Abrikosov vortex lattice is replaced by a distinct new phase of vortex matter, the *Bragg glass* (BrG), which retains topological order but not translational order. There are no unpaired dislocations in the ground state, so in that sense the Abrikosov lattice is not destroyed. On the other hand, the translational order correlation $C_K(r)$ decays beyond R_a, but slowly, only as a power law, very much as in 2D quasi-solids. As a consequence, there should be observable Bragg peaks. At the same time it is a glass. The asymptotic energy exponent is found as $\theta = d-2$.[47,50] The BrG theory, detailed below, is based on both a variational theory and renormalization group calculations, as well as energy arguments which indicate that indeed dislocation lines cost more energy than they allow one to gain.

Most importantly, the hallmark of a true glass being also the existence of diverging barriers between low-lying states, it was argued that to describe creep, the TAFF law should be replaced by:

$$v \sim \rho_{ff} j e^{-U_b(j)/T} \tag{5}$$

with an effective barrier $U_b(j) \sim (j_c/j)^{-\mu}$ at small j. Within the vortex glass picture, this was expected based on the conjecture of a glass phase.[44] Within the elastic approach it could be argued on a more solid basis, within the collective creep theory[48–52] as we will describe in Sec. 4, but only if one assumed the absence of dislocations. The Bragg glass then, with its topological order, guarantees the law (5) with $\mu = 1/2$ asymptotically provided the lowest lying excitations are purely elastic. This implies ultra-slow relaxation at very small current $j \ll j_c$ and zero linear resistivity, i.e a true superconductor. Thus having a glass is essential for true superconductivity, a very new concept as compared to the conventional TAFF picture.[53] Conversely, (5) also implies, in high T_c materials where pinning is weaker, that for larger j near j_c there is a lot of relaxation, accounting for the observations of giant flux creep and relaxation of the Bean profile and of the persistent current[54] with nonlinear logarithmic decay in time $j \sim (T \ln t)^{-1/\mu}$.

2.3.3. *Experiments meet theory: towards a phase diagram*

Since the vortex glass ideas suggested a continuous transition, a phenomenological picture of what a continuous glass transition should look like was developed, based only on the hypothesis of a single diverging length scale at the transition and standard critical scaling.[43–45] It was used then to interpret numerous experiments, but being quite generic, it did not carry information on the detailed nature of the glass phase.

Quite different results were soon obtained, however. In untwinned YBCO single crystals sharp jumps in resistivity were observed[55–58] as shown in Fig. 4, delimiting a first order transition line between a pinned superconducting phase and the liquid. In BSCCO, where anisotropy is larger and pinning weaker, the irreversibility line, i.e. the opening of the magnetization loop, lies significantly below the melting line, hence thermodynamic evidence of first-order equilibrium melting was easier to obtain. The local field B, i.e. the vortex density, measured using microscopic Hall sensors, showed a discontinuous step at the transition,[59] shown in Fig. 5, as in a standard solid. The density of the vortex liquid being *larger* than the solid, as in ice, was interpreted from the entropy gain of letting neighboring vortex lines entangle in the liquid.[60] The phase diagram measured for BSCCO is shown in Fig. 6, where A is a reversible flux liquid, B an ordered vortex solid and C an amorphous or disordered solid. Neutron diffraction[62] showed Bragg peaks in the B phase, but no peaks in the C and A phases, while muon spin rotation experiments[63] showed loss of the triangular lattice lineshape as one goes from B to C. Thermodynamic evidence for a first order transition came also in YBCO[64] as a spike in the specific heat, superposed on the broad maximum, as in Fig. 7. The phase diagram for YBCCO and BSCCO started to look alike, with a FOT at low field and a continuous one at high field, separated by what appeared then as a tri-critical point. Since higher fields (in that range) correspond to stronger disorder as the density of vortices

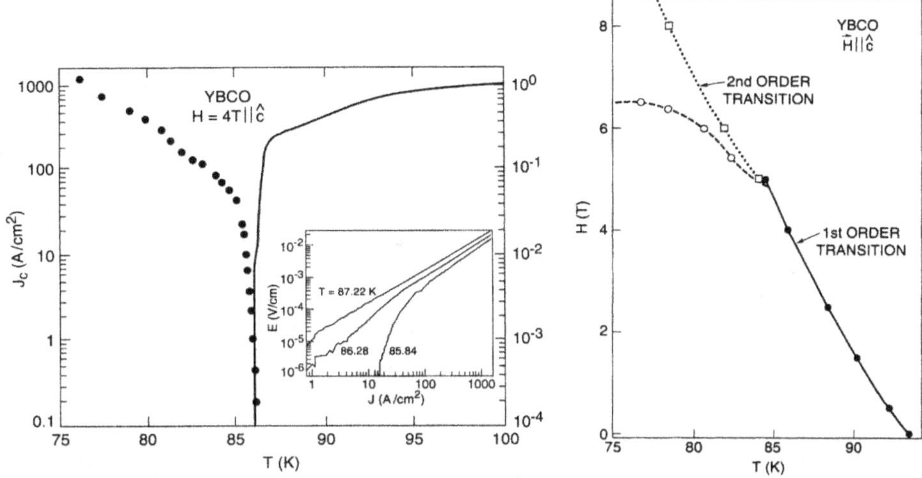

Fig. 4. Left: Linear response resistance (right axis) and critical current as a function of temperature for YBCO at $H = 4T$ near melting. Right: Phase diagram — (solid circle) jumps in resistivity, (squares) continuous transition, (open circles) minimum in v (similar to the maximum of j_c in magnetization). From Ref. 58.

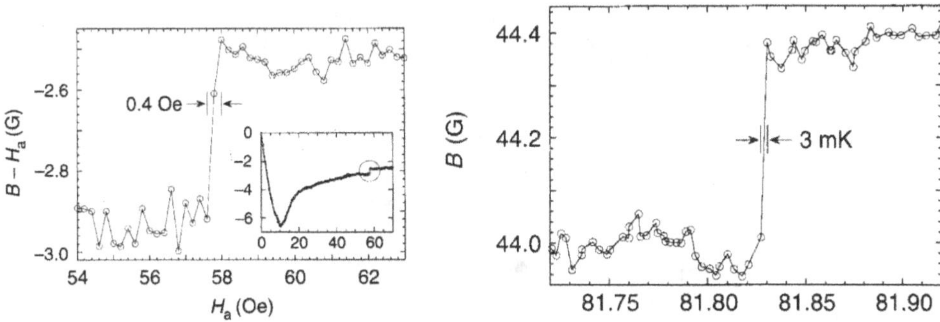

Fig. 5. Sharp jump in local B in BSCCO as H is varied (left) or T (right), evidence of a thermodynamic FOT, from Ref. 59.

is larger, this was in agreement with the idea of two distinct glass phases: (i) a weak disorder phase, the Bragg glass, close to a perfect solid, which would naturally melt through a first order transition (ii) a strong disorder phase which would be amorphous. As seen in Fig. 9, it was also found that the apparent tricritical point is lowered in field upon increasing the weak point disorder by electron irradiation or by changing the doping. An extended Lindemann criterion, $\overline{\langle \delta u(a)^2 \rangle} \sim c_L^2 a^2$, i.e. $R_a \sim a$, which estimates

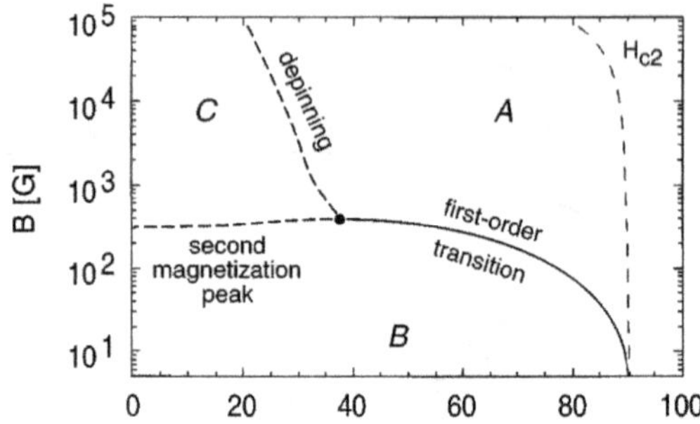

Fig. 6. Phase diagram in BSCCO: An unpinned liquid, B vortex solid (Bragg glass), C pinned amorphous phase, from Ref. 61.

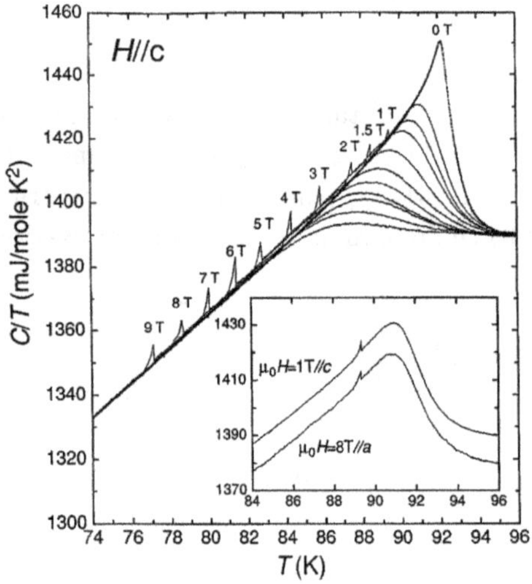

Fig. 7. Specific heat of untwinned YBCO, from Ref. 64.

when the BrG must become unstable to thermal *and* disorder-induced dislocations,[66,67] was consistent with these observations. It furthermore indicated a disorder-induced *transition line* $H_m(T)$ away from the BrG, horizontal at low T, as a simple continuation of the melting line. Strong downward renormalization of disorder was also predicted from flux line thermal wandering,[68] producing in some cases a pronounced maximum in $H_m(T)$ as it merges into the melting line.

There was indeed a feature in the magnetic hysteresis loop at low T, ubiquitous in all high T_c superconductors, a second peak in M(H) also called fishtail (see Fig. 8). Being in the irreversible non-equilibrium region, it could be broad or sharp depending on ramping rate or history, and was the signa-

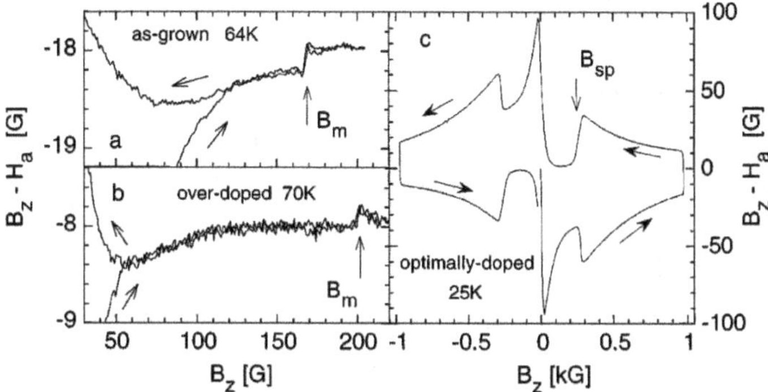

Fig. 8. Magnetization loops in BSCCO. Left top and bottom: High T, reversible jump at melting. Right: Low T, fishtail or second magnetization peak, from Ref. 65.

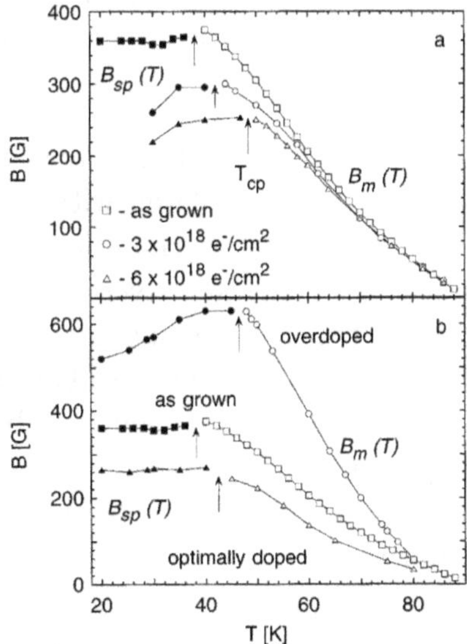

Fig. 9. Phase lines in BSCCO from Ref. 65, as in Fig. 6 upon electron irradiation (top) or changing the doping (bottom): the melting line and second peak line move together. The solid region is reduced upon increasing disorder as expected from an instability line of the Bragg glass to dislocations (correlated disorder has the opposite effect — see below).

ture of a maximum in j_c, or at least in the persistent (i.e. relaxed) current as a function of B. Hence it was unclear whether it was some broad maximum related to field variation of elastic coefficients (due to their non-locality), as often argued to explain the so-called peak effect observed in low T_c, or

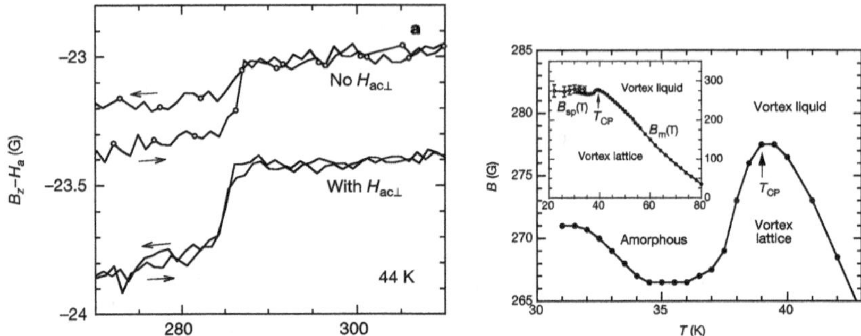

Fig. 10. Left: Hysteretic magnetization is observed below the transition (open circles). Application of $H_{ac\perp}$ removes the residual hysteresis and enhances the magnetization step. Right: Inset, the first-order $B_m(T)$ line (filled circles) along with the second magnetization peak line $B_{sp}(T)$ (open circles). Main panel: The first-order transition was observed at all temperatures around T_{CP} at which fully reversible magnetization was achieved by shaking. From Ref. 69.

some dimensional crossover, etc. Remarkably, it was then discovered that the second peak line was perfectly correlated with the position of the melting curve, appearing to be merely its continuation at low T (see Fig. 9). A proof of that was obtained in a remarkable series of experiments. By shaking the vortex lattice using a transverse *ac* field, the magnetization could be made reversible down to low T. As a result, a FOT was revealed below the T_{cp} point where the melting curve has a reentrant behavior, called inverse melting[69] (see Fig. 10). This made T_{cp} unlikely to be a tricritical point, but rather only a point of rapid variation of the effective disorder due to thermal screening. It was also shown that B_{sp} at low T corresponds to a first order transition line between a low j_c phase and a high j_c one (phases B and C in Fig. 6). Coexistence of these two phases in the same sample near B_{sp}, and their domain wall motions, was revealed by the Bean profile.[70] Note that larger deformations in the high field less ordered glass (or pinned liquid) is expected to result in a higher j_c.[e]

A universal phase diagram was thus emerging, with a low-field weak disorder vortex solid phase, consistent with a Bragg glass (more evidence for that is discussed in Sec. 4), unstable via a first order transition along a unique line which gradually changed from thermal melting to disorder-induced melt-

[e]The E-j characteristics of the two phases are predicted[66] and observed to cross: the persistent currents depends on the observation time, resulting in more or less pronounced peaks.[71]

ing as it crosses the irreversibility line. The BrG instability line was found reentrant at very low fields near H_{c1}. Similar phase diagrams were observed in many high T_c materials with a wide range of anisotropy: thallium compounds, $La_{2-x}Sr_xCuO_4$ (LSCO),[72] YBCO,[73] $Ba_{1-x}K_xBiO_3$ (BKBO),[74] and by a variety of probes (e.g. muons[75]). It was also consistent with numerical simulations.[76]

In BSCCO, layer decoupling of pancake vortices is quite likely to play a role. This is indicated by Josephson Plasma Resonance experiments[77,78] which probe the superconducting phase coherence between adjacent layers. The latter is found to be strongly reduced across the melting line. A decoupling transition was predicted in the absence of disorder.[19,20,108] It should occur already inside the vortex solid, and be triggered by proliferation of vacancy-interstitial defects inside each layer, also called quartets.[79,80] The corresponding picture for more isotropic superconductors is the transition from the usual vortex line crystal to a supersolid, i.e. a solid of entangled vortex lines.[81] Since a vacancy is a *bound* pair of dislocations, this decoupled solid, or supersolid, is still a topologically ordered crystal with long range translational order. In presence of pinning disorder, an analogous decoupling transition is predicted[21,82,83] within the Bragg glass phase, between a coupled BrG (also called a Josephson glass) and a decoupled one. An explanation for the second peak line in terms of such a decoupling was also proposed.[82] Until now, there seems to be no strong evidence for a decoupled solid (or supersolid, or decoupled BrG) so it is quite possible that melting and decoupling occur simultaneously. In terms of the extended Lindemann criterion it means that L_a (the translational order length along z) reaches $\sim d$, the interlayer spacing, before R_a reaches a (i.e. the criterion becomes $\overline{\langle \delta u(z=d)^2 \rangle} \sim c_L^2 a^2$). It is then likely that the unbinding of dislocation loops starts with formation of quartets. In the end, it is the universality of the structure of the phase diagram for all compounds, from the most anisotropic to isotropic one, which is the most convincing argument for a single scenario.

This motivated a reexamination of low T_c superconductors and elucidation of the famous "peak effect", a pronounced but broad maximum in $j_c(T)$ before it vanishes near H_{c2}.[84] It was thus found that in $NbSe_2$, in which thermal fluctuations are very small, the liquid is replaced by a pinned disordered phase with a high j_c.[85] Transport measurements in Corbino geometry (i.e. vortices do not cross sample edges) revealed[86] a first-order jump in j_c (see Fig. 11) in agreement with the Bragg glass to disordered phase universal phase diagram, a feature previously masked by the phase coexistence unavoidable in the conventional stripline geometry commonly used

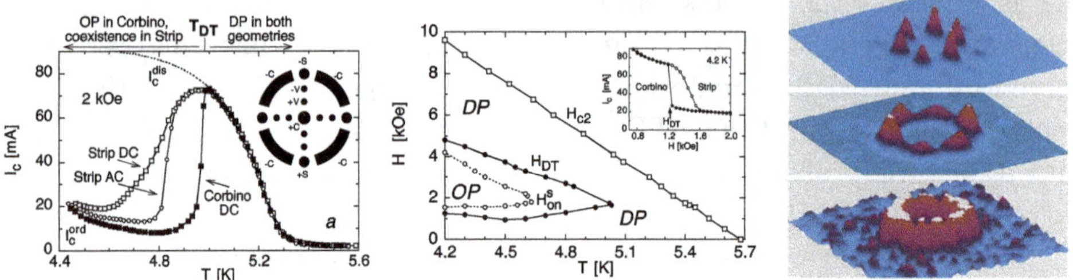

Fig. 11. Left, from Ref. 86: The conventional peak effect in a strip is found to result from coexistence of an ordered solid (OP) and a disordered pinned (DP) phase, the first order jump at the OP/DP disordering transition (DT) (at T_{DT}) is revealed in the Corbino geometry. The phase diagram $H_{DT}(T)$ is plotted and has the universal form, the liquid phase being replaced by the DP phase. H_{c2} is defined as 0.1 the normal resistance. Right, from Ref. 87: Bragg peaks disappearing into a ring in $NbSe_2$ through the peak effect.

for critical current measurements. The topological order of the OP phase over distances $\sim 100a_0$ was confirmed by neutron diffraction,[87] providing a consistent picture in $NbSe_2$ and in Nb.[88]

We close this overview by the effect of *correlated disorder* on the phase diagram. It arises either naturally from twin planes or dislocation lines, and can be created and controlled as columnar disorder from heavy ion irradiation.[89] In the latter case, using the flux line to quantum boson analogy, a new, Bose glass (BoG) phase was predicted[90] where flux lines are localized along columns when these are parallel to the external field. This being a very efficient way to pin the vortices, even a small density of columns results in a dramatic increase in the critical current. A high concentration of columnar defects pushes the melting line to much higher fields and results in a phase diagram with a single glass phase, the BoG, that melts into a liquid phase. In this case, the BoG/liquid transition is predicted to be continuous, very much as the vortex glass, but with anisotropic scaling[90,91] as observed experimentally.[92,160] At sufficiently low concentration of columnar defects, however, the transition was observed to remain first order.[93] Columnar defects are thus of technological importance, and are now included in superconducting tapes in applications, e.g. as nanorods.[94] Their effect can even be increased by introducing a small dispersion in their direction along the c-axis, leading to the splay glass phase,[95] where localization and entanglement effects were predicted and observed[96] to conspire to a slowdown in the creep relaxation. Correlated disorder is developed in Sec. 5.2.

Let us now mention some important open questions. A first one is the nature of the amorphous strongly disordered pinned phase (region C in Fig. 6). As discussed again in Sec. 5.1, the vortex glass scenario is still not fully established. On the theoretical side, the gauge glass phase was found unstable to screening effects (finite London length). On the experimental side, fits to vortex glass scaling in BSCCO seem to fail to lead to universal values for the exponents. In YBCO, the question was raised whether some of the previous vortex glass scaling may have been due to twins, since correlated disorder also leads to similar (though anisotropic) scaling and since the FOT is more visible in untwinned YBCO crystals. Thus other scenarios have been proposed, such as a polymer glass or pinned liquid,[32] a vortex molasses,[97] a decoupled solid,[21] a frozen mesh of dislocations.[98] A second open question stems from recent observations in BSCCO that the disordered solid/liquid depinning line extends as a rather sharp thermal depinning line within the ordered solid.[99] It has been proposed as a true transition,[100] or described in terms of a Josephson glass,[21] however, since a perfect solid is just ruled out by present theories of pinning (in $d = 3$ it would require diverging elastic coefficients), it may also be simply a (rather sharp) thermal crossover within the Bragg glass phase (see e.g. Ref. 194). Even so, a more accurate description of such a sharp crossover is clearly needed.

Finally, we left out new developments in *moving phases* of vortices. A short overview is given in Sec. 5.3.

3. Melting of the Vortex Lattice

Let us recall the GL free energy, $F = F_{\rm kin} + F_\psi + F_{\rm mag}$ in terms of $\psi = |\psi|e^{i\phi}$:

$$F_{\rm kin} = \int d^3x \frac{\hbar^2}{2m}|D\psi|^2 + \frac{\hbar^2}{2m_c}|D_z\psi|^2 \tag{6}$$

$$F_\psi = \int d^3x \alpha|\psi|^2 + \frac{\beta}{2}|\psi|^4, \qquad F_{\rm mag} = \frac{1}{8\pi}\int d^3x B^2 \tag{7}$$

where $D = \nabla + i\frac{2\pi}{\Phi_0}A$, and one also often uses the Gibbs energy $G = F - \frac{1}{4\pi}\int d^3x HB$. One has $|D\psi|^2 = (\nabla|\psi|)^2 + |\psi|^2(D\phi)^2$, with the super-current $j = \frac{\hbar e^*}{m}|\psi|^2 D\phi$.

We are interested in the fluctuation free energy $\mathcal{F} = -T\ln Z$ with $Z = \int D\psi DAe^{-F/T}$ and we denote the free energy density as $\mathcal{F}_d$. We now briefly describe the two main approaches, either from computing the elastic matrix of the vortex lattice from the GL theory (or more simply, the London theory), and applying the Lindemann criterion, or directly from using the lowest

Landau level approximation on GL and comparing the free energies of the liquid and the solid.

3.1. *Interactions between vortices and line energy (isotropic)*

For large κ one can use constant $|\psi|$ and the London equations except in the (small) core region. The kinetic energy becomes $F_{\rm kin} \approx 2\pi \int d^3x (\lambda/c)^2 j^2$ where $\lambda^2 = mc^2/(4\pi(e^*)^2|\psi|^2)$. In presence of vortex lines at positions $x_i(z) = (r_i(z), z)$ the local field $B(r,z)$ is determined from:

$$(1-\lambda^2\nabla^2)B(r,z) = \Phi_0 T(r,z)\,,\quad T(r,z) = \sum_i \left(\frac{dr_i(z)}{dz}, 1\right)\delta^2(r - r_i(z)) \quad (8)$$

with $q \cdot T(q) = 0$ from flux line conservation. The energy, neglecting the core, is the sum of the magnetic and kinetic energy from the currents, using $\nabla \times B = \frac{4\pi}{c} j$:

$$F = \frac{1}{8\pi}\int d^3x[B^2 + \lambda^2(\nabla \times B)^2] = \frac{1}{2}\int \frac{d^3q}{(2\pi)^3} V(q)\,|T(q)|^2 \quad (9)$$

with:

$$V(q) = \frac{\Phi_0^2}{4\pi}\frac{1}{1+\lambda^2q^2} \quad (10)$$

of $3D$ fourier transform $V(x) = \epsilon_0 \frac{e^{-x/\lambda}}{x}$ interaction between line elements. For some quantities, one must use $V_\xi(q)$ (one choice is $V_\xi(q) = e^{-\xi^2q^2}V(q)$) which is cutoff at small distance. One finds the line energy $e_l = F/L$ of a single straight vortex line of length L as $e_l = \frac{1}{2}\int \frac{d^2q_\perp}{(2\pi)^2} V_\xi(q_\perp, 0) \approx \epsilon_0 \ln\kappa = \Phi_0 H_{c1}/(4\pi)$, and the interaction energy $W(r) = 2\epsilon_0 K_0(r/\lambda)$ per unit length, between two straight flux lines separated by r.

3.2. *Elasticity of a vortex line and of the vortex lattice (isotropic)*

For a weakly distorted single vortex line, rotational invariance yields the total energy $F = e_l L \approx F_0 + F_{el}$ and the elastic energy $F_{el} = \frac{e_l}{2}\int dz \left(\frac{dr(z)}{dz}\right)^2$.

For the vortex lattice, one denotes the vortex positions $r_i(z) = R_i + u_i(z)$, where R_i denote the perfect 2D Bravais lattice. One defines $u(q) = a^2 \int dz \sum_i u_i(z) e^{iq_\perp R_i + iq_z z}$ from which one retrieves $u_i(z) = \int_{q,BZ} u(q) e^{-iq_\perp R_i - iq_z z}$, with the limit $\int_{q,BZ} \equiv \frac{a^{-d}}{N}\sum_q \to \int_{BZ} \frac{d^dq}{(2\pi)^d}$ (BZ designates integration over the first Brillouin zone). From (9) the total energy is:

$$F = \frac{1}{2} \int dz \int dz' \sum_{ij} \left(1 + \frac{du_i(z)}{dz} \cdot \frac{du_j(z')}{dz'}\right) V_\xi(x_i(z) - x_j(z')) \quad (11)$$

Expanding to second order one finds:

$$F_{el} = \frac{1}{2} \int_{q,BZ} \Phi_{\alpha\beta}(q) u_\alpha(q) u_\beta(-q) \quad (12)$$

with an elastic matrix:

$$\begin{aligned}\Phi_{\alpha\beta}(q) &= a^{-2} \sum_R \int dz [V''_{\alpha\beta}(R,z)(1 - e^{iq_\perp R + iq_z z}) + \delta_{\alpha\beta} q_z^2 V(R,z) e^{iq_\perp R + iq_z z}] \\ &= a^{-4} \sum_K (M_{\alpha\beta}(K+q) - M_{\alpha\beta}(K)), \quad M_{\alpha\beta}(q) = (q_\alpha q_\beta + \delta_{\alpha\beta} q_z^2) V(q)\end{aligned} \quad (13)$$

using twice Poisson formula $\sum_R f(R) = \sum_K a^{-2} \int d^2 r f(r) e^{iKr}$. For the triangular lattice one finds:

$$\Phi_{\alpha\beta}(q) = \Phi_L(q) P^L_{\alpha\beta}(q_\perp) + \Phi_T(q) P^T_{\alpha\beta}(q_\perp) \quad (14)$$

$$\Phi_L(q) = c_{11}(q) q_\perp^2 + c_{44}(q) q_z^2, \qquad \Phi_T(q) = c_{66}(q) q_\perp^2 + c_{44}(q) q_z^2 \quad (15)$$

with $P^L_{\alpha\beta}(q) = \frac{q_\alpha q_\beta}{q^2}$, $P^T_{\alpha\beta}(q) = \delta_{\alpha\beta} - \frac{q_\alpha q_\beta}{q^2}$ and the bulk and shear moduli, for $H_{c_1} \ll B < 0.4 H_{c2}$:

$$c_{11}(q) \approx c^0_{44}(q) \approx \frac{B^2}{4\pi} \frac{1}{1 + \lambda^2 q^2}, \qquad c_{66} \approx \frac{\Phi_0 B}{(8\pi\lambda)^2} = \frac{\epsilon_0}{4a^2} \quad (16)$$

from $K = 0$ and $K \neq 0$ respectively. The tilt modulus $c_{44}(q) = c^0_{44}(q) + c^c_{44}(q)$ with $c^c_{44}(q)$ from the contribution of $K \neq 0$. One has $c^c_{44}(q) = B(H-B)/(4\pi)$ for $q \to 0$ and $c^c_{44}(\frac{\pi}{a}, q_z) \sim \frac{\epsilon_0}{2a_0^2} \ln(\kappa^2/(1 + \lambda^2(K^2_{BZ} + q_z^2)))$ and $K_{BZ} = 2\sqrt{\pi}/a$ the radius of the circular BZ with same area. An estimate of the density of vortex line times the elastic energy of a single line gives $c_{44} = \frac{\epsilon_0}{a^2} \ln \kappa$ which is valid for $q_\perp \sim \pi/a$. For smaller $q_\perp$ the VL is much stiffer. Note that $c_{11} \gg c_{66}$ and transverse modes are dominating.

The elastic coefficients were also computed for larger fields $B > 0.6 H_{c2}$ using GL beyond the above London approximation.[5,29,30] One effect is that $\lambda \to \lambda' = \lambda/\sqrt{1-b}$ with $b = B/H_{c2}(T)$, with however $c_{66} \approx \frac{\epsilon_0}{4a^2}(1-b)^2$.

At very small fields, near H_{c1}, the vortex lattice is dilute and for $B < H_{c_1}/\ln\kappa$ one has $c_{11} \sim 3c_{66} \sim \epsilon_0 a^{1/2} \lambda^{-3/2} e^{-a/\lambda}$.

3.3. *Anisotropic elasticity*

The above calculation, when extended to the anisotropic case using anisotropic London theory[101] yields an elastic matrix which depends on θ, the angle of H from the c-axis. Here we only discuss $\theta = 0$. c_{66} is found unchanged and (see e.g. Ref. 4):

$$c_{11}(q) \approx \frac{B^2}{4\pi}\frac{1+\lambda_c^2 q^2}{1+\lambda^2 q^2} c_{44}^0(q) , \qquad c_{44}^0(q) \approx \frac{B^2}{4\pi}\frac{1}{1+q_\perp^2\lambda_c^2+q_z^2\lambda^2} \tag{17}$$

while $c_{44}^c(q)$ grows from $\sim \epsilon_0\epsilon^2 \ln(\kappa/\epsilon)/2a_0^2$ for $\lambda q_z \gg 1$ (the Josephson coupling contribution) to $\sim \epsilon_0/(2\lambda^2)$ for $\lambda q_z \ll 1$ (the electromagnetic contribution). For a single vortex line $F_{el} = \frac{1}{2}\int_{q_z} e_l(q_z) q_z^2 |r(q_z)|^2$ where the line tension $e_l(q_z)$ becomes strongly dispersive and increases from a value $\tilde{\epsilon}_1 \approx \epsilon^2\epsilon_0 \ln(\kappa/\epsilon)$ for $\lambda q_z \gg 1$ to $\approx \epsilon_0$ for $\lambda q_z \ll 1$. Anisotropy reduces $\tilde{\epsilon}_1$ and thus strongly enhances the tendency to flux line wandering: in the solid, the individual flux lines are kept place by their neighbors, but in the liquid this results in entanglement of the flux lines.[32]

More generally it was argued[4] that for Q a quantity such as volume V, energy E, temperature T, or field, one can relate the anisotropic system to the isotropic one using:

$$Q(\theta, H, T, \xi, \lambda, \Delta) = s_Q Q_{\text{iso}}(\epsilon_\theta H, T/\epsilon, \xi, \lambda, \Delta/\epsilon) \tag{18}$$

where $\epsilon_\theta^2 = \epsilon^2 \sin^2\theta + \cos^2\theta$ and θ is the angle of H from the c-axis. One has $s_V = s_E = s_T = \epsilon$, $s_B = s_H = 1/\epsilon_\theta$. Here Δ is the disorder strength (see below). This again shows that fluctuations, both thermal and from disorder, are enhanced by anisotropy.

3.4. *Melting of the elastic vortex lattice*

We now consider the thermal fluctuations, described by the canonical partition function $Z = \int \mathrm{D}u e^{-F_{el}[u]/T}$. To estimate the melting line from the Lindemann criterion one usually neglects the longitudinal mode since $c_{11}(k) \gg c_{66}$ and write:

$$\langle u^2 \rangle \approx T \int_{BZ} \frac{d^2 q_\perp}{(2\pi)^2} \int \frac{dq_z}{2\pi} \frac{1}{c_{66}q_\perp^2 + c_{44}(q) q_z^2} \tag{19}$$

If one neglects the dispersion of c_{44} and uses a circular BZ:

$$\langle u^2 \rangle \approx T \frac{K_{BZ}}{4\pi\sqrt{c_{44}c_{66}}} , \qquad T_m \approx 2\sqrt{\pi}\sqrt{c_{44}c_{66}} c_L^2 a^3 \tag{20}$$

If one instead uses the expression:

$$c_{44}(q) = \frac{B^2}{4\pi(1+\lambda'^2 q_z^2 + \lambda_c'^2 q_\perp^2)} \tag{21}$$

one finds (20) with $c_{44} \to c_{44}^{\text{eff}} = c_{44}(0)/F(y_1, y_2)^2$ with $y_1 = \lambda_c'^2 K_{BZ}^2$ $y_2 = \frac{c_{66}}{c_{44}(0)}\lambda'^2 K_{BZ}^2$ with $F(y_1, y_2) = \int_0^1 dx(1+y_1x^2)^{1/2}(1+y_2x^2)^{-3/2}$. The combination $y_1 = \lambda_c'^2 K_{BZ}^2$ is large and using $c_{66} = \epsilon_0(1-b)^2/4a^2$ with $b = B/H_{c2}(T) = 2\pi\xi^2/a^2$ one finds $y_2 = \frac{1-b}{4}$ and

$$\langle u^2 \rangle \approx a^2 \frac{kT}{\epsilon\epsilon_0 a} \frac{1}{\sqrt{\pi}(1+y_2+\sqrt{1+y_2})}(1-b)^{-3/2} \tag{22}$$

Note that thermal fluctuations are large since T and λ_c are large. It is the region $q_\perp \approx K_{BZ}$ that contributes most. A slightly more detailed calculation[29,30] gives the melting curve $b_m(T)$ as the solution of:

$$\frac{\sqrt{b}}{1-b}\frac{t}{\sqrt{1-t}}\left(\frac{4(\sqrt{2}-1)}{\sqrt{1-b}}+1\right) = \frac{2\pi c_L^2}{\sqrt{Gi}} \tag{23}$$

Near T_c $b_m(T) \ll 1$ and one finds:

$$B_m(T) = 5.6\frac{c_L^4}{Gi}H_{c2}(0)(1-t)^2 \tag{24}$$

Although we did not discuss here melting in 2D geometries, such as films, we may indicate a few among numerous studies.[102,103]

3.5. *Landau–Ginzburg approach to melting*

Another approach is the lowest Landau level approximation (LLL) within GL. Here we follow the presentation in Ref. 40, to which we refer for details (see also Ref. 36). Consider (6) with $\alpha(T) = -\alpha_0(1-t)$ and rescale $r \to \xi(0)r$, $z \to \epsilon\xi(0)z$ where $\xi(0) = \frac{\hbar^2}{2m\alpha_0}$ and $\psi^2 \to \frac{2\alpha_0}{\beta}\psi^2$, $b = \frac{B}{H_{c2}(0)}$ with $H_{c2}(0) = \frac{\Phi_0}{2\pi\xi(0)^2}$. One then obtains:

$$\frac{F}{T} = \frac{1}{\omega}\int d^3x\left[\frac{1}{2}|D\psi|^2 + \frac{1}{2}|\partial_z\psi|^2 - \frac{1-t}{2}|\psi|^2 + \frac{1}{2}|\psi|^4 + \frac{\kappa}{4}(b-h)^2\right] \tag{25}$$

with $\omega = T\beta/(4\epsilon\alpha_0^2\xi(0)^4) = \pi t\sqrt{2Gi}$. For large κ one can neglect the fluctuations of the electromagnetic field, hence we can consider a uniform B and use the Landau gauge, i.e. $D_x = \partial_x - iby$, $D_y = \partial_y$. The LLL condition $-D^2\psi = b\psi$ implies $\int_x |D\psi|^2 = b\int_x |\psi|^2$. Further rescaling $r \to r/\sqrt{b}$, $z \to z\left(\frac{b\omega}{4\pi\sqrt{2}}\right)^{-1/3}$, $\psi^2 \to \left(\frac{b\omega}{4\pi\sqrt{2}}\right)^{2/3}\psi^2$ and neglecting the last term yields the fluctuation free energy density $\mathcal{F}_d = \frac{H_{c2}(0)^2}{2\pi\kappa^2}\left(\frac{b\omega}{4\pi\sqrt{2}}\right)^{4/3} f_d$ in terms of the reduced free energy density f_d:

$$f_d = -\frac{4\pi\sqrt{2}}{V'} \ln \int D\psi D\psi^* \exp\left(-\frac{f[\psi]}{4\pi\sqrt{2}}\right) \tag{26}$$

$$f[\psi] = \int d^3x \left[\frac{1}{2}|\partial_z\psi|^2 + a_T|\psi|^2 + \frac{1}{2}|\psi|^4\right], \quad a_T = -\left(\frac{b\omega}{4\pi\sqrt{2}}\right)^{-2/3} \frac{1-t-b}{2} \tag{27}$$

V' being the volume in the rescaled units. Let us start with the mean field theory[3] which simply minimizes (27), i.e. $f_d = f[\psi_{\min}]/V'$. For $a_T > 0$ the minimum is $\psi = 0$ corresponding to the normal phase. For $a_T < 0$, i.e. $B < H_{c2}(T) = H_{c2}(0)(1-t)$, a non-trivial minimum is searched for in the LLL (exact at $a_T \approx 0$) i.e. it obeys:

$$((\partial_x - iy)^2 + \partial_y^2 + 1)\psi = 0 \tag{28}$$

hence it must be a superposition of the form $\psi(r) = \sum_n C_n \phi_{k_n}(r)$ with $\phi_{k_x}(r) = \pi^{-1/4} e^{ik_x x - \frac{1}{2}(y-k_x)^2}$. The triangular vortex lattice is described choosing x along lattice direction $k_x = k_n = 2\pi n/a_0$ with $C_{n+2} = C_n$ and $C_1 = iC_0$ leading to the Abrikosov lattice (AL) wave function $\psi(r) = C\phi(r)$ where

$$\phi(r) = 3^{1/8} \sum_{\ell=-\infty}^{\infty} e^{i(\frac{\pi}{2}\ell^2 + 3^{1/4}\pi^{1/2}\ell x) - \frac{1}{2}(y - 3^{1/4}\pi^{1/2}\ell)^2} \tag{29}$$

is normalized in the unit cell, and $f_d = \min_C[|C|^2 a_T + \frac{1}{2}|C|^4 \beta_A] = -\frac{a_T^2}{2\beta_A}$ where:

$$\beta_A = \langle|\phi(r)|^4\rangle_{u \cdot c} = \sum_{n_1,n_2} e^{-\frac{1}{2}(n_1 d_1 + n_2 d_2)^2} \approx 1.16 \tag{30}$$

for any lattice[40] with vectors d_1, d_2, is minimum for the triangular lattice, hence selected. Since $f_d = 0$ for $a_T < 0$, it leads to the mean field specific heat jump $\Delta C/T = H_{c2}^2/8\pi\kappa^2\beta_A T_c^2$. The (reduced) structure factor of the perfect AL, associated to the density $|\psi(r)|^2$ is found as[40] $S(q_\perp, z = 0) = \left(\frac{a_T}{\beta_A}\right)^2 4\pi^2 \sum_K e^{-K^2/2}\delta(q_\perp - K)$.

From (27) the fluctuation free energy of the liquid was obtained in an improved perturbation theory[40] in the $|\psi|^4$ term around $\psi = 0$ leading to:

$$f_d^{\text{liq}} = 4\mu(1 + g(x)) \approx_{a_T \to -\infty} 4a_T^{1/2} + \frac{4}{a_T} + \cdots \tag{31}$$

$$\mu^3 - a_T\mu - 4 = 0, \qquad x = 1/(2\mu^3) \tag{32}$$

and $g(x)$ is the function (called $f(x)$ there) computed in Ref. 36. For the solid a two-loop calculation perturbing around the perfect AL gives:

$$\tilde{f}_d^{\rm sol} = -\frac{a_T^2}{2\beta_A} + 2.848|a_T|^{1/2} + \frac{2.4}{a_T} + \cdot\cdot \tag{33}$$

It is found that they intersect at $a_T \approx -9.5$, leading to a first order VL melting transition.[104] It compares reasonably with numerics[37] at high fields (the same method in 2D gives $a_T \approx -13.2$ in very good agreement with numerics). The Debye–Waller factor is found $e^{-2W} \approx 0.50$ to one loop which predicts the Lindemann number through $2W = \frac{1}{2}K^2 a_0^2 c_L^2$. From the scaled magnetization $m = -f'(a_T)$ one gets a jump $\Delta M/M = 0.018$ consistent with experiments[105] if one takes into account additional weak pinning disorder.[104] This jump is responsible for the spike in the specific heat[40] which is also observed to be accompanied by a jump.[64] The prediction $\Delta c = 0.0075((2 - 2b + t)/t)^2 - 0.20 Gi^{1/3}(b - 1 - t)(b/t^2)^{2/3}$ is about twice larger than the one observed in Ref. 64. The LLL approximation should hold for $4B\xi^2(0)/\Phi_0 > Gi$ or $1 - t - b < 2b$. Extension of the theory including weak disorder has been studied,[40,104] however it does not, at present day, provide a detailed description of the Bragg glass (see below).

3.6. *Superconducting phase coherence and decoupling*

Thermal fluctuations of the vortex positions lead to strong phase fluctuations:

$$\delta\phi(k) \approx \frac{2\pi}{a^2}\frac{ik_\perp \times u(k)}{k^2} \tag{34}$$

Since $\delta\phi(k) \sim u(k)/k$, it leads to suggestion of breakdown of superconducting phase order for $d \leq 4$.[106] There is still some debate on the validity and consequences of this argument (see Refs. 4 and 6 for review). In particular, the consideration of the gauge invariant phase $\tilde{\phi}(x) = \phi(x) - \frac{2\pi}{\Phi_0}\int^x A \cdot d\ell$ led to the opposite conclusion that there is phase long range order in $d = 3$, with cancellation of the divergences.[40,107]

The question of the phase fluctuations was much discussed for highly anisotropic superconductors, described by the Laurence–Doniach model, a generalization of GL[4] with discrete 2D layers indexed by $z = n$. The layers are coupled via (i) the Josephson coupling energy $-E_J \sum_n \int d^2r \cos(\tilde{\phi}(r, n+1, n))$ with $\tilde{\phi}(r, m, n) = \tilde{\phi}(r, m) - \tilde{\phi}(r, n)$ and $E_J = \epsilon^2\epsilon_0^2/\pi d$ (also noted J/ξ^2). (ii) the magnetic energy (the term B^2). Glazman and Koshelev[19] computed, within the London theory, the quadratic fluctuations of the gauge invariant phase. Their calculation suggests that the gauge invariant correlation

$\langle\tilde{\psi}(x)\tilde{\psi}^*(x')\rangle$ with $\tilde{\psi}(x) = |\psi_0|e^{i\tilde{\phi}(x)}$ does not always decay to zero along z. Along z, they found that for $B > B_{cr} \sim \Phi_0/\gamma^2 d$ the melting transition is given by its 2D limit $T_m = T_m^{2D} = \frac{0.62}{8\pi\sqrt{3}} d\epsilon_0$ and the phase fluctuations are $\langle\tilde{\phi}(0, n+1, n)^2\rangle \sim T/T_0(B)$ with $T_0(B) \approx T_m^{2D}(B_{cr}/B)^{1/2}$. Hence they concluded that for $B > B_{cr}$ there is a range of temperatures $T_0(B) < T < T_m$ where phase fluctuations are large. This result, and the further Gaussian self-consistent treatment[20] suggests the vanishing of critical current along z at a decoupling transition $T_{\rm dec}$. In this transition, the effective Josephson coupling between layers vanishes while the lattice can be maintained by the electromagnetic (EM) coupling between layers.

To predict more definitely a decoupling transition beyond the Gaussian (self-consistent) approximation, one needs to (i) use the renormalization group which predicts that the renormalized Josephson coupling flows to zero,[21] i.e. a decay of long range order in the phase, i.e. a decay of $\langle\cos(\tilde{\phi}(r, n, n+1))\cos(\tilde{\phi}(r, 1, 2))\rangle$ at large n, while a finite Josephson coupling is maintained locally between adjacent layers (and, in fact accounts for the experimentally observed plasma resonance[77]) (ii) a true transition requires topological defects. In the case of decoupling, these are interstitial and vacancies in a single layer.[80] These defects have also been called quartets of dislocations.[79] Indeed, in a layered superconductor one can think of the vortices as pancake vortices in each layer, joined by a Josephson string. The component of this string parallel to the layer is the Josephson vortex or fluxon. Such a fluxon excitation can also exist as a loop in between two layers, within which the phase difference between the layers varies by 2π. A vacancy-interstitial VI pair in a single layer can then also be seen as a closed Josephson loop connecting them. In the presence of a Josephson coupling J these pairs are bound. At the decoupling transition these defects proliferate, with loops of arbitrary sizes.

There are two limits in which the decoupling transition problem is simpler. (i) When $J \to 0$ and the layers are coupled only magnetically: then there is an unbinding transition of the VI pairs, from a perfect VL where phase coherence is maintained between the layers and a defected VL with vacancies and interstitials where phase coherence is lost. In the presence of pinning disorder this limit was studied in Ref. 83, (ii) the more isotropic superconductor: one can keep the description in terms of vortex *lines*: a VI pair then corresponds to a flux line which wanders of order a in a height $z = d$, or two flux lines which exchange positions. It is thus an entangled solid, the analogous of a super-solid in quantum problems.[81] In between these two extremes the topological transition merges with the above thermal decoupling once the Josephson coupling is finite, being two anisotropic limits of the same transition,[108] at which superconducting order is destroyed while the flux line positional correlations are maintained.

4. From the Abrikosov Lattice to the Bragg Glass

4.1. *Pinning disorder*

Here we consider point disorder due to point defects perturbing the superconductor on a scale smaller than ξ. This is the case of oxygen defects in high-T_c materials, which are found to account for the pinning disorder.[4,109] The coupling energy to the disorder is modeled as:

$$F_{\text{dis}} = \int d^3x \tilde{\rho}(x)\tilde{U}(x) \tag{35}$$

where $x = (r, z)$ and $\tilde{\rho}(r, z) = \sum_i p(r - R_i - u_i(z))$. Within the GL model, one can consider uncorrelated δT_c-disorder, $\tilde{U}(x) = |\psi_0|^2\alpha(x)$ and a form factor equal to $p(r) = 1 - |\psi_v(r)|^2$ from the solution $\psi_v(r)$ of the GL equation for a vortex normalized to one at infinity. One may also consider mean free path disorder, i.e. random mass. Equation (35) can be rewritten as a coupling to the vortex density $\rho(x)$:

$$F_{\text{dis}} = \int d^3x \rho(x) U(x) , \qquad \rho(r, z) = \sum_i \delta^{(2)}(r - R_i - u_i(z)) \tag{36}$$

with, in Fourier $U(q) = p(q_\perp)\tilde{U}(q)$. The correlator of the random potential is then

$$\overline{U(x)U(x')} = R^U(r - r')\delta(z - z') \tag{37}$$

where here and below $\overline{\cdots}$ denote average over the random potential, i.e. sample averages. Here $R^U(q) = \tilde{\gamma}\ p(q_\perp)^2$, which leads to $R^U(r) = \gamma\xi^2 k(r/\xi)$ with $k(0) = 1$ and $k(y) \sim y^{-2}\ln y$ at large y. For simplicity, one often considers $R^U(r) = dU_p^2 e^{-r^2/r_f^2}$ where d is the layer separation, and U_p typical pinning energy per unit length along z. The correlation length of the disorder seen by a vortex is $r_f \approx \xi$ since this is the smallest length that it can resolve. For a single vortex line one obtains:

$$F_{\text{dis}} = \int dz\ U(u(z), z) \tag{38}$$

A phenomenological description in terms of a 3D density n_i of independent impurities with random individual 3D pinning forces of r.m.s f_p can also be used. If there are a few defects within the volume $\xi^2 d$, as typically for oxygen vacancies, the resulting random potential is nearly Gaussian with $dU_p^2 = \gamma\xi^2 = f_p^2 n_i \xi^4$.

Pinning disorder deforms the vortex lattice and one first describes these deformations using elastic theory. The elastic description assumes small nearest neighbor displacements, $|u(x) - u(x+a)| \ll a$ and is *a priori* guaranteed to work at weak disorder and short scale (i.e. small system size). Even for weak disorder, there is no guarantee that it holds at large scales, and indeed we discuss later the possibility of more violent plastic deformations such as dislocations. An important correlation function, which probes translational order (also called positional order) is defined as

$$C_K(x) = \overline{\langle e^{iK\cdot u(x)} e^{-iK\cdot u(0)} \rangle} \tag{39}$$

measured in practice as a translational average over a large given sample.

4.2. *Larkin Ovchinnikov theory*

The first assault on the Abrikosov lattice came from Larkin who carried perturbation theory in the disorder[13] and found that the deformations of the vortex lattice must grow with scale in space dimension $d \leq 4$. His more involved original calculation can be summarized in terms of the beautifully simple, random force Larkin model:

$$F = F_{el}[u] - \int d^d x f(x) u(x) \tag{40}$$

which assumes a linear coupling of the displacement field to the disorder. Since this is a quadratic energy it can be solved for any realization of the disorder:

$$\langle u_\alpha(q) \rangle = \Phi(q)^{-1}_{\alpha\beta} f_\beta(q) \tag{41}$$

where $\Phi(q)$ is the elastic matrix (12 and 14). For a short range isotropic correlated random force $f(x)$ of variance W, with $\overline{f(x)f(x')} = W\delta^d(x-x')$, one finds, upon disorder averaging:

$$\overline{\langle u_\gamma(-q) u_\gamma(q) \rangle} = \frac{T}{\Phi_\gamma(q)} + \frac{W}{\Phi_\gamma^2(q)} \tag{42}$$

with $\gamma = L, T$ the longitudinal and transverse components. For isotropic elasticity $\Phi(q) = cq^2$ and at $T = 0$ this leads to:

$$\begin{gathered} \tilde{B}(x) = \overline{(u(x) - u(0))^2} = 4W \int_{q,BZ} \frac{1}{c^2 q^4}(1 - \cos(qx)) = \frac{4W}{c^2} b(x) \\ C_K(x) = e^{-K^2 \tilde{B}(x)/4}, \quad b(x) \sim |x|^{2\zeta_L}, \quad \zeta_L = \frac{4-d}{2} \end{gathered} \tag{43}$$

hence relative distortions grow as $u \sim |x|^{1/2}$ in 3D and translational order decays exponentially. Note that thermal fluctuations are subdominant.

Although solvable and instructive, this linear model has seriously unwanted features (i) it is not invariant by a lattice translation $u(r) \to u(r) + a_0$, (ii) if an external force is applied, the vortex lattice flows freely, i.e. there is no barrier to motion. One may wonder why it is relevant at all to the problem. It was understood by Larkin that linearizing the random potential seen by the vortices is a valid approximation for scales smaller than a characteristic length R_c. Since the random potential is smooth and varies on scale r_f, one expects this approximation to break down when the deformations reach $u \sim r_f$, i.e. at scale such that $\tilde{B}(x = R_c) \sim r_f^2$ which, from (43) gives:

$$R_c = (c^2 r_f^2 / W)^{1/(4-d)} \tag{44}$$

a scale which, for weak disorder $W \ll W_c = c^2 r_f^2 / a^{4-d}$ can be much larger than a. The case $R_c < a$, called single vortex pinning, is mentioned below. From Sec. 4.1 one estimates $W = \frac{1}{2a^4} \sum_K K^2 R^U(K)$ for the 3D vortex lattice, as $W \approx \frac{1}{2a^2} tr R^{U\prime\prime}(0) \sim dU_p^2 / a^2 r_f^2$ when $r_f \ll a$, using $a^{-2} \sum_K \equiv \int_k$ in that case.

With remarkable insight, Larkin and Ovchnnikov[14] (LO) understood that the scale R_c also determines the threshold depinning force. In their picture, the system breaks into domains of size R_c which are *independently pinned.* This leads to the collective pinning theory. To unpin the system one must apply a total external force FR_c^d on a Larkin domain at least equal to the typical pinning force acting on the same domain, which scales as $\int d^d x f(x) \sim W^{1/2} R_c^{d/2}$. This leads to the LO estimate for the depinning threshold force per unit volume:

$$F_c = \frac{f_c}{a^2} \approx \frac{W^{1/2}}{R_c^{d/2}} = \frac{c r_f}{R_c^2} \tag{45}$$

the last equality means that the elastic, disorder and external force energies within a Larkin domain are of the same order, defined as $U_c = c r_f^2 R_c^{d-2} \sim W^{1/2} r_f R_c^{d/2} \sim F_c r_f R_c^d$. Although sometimes termed "non-universal physics" since it occurs at short scale, the validity of (45) is by now proved by more sophisticated methods (see below) and has been incredibly useful in numerous experimental realizations of pinned systems. In superconductors, it determines quite well the magnitude of the critical current density $j_c = \mathsf{c} f_c / \Phi_0 = \mathsf{c} F_c / B$.[110]

For the triangular VL in $d = 3$ the deformation splits into $\tilde{B}(x) = \tilde{B}_T(x) + \tilde{B}_L(x)$ and there are two Larkin lengths, R_c in ab plane, and L_c along z. The dominant contribution arises from transverse modes:

$$\tilde{B}_T(x)=2W\int_{q,BZ}\frac{(1-\cos(qx))}{(c_{66}q_\perp^2+c_{44}(q)q_z^2)^2}\approx\frac{2W}{c_{66}^{3/2}c_{44}^{1/2}}b(\rho)\,,\quad \rho=\left(r^2+\frac{c_{66}}{c_{44}}z^2\right)^{1/2}$$

for large enough R_c (weak disorder) when one can neglect non-locality and use $c_{44}=c_{44}(0)$. In that case $b(\rho)\sim\frac{1}{4\pi^2}\rho$ and R_c is given by (44) with $c^2\to c_{66}^{3/2}c_{44}^{1/2}$. The Larkin length along z is $L_c=R_c\sqrt{c_{44}/c_{66}}\sim\frac{\lambda}{a}R_c$. For $R_c<\lambda/\epsilon$ one cannot neglect non-locality: since $c_{44}\sim 1/q_\perp^2$ one finds much slower logarithmic growth, i.e. $b(\rho)\to\lambda_c\left(\frac{r^2}{a^2}+\frac{z}{\epsilon a}\right)$. The Larkin length thus varies exponentially with the disorder for $a<R_c<\lambda/\epsilon$ and one has $L_c=\epsilon R_c^2/a$.

The energy scale $U_c\sim r_f^2c_{66}L_c\sim r_fF_cL_cR_c^2$ determines the critical force per unit volume F_c and the critical current j_c as:

$$F_c=\frac{j_c}{\mathsf{c}}B\approx c_{66}\frac{r_f}{R_c^2}\tag{46}$$

The above arguments also apply to a single flux line, setting $d=1$ and $c=e_l$ and $f(z)=\nabla U(0,z)$ in (38). This gives the single vortex Larkin length (along the field direction z), $L_c^{sv}=(e_l^2r_f^2/W_{sv})^{1/3}$, with $W_{sv}=\frac{1}{2}\nabla^2R^U(0)\approx\gamma$, leading to $f_c=(\Phi_0/\mathsf{c})j_c\approx e_lr_f/(L_c^{sv})^2$ (in case of a dispersive line tension the value $e_l(q_z\sim 1/L_c^{sv})$ should be used). The interactions with the neighboring vortex lines can be neglected if $L_c^{sv}<\epsilon a$. One checks from above and the previous sections that the condition $L_c^{sv}<\epsilon a$ is equivalent to $R_c<a$, that the term $c_{66}q_\perp^2\ll c_{44}(q)q_z^2$ for $q_\perp\sim K_{BZ}$ and that j_c is then determined by single vortex pinning.

To summarize, the LO theory shows that disorder makes the Abrikosov vortex lattice unstable. It does not predict however what is the new phase of vortex matter at scales larger than the Larkin length R_c. Beyond that scale one expects multiple metastable states with barriers between them, and to describe the system one must use modern tools which have emerged in nonlinear physics of glasses and glassy systems.

4.3. *The Bragg glass*

To describe the vortex system beyond the LO theory one must study the full model:

$$F[u]=F_{el}[u]+F_{\rm dis}[u]\tag{47}$$

As the Larkin model, it leads to a competition between the elastic energy (12) — which tends to reduce deformations — and the disorder energy (36)

— which tends to favor them — but where now the full nonlinear dependence in $u(x)$ is kept. The third ingredient is the periodicity of the lattice, i.e. the exact invariance of $F[u]$ by a lattice translation $u_i(z) \to u_i(z) + a_0$. This turns out to be crucial: since there is no energy gain to shift the lattice by a_0, large deformations will *not* be favored, by contrast to Larkin model which, it turns out, *overestimates them.* As discussed below, this simple fact leaves room for the existence of a Bragg glass.

Since periodicity is so important, we now examine the coupling of a periodic object to a random potential. To simplify the model (47) we need to trade the vortex position variables $u_i(z)$ for a smooth continuum deformation field $u(r, z)$. Since the density however is not a smooth function, this raises some subtle issues.

4.3.1. *The model: coupling a lattice to disorder*

It turns out that the density of a periodic object can be written:

$$\rho(x) \approx \rho_0 \left(1 - \partial_\alpha u_\alpha(x) + \sum_{K \neq 0} e^{iKx - iKu(x)} \right) \tag{48}$$

where ρ_0 is the average density, in terms of the smooth field $u(x) = \int_{q,BZ} u(q) e^{iqx}$ such that $u(R_i, z) = u_i(z)$. In the absence of topological defects an exact version of this decomposition can be given,[47,111] introducing a labeling phase field $\phi(R_i + u_i(z), z) = R_i$, from which corrections to (48), mainly higher gradients, can be computed. The gradient term is the change in local density due to compression. For a slowly varying $u(x)$ it contains Fourier components only near $q \approx 0$, while the other terms oscillate fast, with the periodicity of the lattice, i.e. $q_\perp \approx K$. They correspond to a translation of the crystal by u, with little change in the averaged density over a period, and can be viewed as a local shift in the phase of the periodic density wave, as in charge density wave theory, where usually a single K is retained. The coupling to the random potential (36) thus splits into two very different contributions[47,50,112]:

(i) *Long wavelength disorder*: The modes of the disorder $U(q)$ with small q give rise to a coupling, in general dimension d:

$$F^{\text{dis}}_{q \sim 0} = -\rho_0 \int d^d x U_0(x) \nabla \cdot u(x) \tag{49}$$

where $U_0(x) = \int_{q \approx 0} U(q) e^{iqx}$, which mostly compresses or dilates the lattice locally. Although reminiscent of Larkin's model it is down by a gradient and is thus much less efficient in destroying the lattice. For point impurities and standard elasticity, it results (by itself) in displacements $u \sim L^{2-d}$ which are unbounded only for $d \leq 2$ (and as $\sim \ln L$ in $d = 2$). Furthermore, *it does not lead to pinning*, since it is invariant by an arbitrary global translation of the lattice. It is a particular case of the random stress disorder with linear coupling energy $F \sim \sigma_{ij}(r) u_{ij}(r)$ to the strain matrix, a type of disorder also generated by *internal disorder*, e.g. a fixed connectivity crystal with atoms of different sizes.[103,113] In $d = 2$, it generates topological defects only above a critical strength, if alone,[114] or for any strength if combined with pinning disorder[115,116] (see below). It can however be neglected in $d = 3$.

(ii) *Pinning (or commensurate) disorder*: More dangerous are the modes $U_{q\approx K}$ close in periodicity to the lattice, which couple as:

$$F^{\rm dis}_{q\sim K} = \rho_0 \int d^d x U_{-K}(x) e^{-iKu(x)} \tag{50}$$

where $U_K(x) = \int_{q\approx K} U(q) e^{iqx}$. Their effect is to shift locally the crystal. As a result, an important new length scale appears, R_a, defined as the scale over which u varies by $\sim a$. It can be estimated as follows. The phase u, roughly constant on scale R_a, adjusts to cancel the phase of sum of random complex numbers $-\int d^d r U_K(r)$ over a volume R_a^d. A complex plane random walk argument yields that the gain in energy *density* from disorder as $\sim \rho_0 R^U(K)^{1/2}/R_a^{d/2}$. On the other hand, the cost in energy density of such a deformation $\sim a$ on scale R_a scales as $\sim c(a/R_a)^2$. The optimum occurs at a length scale:

$$R_a \sim (c^2 a^8 / R^U(K))^{1/(4-d)} \tag{51}$$

also called the Fukuyama–Lee length and was studied in the context of charge density waves,[112] a particular case of a more general argument for disordered system.[117] Note that in systems where $r_f \gg a$ and $R^U(K) \sim e^{-r_f^2 \frac{K^2}{2}}$ pinning disorder can be much suppressed. This occurs in some 2D systems (such as electrons on the surface of helium).

Keeping only pinning disorder (50), and using $\overline{U_K(x) U_{-K'}(x')} \approx R^U(K) \delta_{K,K'} \delta^d(x - x')$, the model (36) can thus be replaced by:

$$F[u] = F_{el}[u] + \int d^d x V(u(x), x)\,, \quad \overline{V(u,x)V(u',x')} = R(u-u')\delta^d(x-x') \tag{52}$$

where, for a periodic object the correlator of the disorder reads:

$$R(u) = \sum_{K\neq 0} \rho_0^2 R^U(K) \cos(Ku) \tag{53}$$

hence it is a periodic function with the periodicity of the lattice.

It turns out that (52) has emerged as *the universal model for pinning*, i.e. depending on the choice of space dimension d, number of components N of $u(x)$, and function $R(u)$ it can model a variety of disordered elastic systems — without topological defects. For $d = D - 1$, $N = 1$, it models a magnetic interface (without large scale overhangs) in a D-dimensional magnet with either random bond (RB) type disorder (a short range function $R(u) \sim e^{-u^2/r_f^2}$) or random field (RF) (a long range function $R(u) \sim |u|$), contact line of fluids $d = 1$, $N = 1$ and RF disorder, as well as various periodic systems $R(u)$ periodic such as charge density waves $N = 1$, magnetic bubbles ($d = 2$, $N = 2$) and of course vortex lattices in $d = 3$, $N = 2$ or in $d = 2$, $N = 2$ (field perpendicular to plane) $N = 1$ field along the plane.

To handle disorder, it is customary to use replica, i.e. introduce n copies of the displacement field $u_a(r)$, $a = 1, \cdots n$ and average their Gibbs measure over disorder, defining the replicated Hamiltonian through $e^{-\beta F_{\text{rep}}[u]} = \overline{e^{-\beta \sum_a F[u_a]}}$. One can then show that all disorder averaged correlation functions can be expressed as suitable averages of u_a, e.g. $\overline{\langle u(x)u(x')\rangle} = \langle u_a(x)u_a(x')\rangle_{H_{\text{rep}}}$ in the limit $n = 0$. For model (52)

$$F_{\text{rep}}[u] = \sum_{a=1}^{n} F_{el}[u_a] - \frac{1}{2T}\int d^dx \sum_{a,b=1}^{n} R(u^a(x) - u^b(x)) \tag{54}$$

where the disorder term is highly nonlinear and couples the replica. Expanding (54) to quadratic order at small $u^a - u^b < r_f$ one nicely recovers the Larkin model, in its replicated version: the Larkin regime $x < R_c$ thus exists for all pinned systems.

The Bragg glass, however, is characterized by two scales R_c and R_a, and that leads to an interesting scale dependence. When $H \ll H_{c2}$ (and temperature is not too high) the correlation length of the disorder $r_f \ll a$, thus $R_c \ll R_a$. In that case the function $R(u)$ is the sum of many harmonics and in fact, for $r_f \ll u \ll a$ it looks like a short range function, e.g. if $R_K^U \sim e^{-K^2 r_f^2/2}$ then $R(u) \sim e^{-u^2/2r_f^2}$, obtained by replacing the sum in (53) by an integral. Thus one may expect, as confirmed below, that in the range of scales $R_c < x < R_a$, the so-called "random manifold" regime, the problem looks the same as the $N = 2$, $d = 3$ random bond (non-periodic) elastic manifold: relative displacements between vortices being $< a$, each vortex sees in effect an independent disorder from its neighbor. Only at scales $x \gtrsim R_a$ does it feel the periodicity of the system and one expects another behavior.

There are at present two main analytical methods to treat the pinning model (52) and its replicated version (54), and to describe the vortex lattice beyond R_c. We first describe each in some generality, giving only the main idea. Both predict that the pinning problem is characterized by two exponents:

$$\tilde{B}(x) = \overline{\langle (u(x) - u(0))^2 \rangle} \sim x^{2\zeta}\,, \qquad \overline{F^2}^c \equiv \overline{F^2} - \overline{F}^2 \sim R^{2\theta} \tag{55}$$

where ζ describes the roughness of the pinned elastic object, and $\theta = d - 2 + 2\zeta$ the sample to sample fluctuations of its free energy F, with only a few universality classes depending on d, N, the type of disorder, i.e. of $R(u)$, and the type of elasticity (local, non-local, ...). Finally, we present the main results specific to the Bragg glass.

4.3.2. *Variational method: a mean field theory*

The idea,[47,51,118,119] of the Gaussian variational method (GVM) is to look for the best trial Gaussian Hamiltonian F_0 which approximates (54):

$$F_0 = \frac{1}{2} \int_q [\Phi_{\alpha\beta}(q)\delta_{ab} + \sigma_{ab}\delta_{\alpha\beta}] u^a_\alpha(q) u^b_\beta(-q) \tag{56}$$

By extremization of the variational free energy $\mathcal{F}_{\text{var}} = \mathcal{F}_0 + \langle F_{\text{rep}} - F_0 \rangle_{F_0}$ one obtains the saddle point equation for $n \times n$ matrix of variational parameters σ_{ab}. This method is known to give a reasonable approximation in the case of the pure one component Sine Gordon model, predicting a flat (massive) phase in $d = 3$, and the roughening transition in $d = 2$. Here, two types of solutions exist. The simplest one of the form $\sigma_{ab} = \sigma_c \delta_{ab} + \sigma$, mimics the distribution (thermal and over disorder) of each displacement mode $u(q)$ by a simple Gaussian, and leads to Larkin model behavior. The second type, which is a better variational solution above the Larkin scale, i.e. for system size $R > R_c$, exhibits spontaneous replica permutation symmetry breaking (RSB). This feature corresponds to the phase space breaking into separate ergodic components and was found for the first time in the mean field description of spin glasses. In a nutshell, the RSB solution approximates the distribution of displacements by a (hierarchical) superposition of Gaussians centered at different randomly located points in space[118]: all modes $u(q)$ are in effect decoupled and in a given sample their Gibbs distribution takes the form (for isotropic elasticity $\Phi(q) = cq^2$):

$$P(u(q)) \propto \sum_i e^{-f^{(i)}/T} e^{-\frac{c}{2T}(q^2 + R_c^{-2})|u(q) - u^{(i)}(q)|^2} \tag{57}$$

Each mode thus fluctuates around preferred configurations $u^{(i)}(q)$, i.e. metastable states with an associated free energy $f^{(i)}$. One recovers qualitatively the LO picture: the Larkin length R_c sets the internal size of the elastically correlated domains. In the simplest type of RSB solution, there is only one level to the hierarchy: the f_i are distributed according to an exponential distribution $P(f) \sim \exp(u_c f/T)$, where $u_c = T/T_g$, and the $u^{(i)}(q)$ according to:

$$P(u^{(i)}(q)) \sim \prod_i e^{-\frac{c}{4T_g}q^2|u^{(i)}(q)|^2} \tag{58}$$

where T_g is a temperature scale corresponding to thermal depinning. This so-called one-step RSB holds for a periodic object in $d = 2$: there, the elastic model (with topological defects forbidden) presents a genuine glass transition at T_g, the glassy analog of the roughening transition (in a real 2D crystal however, this transition is usually preempted by melting). For the vortex lattice in $d = 3$ however, no such transition exists and one needs the infinite-step RSB solution. There the states $u^{(i)}(q)$ are organized in a hierarchical manner (clusters of states inside larger clusters and so on) generalizing the above construction. The GVM gives the correlation function as[118]:

$$\overline{\langle u(-q)u(q)\rangle} = \frac{T}{\Phi(q)}\left(1 + \int_0^1 \frac{dv}{v^2}\frac{[\sigma](v)}{\Phi(q)+[\sigma](v)}\right) \sim_{q\to 0} \frac{Z}{q^{d+2\zeta}} \tag{59}$$

where $[\sigma](v) \sim (v/T)^{2/\theta}$ at small v, with $\theta = [d-2+2\zeta]$ and Z finite as $T \to 0$. For the "random manifold" problem, i.e. a short range correlator $R(u)$, and $\Phi(q) = cq^2$, the GVM predicts $\zeta = \zeta_F = \frac{4-d}{4+N}$. This is the same result as the "Flory approximation" which implies the balance $u^2R^{d-2} \sim u^{-N/2}R^{d/2}$ of elastic energy and disorder energy in (52) estimated from a dimensional estimate $R(u) \sim \delta^{(N)}(u)$. It predicts $\zeta = 3/5$ in $d = 1, N = 1$ a case where $\zeta = 2/3$ is known to be exact. Being a mean field method, the GVM cannot be expected to be exact, but it does provide a quite consistent qualitative picture and reasonable approximations. For a periodic object it correctly predicts $\zeta = 0$ and an amplitude Z not so far off from the correct value (see below). The extension of the GVM to the dynamics provides an interesting approach to aging phenomena[121] but is not at present the most convenient or accurate method to describe creep and depinning.[122]

We now turn to the more powerful FRG method, which is exact in a $d = 4 - \varepsilon$ expansion and treats more accurately the coupling between the modes.

4.3.3. *The functional renormalization group (FRG)*

The method, pioneered in Refs. 47 and 120, is presented here it in its modern form.[123–125] There, the central object, the *renormalized disorder correlator*, is not an abstract but a perfectly measurable quantity. To define it, imagine a gedanken experiment where one puts the elastic system of size R in an harmonic well:

$$F_{V,w}[u] = F_{el}[u] + \int d^d x \, V(u(x), x) + \int d^d x \, \frac{m^2}{2}(u(x) - w)^2 \qquad (60)$$

and denote $u(x, w)$ the configuration which minimizes $F_{V,w}$ and $u(w) = R^{-d} \int d^d x \, u(x, w)$ its barycenter. The role of the harmonic well is to fix the global translation mode of the system at $u(x, w) = w$ in absence of disorder, and, on average, at $\overline{u(x, w)} = w$ in presence of disorder. It adds to the elastic energy and provides an additional length scale R_m such that the system feels the well only on scales $x \geq R_m$: in the simplest case of local isotropic elasticity $cq^2 \to cq^2 + m^2$ and $R_m = \sqrt{c}/m$.

A remarkable feature is that the minimum energy configuration $u(x, w)$ experiences discontinuous jumps as w is varied. These jumps, also called shocks, occur each time a distinct local energy minimum becomes more favorable. The total force density exerted by the harmonic spring $f = m^2(w - u(w))$ is zero on average but its fluctuations are measured by the second cumulant[123]:

$$m^4 \overline{(u(w) - w)(u(w') - w')}^c = R^{-d} \Delta(w - w') \qquad (61)$$

which defines the function $\Delta(w)$, called *renormalized force correlator.* For notational simplicity we focus here on $N = 1$ (interfaces), but for an N component displacement field, $\Delta(w) \to \Delta_{\alpha\beta}(\vec{w})$. The factor R^{-d} is natural there from the central limit theorem, as the effect of the well is to split the system into $\sim (R/R_m)^d$ regions which may be viewed as independent. Accordingly, roughness $u \sim x^\zeta$ is expected to saturate for scales $x > R_m$. This leads to define a rescaled force correlator through:

$$\Delta(w) = A_d R_m^{2\zeta - (4-d)} \tilde{\Delta}(w/R_m^\zeta) \qquad (62)$$

where A_d is a known constant. The main result of the FRG is that $\Delta(w)$ satisfies a nonlinear differential equation, which can be put in the form:

$$-m\partial_m \tilde{\Delta}(w) = (4-d-2\zeta)\tilde{\Delta}(w) + \zeta w \tilde{\Delta}'(w) - \left[\frac{1}{2}\tilde{\Delta}(w)^2 - \tilde{\Delta}(0)\tilde{\Delta}(w)\right]'' + \sum_{n\geq 2} \beta_n[\tilde{\Delta}] \qquad (63)$$

for $N = 1$, and similar generalizations for $N > 1$, where the nonlinear term can be computed to successive orders $\beta_n \sim O(\tilde{\Delta}^{n+1})$ in a so-called loop expansion. Hence the original complicated, many degree of freedom pinning problem has been reduced to solving a (relatively) simple differential equation.

Analyzing Eq. (63) reveals striking features. First, one finds that in the limit of interest R_m large ($m \to 0$) $\tilde{\Delta}(w)$ flows to some fixed point (FP) $\tilde{\Delta}^*(w)$ which is uniformly $O(\varepsilon) = 4 - d$) near $d = 4$, hence justifying the small $\tilde{\Delta}$ expansion in (63), and justifying the form (62) in the small m limit. One finds several possible FP,[47,120] each associated to a distinct value of the roughness exponent ζ and corresponding to the different universality classes of the pinning model (52). The most striking feature is that these FP are *non-analytic* and exhibit a linear cusp $\tilde{\Delta}^*(w) - \tilde{\Delta}^*(0) \sim |w|$ at small w. In fact, this feature becomes pronounced at scale $R_m \geq R_c$, the *Larkin length*, and directly follows from the fact that $u(w)$ exhibits jumps $S_i = u(w_i^+) - u(w_i^-)$ at some discrete set of w_i. More precisely one shows[126] from (61) that the right derivative is:

$$|\Delta'(0^+)| = m^4 \langle S^2 \rangle / 2 \langle S \rangle \tag{64}$$

i.e. it contains information on the moments of the distribution $P(S)$ of jump sizes.

Until now, we focused on the minimal energy state, but the method can be extended to Gibbs *equilibrium* at non-zero temperature $T > 0$, simply replacing everywhere $u(x, w) \to \langle u(x) \rangle_{F_{V,w}}$ i.e. thermal averages at fixed V and w. The resulting FRG equation is discussed in Sec. 4.4(c).

It can also be extended to the *dynamics*. There w is changed with time, e.g. $w = vt$. The limit $v = 0^+$ at $T = 0$ allows to study the depinning transition. There, as w increases adiabatically, the configuration $u(r, w)$ follows a sequence of metastable states: at a given w the system is stuck in a local energy minimum, distinct in general from the global minimum. When w is increased, nothing happens until the barrier seen by the system vanishes at some w_i. At $w = w_i^+$ the system jumps to the next metastable state, and so on. The attentive reader will guess that the average force with which the harmonic spring drives the system is now non-zero and behaves as:

$$m^2 \overline{w - u(w)} = F_c(m) \approx F_c - C\, m^{2-\zeta_{dep}} \tag{65}$$

at small m, where F_c is the depinning threshold force density in the absence of the harmonic well. Equation (61) still defines the renormalized force correlator, and (64) also holds for these dynamical jumps, called avalanches. Thanks to the FRG one can again compute $\Delta(w)$. Curiously it is found to obey, to lowest (one-loop) order, *the same* Eq. (63) as for the statics, the difference only starts in the two-loop term[125]:

$$\beta_2[\tilde{\Delta}] = \left[\frac{1}{2} (\tilde{\Delta}'(w)^2 - \lambda \tilde{\Delta}'(0^+)^2)(\tilde{\Delta}(w) - \tilde{\Delta}(0)) \right]'' \tag{66}$$

with $\lambda = 1$ for the statics and $\lambda = -1$ for depinning. Solving for the FP one finds $\zeta_{RB} = 0.20829804\varepsilon + 0.006858\varepsilon^2 + \cdot\cdot$ and $\zeta_{\rm dep} = \frac{\varepsilon}{3}(1 + 0.14331\varepsilon + \cdot\cdot)$ for the roughness of non-periodic objects with short range disorder (e.g. an Ising interface with random bond disorder) in equilibrium and at the depinning transition, respectively.

Hence thanks to the FRG, the function $\Delta(w)$ defined in (61) has been *computed* order by order in the $\varepsilon = 4 - d$ expansion. It has also been measured in numerical simulations for interfaces, both for the statics and for the depinning transition.[127] The agreement was found remarkable. Recently, $\Delta(w)$ was measured in wetting experiments on contact line depinning,[128] and should be accessible in cracks and magnets. Whether it can be directly measured in experiments on vortex systems remains open.

Knowing $\Delta(w)$ the FRG method allows to compute other quantities in an expansion in $\tilde{\Delta}^*$, i.e. in a $\varepsilon = 4 - d$ expansion, such as the correlation function $\overline{u(-q)u(q)} = q^{-(d+2\zeta)} f(qL_m)$, the avalanche size distribution $P(S)$, or the dynamical exponent z. It also allows to obtain the v–f characteristics above and near the depinning transition, as $v \sim (f - f_c)^\beta$ where $\beta = (z - \zeta)/(2 - \zeta)$, with generally $\beta < 1$ for the depinning of an elastic object. This differs from the conventional picture, (3) and Fig. 2, and leads to the $T = 0$ characteristics of Fig. 14.

Let us close by indicating that while the setting using a harmonic well makes the FRG well defined, and allows e.g. reliable higher loop calculations, if one only wants one-loop approximation the Wilson shell method, where the small scale degrees of freedom $\Lambda_{\ell+d\ell} < q < \Lambda_\ell = \Lambda e^{-\ell}$ are iteratively eliminated, can also be used.[120] One recovers (63) with $-m\partial_m \to \partial_\ell$.

4.4. *Correlations and dynamics in the Bragg glass*

We now present the main properties of the vortex lattice in presence of pinning disorder which can be obtained using the above methods and the resulting physical picture for the Bragg glass phase.

(a) *Quasi-long range order:*

Let us first discuss the static, i.e. equilibrium correlations:

$$\tilde{B}_{\alpha\beta}(x) = \overline{\langle (u_\alpha(x) - u_\alpha(0))(u_\beta(x) - u_\beta(0)) \rangle} \tag{67}$$

The Gaussian variational method (GVM), complemented by the FRG, leads to three main regimes in the case $r_f \ll a$ and allows to describe the crossover between them.[47] For simplicity, we start with isotropic elasticity

$\Phi(q) = cq^2$, where $\tilde{B}_{\alpha\beta}(x) = \frac{1}{N}\delta_{\alpha\beta}\tilde{B}(x)$ and describe below the more realistic case.

(i) *Larkin regime* $\tilde{B}(x) \ll r_f^2(T)$: The results of Sec. 4.2 are recovered. In addition, although thermal fluctuations are subdominant in (48), they also average out the disorder locally, as captured by the variational method which in effects replaces $r_f^2 \to r_f^2(T) = r_f^2 + \langle u^2 \rangle$, i.e. the thermal width of the vortex line, leading to an effectively temperature dependent Larkin length $R_c(T)$, a useful concept to study thermal depinning phenomena (discussed in details e.g. in Ref. 4).

(ii) *Random manifold regime* $r_f(T)^2 \ll \tilde{B}(x) \ll a^2$: There $\tilde{B}(x) \approx r_f^2(x/R_c)^{2\zeta}$, i.e. displacements grow as a power law with $\zeta = \zeta_{RM} \approx 0.2$ ($\zeta_F = \frac{1}{6}$) for 3D vortex lattices ($N = 2$). The decay of the translational order in that regime is a stretched exponential $C_K(x) \sim e^{-Cx^{2\zeta}}$. This regime was anticipated in Refs. 48 and 51.

(iii) *Asymptotic regime* $\tilde{B}(x) \geq a^2$, i.e. $x > R_a$. The periodicity of the lattice becomes important and one finds that displacements should grow only logarithmically at large scale.[50,129] More precisely[47,119]:

$$\tilde{B}(x) \approx \frac{2NA}{K_0^2} \ln |x| \tag{68}$$

with $\eta = A = 4 - d$ in the GVM and K_0 the lowest reciprocal lattice vector. Within the GVM the decay of translational order is found to be a power law:

$$C_{K_0}(x) \sim \frac{1}{|x|^\eta} \tag{69}$$

as confirmed by the one-loop FRG, which gives $\eta = A = \frac{\pi^2}{9}\varepsilon \approx 1.096$ and for the $N = 2$ triangular lattice found within 5% of that value.[47] A more accurate one loop estimate for the triangular lattice, $A = \eta = 1.143$, was later obtained.[130] Higher loop corrections to (68), computed only for $N = 1$:[125] $A = \varepsilon(1 + \frac{\varepsilon}{6} + \cdot\cdot)$, are likely to increase again this result slightly.

The conclusion is thus that the translational order decays as a power law, quite reminiscent of the quasi-long range order exhibited by 2D solids. The value of the exponent $\eta \approx 1$ means that the structure factor:

$$S_K(q) = \int d^d x e^{iqx} C_K(x) \approx \int d^d x e^{iqx} e^{-\frac{1}{2}K_\alpha K_\beta \tilde{B}_{\alpha\beta}(x)} \sim \frac{1}{q^{3-\eta}} \tag{70}$$

exhibits *divergent Bragg peaks*[47] as $q \to 0$ (it does so as long as $\eta < 3$).

Note that for $r_f \ll a$, the scale R_a is obtained by matching $\tilde{B}(x)$ between regimes (ii) and (iii), which predicts $R_a \sim R_c(a/r_f)^{1/\zeta_{RM}}$. The variational method thus reproduces the Flory approximation result $R_a \sim R_c(a/r_f)^6$, which is also what (51) predicts for $K = K_0$. In principle, the FRG allows to predict more accurately the crossover function with the correct value for ζ but this has not yet been attempted. When $r_f(T) \sim a$ (high fields or temperatures) the crossover occurs directly from the Larkin regime to the asymptotic one, with $R_a \sim R_c$.

These results are not much changed by considering realistic elasticity matrix $\Phi_{\alpha\beta}(q)$ for the vortex lattice. The variational method allows to compute the correlation $\tilde{B}_{\alpha\beta}(x) = P^T_{\alpha\beta}(r)\tilde{B}_T(x)+P^L_{\alpha\beta}(r)\tilde{B}_L(x)$, which takes the form, in the limit $c_{66} \ll c_{11}$:

$$\tilde{B}_{L,T}(r,z) = \frac{a^2}{\pi^2} b_{L,T}\left(\frac{r}{R_a}, \frac{z}{L_a}\right), \tag{71}$$

with $L_a = \sqrt{\frac{c_{44}}{c_{66}}} R_a$, and $R_a \sim a^4 c_{66}^{3/2} c_{44}^{1/2} / \rho_0^2 U_p^2 d r_f^2$, where $\rho_0 = a^{-2}$ is the average vortex density. It can also be written in a form similar to the Larkin length, as $R_a = a^2 c_{66}^{3/2} c_{44}^{1/2} / W_{BrG}$ where the disorder parameter W_{BrG} is much weaker than W defined in Sec. 4.2, as $W_{BrG} = W(r_f/a)^4$, leading to the Flory approximation $R_a/R_c = (a/r_f)^6$. If we use local elastic constants $\sim 1/a^2$ it leads to $R_c \sim 1/a^2$ while $R_a \sim a^4$, hence as the field is increased the Larkin length increases, while R_a decreases leading eventually to the instability of the Bragg glass at some "disorder induced" melting field. The qualitative statement that "increasing the field increases the disorder" is thus justified in the Bragg glass regime (for $R \gtrsim R_a$): scaling the field as $u = a\tilde{u}$, then results in a theory with a-independent elastic constants and a appearing only as ρ_0^2 in the disorder strength in front of the cosine term in (53).

The in-plane anisotropy of the deformations is measured by the ratio $\sigma = \tilde{B}_T(r,0)/\tilde{B}_L(r,0)$ in decoration experiments in Fig. 14 (left). It is predicted to cross over, as r increases, from the universal value[51] $\sigma = 2\zeta_{RM} + 1 \approx 1.4$, in the RM regime, to $\sigma = 1$, i.e. isotropic displacements, at large distance.[47] Within the one loop FRG the exponents ζ_{RM} and η are found to depend slightly on the ratio c_{66}/c_{11}, however the effect is very weak[130] (smaller than the two-loop correction): η varies between 1.143 and 1.159. Finally, consideration of non-local elastic moduli yields further regimes and crossovers at intermediate scales. If anything, the behavior $c_{44} \sim 1/q_\perp^2$ for $q_\perp > \lambda_c^{-1}$ makes the Bragg glass more stable, i.e. the deformations grow more slowly as $\tilde{B}(x) \sim \ln|x|$ in the Larkin and RM regimes, and as $\tilde{B}(x) \sim \ln\ln|x|$ for $r > R_a$. Of course, the asymptotic regime (iii) remains unchanged at the largest scales $r > \lambda_c$ and is dominated by local elasticity with $c_{44} \equiv c_{44}(0)$.

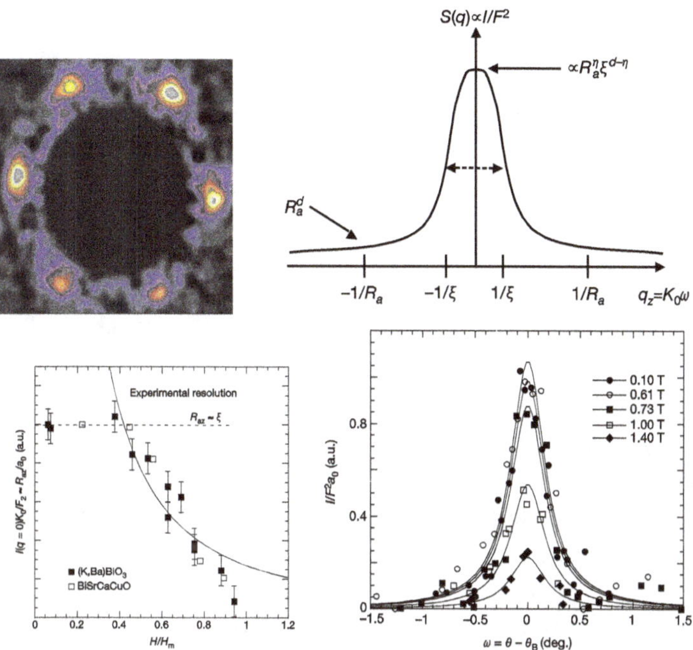

Fig. 12. Top: (left) Neutron diffraction peaks in the vortex solid phase of a (K,Ba)BiO3 superconductor at $T = 2K$ (right) predicted angular dependence of the intensity, taking into account the (fixed) experimental resolution (called here ξ, distinct from the coherence length). For $R_a > \xi$, as R_a decreases (H increases), the peak width at half-maximum should remain constant $\sim 1/\xi$. Bottom (right): The diffracted intensity (rocking curves) indeed collapses without any broadening above 0.7 T an evidence for the Bragg glass (left): $L_a \equiv R_{az}$ as a function of H obtained from the maximum of the peak: resolution limited $\xi \sim 50a_0$ at small field, it decreases near H_m which coincides with the position of the second peak in magnetization curves. From Ref. 74.

The prediction of divergent Bragg peaks was compared with experiments,[131] and most directly verified[74] on $(K, Ba)BiO_3$, an isotropic material where no complications due to anisotropy or dimensional crossover can be invoked. Taking into account the finite experimental resolution ξ, the divergent Bragg peak (70) are predicted to take the shape shown in Fig. 12 (top right) if $R_a > \xi$: (i) the peak width is determined only by ξ (ii) the peak height allows to measure the positional correlation length R_a. Thus if disorder (i.e. here the magnetic field) is increased, R_a decreases and the observed peaks should *collapse without broadening.* This is what they do in Fig. 12 (bottom right), a direct evidence of the Bragg glass phase and its algebraic positional order.

(b) *Topological order*

The striking result that quasi-long range order survives was derived within an elastic theory, assuming the absence of dislocations. However, it was argued early on[45,129] that dislocation loops of size L could gain an energy $L^{d/2}$ from disorder, larger than their core energy cost L^{d-2} in $d \leq 4$, hence that they would always be generated even at weak disorder. The argument is similar to the Imry Ma argument showing the destruction of a ferromagnetic phase by random fields.[117] Although this estimate was implicitly based on the Larkin model, it remained unchallenged, leading to the belief that the elastic theory would hold only below some finite scale L_{disloc}, beyond which the lattice was destroyed, presumably leading to a vortex glass. Indeed, such dislocations were always seen in early decoration pictures,[27] although it was questioned whether these were a true equilibrium feature or simply the result of the quench process (field cooling).[131] These also suggested some *hexatic glass* regime, where translational order decays exponentially beyond L_{disloc}, but disclinations being absent, a bond orientational order could survive.[132] Ultimately however, and in absence of a locking effect to a lattice direction,[133] Larkin and Bragg glass type arguments would also apply and perfect hexatic order would also be unstable and lead to quasi-order, also potentially unstable to disclinations at large scale.

The energy argument was then re-examined in light of the more detailed theory for lattice deformations beyond the Larkin scale.[47] One compares the core energy cost $\sim L \ln L$ of a dislocation loop in $d = 3$ on scale L, with the cost of an elastic deformation, of the order of $L^{d-2+2\zeta}$, which can be typically relaxed by allowing for a dislocation loop, provided $u \sim L^{\zeta}$. So if translational order is destroyed $\zeta > 0$, i.e. if the Larkin or the random manifold regime were true up to infinite scales, it would indeed be favorable to create dislocations. However, if quasi long range order persists $\zeta = 0$ and for weak disorder, dislocations may not be generated by disorder and the elastic solution may thus be self-consistently stable. It was thus proposed[47] that a *distinct*, topologically ordered, phase with no large unbound dislocation, the Bragg glass, survives at weak disorder, and that the vortex glass, if it exists, would only occur at stronger disorder, i.e. higher fields in the phase diagram. The original energy argument was strengthened in subsequent works[134,135] and received support from numerical simulations.[136] Note that by contrast, $d = 2$ lattices (e.g. a single layer with H perpendicular to the plane) are ultimately unstable to dislocations at large scales.[115,116] One shows that pinning disorder, if present, renormalizes upward the long wavelength disorder, which grows unboundedly and generates dislocations at a

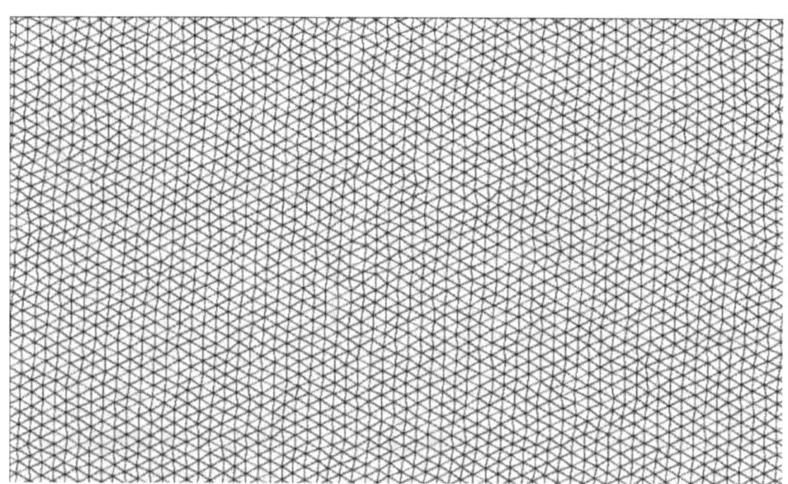

Fig. 13. Decoration picture from Ref. 138 on BSCCO: the Delaunay triangulation shows that each vortex has six neighbors and reveals the absence of dislocations.

scale $L_{\text{disloc}} \sim R_a e^{\sqrt{\ln(R_a/a)}} \sim a e^{(E_c/E^*)^{2/3}}$ which however can be large for large bare core energy E_c, or weak disorder. A topologically ordered glass however exists in $d = 2$ for lines directed *along* the plane, with however a faster decay $C_{K_0}(r) \sim e^{-B(\ln r)^2}$ of translational order than in the 3D Bragg glass, and was studied in Ref. 137.

The most vivid illustration of what the Bragg glass looks like, i.e. a distorted but topologically ordered lattice, is probably the decoration picture shown in Fig. 13. In this later experiment non-equilibrium dislocations were eliminated by tilting back and forth the external field, and *equilibrated* regions containing $\sim 10^5$ vortices without dislocations were obtained in fields of 70–120 G. The mean square distortions $\tilde{B}_{L,T}(r, z = 0)$ were measured and Fig. 14 shows a remarkable agreement with the predictions of the random manifold regime (ii). The data also shows saturation of growth where it is expected from the asymptotic regime (iii) but it was not possible in that experiment to equilibrate the system at such large scales.

(c) *Creep and glassy features*

Although the Bragg glass retains the topological order of the Abrikosov lattice, it differs from it by being *a glass*, with many metastable states and barriers to motion. At zero temperature, $T = 0$, this implies a non-zero critical current j_c (i.e. depinning threshold force f_c) with a depinning transition, and the $v - f$ or $E - j$ curve represented in Fig. 14. At $T > 0$ flux creep occurs by thermal activation over barriers $\sim U_b(j)$, even at very small

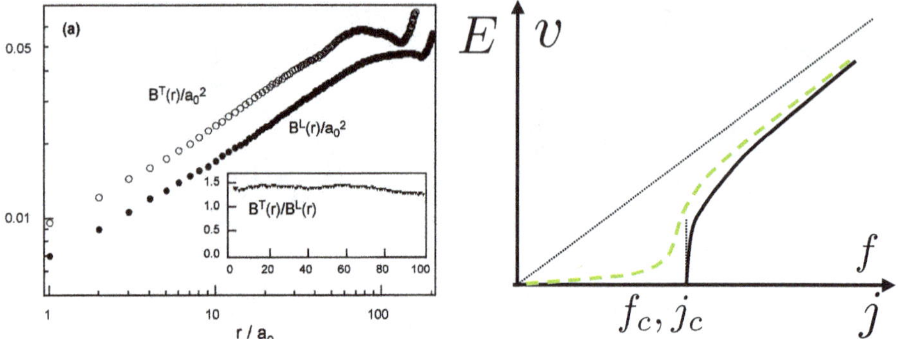

Fig. 14. Left: Mean squared deformation $\tilde{B}_{L,T}(r, z = 0)$ as a function of r in the decoration picture of Fig. 13. The exponent $\zeta = 0.22$ of the random manifold regime is the best fit. Inset: Anisotropy ratio as a function of scale, in agreement with the predicted value $1 + 2\zeta$. From Ref. 138. Right: v–f curve for an elastic object. Elastic depinning $v \sim (f - f_c)^\beta$ at $T = 0$ (solid line) with $\beta < 1$, i.e. a vertical slope at f_c, and creep $v \sim e^{-A/(Tf^\mu)}$ at small f and $T > 0$ (dashed line) with zero linear resistivity. Both can be derived using the FRG theory.

$j \ll j_c$ ($f \ll f_c$). The salient feature of a glass is the divergence of the barrier $U_b(j) \sim j^{-\mu}$ at small j. The classical argument[48,49] assumes that for small $j \ll j_c$, most of the time is spent near equilibrium and that the motion proceeds via thermal nucleation of an excitation involving displacements of order $u(R) \sim r_f(R/R_c)^\zeta \gg r_f$ over an internal scale R, which costs a typical energy $F_{\text{dis}} \sim U_c(R/R_c)^\theta$. Taking into account the energy gain from the external force, one sees that in order to move, the system must overcome the barrier:

$$U_b(f) \sim \max_R[F_{\text{dis}} - f u(R) R^d]\,, \quad U_b(f) \sim U_c\left(\frac{f_c}{f}\right)^\mu, \quad \mu = \frac{\theta}{d + \zeta - \theta} \tag{72}$$

given by the optimal nucleus of size $R_{\text{opt}} \sim f^{1/(d+\zeta-\theta)}$. This leads to the creep law:

$$v \sim v_{ff} e^{-U_b(f)/T} \sim v_{ff} e^{-\frac{U_c}{T}\left(\frac{f_c}{f}\right)^\mu}, \qquad \mu = \frac{d - 2 + 2\zeta_{\text{eq}}}{2 - \zeta_{\text{eq}}} \tag{73}$$

where in the latter we have assumed that $\theta = \theta_{\text{eq}} = d - 2 + 2\zeta_{\text{eq}}$, i.e. equilibrium values for the exponents.

Although this argument has been quite successful, fascinating questions remain, still mostly under investigation (i) do relevant barriers indeed scale as fluctuations of energy minima, i.e. $\theta = \theta_{\rm eq}$ as assumed above, and is the energy landscape characterized by a unique energy scale, (ii) is the use of the Arrhenius law $v \sim \exp(-U_b/T)$ for a single barrier fully justified or does the distribution of barriers (eventually broad) have an effect, (iii) after the optimal barrier has been passed, what is the subsequent motion, is it deterministic and looks like avalanches near $T = 0$ depinning? The closely related question of the thermal rounding of the depinning transition is also not fully understood.[139] The one-loop FRG in the presence of temperature amounts to adding the term $\tilde{T}\tilde{\Delta}''(u)$ in Eq. (63). In the presence of this term, the correlator $\tilde{\Delta}(u)$ remains a smooth function in a very narrow boundary layer $u \sim \tilde{T} \sim TR^{-\theta}$. The dynamical exponent $t \sim R^z$ is controlled by the curvature, $z = 2 - \tilde{\Delta}''(0) \sim 1/\tilde{T}$, and blows up leading to diverging barriers. The FRG thus leads to a derivation of the creep law with $\theta = \theta_{\rm eq}$.[52] It also allows to study the effect of barrier distributions[140] and to predict that the behavior at intermediate scale in the creep process is indeed governed by $T = 0$ depinning.[52] Novel algorithms,[142] exploiting the so-called no passing rule for interfaces,[141] have confirmed that the roughness exponent is $\zeta = \zeta_{\rm dep}$ in the critical region above the scale $R_{\rm opt}$ (and below a much larger scale set by the velocity).

Since for $r_f \ll a$ the Bragg glass is characterized by two distinct scales R_c and R_a, the creep barrier crosses over from the random manifold (RM) regime, $U_b(f) \sim U_c(f_c/f)^{\mu_{\rm RM}}$ for $f_a \ll f \ll f_c$ to $U_b(f) \sim U_a(f_c/f)^{\mu_{\rm BrG}}$ for $f \ll f_a$ with $U_a = U_c(R_a/R_c)^{\theta_{\rm RM}}$ and $f_c/f_a = (R_a/R_c)^{2-\zeta_{\rm RM}}$. The asymptotic exponent for the divergence of barriers in the Bragg glass is thus predicted to be $\mu_{\rm BrG} = 1/2$, corresponding to $\zeta = 0$ and $\theta = d - 2$.[47,50]

Although in other pinned systems, such as domain walls in magnetic films,[143] or in ferroelectric films,[144] the creep law has been nicely confirmed over many orders of magnitude, precise and unambiguous transport measurements for vortices are a challenge. This is due to the many length scales and complicated effects in the spatial distribution of the current due to surface barriers.[145] Indeed in BSCCO crystals near melting T_m, the bulk pinning is weak (an ideal situation to measure creep) and surface barriers dominate. In a remarkable experiment this effect was circumvented by positioning the contacts far from the edge.[146] The results are represented in Fig. 15. It demonstrates a pure Arrhenius law for activated motion just below the melting transition in the vortex solid. The measured barrier $U_b(j)$ (inset) shows a dependence in current consistent with the exponent $\mu = 1/2$ expected for the Bragg glass in its asymptotic regime.

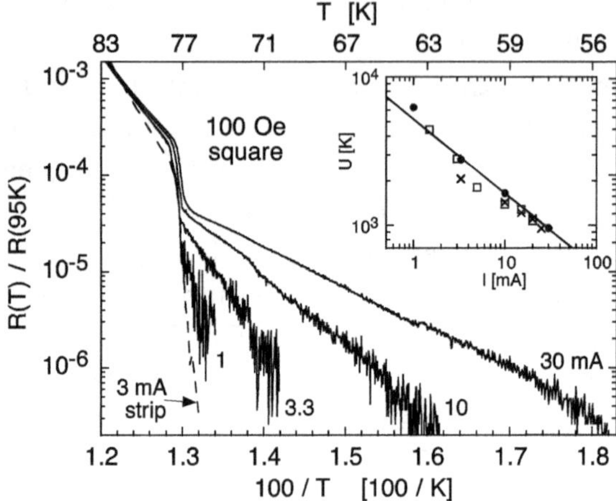

Fig. 15. Arrhenius plot of the normalized resistance at 100 G and various currents between 1 and 30 mA for a BSCCO strip. The dashed curve is the normalized resistance of the strip at 3 mA for comparison. Inset: Activation energy as a function of current in the vortex solid phase at applied fields of 100 G (dots), 300 G (crosses) and 500 G (squares). Solid line: $U = CI^{1/2}$. From Ref. 146.

5. Other Glasses

5.1. *Vortex glass and gauge glass*

It was proposed early on,[44,45] by analogy with spin glasses, that in presence of disorder the Abrikosov lattice was replaced by a phase, called the vortex glass, characterized by random but frozen phase coherence, with correlation defined as:

$$C_{VG}(x) = \overline{|\langle \tilde{\psi}^*(x)\tilde{\psi}(0)\rangle|^2} \tag{74}$$

with non-zero limit $C_{VG}(x \to \infty)$, where $\tilde{\psi}(x) = \psi(x)e^{-i\frac{2\pi}{\Phi_0}\int^x A\cdot d\ell}$ to maintain gauge invariance. Besides some mean field calculations,[152] additional material was put on this proposal by considering the *gauge glass model*, i.e. an XY model with random gauge. In this approach, more suitable to strong disorder, the vortex lattice is fully ignored and the superconductor is represented by grains with superconducting phase ϕ_i at positions x_i on the sites of a d-dimensional lattice

$$F = -\sum_{\langle ij\rangle} \cos\left(\phi_i - \phi_j - A_{ij} - \frac{1}{\lambda_0}\int_{x_i}^{x_j} a(x)dx\right) + \frac{1}{2}\sum_{\square}(\nabla \times a)^2 \tag{75}$$

where A_{ij} is a random variable uniformly distributed in $[0, 2\pi]$ and a denotes the fluctuations of the vector potential which are limited by the bare screening length λ_0. This model was found numerically to exhibit a vortex glass phase in $d = 3$ for $\lambda_0 = \infty$, with a continuous transition to a high temperature phase.[46] This $d = 3$ vortex glass phase was, however, found to be unstable at non-zero temperature in the more realistic case of finite λ_0.[153]

A continuous vortex liquid to glass transition can be described phenomenologically assuming a single diverging length scale $\xi \sim |T - T_g|^{-\nu}$, and time scale $\tau \sim \xi^z$, as:

$$E \sim \xi^{-z-1} e_{\pm}(j\xi^{d-1}), \quad \sigma(\omega) \sim \xi^{2+z-d} s_{\pm}(\omega\xi^z) \tag{76}$$

with $e_+(x) \sim x$ and $e_-(x) \sim \exp(-a/x^{\mu})$ at small x, tailored to match linear response and creep, respectively. At the transition, consistency requires $E \sim j^{\frac{z+1}{d-1}}$ and $\sigma(\omega) \sim (-i\omega)^{(d-z-2)/z}$, and above the transition, the resistivity vanishes as $\rho \sim (T - T_c)^{\nu(z+2-d)}$. Note that this phenomenological theory does not say much about the detailed nature of the glass phase, and should hold as long as the transition is continuous and scaling holds (look for a simple but instructive toy example Ref. 154). Many experiments were analyzed using (76) leading to values $\nu = 1.1 \pm 0.4$ and $z = 5.2 \pm 0.6$, however there has been much debate on the interpretation in terms of a transition to a vortex glass. Pinning by dislocation lines in films, or by twin boundaries in bulk samples, lead to a different Bose glass phase (see below). Some of the more clear-cut experiments have been done on the isotropic BaKBiO superconductor.[155] In BSCCO similar fits fail to produce reasonable (i.e. universal) values for the exponents.[156]

5.2. *Bose glass and splay glass*

Correlated disorder occurs either naturally as dislocation lines or twin boundary planes, or can be induced through irradiations by heavy ions. The latter produce lines of damage in the material which act as *columnar defects* very efficient to pin the flux lines, hence of technological importance. Figure 16 shows the phase diagram of BSCCO and the irreversibility line before and after irradiation producing columnar defects of density d_0^{-2} and matching field $B_\phi = \Phi_0/d_0^2$. The irreversible regime where the vortex solid is pinned is considerably enhanced in all compounds,[89] and predicted to form a distinct phase, the Bose glass.[90] The simplest description uses the mapping from flux lines to quantum mechanics. The partition function of a single flux line at temperature T obeys, as a function of its endpoint r,

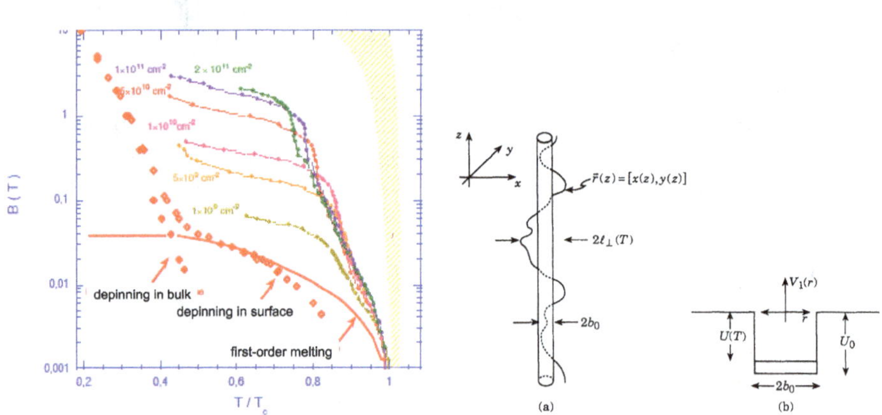

Fig. 16. Left: Upward shift in irreversibility line in BSCCO after irradiations by 5.8 GeV Pb ions at various fluences. A saturation of the maximum temperature with d^{-2} signals diffusion of two-dimensional pancake vortices between vortex lines and lost linelike character of vortices. From Ref. 159 and courtesy C. J. van der Beek. Right: A 3D thermally fluctuating flux line around a columnar defect map onto a quantum particle in a 2D well.

the Schrodinger equation with imaginary time replaced by z and $\hbar$ by T. Its binding energy to a columnar defect is obtained from solving the quantum problem in dimension $d-1$ for a particle in a well (see Fig. 16), which yields $U(T) = U_0 f(T/T^*)$ with $T^* = b_0\sqrt{\tilde{\epsilon}_1 U_0}$ and $U_0 \sim \epsilon_0 (c_0/2\xi)^2$ where $b_0 = \max(c_0, \sqrt{2}\xi)$ is the effective width of the well and c_0 the defect radius. Here $\tilde{\epsilon}_1$ is the local tilt modulus, see Sec. 3.3. It shows that a single vortex line in a forest of random defects has bounded wandering $\delta r(z) \sim \ell_\perp(T)$ and obtains estimates of its *localization length* $\ell_\perp(T)$ and (effective) critical current $j_c(T) \approx cU(T)/\Phi_0\ell_\perp(T)$. The many vortex line 3D problem maps to repulsive bosons in a $2D$ random potential, known from numerics to exhibit a Bose glass phase with finite localization length.[157] This mapping together with numerics and phenomenological arguments predicts[90] (i) *continuous Bose glass to vortex liquid transition* with a diverging localization length $\ell_\perp(T) \sim |T - T_{BG}|^{-\nu_\perp}$ with $\nu_\perp \approx 1$, $\ell_z \sim \ell_\perp^2$ and anisotropic vortex glass-like scaling at the transition,[91] (ii) *divergent barriers* in the glass phase with two regimes of creep as $E \sim \exp\left(-\frac{E_k}{T}(j_1/j)\right)$ for intermediate $j_1 = cU_0/\Phi_0 d_0 \ll j \ll j_c$ where the nucleation barrier arises from a half-loop kinks away from a single column, and $E \sim \exp\left(-\frac{E_k}{T}(j_0/j)^{1/3}\right)$ at very small j, when optimization of the nucleation barrier over many defects can occur, by an extension of Mott's variable range hopping argument for bosons,

(iii) *transverse Meissner effect*: in the BoG phase the effective tilt modulus is infinite and it requires a finite transverse threshold field $H_\perp(T) \sim E_k(T)/\Phi_0 d_0$, with $E_k(T) = \sqrt{\tilde{\epsilon}_1 U(T)}d$, to tilt the vortex lines away from the columns and produce a non-zero bulk $B_\perp$. Many of the predicted properties of the BoG have been checked in transport and magnetization experiments.[92,160] Remarkable double-sided decoration experiments[158] show localization and Coulomb gap effects very much as in semiconductors. Surprisingly though, detecting the transverse Meissner effect has been a challenge.[161] Finally, in very anisotropic compounds, localization of pancakes by columns can be studied from some discrete extension of the mapping to quantum mechanics,[162] and phenomena such as recoupling by columnar defects can occur.

One can also describe the Bose glass within an elastic theory, starting from the vortex lattice and adding correlated disorder along z. The GVM and the FRG can both be used,[163] very much as in Sec. 4.4 above, and also lead to predictions of localization and transverse Meissner effect. However in the case of heavy ion irradiations, the columnar disorder is very strong and immediately creates dislocations: the elastic description works then only in some high temperature region just below the transition where the effective disorder is weakened and L_{disloc} is large (e.g. $R_a \sim 30a_0$ was estimated in Ref. 90). In fact, if R_a is large enough, the BoG-liquid transition should become first order, as is indeed observed for very small ion fluence.[93] One notes that the ground state with columnar disorder is *identical*, up to translations along z, to the ground state in $d-1$ dimension with *point* disorder. Hence, even weak columnar disorder in $d=3$ produces dislocations at large scales, and exponential decay of $C_K(r)$. A Bragg–Bose glass phase should however exist in the $d=3+1$ problem, which has no realization in real superconductors, but could be realized in a genuine 3D quantum crystal with substrate quenched point disorder.

It was proposed[95] that a moderate dispersion in the directions of the irradiation tracks would keep the vortices localized along the splayed columns, leading to yet another glass phase, the splay glass, with further improved transport properties: (i) the vortices being now entangled, the energy cost of flux cutting increases the barrier for activated motion, (ii) the variable range vortex creep should be suppressed since the dominant excitations occur within the families of similarly inclined tracks, leading e.g. to $E \sim \exp\left(-\frac{E_k}{T}(j_0/j)^{3/5}\right)$. This was checked in experiments, which did show cases of reduced relaxation, see Fig. 17, the effect being more pronounced in the less anisotropic compounds, such as YBCCO.[96,164]

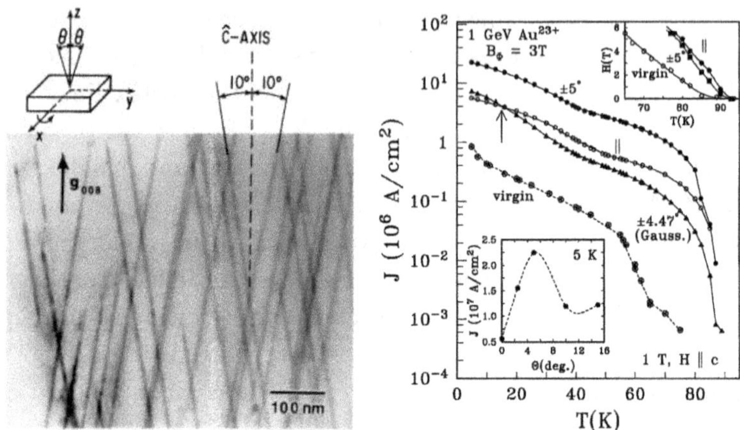

Fig. 17. Left: TEM image of YBCO irradiated with 1.08 GeV Au ions in two splayed family. Right: Persistent current density $J(T)$ for splayed configuration $\pm 5^0$ (solid circles), Gaussian distribution (triangles) compared to parallel columns (open circles). From Ref. 96.

5.3. *Moving glasses*

Finally, we now briefly discuss moving phases of vortices. Consider the vortex lattice undergoing overdamped dynamics under the action of the Lorentz force produced by an external current, which, for simplicity we take as uniform in space. If $j \gg j_c$, i.e. $f \gg f_c$ the system is in the flux flow regime $v \sim f/\eta$ at large velocity. There, it is natural to argue that disorder is weak, being averaged out by fast motion, hence the elastic description is a reasonable starting point:

$$\eta(\partial_t + v \cdot \nabla) u^\alpha(x,t) = \int_{x'} \Phi_{\alpha\beta}(x-x') u^\beta(x',t) + f^\alpha_{\rm pin}(u,x,t) + f^\alpha - \eta v^\alpha \quad (77)$$

where $vt + u(x,t)$ is the deformation expressed in laboratory frame coordinates, and the pinning force is:

$$f^\alpha_{\rm pin}(u,x,t) = -\frac{\delta F_{\rm dis}[u]}{\delta u^\alpha(x,t)} = U(x)\rho_0 \sum_K iK^\alpha e^{iK\cdot(x-vt-u(x,t))} \quad (78)$$

At large velocity (78) oscillates very fast and one can perform a weak disorder/large velocity expansion[165] and compute the deformations u in powers of $1/v$. This was done in Ref. 166 and it was concluded that above some velocity the VL becomes a perfect crystal again at an effective temperature $T' = T + T_{\rm sh}$, the effect of disorder being described by an effective *shaking temperature* $T_{\rm sh} \sim 1/v^2$. This approach would suggest bounded displacements in $d > 2$ and the absence of glassy properties in the moving solid.

The moving vortex lattice turns out to be a much more subtle and interesting object. An examination of (78) immediately shows what is missing[167]: a part of the pinning force along the direction (denoted by y) perpendicular to motion, is *static*! This part comes from the contribution of the vectors $K = (0, K^y)$, such that $K \cdot v = 0$, and is *not* averaged by motion. Hence $u^y(x,t)$ sees a resulting *static random potential* and as a result exhibits glassy properties. This is also apparent in the weak disorder perturbation theory which leads to infrared divergences for corrections to u^y. These can again be treated using the dynamical FRG on the equation retaining only transverse deformations, the so-called moving glass equation[167]:

$$\eta\partial_t u^y(x,t)+\eta v\cdot\nabla u^y(x,t) = c\nabla^2 u^y(x,t)+U(x)\rho_0 \sum_{K^y} iK^y e^{iK^y(y-u^y(x,t))} \quad (79)$$

Apart from the convective term, which retains the memory that the system is moving and dissipates, it is very similar to the standard pinning equation for a periodic elastic object. Analysis[167] shows that $u^y(x,t)$ is pinned, hence (i) motion occurs along static channels directed along the average velocity, in which the vortices flow like beads on a string (see Fig. 18 right), (ii) these channels are rough, hence translational order decays for $d \leq 3$ (analogous to Larkin's $d \leq 4$ for static, with here anisotropic scaling $L_{\|} \sim vL_y^2/c$). In $d = 3$ smectic type quasi-long range translational order persists in the y direction, (iii) there is *transverse pinning*, i.e. the system resists if an additional force is applied transverse to motion, there are *transverse barriers*. Two dynamical phases with these properties are predicted[167,168] (i) a *moving Bragg glass* (MBrG) where topological order is retained in all directions (transverse and longitudinal), (ii) a flowing smectic glass (MSG), where dislocation exists in between the channels (decoupled channels) but transverse (i.e. smectic) topological order survives. All these predictions have been verified in numerics.[169] Decoration in motion have shown flow along channels[170,171] and both predicted phases can be seen in the experiment in Fig. 18. At weak disorder the BrG should depin elastically and transform continuously into the MBrG: a creeping moving glass was seen in STM experiments.[172] Another signature of a topologically ordered MBrG, narrow band noise at the washboard frequency $\omega = a/v$, has been reported.[173] The detection of the transverse critical force f_c^t in vortex lattices has been a challenge, although it was reported in superconducting metal glasses $Fe_xNi_{1-x}Zr_2$.[174] It has been neatly observed in Wigner crystals.[175] CDW in presence of a transverse current is also described by Eq. (79) and also exhibits the moving glass physics.[176]

The fact that moving phases may be glassy and exhibit barriers came as a surprise. In fact, one shows that with point disorder the MBrG and MSG

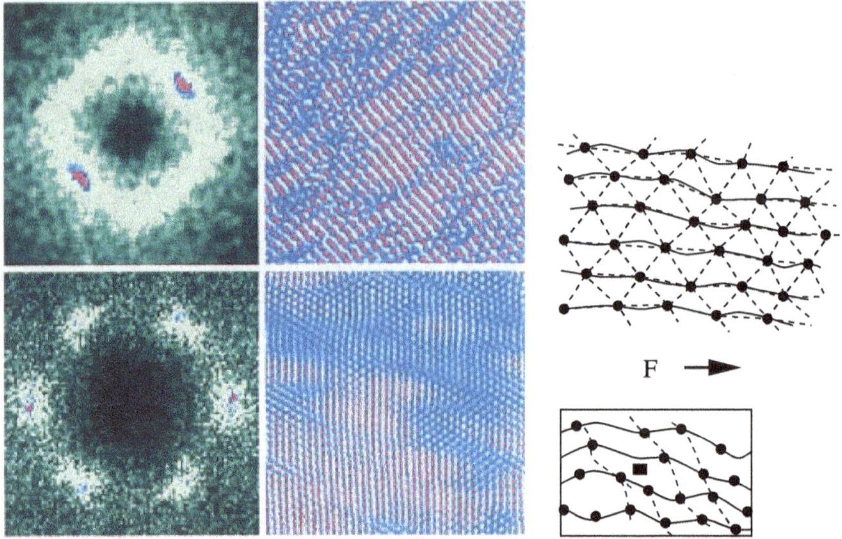

Fig. 18. Left: Decoration in motion images in $NbSe_2$ at 4 K from Ref. 170. Right column: Real space image (Fourier filtered) which shows the static channels along direction of motion. Left column: Its Fourier transform, which shows transverse order (two peaks) or full triangular lattice order (six peaks). Top: Moving smectic (low-field, higher-velocity). Bottom: Moving Bragg glass (high-field, lower-velocity). Right: (top) Moving Bragg glass on static channels, (bottom) moving smectic, channels are decoupled by dislocations (black square).

do not possess diverging barriers,[167,168] hence they are weaker glasses than their static parent. However, in the presence of columnar disorder, a truly glassy moving phase with diverging barriers was predicted,[177] and seen in numerics,[178] the moving Bose Glass. Its detection in experiments remains a challenge.

Let us close by mentioning that dynamical phases and effects in moving vortices are much richer, there are also strong disorder or nonlinear physics situations such as plastic depinning, filamentary flow and chaos,[179] incommensurate easy flow channels,[180] avalanche motion,[181] dendritic and fingering phenomena,[182] helical instabilities,[5,183] turbulent flow,[185] vortex twisters,[184] which are often more difficult to model.

6. Conclusion

We close this chapter by (very few) examples of topics of current research. First, some effects require a description of the vortex lattice beyond the

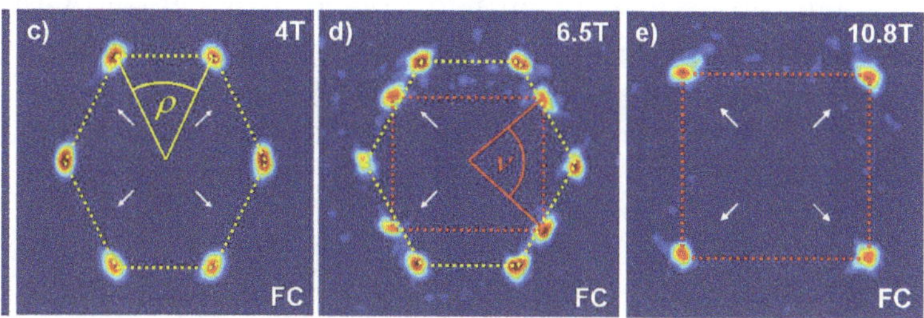

Fig. 19. The VL diffraction patterns for untwinned YBCO at $T = 2K$ and (c) $B = 4T$, (d) $B = 6.5T$, (e) $B = 10.8T$ after field cooling. The transition is between hexagonal and rhombic structures with distortions of opposite sign, providing evidence for a $d + s$-wave admixture for the order parameter, from Ref. 189.

s-wave Ginzburg–Landau model. It has been predicted that the d-wave symmetry of the order parameter should favor a square vortex lattice over a triangular one at high field, leading to a first order structural transition as the field is increased.[186,187] Small-angle neutron-scattering experiments on YBCO have indeed revealed a structural transition.[188] More recent experiments on untwinned samples[189] are shown in Fig. 19: there, the coexistence region suggests a first order transition around $6T$. A distinct transition at lower fields around $2T$ is also seen[189] between a low-field hexagonal to an intermediate-field distorted hexagonal phase and is claimed to be driven by Fermi surface effects. Recent specific heat experiments, performed at higher temperature, do find signature of first order transition at about this field which suggest a FOT structural transition line in the phase diagram. Note that these transitions occur very deep within the vortex solid, i.e. within the Bragg glass.

Next, we mention that the FRG and concepts such as quasi translational order, topological order and Bragg glass phases have been useful in other problems with quenched disorder[150]: e.g. liquid crystals[147,149] confined in a random matrix, the spontaneous vortex lattice in ferromagnetic superconductors (such as $ErNi_2B_2C$),[148] and random field magnets.[151] A direction of present research, mentioned in Sec. 4.3.3, is to study avalanches phenomena in other pinned systems, as well as relaxation from a prescribed initial state,[191] with close contact between theory[126] and experiments.[128] Remarkable recent developments[192] allow to image and manipulate individual vortices using magnetic force microscopy. It allows to directly measure the interaction of a moving vortex with the local disorder potential. Being near the realization of the quadratic well setting described in Sec. 4.3.3 it could be a useful probe to perform more precise comparison with theory, e.g. for processes such as depinning, vortex creep and vortex avalanches[181] and to study relaxation from a chosen at will initial configuration.

Acknowledgments

I am particularly indebted to T. Giamarchi and K. Wiese for multiple and fruitful collaborations. I am grateful to B. Horovitz, D. S. Fisher, T. Nattermann, D. R. Nelson and V. Vinokur, as well as L. Balents, P. Chauve, D. Carpentier, L. Cugliandolo, A. Fedorenko, A. Koshelev, C. Marchetti, A. Middleton, E. Olive, A. Rosso, G. Schehr, J. C. Soret, T. Hwa for collaborations, and numerous other colleagues for useful discussions. I have learned a lot about experiments from C. J. van der Beek, M. Charalambous, T. Klein, C. Simon, A. Kapitulnik, E. Zeldov and many others. I am very grateful to C. J. van der Beek, D. Feldman, B. Horovitz, T. Giamarchi and E. Zeldov for careful reading of the manuscript and useful suggestions.

References

1. M. Tinkham, *Introduction to Superconductivity* (McGraw Hill, New York, 1975).
2. P. G. De Gennes, *Superconductivity of Metals and Alloys* (W. A. Benjamin, New York, 1966); P. G. de Gennes and J. Matricon, *Rev. Mod. Phys.* **36**, 45 (1964).
3. A. A. Abrikosov, *Sov. Phys. JETP* **5**, 1174 (1957).
4. G. Blatter *et al.*, *Rev. Mod. Phys.* **66**, 1125 (1994).
5. H. Brandt, *Rep. Prog. Phys.* **58**, 1465 (1995), arXiv:0806.1058.
6. T. Nattermann and S. Scheidl, *Adv. Phys.* **49**, 607 (2000).
7. T. Giamarchi and S. Bhattacharya, arXiv:cond-mat/0111052.
8. U. C. Tuber and D. R. Nelson, *Phys. Rep.* **289** (1997) 157.
9. D. A. Huse and L. R. Radzihovsky, arXiv:cond-mat/9310047.
10. M. Kardar, cond-mat/9704172 and cond-mat/9507019.
11. T. Giamarchi and P. Le Doussal, in *Spin Glasses and Random Fields*, ed. A. P. Young (World Scientific, Singapore, 1998), p. 321, cond-mat/9705096.
12. G. Eilenberger, *Phys. Rev.* **164**, 928 (1967).
13. A. I. Larkin, *Sov. Phys. JETP* **31**, 784 (1970).
14. A. I. Larkin and Y. N. Ovchinnikov, *J. Low Temp. Phys.* **34**, 409 (1979).
15. P. W. Anderson and Y. B. Kim, *Rev. Mod. Phys.* **36**, 39 (1964).
16. M. R. Beasley, R. Labush and W. W. Webb, *Phys. Rev.* **181**, 682 (1969).
17. Y. B. Kim, C. F. Hempstead and A. R. Strnad, *Phys. Rev. Lett.* **9**, 306 (1962).
18. C. C. Tsuei *et al.*, *Phys. Rev. Lett.* **73**, 593 (1994).
19. L. I. Glazman and A. E. Koshelev, *Phys. Rev. B* **43**, 2835 (1991).

20. L. L. Daemen *et al.*, *Phys. Rev. Lett.* **70**, 1167 (1993); *Phys. Rev. B* **47**, 11291 (1993).
21. B. Horovitz and T. R. Goldin, *Phys. Rev. Lett.* **80**, 1734 (1998); B. Horovitz, *Phys. Rev. B* **72**, 024518 (2005).
22. C. J. van der Beek *et al. Phys. Rev. B* **72**, 214504 (2005).
23. K. Hikada *et al.*, *Jpn. J. Appl. Phys.* **27**, L538 (1989); T. T. M. Palstra *et al.*, *Phys. Rev. Lett.* **61**, 1662 (1988).
24. P. L. Gammel *et al.*, *Phys. Rev. Lett.* **61**, 1666 (1988); D. E. Farrell, J. P. Rice and D. M. Ginsberg, *Phys. Rev. Lett.* **67**, 1165 (1991).
25. A. P. Malozemoff *et al.*, *Phys. Rev. B* **38**, 7203 (1988).
26. P. L. Gammel *et al.*, *Phys. Rev. Lett.* **59**, 2592 (1987).
27. D. G. Grier *et al.*, *Phys. Rev. Lett.* **66**, 2270 (1991).
28. Y. Yeshurun and A. P. Malozemoff, *Phys. Rev. Lett.* **60**, 2202 (1988).
29. A. Houghton, R. Pelcovits and A. Sudbo, *Phys. Rev. B* **40**, 6763 (1989).
30. E. H. Brandt, *Phys. Rev. Lett.* **63**, 1106 (1989).
31. T. R. Goldin and B. Horovitz, *Phys. Rev. B* **58**, 9524 (1998).
32. D. R. Nelson, *Phys. Rev. Lett.* **69**, 1973 (1988); D. R. Nelson and S. Seung, *Phys. Rev. B* **39**, 9153 (1989); D. R. Nelson and P. Le Doussal, *Phys. Rev. B* **42**, 10113 (1990).
33. E. Brezin, D. R. Nelson and A. Thiaville, *Phys. Rev. B* **31**, 7124 (1985).
34. M. A. Moore and T. J. Newman, *Phys. Rev. Lett.* **75**, 533 (1995); L. Balents and L. Radzihovsky, *Phys. Rev. Lett.* **76**, 3416 (1996).
35. Z. Tesanovic *et al.*, *Phys. Rev. Lett.* **69**, 3563; Z. Tesanovic and A. V. Andreev, *Phys. Rev. B* **49**, 4064 (1994).
36. S. Hikami, A. Fujita and A. I. Larkin, *Phys. Rev. B* **44**, 10400 (1991).
37. R. Sasik and D. S. Stroud, *Phys. Rev. Lett.* **75**, 2582 (1995).
38. D. Li and B. Rosenstein, *Phys. Rev. Lett.* **86**, 3618 (2001); B. Rosenstein, *Phys. Rev. B* **60**, 4268 (1999).
39. D. Li and B. Rosenstein, *Phys. Rev. B* **65**, 220504(R) (2002).
40. B. Rosenstein and D. Li, arXiv:0905.4224 (2009).
41. R. Hetzel, A. Sudbo and D. Huse, *Phys. Rev. Lett.* **69**, 518 (1992).
42. M. C. Marchetti and D. R. Nelson *Phys. Rev. B* **41**, 1910 (1990).
43. R. Koch *et al.*, *Phys. Rev. Lett.* **63**, 1511 (1989).
44. M. P. A. Fisher, *Phys. Rev. Lett.* **62**, 1415 (1989).
45. D. S. Fisher, M. P. A. Fisher and D. A. Huse, *Phys. Rev. B* **43**, 130 (1990).
46. See, e.g. T. Olson and A. P. Young, *Phys. Rev. B* **61**, 12467 (2000) and references therein.
47. T. Giamarchi and P. Le Doussal, *Phys. Rev. Lett.* **72**, 1530 (1994); *Phys. Rev. B* **52**, 1242 (1995).
48. M. Feigelman, V. B. Geshkenbein, A. I. Larkin and V. Vinokur, *Phys. Rev. Lett.* **63**, 2303 (1989).
49. T. Nattermann, *Europhys. Lett.* **4**, 1241 (1987); L. B. Ioffe and V. M. Vinokur, *J. Phys. C* **20**, 6149 (1987).
50. T. Nattermann, *Phys. Rev. Lett.* **64**, 2454 (1990).
51. J. P. Bouchaud, M. Mezard and J. S. Yedidia, *Phys. Rev. Lett.* **67**, 3840 (1991); *Phys. Rev. B* **46**, 14686 (1992).

52. P. Chauve, T. Giamarchi and P. Le Doussal, *Europhys. Lett.* **44**, 110 (1998); *Phys. Rev. B* **62**, 6241 (2000).
53. D. A. Huse, M. P. A. Fisher and D. S. Fisher, *Nature* **358**, 553 (1992).
54. See, e.g. P. Svedlindh *et al.*, *Physica C* **176**, 336 (1991).
55. M. Charalambous, J. Chaussy and P. Lejay, *Phys. Rev. B* **45**, 5091 (1992).
56. H. Safar *et al.*, *Phys. Rev. Lett.* **69**, 824 (1992) and 70, 3800 (1993).
57. W. K. Kwok *et al.*, *Phys. Rev. Lett.* **69**, 3370 (1992).
58. H. Safar *et al.*, *Phys. Rev. B* **52**, 6211 (1995).
59. E. Zeldov *et al.*, *Nature* **375**, 373 (1995).
60. D. R. Nelson, *Nature* **375**, 356 (1995).
61. B. Khaykovich *et al.*, *Phys. Rev. B* **56**, R517 (1997).
62. R. Cubitt *et al.*, *Nature* **365**, 407 (1993).
63. S. L. Lee *et al.*, *Phys. Rev. Lett.* **71**, 3862 (1993); C. Bernhard *et al.*, *Phys. Rev. B* **52**, R7050 (1995).
64. A. Schilling *et al.*, *Nature* **382**, 791 (1996); *Phys. Rev. Lett.* **78**, 4833 (1998); M. Roulin *et al.*, *Science* **273**, 1210 (1996); F. Bouquet *et al.*, *Nature* **411**, 448 (2001).
65. B. Khaykovich *et al.*, *Phys. Rev. Lett.* **76**, 2555 (1996).
66. T. Giamarchi and P. Le Doussal, *Phys. Rev. B* **55**, 6577 (1997).
67. G. Mikitik and E. H. Brandt, *Phys. Rev. B* **64**, 184514 (2001); ibid *Phys. Rev. B* **68**, 054509 (2003).
68. D. Ertas and D. R. Nelson, *Physica C* **272**, 79 (1996); J. Kierfeld, *Physica C* **300**, 171 (1998).
69. N. Avraham *et al.*, *Nature* **411**, 451 (2001).
70. C. J. van der Beek *et al.*, *Phys. Rev. Lett.* **84**, 4196 (2000); D. Giller *et al.*, *Phys. Rev. Lett.* **84**, 3698 (2000); M. Konczykowski *et al.*, *Physica C* **341** (2000); C. J. van der Beek *et al.*, *Physica C* **341**, 1279 and 1319 (2000).
71. M. Konczykowski *et al.*, *Physica C* **332**, 219 (2000).
72. Y. Radzyner, A. Shaulov and Y. Yeshurun, *Phys. Rev. B* **65**, 100513 (2002).
73. Deligiannis *et al.*, *Phys. Rev. Lett.* **79**, 2121 (1997); T. Nishizaki *et al.*, *Phys. Rev. B* **61**, 3649 (2000).
74. T. Klein *et al.*, *Nature* **413**, 404 (2001).
75. U. Divakar *et al.*, *Phys. Rev. Lett.* **92**, 237004 (2004).
76. A. V. Otterlo, R. Scalettar and G. Zimanyi, *Phys. Rev. Lett.* **81**, 1497 (1998).
77. Y. Matsuda *et al.*, *Phys. Rev. Lett.* **75**, 4512 (1995); T. Shibauchi *et al.*, *Phys. Rev. Lett.* **83**, 1010 (1999); M. B. Gaifullin *et al.*, *Phys. Rev. Lett.* **84**, 2945 (1999); S. Colson *et al.*, *Phys. Rev. Lett.* **90**, 137002 (2003).
78. C. J. van der Beek, "Thermodynamique des vortex dans les supraconducteurs dsordonns", Open archive http://tel.archives-ouvertes.fr/ under number tel-00483670 (in French).
79. M. V. Feigelman, V. B. Geshkenbeim and A. I. Larkin, *Physica C* **167**, 177 (1990).
80. M. J. W. Dodgson, V. B. Geshkenbein and G. Blatter, *Phys. Rev. Lett.* **83**, 5358 (1999).
81. E. Frey, D. R. Nelson and D. S. Fisher, *Phys. Rev. B* **49**, 9723 (1994).
82. B. Horovitz, *Phys. Rev. B* **60**, R9939 (1999).

83. B. Horovitz and P. Le Doussal, *Phys. Rev. Lett.* **84**, 5395 (2000); *Phys. Rev. B* **71**, 134202 (2005).
84. P. Koorevaar, J. Aarts, P. Berghuis and P. H. Kes, *Phys. Rev. B* **42**, 1004 (1990).
85. A. M. Troyanovskii, M. van Hecke, N. Saha, J. Aarts and P. H. Kes, *Phys. Rev. Lett.* **89**, 147006 (2002).
86. Y. Paltiel *et al.*, *Phys. Rev. Lett.* **85**, 3712 (2000), arXiv:cond-mat/0008092.
87. P. L. Gammel *et al.*, *Phys. Rev. Lett.* **80**, 833 (1998).
88. S. R. Park *et al.*, *Phys. Rev. Lett.* **91**, 167003 (2003).
89. L. Civale *et al.*, *Phys. Rev. Lett.* **67**, 648 (1991); M. Konczykowski *et al.*, *Phys. Rev. B* **44**, 7167 (1991); R. C. Budhani, M. Suenaga and S. H. Liou, *Phys. Rev. Lett.* **69**, 3816 (1992); V. Hardy, C. Simon, J. Provost and D. Groult, *Physica C* **205**, 371 (1993).
90. D. R. Nelson and V. M. Vinokur, *Phys. Rev. Lett.* **68**, 2398 (1992); *Phys. Rev. B* **48**, 13060 (1993).
91. J. Lidmar and M. Wallin, *Europhys. Lett.* **47**, 494 (1999).
92. W. Jiang *et al.*, *Phys. Rev. Lett.* **72**, 550 (1994); S. A. Grigera *et al.*, *Phys. Rev. Lett.* **81**, 2348 (1998); J. C. Soret *et al.*, *Phys. Rev. B* **61**, 9800 (2000).
93. S. S. Banerjee *et al.*, *Phys. Rev. Lett.* **90**, 087004 (2003); Menghini *et al.*, *Phys. Rev. Lett.* **90**, 147001 (2003); T. Verdene *et al.*, *Phys. Rev. Lett.* **101**, 157003 (2008).
94. W. D. Huang, *Superconductor Science and Technology*, **13**, 1499 (2000); A. Ibi, *Physica C* **468**, 1514 (2008); B. Maiorov *et al.*, *Nature Materials* **8**, 398 (2009); S. R. Foltyn *et al.*, *Nature Materials* **6**, 631 (2007).
95. T. Hwa, P. Le Doussal, D. R. Nelson and V. M. Vinokur, *Phys. Rev. Lett.* **71**, 3545 (1993); P. Le Doussal and D. R. Nelson, *Physica C* **232**, 69 (1994).
96. L. Krusin-Elbaum *et al.*, *Phys. Rev. Lett.* **76**, 2563 (1996).
97. C. Reichhardt, A. van Otterlo and G. T. Zimnyi, *Phys. Rev. Lett.* **84**, 1994 (2000).
98. J. Kierfeld and V. M. Vinokur, *Phys. Rev.* **61**, R14928 (2000); ibid. **69**, 024501 (2004).
99. H. Beidenkopf *et al.*, *Phys. Rev. Lett.* **95**, 257004 (2005).
100. H. Beidenkopf *et al.*, *Phys. Rev. Lett.* **98**, 167004 (2007).
101. V. G. Kogan, *Phys. Rev. B* **24**, 1572 (1981).
102. P. Berghuis, A. L. F. van der Slot and P. H. Kes, *Phys. Rev. Lett.* **65**, 2583 (1990).
103. D. Carpentier and P. Le Doussal, *Phys. Rev. Lett.* **81**, 1881 (1998), and reference therein.
104. D. Li and B. Rosenstein, *Phys. Rev. Lett.* **90**, 167004 (2003).
105. M. Willemin *et al.*, *Phys. Rev. Lett.* **81**, 4236 (1998); T. Nishizaki *et al.*, *Physica C* **362**, 121 (2001).
106. M. A. Moore, *Phys. Rev. B* **39**, 136 (1989).
107. A. Houghton, R. A. Pelcovits and A. Sudb, *Phys. Rev. B* **42**, 906908 (1990).
108. For a related crossover in the $B = 0$ transition, see B. Horovitz, *Phys. Rev. Lett.* **67**, 378 (1991); *Phys. Rev. B* **47**, 5947 (1993).
109. M. Tinkham, *Helv. Phys. Acta* **61**, 443 (1988).

110. See, e.g. P. H. Kes and C. C. Tsuei, *Phys. Rev. Lett.* **47**, 1930 (1981).
111. F. D. M. Haldane, *Phys. Rev. Lett.* **47**, 1840 (1981).
112. H. Fukuyama and P. A. Lee, *Phys. Rev. B* **17**, 535 (1978).
113. D. R. Nelson, *Phys. Rev. B* **27**, 2902 (1983); D. Carpentier and P. Le Doussal, *Phys. Rev. B* **55**, 12128 (1997).
114. T. Nattermann *et al.*, *J. Phys. I (France)* **5**, 565 (1995); M. Cha and H. A. Fertig, *Phys. Rev. Lett.* **74**, 4867 (1995).
115. C. Zeng, P. L. Leath and D. S. Fisher, arXiv:cond-mat/9807281.
116. P. Le Doussal and T. Giamarchi, *Physica C* **331**, 233 (2000), arXiv:cond-mat/9810218.
117. Y. Imry and S. K. Ma, *Phys. Rev. Lett.* **35**, 1399 (1975).
118. M. Mezard and G. Parisi, *J. Phys. I (France)* **1**, 809 (1991); E. I. Shakhnovich and A. M. Gutin, *J. Phys. A* **22**, 1647 (1989).
119. S. E. Korshunov, *Phys. Rev. B* **48**, 3969 (1993).
120. D. S. Fisher, *Phys. Rev. Lett.* **56**, 1964 (1986); L. Balents and D. S. Fisher, *Phys. Rev. B* **48**, 5959 (1993); T. Nattermann, S. Stepanow, L. H. Tang and H. Leschhorn, *J. Phys. II France* **2**, 1483 (1992); O. Narayan and D. S. Fisher, *Phys. Rev. B* **48**, 7030 (1993); O. S. Wagner *et al.*, *Phys. Rev. B* **59**, 11551 (1999).
121. L. F. Cugliandolo, J. Kurchan and P. Le Doussal, *Phys. Rev. Lett.* **76**, 2390 (1996); L. F. Cugliandolo, T. Giamarchi and P. Le Doussal, *Phys. Rev. Lett.* **96**, 217203 (2006), and references therein.
122. H. Horner, arXiv:cond-mat/9508049.
123. P. Le Doussal, *Europhys. Lett.* **76**, 457 (2006); *Ann. Phys.* **325**, 49 (2010).
124. K. Wiese and P. Le Doussal, *Markov Processes Relat. Fields* **13**, 777 (2007).
125. P. Le Doussal, K. J. Wiese and P. Chauve, *Phys. Rev. B* **66**, 174201 (2002); *Phys. Rev. E.* **69**, 026112 (2004).
126. P. Le Doussal and K. J. Wiese, *Phys. Rev. E* **79**, 051106 (2009); P. Le Doussal, A. A. Middleton and K. J. Wiese, *Phys. Rev. E* **79**, 050101 (2009).
127. A. A. Middleton, P. Le Doussal and K. J. Wiese, *Phys. Rev. Lett.* **98**, 155701 (2007); A. Rosso, P. Le Doussal and K. J. Wiese, *Phys. Rev. B* **75**, 220201 (2007).
128. P. Le Doussal, K. J. Wiese, S. Moulinet and E. Rolley, *Europhys. Lett.* **87**(5), 56001 (2009).
129. J. Villain and J. F. Fernandez, *Z. Phys. B* **54**, 139 (1984).
130. T. Emig, S. Bogner and T. Nattermann, *Phys. Rev. Lett.* **83**, 400 (1999); S. Bogner, T. Emig and T. Nattermann, *Phys. Rev. B* **63**, 174501 (2001).
131. T. Giamarchi and P. Le Doussal, *Phys. Rev. Lett.* **75**, 3372 (1995).
132. E. M. Chudnovsky, *Phys. Rev. B* **40**, 11355 (1989).
133. J. Toner, *Phys. Rev. Lett.* **66**, 2523 (1991).
134. J. Kierfeld, T. Nattermann and T. Hwa, *Phys. Rev. B* **55**, 626 (1997); D. Carpentier, P. Le Doussal and T. Giamarchi, *Europhys. Lett.* **35**, 379 (1996).
135. D. S. Fisher, *Phys. Rev. Lett.* **78**, 1964 (1997).
136. M. J. P. Gingras and D. A. Huse, *Phys. Rev. B* **53**, 15193 (1996); S. Ryu, A. Kapitulnik and S. Doniach, *Phys. Rev. Lett.* **77**, 2300 (1996); D. McNamara, A. A. Middleton and C. Zeng, *Phys. Rev. B* **60**, 10062 (1999).

137. J. Cardy and S. Ostlund, *Phys. Rev. B* **25**, 6899 (1982); J. Toner and D. DiVincenzo, *Phys. Rev. B* **41**, 632 (1990); T. Nattermann, I. Lyuksyutov and M. Schwartz, *Europhys. Lett.* **16**, 295 (1991); J. Toner, *Phys. Rev. Lett.* **67**, 2537 (1991); M. Bauer and D. Bernard, *Europhys. Lett.* **33**, 255 (1996); C. Zeng, A. A. Middleton and Y. Shapir, *Phys. Rev. Lett.* **77**, 3204 (1996); P. Le Doussal and G. Schehr, *Phys. Rev. B* **75**, 184401 (2007).
138. P. Kim, Z. Yao and C. A. Bolle, *Phys. Rev. B* **60**, R12589 (1999).
139. S. Bustingorry, A. B. Kolton and T. Giamarchi, *Europhys. Lett.* **81**, 26005 (2008), and reference therein.
140. L. Balents and P. Le Doussal, *Phys. Rev. E* **69**, 061107 (2004).
141. A. A. Middleton, *Phys. Rev. Lett.* **68**, 670 (1992).
142. A. B. Kolton, A. Rosso and T. Giamarchi, *Phys. Rev. Lett.* **94**, 047002 (2005); A. B. Kolton, A. Rosso, T. Giamarchi and W. Krauth, *Phys. Rev. B* **79**, 184207 (2009).
143. S. Lemerle *et al.*, *Phys. Rev. Lett.* **80**, 849 (1998).
144. P. Paruch, T. Giamarchi and J.-M. Triscone, *Phys. Rev. Lett.* **94**, 197601 (2005).
145. D. T. Fuchs *et al.*, *Nature* **391**, 373 (1998).
146. D. T. Fuchs *et al.*, *Phys. Rev. Lett.* **81**, 3944 (1998).
147. B. Jacobsen *et al.*, *Phys. Rev. Lett.* **83**, 1363 (1999); K. Saunders, L. Radzihovsky and J. Toner, *Phys. Rev. Lett.* **85**, 4309 (2000); L. Radzihovsky and J. Toner, *Phys. Rev. B* **60**, 206 (1999).
148. A. M. Ettouhami, K. Saunders, L. Radzihovsky and J. Toner, *Phys. Rev. B* **71**, 224506 (2005).
149. D. E. Feldman, *Phys. Rev. Lett.* **84**, 4886 (2000); D. E. Feldman, *Phys. Rev. B* **61**, 382 (2000).
150. D. E. Feldman, *Phys. Rev. E* **70**, 040702(R) (2004); *Phys. Rev. B* **62**, 5364 (2000).
151. D. S. Fisher, *Phys. Rev. B* **31**, 7233 (1985); D. E. Feldman, *Phys. Rev. Lett.* **88**, 177202 (2002); P. Le Doussal and K. J. Wiese, *Phys. Rev. Lett.* **96**, 197202 (2006); M. Tissier and G. Tarjus, *Phys. Rev. Lett.* **96**, 087202 (2006).
152. A. T. Dorsey, M. Huang and M. P. A. Fisher, *Phys. Rev. B* **45**, 523 (1992); R. Ikeda, *J. Phys. Soc. Japan* **65**, 1170 and 3398 (1996).
153. H. S. Bokil and A. P. Young, *Phys. Rev. Lett.* **74**, 3021 (1995); F. Pfeiffer and H. Rieger, *Phys. Rev. B* **60**, 6304 (1999).
154. P. Le Doussal and V. M. Vinokur, *Physica C* **254**, 63 (1995).
155. T. Klein *et al.*, *Phys. Rev. B* **58**, 12411 (1998).
156. C. J. van der Beek *et al.*, *Physica C* **195**, 307 (1992).
157. M. Wallin and S. M. Girvin, *Phys. Rev. B* **47**, 14642 (1993).
158. U. C. Tauber, H. Dai, D. R. Nelson and C. M. Lieber, *Phys. Rev. Lett.* **74**, 5132 (1995), and references therein.
159. C. J. van der Beek *et al.*, *Phys. Rev. Lett.* **86**, 5136 (2001).
160. C. J. van der Beek *et al.*, *Phys. Rev. Lett.* **74**, 1214 (1995); C. J. van der Beek *et al.*, *Phys. Rev. B* **51**, 15492 (1995).
161. V. Ta Phuoc *et al.*, *Phys. Rev. Lett.* **88**, 187002 (2002).
162. A. E. Koshelev, P. Le Doussal and V. M. Vinokur, *Phys. Rev. B* **53**, R8855 (1996).

163. T. Giamarchi and P. Le Doussal, *Phys. Rev. B* **53**, 15206 (1996); L. Balents, *Europhys. Lett.* **24**, 489 (1993); T. Giamarchi, P. Le Doussal and E. Orignac, *Phys. Rev. B* **64**, 245119 (2001); A. Petkovic, T. Emig and T. Nattermann, *Phys. Rev. B* **79**, 224512 (2009).
164. V. Hardy *et al.*, *Physica C* **257**, 16 (1996); Th. Schuster *et al.*, *Phys. Rev. B* **53**, 2257 (1996).
165. A. I. Larkin and Y. N. Ovchinnikov, *Sov. Phys. JETP* **38**, 854 (1974); A. Schmidt and W. Hauger, *J. Low Temp. Phys.* **11**, 667 (1973).
166. A. E. Koshelev and V. M. Vinokur, *Phys. Rev. Lett.* **73**, 3580 (1994).
167. T. Giamarchi and P. Le Doussal, *Phys. Rev. Lett.* **76**, 3408 (1996); P. Le Doussal and T. Giamarchi, *Phys. Rev. B* **57**, 11356 (1998).
168. L. Balents, C. Marchetti and L. Radzihovsky, *Phys. Rev. B* **57**, 7705 (1998).
169. K. Moon *et al.*, *Phys. Rev. Lett.* **77**, 2378 (1997); C. J. Olson *et al.*, *Phys. Rev. B* **61**, R3811 (1999); H. Fangohr, P. A. J. de Groot and S. J. Cox, *Phys. Rev. B* **63**, 064501 (2001).
170. F. Pardo *et al.*, *Nature* **396**, 348 (1998).
171. M. Marchevsky *et al.*, *Phys. Rev. Lett.* **78**, 531 (1997).
172. A. M. Troyanovski *et al.*, *Nature* **399**, 665 (1999).
173. Y. Togawa *et al.*, *Phys. Rev. Lett.* **85**, 3716 (2000).
174. J. Lefebvre, M. Hilke and Z. Altounian, *Phys. Rev. B* **78**, 134506 (2008); J. Lefebvre, M. Hilke, R. Gagnon and Z. Altounian, *Phys. Rev. B* **74**, 174509 (2006).
175. F. Perruchot *et al.*, *Physica B* **256**, 587 (1998).
176. N. Markovic, N. Dohmen and H. van der Zant, *Phys. Rev. Lett.* **84**, 534 (2000).
177. P. Chauve, P. Le Doussal and T. Giamarchi, *Phys. Rev. B* **61**, R11906 (2000).
178. Y. Fily, E. Olive and J. C. Soret, *Phys. Rev. B* **79**, 212504 (2009); E. Olive, J. C. Soret, P. Le Doussal and T. Giamarchi, *Phys. Rev. Lett.* **91**, 037005 (2003).
179. E. Olive and J. C. Soret, *Phys. Rev. Lett.* **96**, 027002 (2006); *Phys. Rev. B* **77**, 144514 (2008).
180. R. Besseling, R. Niggebrugge and P. H. Kes, *Phys. Rev. Lett.* **82**, 3144 (1999).
181. E. Altshuler *et al.*, arXiv:cond-mat/0208266.
182. T. H. Johansen *et al.*, *Europhys. Lett.*, **59**(4), 599 (2002).
183. M. Kohandel and M. Kardar, arXiv:cond-mat/9912056.
184. See e.g. G. Danna *et al.*, *Physica C* **281**, 278 (1997).
185. T. Hwa, arXiv:cond-mat/9206008.
186. M. Ichioka, A. Hasegawa and K. Machida, *Phys. Rev. B* **59**, 8902 (1999).
187. M. Franz, I. Affleck and M. H. S. Amin, *Phys. Rev. Lett.* **79**, 1555 (1997).
188. S. P. Brown *et al.*, *Phys. Rev. Lett.* **92**, 067004 (2004).
189. J. S. White *et al.*, *Phys. Rev. Lett.* **102**, 097001 (2009).
190. M. Reibelt, S. Weyeneth, A. Erb and A. Schilling, arXiv:1003.4688.
191. A. B. Kolton, G. Schehr and P. Le Doussal, *Phys. Rev. Lett.* **103**, 160602 (2009); A. B. Kolton *et al.*, *Phys. Rev. B* **74**, 140201(R) (2006).
192. E. H. Brandt, G. P. Mikitik and E. Zeldov, *Phys. Rev. B* **80**, 054513 (2009); O. M. Auslaender *et al.*, *Nature Physics* **5**, 35 (2009).
193. M. Kohandel and M. Kardar, *Phys. Rev. B* **59**, 9637 (1999).
194. M. I. Dolz, A. B. Kolton and H. Pastoriza, *Phys. Rev. B* **81**, 092502 (2010).

BREAKING TRANSLATIONAL INVARIANCE BY POPULATION IMBALANCE: THE FULDE–FERRELL–LARKIN–OVCHINNIKOV STATES

Gertrud Zwicknagl

Institut für Mathematische Physik,
Technische Universität Braunschweig, 38106 Braunschweig, Germany

Jochen Wosnitza

Hochfeld-Magnetlabor Dresden,
Forschungszentrum Dresden-Rossendorf, 01314 Dresden, Germany

An overview is given of our present understanding of superconductivity with spontaneously broken translation symmetry in polarized Fermi systems. The existence of "crystalline" superconducting phases is considered in a wide range of systems, prominent examples being conduction electrons in metals, ultra-cold atoms in a trap, nuclear matter and dense quark systems. The underlying physics is delineated and theoretical approaches to the inhomogeneous phases and their properties are discussed. From the experimental side, it is argued that superconductivity with imbalance-induced order parameters is realized in layered organic compounds and potentially in heavy-fermion systems.

1. Introduction

Superfluid phases with unequal numbers of particles forming Cooper pairs have been the focus of interest for many decades. Of particular interest has been the conjecture by Fulde and Ferrell (FF)[1] and Larkin and Ovchinnikov (LO)[2] that at low temperatures, superfluid states with spontaneously broken translational symmetry should form. These FFLO states have been under theoretical investigation for a variety of systems including metals,[3,4] ultra-cold atoms,[5] nuclear matter and dense quark systems.[6]

In the present review, we shall restrict ourselves to non-relativistic two-component Fermi systems, well-known examples being the conduction electrons in metals. In an imbalanced Fermi liquid there will be a majority and a minority component which we shall denote as $\uparrow$ and $\downarrow$, respectively. The system is characterized by the corresponding densities $n_\uparrow \geq n_\downarrow$. For independent particles with no interconversion, the densities of the majority and minority degrees of freedom, $n_\uparrow$ and $n_\downarrow$, can alternatively be specified by chemical potentials $\mu_\uparrow$ and $\mu_\downarrow$. The latter are expressed by the average value and the difference

$$\mu_\uparrow = \mu + \delta\mu \, ; \qquad \mu_\downarrow = \mu - \delta\mu \, . \tag{1}$$

In spin-1/2-fermion systems, such as the conduction electrons in metals, the difference in the chemical potentials, $\delta\mu$, is tuned by an external magnetic field. In trapped atomic systems, on the other hand, imbalance can be generated by preparing mixtures with different numbers of partners and the natural variable is the polarization, i.e. the difference in the corresponding particle numbers.

Due to the formation of Cooper pairs the susceptibility of the superconductor is reduced as compared to the normal state. For high differences in the chemical potentials the polarized normal state has a lower energy than the superconducting phase. Qualitatively, the transition from the paired superconducting to the polarized normal phase will occur when the polarization energy equals the condensation energy of the Cooper pairs. This condition defines the limiting chemical-potential mismatch, $\delta\mu_c\,(T)$, beyond which homogeneous superconductivity will not exist. The low-temperature limit $\delta\mu_c\,(0)$ for a conventional superconductor was first given by Chandrasekhar[7] and Clogston[8]

$$\delta\mu_c\,(0) = \frac{\Delta_0}{\sqrt{2}} \, ,$$

where Δ_0 is the energy gap of the uniformly balanced reference system. The $\delta\mu$-T phase diagram for homogeneous superconductors was studied in detail by Sarma.[9,10] In addition to the normal phase and the homogeneous superconductor, he considered a compromise state which embodies "phase separation in momentum space". The breached-pair (BP) states[11] are characterized by the coexistence of paired superfluid and normal polarized components which reside in different regions of momentum space. Unfortunately, these gapless states turn out to be unstable in the "classical" situation of a superconductor in an external magnetic field acting on the spins of the electrons. The interest in BP states, however, has been recently rekindled since they could be realized in dense quark systems or ultra-cold atoms. FFLO predicted a spatially inhomogeneous partially gapped compromise state with both paired superfluid and normal polarized regions to be the stable ground state in a (narrow) range above the Chandrasekhar–Clogston limit.

The idea of forming a textured state with paired and normal regions closely parallels the concept of the vortex phase which appears in type-II superconductors below the upper critical field. This phase gains its stability from the fact that it allows magnetic flux to penetrate into the superconductor. Like the FFLO state, it was predicted theoretically by Abrikosov.[12] Shortly afterwards, it was confirmed by numerous experiments employing a variety of techniques. Although the existence of inhomogeneous compromise states in imbalanced superfluids seems plausible and despite intensive search, these phases have not yet been identified unambiguously. There are, however, indications that they are realized in some highly anisotropic materials, as will be discussed below.

There are several reasons for the difficulties in the experimental observation. First, the FFLO state requires very stringent conditions which strongly restrict the appropriate classes of superconducting materials. Soon after the FFLO phases were conjectured, it became clear that impurities act as pair breakers suppressing the corresponding transition temperatures.[13] A necessary criterion for the realization of an FFLO state is a long mean-free path exceeding the superconducting coherence length. An additional technical difficulty encountered in metals is related to the fact that electrons are charged particles. An external magnetic field required to create the imbalance in the conduction electron system will inevitably act on their orbits and suppress superconductivity. To observe imbalance-induced high-field phases the orbital pair breaking must be strongly reduced as pointed out by Maki.[13] The early search for the FFLO phases concentrated on thin superconducting films in parallel magnetic fields.[14] Although polarization effects were clearly seen, it was not possible to stabilize the FFLO state. Promising candidates are recently discovered heavy-fermion compounds and the layered organic superconductors. Second, there are no simple criteria which could help to identify the phases. The order parameter as such is a quantum mechanical object and hence cannot be observed directly. The generally accepted procedure in this situation is to calculate observable properties which depend on the order parameter and hence reflect its structure. It turns out, however, that the properties of FFLO states are non-universal depending sensitively on the normal state, i.e. the dispersion of the quasiparticles and their interactions. The latter strongly affect the transition line separating the normal and the superfluid phases as well as the order of the phase transitions. In addition, the anisotropy of the order parameter in the homogeneously balanced phase which is related to the presence or absence of low-energy excitations plays an important role.

Considering these complications, the present article does not attempt at providing a final answer to the existence and potential realization of imbalanced-induced spontaneous breaking of translational symmetry. The central goal is to summarize basic concepts and discuss the present state-of-the-art emphasizing what has been solved and what should be solved.

The article is organized as follows: in Sec. 2, we briefly review the ground-state properties of a superconductor with population imbalance. The properties of inhomogeneous phases are summarized in Sec. 3. In Sec. 4, we comment on the experimental situation with an emphasis on condensed-matter systems. The article closes with a summary and an outlook in Sec. 5.

2. Theoretical Background

2.1. *Order parameter*

The highly correlated superfluid phases in a multi-component Fermi liquid are characterized by long-range order in the two-particle density matrix which allows us to define order parameters

$$\Psi_{AB}(x,x') = \langle \psi_A(x)\psi_B(x') \rangle \,. \tag{2}$$

The indices A and B denote (pseudo) spins in the case of electron systems or ^{3}He, for ultra-cold atoms they refer to hyperfine states, whereas they include color, spin, and flavor indices in quark condensates.

In the two-component systems to be considered here, the pair amplitudes Eq. (2) depend upon the fermion spins and positions which are conveniently expressed in terms of the center-of-mass and relative coordinates $\mathbf{R} = \frac{1}{2}(\mathbf{x}+\mathbf{x}')$ and $\mathbf{r} = \mathbf{x}-\mathbf{x}'$, $\Psi_{\sigma\sigma'}\left(\mathbf{R}+\frac{1}{2}\mathbf{r}, \mathbf{R}-\frac{1}{2}\mathbf{r}\right)$. We focus on the BCS regime where the characteristic time and length scales in superconducting phenomena, set by the inverse gap function Δ^{-1} and the coherence length $\xi_0 = \frac{\hbar v_F}{\pi\Delta}$, respectively, are very large on the atomic scale determined by the inverse Fermi energy E_F and Fermi wavelength λ_F. This separation of length and time scales allows us to classify the order parameter of inhomogeneous systems in close analogy to their homogeneous counterparts.

In homogeneous superconductors, the order parameter is a function of the relative variable only. It varies on the scale of the Fermi wavelength which follows from the short-ranged character of the attractive quasiparticle interaction. Performing a Fourier transformation with respect to $\mathbf{r}$ with $\Psi_{\sigma\sigma'}(\mathbf{r}) \rightarrow \Psi_{\sigma\sigma'}(\mathbf{k})$ and restricting the wave vector to the Fermi surface $\mathbf{k}_F$

yields a pairing amplitude $\Psi_{\sigma\sigma'}(\mathbf{k}_F)$ which depends upon the direction $\mathbf{k}_F$ on the Fermi surface.

The fundamental property of $\Psi_{\sigma\sigma'}$ is its behavior as a two-fermion wave function in many respects. This expresses the fact that a superconducting order parameter is not the thermal expectation value of a physical observable, but rather a complex (pseudo) wave function describing quantum phase correlations on the macroscopic scale of the superconducting coherence length. Its phase is a direct signature of the broken gauge invariance in the superconducting condensate.

The Pauli principle requires $\Psi_{\sigma\sigma'}$ to be antisymmetric under the interchange of particles $\Psi_{\sigma\sigma'}(\mathbf{k}_F) = -\Psi_{\sigma'\sigma}(-\mathbf{k}_F)$. In addition, it transforms like a two-fermion wave function under rotations in position and spin space and under gauge transformations. The transformation properties yield a general classification scheme for the superconducting order parameter which is represented by a 2×2 matrix in (pseudo) spin space. It can be decomposed into an antisymmetric "singlet" (s) and a symmetric "triplet" (t) contribution according to $\mathbf{\Psi}(\mathbf{k}_F) = \mathbf{\Psi}_s(\mathbf{k}_F) + \mathbf{\Psi}_t(\mathbf{k}_F)$ with $\mathbf{\Psi}_s(\mathbf{k}_F) = \phi(\mathbf{k}_F)i\sigma_2$ and $\mathbf{\Psi}_t(\mathbf{k}_F) = \sum_{\mu=1}^{3} d_\mu(\mathbf{k}_F)\sigma_\mu i\sigma_2$, where σ_μ with $\mu = 1, 2, 3$ denote the Pauli matrices. Antisymmetry, $\mathbf{\Psi}(\mathbf{k}_F) = -\mathbf{\Psi}^T(-\mathbf{k}_F)$, requires $\phi(\mathbf{k}_F) = \phi(-\mathbf{k}_F)$ and $d_\mu(\mathbf{k}_F) = -d_\mu(-\mathbf{k}_F)$ for the complex orbital functions $\phi(\mathbf{k}_F)$ and $d_\mu(\mathbf{k}_F)$.

The general classification scheme for superconducting order parameters proceeds from the behavior under the transformations of the symmetry group $\mathcal{G}$ of the Hamiltonian. It consists of the crystal point group G or the SO(3) in the case of an isotropic liquid, the spin-rotation group SU(2), the time-reversal symmetry group $\mathcal{K}$, and the gauge group U(1). A comprehensive description is given e.g. in Ref. 15 and references therein.

Population imbalance affects only opposite "spin" pairs, i.e. pairs formed by different species which occur in the even-parity (spin-singlet) superconductors as well as the d_z-component of a triplet state. In the following discussion, we shall focus on even-parity (spin-singlet) superconductors belonging to a one-dimensional representation Γ of the symmetry group. These states are characterized by a complex scalar function, $g(\mathbf{k}_F)$, which can have zeros, symmetry-imposed or accidental, on the Fermi surface. If Γ is the unity representation, the superconductor is called "conventional", otherwise we call it "unconventional".

The classification scheme described above refers to the symmetry properties of the pair amplitudes with respect to the relative variable which reflects the short-ranged quasiparticle interaction varying on the atomic scale.

Considering the separation of length scales, we adopt the same scheme to classify the variation with the relative variable of inhomogeneous order parameters speaking of s- and d-wave, conventional and unconventional FFLO states. A textured order parameter which varies periodically with the center-of-mass variable $\mathbf{R}$ on the scale of the coherence length may break both translational and rotational symmetry. The general classification scheme for a periodic order parameter has not yet been worked out. An interesting phenomenon is the potential spontaneous breaking of relative symmetries in FFLO states. The orientation of the "crystalline" order parameter is determined by anisotropies of the normal-state properties but also by anisotropies in the order parameter of the homogeneous reference system.

2.2. *Population imbalance and breaking of translational symmetry*

To start with, we consider a two-component Fermi liquid at $T = 0$ where the two species are denoted by $\uparrow$ and $\downarrow$, respectively. The elementary excitations in the normal state are weakly interacting quasiparticles with dispersion $\epsilon(\mathbf{k})$ where the energies are conveniently measured relative to the average chemical potential μ. The quasiparticles feel a short-ranged attraction V which causes the instability of the normal state. Adopting mean-field approximation and introducing the abbreviation

$$\Delta\left(\mathbf{k};\mathbf{q}\right) = \sum_{\mathbf{k}'}{}' V(\mathbf{k},-\mathbf{k};\mathbf{k}',-\mathbf{k}') \left\langle c_{-\mathbf{k}'+\frac{\mathbf{q}}{2}\downarrow} c_{\mathbf{k}'+\frac{\mathbf{q}}{2}\uparrow} \right\rangle \tag{3}$$

for the effective pair field leads to the separable model Hamiltonian[16]

$$H = \sum_{\mathbf{k}} H(\mathbf{k}) + \text{const}, \tag{4}$$

where

$$\begin{aligned} H(\mathbf{k}) = & \left(\epsilon\left(\mathbf{k}+\frac{\mathbf{q}}{2}\right) - \delta\mu\right) c^{\dagger}_{\mathbf{k}+\frac{\mathbf{q}}{2}\uparrow} c_{\mathbf{k}+\frac{\mathbf{q}}{2}\uparrow} + \left(\epsilon\left(-\mathbf{k}+\frac{\mathbf{q}}{2}\right) + \delta\mu\right) c^{\dagger}_{-\mathbf{k}+\frac{\mathbf{q}}{2}\downarrow} c_{-\mathbf{k}+\frac{\mathbf{q}}{2}\downarrow} \\ & + \Delta^{*}\left(\mathbf{k};\mathbf{q}\right) c_{-\mathbf{k}+\frac{\mathbf{q}}{2}\downarrow} c_{\mathbf{k}+\frac{\mathbf{q}}{2}\uparrow} + \Delta\left(\mathbf{k};\mathbf{q}\right) c^{\dagger}_{\mathbf{k}+\frac{\mathbf{q}}{2}\uparrow} c^{\dagger}_{-\mathbf{k}+\frac{\mathbf{q}}{2}\downarrow} . \end{aligned} \tag{5}$$

The operators $c^{\dagger}_{\mathbf{k}\sigma}$ ($c_{\mathbf{k}\sigma}$) create (annihilate) a quasiparticle in state $\mathbf{k}\sigma$, $\delta\mu$ denotes the chemical-potential mismatch. In Eq. (4), the constant corrects for double counting of interactions. An ultraviolet cut-off is introduced as indicated by the prime in the sum in Eq. (3). As written, Eq. (5) is the usual Bardeen-Cooper-Schrieffer (BCS) Hamiltonian for an order parameter with finite center-of-mass momentum $\mathbf{q}$ varying in real space $\sim e^{i\mathbf{q}\cdot\mathbf{R}}$.[17,18] We assume that the corresponding length scale is set by the coherence length ξ_0 which implies $q \sim 1/\xi_0 \ll k_F$. In that respect, the modulated order parameters considered here differ from those of antiferromagnetic superconductors[19,20] where the staggered field induces contributions $\langle c_{-\mathbf{k}'+\frac{\mathbf{q}}{2}\downarrow} c_{\mathbf{k}'+\frac{\mathbf{q}}{2}\uparrow}\rangle$ with $q \sim k_F$.[21]

The thermodynamic properties of the model Eq. (4) can be deduced from the free energy. For a given chemical-potential mismatch the latter is calculated by treating the center-of-mass momentum $\mathbf{q}$ and the pair potential $\Delta(\mathbf{k};\mathbf{q})$ as variational parameters. The values of the latter are subsequently determined so as to minimize the free energy. Proceeding along these lines yields directly the (meta-) stable states. The traditional approach based on solving the self-consistency equation for the order parameter yields the stationary points of the free energy which include (meta-) stable as well as unstable phases.

Following this line of thought we begin by calculating the eigenvalues and eigenstates of the model Hamiltonian for a given chemical-potential mismatch $\delta\mu$, center-of-mass momentum $\mathbf{q}$, and pair potential $\Delta(\mathbf{k};\mathbf{q})$.

Since all pair Hamiltonians $H(\mathbf{k})$ commute the eigenfunctions of the BCS Hamiltonian Eq. (4) will be products of the eigenstates of $H(\mathbf{k})$. The operators $H(\mathbf{k})$ are easily diagonalized in the basis $c^\dagger_{\mathbf{k}+\frac{\mathbf{q}}{2}\uparrow}|0\rangle$, $c^\dagger_{-\mathbf{k}+\frac{\mathbf{q}}{2}\downarrow}|0\rangle$, $c^\dagger_{\mathbf{k}+\frac{\mathbf{q}}{2}\uparrow}c^\dagger_{-\mathbf{k}+\frac{\mathbf{q}}{2}\downarrow}|0\rangle$, and $|0\rangle$ where the single-particle states $|\mathbf{k};\uparrow\rangle$ and $|\mathbf{k};\downarrow\rangle$ are eigenstates with energies $E_{\mathbf{k}\uparrow} = \epsilon\left(\mathbf{k}+\frac{\mathbf{q}}{2}\right) - \delta\mu$ and $E_{\mathbf{k}\downarrow} = \epsilon\left(-\mathbf{k}+\frac{\mathbf{q}}{2}\right) + \delta\mu$, respectively. The characteristic feature of the superconducting state are the pair states $|\mathbf{k};-\rangle$ and $|\mathbf{k};+\rangle$ with energies $E_{\mp}(\mathbf{k}) = \epsilon(\mathbf{k}) \mp \sqrt{(\epsilon(\mathbf{k}))^2 + |\Delta(\mathbf{q};\mathbf{k})|^2}$, which are obtained as coherent superpositions of the two-particle states and the empty state in close analogy to bonding and anti-bonding states in molecular physics. We explicitly used the fact that the anticipated center of mass momentum of the Cooper pairs is small on the scale of the Fermi momentum setting $\frac{1}{2}(\epsilon(\mathbf{k}+\frac{\mathbf{q}}{2})+\epsilon(-\mathbf{k}+\frac{\mathbf{q}}{2})) \simeq \epsilon(\mathbf{k})$. Transitions between the different branches labeled by $\nu = \pm, \uparrow, \downarrow$ are described in terms of ladder operators, the Bogoliubov operators.

Figure 1 displays the eigenvalues for the unperturbed balanced superconductor (a) and one with imbalance and finite center of mass momentum (b) and (c). In the unperturbed superconductor, the bound pair states $|\mathbf{k};-\rangle$ are energetically more favorable than the single-particle states. Both the chemical-potential mismatch $\delta\mu$ and finite center-of-mass momentum $\mathbf{q}$, shift the single-particle state relative to the bound pair states which remain unaffected. For a fixed direction on the Fermi surface, $\delta\mu$ and $\frac{\hbar}{2}\mathbf{v}_F(\mathbf{k}_F)\cdot\mathbf{q}$ can

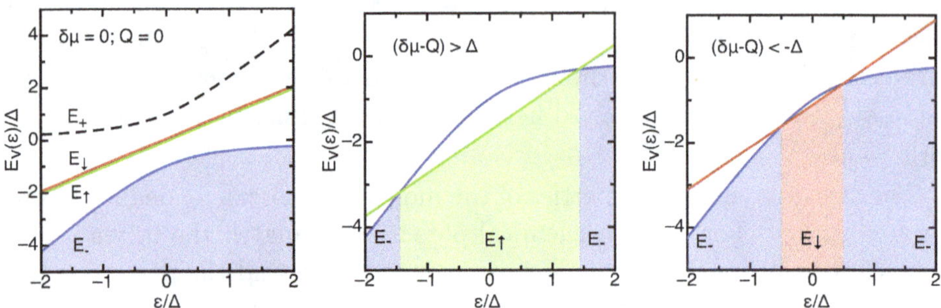

Fig. 1. Eigenvalues E_ν with $\nu = -, \uparrow, \downarrow, +$ of the pair Hamiltonian Eq. (5) for $\Delta > 0$ versus kinetic energy $\epsilon(\mathbf{k})$ of the normal-state quasiparticles with Q as defined in the text. In a homogeneous superconductor without imbalance ($\delta\mu = 0$, $\mathbf{q} = 0$, i.e. $Q = 0$) (left panel) the ground state is formed as a product of bound-pair states, i.e. $\nu(\mathbf{k}) = -$ or all momenta $\mathbf{k}$. Population imbalance $\delta\mu > 0$ in a homogeneous superconductor as well as finite center-of-mass momenta $\mathbf{q} \neq 0$, i.e. $Q \neq 0$ (middle and right panel) may reduce the energies of the single particle states $E_{\uparrow(\downarrow)}$ below those of the bound-pair states. The Zeeman splitting may be partially compensated by finite center-of-mass momentum. Forming a many-particle state with single-particle states in the vicinity of the Fermi energy $\xi = 0$ reduces the phase space for pairing. At the intersection of the branches, the character of the compromise state changes abruptly. This leads to singularities in the ground-state energy as function of Δ, $\delta\mu$, and $\mathbf{q}$.

be combined to mutually compensate the shifts. The appropriate $\mathbf{q}$-vector depends, of course, on the Fermi velocity $\mathbf{v}_F(\mathbf{k}_F)$ which varies on the Fermi surface. As a consequence, it is impossible to fully compensate a mismatch $\delta\mu$ for all fermions at the Fermi surface in two (2D) and three-dimensional (3D) superconductors by choosing a single wave vector $\mathbf{q}$. Depending on the relative size of $\left|\delta\mu - \frac{\hbar}{2}\mathbf{v}_F(\mathbf{k}_F) \cdot \mathbf{q}\right|$ and $\Delta(\mathbf{k}; \mathbf{q})$ we may have partial compensation (see Fig. 1(b) and Fig. 1(c)). In a one-dimensional (1D) superconductor, on the other hand, the mismatch $\delta\mu$ can be fully compensated resulting in the absence of a limiting $\delta\mu$ value.

Having diagonalized the mean-field Hamiltonian for arbitrary $\Delta(\mathbf{k}; q)$, $\delta\mu$ and $\mathbf{q}$ we now turn to a discussion of the ground state of the system which — like any eigenstate of the model Hamiltonian Eq. (4) — is given as a product $|\Psi\{\nu(\mathbf{k})\}\rangle = \prod_{\mathbf{k}} |\mathbf{k}; \nu(\mathbf{k})\rangle$. The index function $\nu(\mathbf{k}) = \mp, \uparrow, \downarrow$ specifies the state to be selected for a given $\mathbf{k}$-vector. At this point, we would like to emphasize that the product must run over all $\mathbf{k}$ states in order to obtain an eigenfunction of the model Hamiltonian. This feature reflects the highly correlated nature of the states. Any eigenstate is characterized by the momentum-space function $\nu(\mathbf{k})$ which can be viewed as an analogue of the momentum-distribution function. The corresponding energy, given by

$$E(\{\nu(\mathbf{k})\}) = \sum_{\mathbf{k}} E_{\nu(\mathbf{k})}(\mathbf{k}), \tag{6}$$

is minimal when the state $\nu_0(\mathbf{k})$ with the lowest value for every $\mathbf{k}$ is selected. Comparing the energies of the bound pair states $E_-(\mathbf{k})$ to those of the single-particle states $E_\uparrow(\mathbf{k})$ and $E_\uparrow(\mathbf{k})$ in Fig. 1 shows that the BCS wave function built as a product of bound pair states $|\Psi_{0BCS}\rangle = \prod_{\mathbf{k}} |\mathbf{k}; -\rangle$ gives the lowest energy for weak imbalance ($\delta\mu < \Delta$) and long-wavelength modulation ($\frac{\hbar}{2} v_F q < \Delta$).

The central focus of the present article are compromise states built from a combination of bound pair states and single-particle states. These states include the gapless homogeneous BP phase where the normal majority component is found in the immediate vicinity of the Fermi surface. Allowing for finite center-of-mass momentum $\mathbf{q}$, we arrive at partially gapped states where quasiparticle excitations of both the majority and the minority components may be present at $\mathbf{k}_F$. The gapless regions do not contribute to pairing, they are "blocked". In general, blocking reduces the energy gain due to pair formation and, consequently, destabilizes the superfluid state. Treating Δ and $\mathbf{q}$ as independent variables, the energy gain due to pair formation for given $\delta\mu$ is easily evaluated from

$$\Delta E_S(\Delta, \delta\mu, \mathbf{q}) = \sum_{\mathbf{k}}{}' \min_\nu E_\nu(\mathbf{k}) - \sum_{\mathbf{k},\sigma} (\epsilon(\mathbf{k}) + \sigma\delta\mu)\, \Theta(-(\epsilon(\mathbf{k}) + \sigma\delta\mu)) + \text{const}, \tag{7}$$

with the Heaviside function $\Theta(x) = 0$ for $x < 0$ and $\Theta(x) = 1$ otherwise. The first two terms are the quasiparticle contribution while the constant corrects for double counting of interaction energies in the quasiparticle contribution.

Before we turn to the solution we simplify the notation and introduce dimensionless variables. We explicitly account for the fact that the relevant degrees of freedom are restricted to the immediate vicinity of the Fermi surface. As a result, the summation over the quasiparticle states separate into products of integrals over the Fermi surface parameterized by $\mathbf{k}_F$ and energy integrations over ξ

$$\sum_{\mathbf{k}} \cdots = N(0) \int w(\mathbf{k}_F)\, d\mathbf{k}_F \int d\xi \cdots = N(0) \left\langle \int d\xi \cdots \right\rangle_{FS}. \tag{8}$$

We introduced the density of states at the Fermi energy, $N(0)$, and $\langle \cdots \rangle_{FS}$ as a short-hand notation for the Fermi-surface integral $\int w(\mathbf{k}_F)\, d\mathbf{k}_F \cdots$ with the normalized weight function $\langle w(\mathbf{k}_F)\rangle_{FS} = 1$ given by the Fermi velocity $w(\mathbf{k}_F) \propto \frac{1}{|\mathbf{v}_F(\mathbf{k}_F)|}$. This allows us to fully account for anisotropies in the Fermi wave vector and in the effective mass. The anisotropic gap function is given by $\Delta_{\mathbf{q}} g(\mathbf{k}_F)$ with the normalized function $\left\langle |g(\mathbf{k}_F)|^2 \right\rangle = 1$. All energies are measured in units of the order-parameter amplitude Δ_0 which minimizes the energy difference Eq. (7) of the homogeneous reference system $\mathbf{q} = 0$ without imbalance $\delta\mu = 0$. We define $Q(\mathbf{k}_F) = \frac{\hbar \mathbf{v}_F(\mathbf{k}_F)\cdot \mathbf{q}}{2}$ and measure the center-of-mass momentum in units of the inverse coherence length. Finally, the coupling constant and the ultra-violet cut-off are eliminated in favor of the energy gap Δ_0 of the balanced reference system and we arrive at

$$
\begin{aligned}
&\Delta E_S(\Delta_{\mathbf{q}}, \delta\mu, \mathbf{q})/N(0) \\
&= -\frac{1}{2}|\Delta_{\mathbf{q}}|^2 + \left\langle (\delta\mu + Q(\mathbf{k}_F))^2 \right\rangle_{FS} \\
&\quad - \left\langle \mathrm{Re}\sqrt{(\delta\mu + Q(\mathbf{k}_F))^2\left((\delta\mu + Q(\mathbf{k}_F))^2 - |\Delta_{\mathbf{q}}|^2 |g(\mathbf{k}_F)|^2\right)} \right\rangle_{FS} \\
&\quad + |\Delta_{\mathbf{q}}|^2 \left\langle |g(\mathbf{k}_F)|^2 \,\mathrm{Re}\log \frac{\sqrt{(\delta\mu + Q(\mathbf{k}_F))^2} + \sqrt{(\delta\mu + Q(\mathbf{k}_F))^2 - |\Delta_{\mathbf{q}}|^2 |g(\mathbf{k}_F)|^2}}{\Delta_0 |g(\mathbf{k}_F)|} \right\rangle_{FS} .
\end{aligned}
\tag{9}
$$

The expression has singularities which correspond to the bounds of the broken-pair regions. This fact already indicates that an expansion of the ground-state energy in terms of the order-parameter amplitude Δ is likely to fail. The radius of convergence of such an expansion will depend upon the dimensionality of the system under consideration. The consequences were discussed in detail by Combescot and Mora.[22]

The evolution of the ground-state energy of a homogeneous superconductor with imbalance is displayed in Fig. 2 for a homogeneous isotropic s-wave superconductor and its $d_{x^2-y^2}$-wave-counterpart. The characteristic features for the isotropic s-wave superconductor can be summarized as follows: for large values of the chemical-potential mismatch $\delta\mu > \Delta_0$, the energy increases monotonically with Δ and the normal state $\Delta = 0$ is the only stable phase. As $\delta\mu$ is lowered below $\delta\mu = \Delta_0$, the BCS ground state $\Delta = \Delta_0$ appears as a metastable state and we enter the two-phase regime. The latter extends down to the supercooling field $\delta\mu = \frac{\Delta_0}{2}$ below which the system must be superconducting. In the two-phase regime, the local minima,

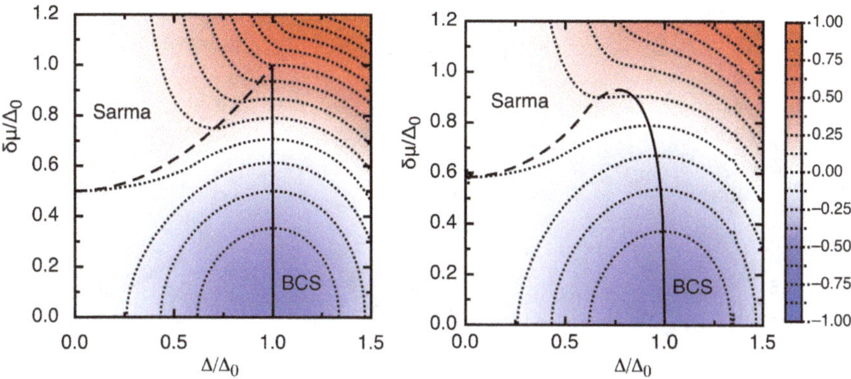

Fig. 2. Contour plot of variation with $\delta\mu$ and Δ of the energy gain $\Delta E_S(\Delta_{\mathbf{q}=0}, \delta\mu, \mathbf{q} = 0)/(N(0)\Delta_0^2)$ (see Eq. (7)) for the homogeneous s-wave (left panel) and $\mathrm{d}_{x^2-y^2}$-wave (right panel) superconductor. Solutions of the selfconsistency equation correspond to stationary points of ΔE_S with respect to variations of $\Delta_{\mathbf{q}=0}$. The order-parameter amplitude $\Delta_{\mathbf{q}=0}$ and the chemical-potential mismatch $\delta\mu$ are measured in units of the order-parameter amplitude Δ_0 in the BCS ground state of the corresponding balanced reference system. The latter corresponds to the minimum of ΔE_S along the $\delta\mu = 0$-line. For the homogeneous s-wave state, ΔE_S has a (local) minimum at $\Delta/\Delta_0 = 1$ as long as $\delta\mu/\Delta_0 < 1$. The BCS state (solid line) therefore persists as a (meta-) stable state up to the superheating "field" $\delta\mu/\Delta_0 = 1$. For small values of the mismatch $\delta\mu/\Delta_0 < 0.5$, the BCS state is the only minimum of ΔE_S. At the supercooling "field" $\delta\mu/\Delta_0 = 0.5$, the normal state $\Delta/\Delta_0 = 0$ becomes a (local) minimum. The homogeneous systems exhibit a rather broad two-phase regime for $0.5 \le \delta\mu/\Delta_0 \le 1$. In this $\delta\mu$-range, ΔE_S has a local maximum which corresponds to Sarma's unstable solution of the selfconsistency equation (dashed line). The qualitative behavior of ΔE_S for a homogeneous $\mathrm{d}_{x^2-y^2}$-wave order parameter closely parallels that of the homogeneous isotropic s-wave superconductor. The quantitative differences, however, are due to the presence of low-energy quasiparticle excitations.

i.e. the normal state $\Delta = 0$ and the usual BCS state $\Delta = \Delta_0$, are separated by a well which is centered at the line $\Delta = \Delta_0\sqrt{2\frac{\delta\mu}{\Delta_0} - 1}$ which was found by Sarma.[9] At the Chandrasekhar–Clogston limit, $\delta\mu = \frac{\Delta_0}{\sqrt{2}}$, the normal phase and the BCS state become degenerate and a first-order transition between the normal and the superconducting state occurs. It is important to note that there is no stable BP state, superconducting (meta-) stable states occur only for $\Delta = \Delta_0 \ge \delta\mu$. The main features are also present in the case of a $\mathrm{d}_{x^2-y^2}$-wave order parameter which is likely to occur in many superconductors with strongly correlated electrons. The presence of low-energy

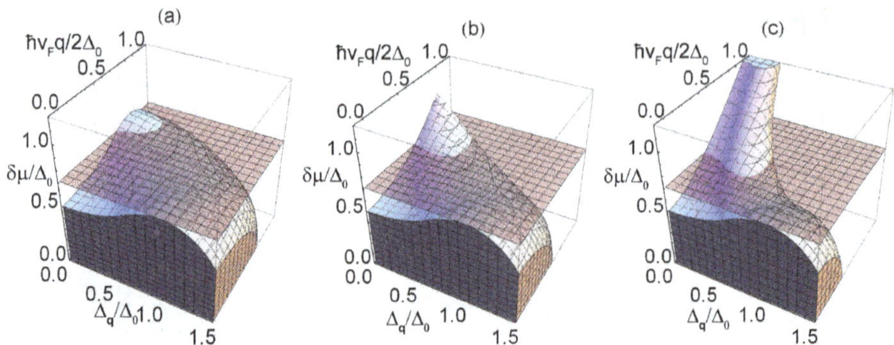

Fig. 3. Variation with chemical-potential mismatch $\delta\mu/\Delta_0$, order-parameter amplitude $\Delta_{\mathbf{q}}/\Delta_0$, and length of modulation vector $Q = \frac{\hbar}{2}\frac{v_F q}{\Delta_0}$ of the energy difference ΔE_S from Eq. (9) for (a) an isotropic three-dimensional (3D), (b) an isotropic two-dimensional (2D), and (c) one-dimensional (1D) Fermi surface. All energies are measured in units of the order-parameter amplitude Δ_0 characterizing the BCS ground state of the homogeneous ($\mathbf{q} = 0$) balanced ($\delta\mu = 0$) reference system. In the filled regions, the superconducting solution is energetically more favorable than the normal state ($\Delta E_S < 0$). The boundary surface corresponds to combinations of $\delta\mu/\Delta_0$, $\Delta_{\mathbf{q}}/\Delta_0$, and $Q = \frac{\hbar}{2}\frac{v_F q}{\Delta_0}$ for which the energy of the superconducting solution equals that of the normal state, i.e. $\Delta E_S = 0$. The plane $\delta\mu/\Delta_0 = 1/\sqrt{2}$ corresponds to the Chandrasekhar–Clogston limit for the first-order transition between the homogeneous superconducting and normal states. For modulated states $Q \neq 0$, pair formation can lead to energy gain for chemical-potential mismatch $\delta\mu/\Delta_0$ exceeding the Chandrasekhar–Clogston limit.

excitations, however, reduces the energy-gap opening at the superheating field.

We next turn to the case of finite-q pairing. As pointed out before, the possible compensation of the chemical-potential mismatch depends on the dimensionality of the system. This is clearly evident from Fig. 3 which shows the regions in $\delta\mu$-$\mathbf{q}$-Δ space where the energy of the superconducting state is lower than that of the normal phase, $\Delta E_S\left(\Delta_{\mathbf{q}}, \delta\mu, \mathbf{q}\right) \leq 0$, for isotropic 3D and 2D as well as for 1D superconductors with s-wave order parameter. The limiting value of the chemical-potential mismatch $\delta\mu/\Delta_0$ is given by the maximum which only slightly exceeds the Chandrasekhar–Clogston limit in 3D. In 2D, the limiting value is enhanced and coincides with the beginning of the two-phase regime. The singularities for small $\Delta q/\Delta_0$ give rise to highly anomalous properties to be discussed in the subsequent section. The absence of a limiting mismatch in 1D is obvious.

At finite temperatures $T > 0$, the free-energy gain due to pairing is given by

$$\Delta F(\Delta_{\mathbf{q}}, \delta\mu, \mathbf{q}; T) = -N(0)\Bigg[-\left\langle(\delta\mu - Q(\mathbf{k}_F))^2\right\rangle_{FS} + |\Delta_{\mathbf{q}}|^2 \ln\frac{T}{T_{c0}} + 2\pi k_B T \sum_{\ell=0}^{\infty}\left(2\mathrm{Re}\left\langle\sqrt{(\omega_\ell + i(\delta\mu - Q(\mathbf{k}_F)))^2 + |\Delta_{\mathbf{q}}|^2 g^2(\mathbf{k}_F)}\right\rangle_{FS} - 2\omega_\ell - \frac{|\Delta_{\mathbf{q}}|^2}{\omega_\ell}\right)\Bigg], \quad (10)$$

where $\omega_\ell = \pi k_B T(2\ell + 1)$ are the fermionic Matsubara frequencies and T_{c0} denotes the superconducting transition temperature of the balanced reference system. Of particular interest is the second-order transition from the normal phase to a superconducting state $\Delta_{\mathbf{q}} e^{i\mathbf{q}\cdot\mathbf{r}}$ which occurs at a critical mismatch $\delta\mu(T)$ given by the linearized selfconsistency equation (linearized stationarity condition) for $\Delta F(\Delta_{\mathbf{q}}, \delta\mu, \mathbf{q}; T)$. The latter is given by the implicit equation

$$\ln\frac{T}{T_{c0}} = \psi\left(\frac{1}{2}\right) - \min_Q \left\langle g^2(\mathbf{k}_F)\,\mathrm{Re}\,\psi\left(\frac{1}{2} + i\frac{1}{2\pi k_B T}(\delta\mu(T) + Q(\mathbf{k}_F))\right)\right\rangle_{FS}, \quad (11)$$

where $\psi(z)$ is Euler's digamma function. The optimal modulation vector depends on the anisotropies of the normal-state quasiparticle dispersion and the superconducting order parameter as given by $\mathbf{v}_F(\mathbf{k}_F)$ and $g(\mathbf{k}_F)$. The variation with T has been analyzed in great detail by Saint-James and Sarma[10] for the s-wave case. The formal key to the phase diagram is the variation with x of the curvature $-\mathrm{Re}\,\psi''\left(\frac{1}{2} + ix\right)$ which is negative for small values of the argument and changes sign at $x = 0.304$. This implies that close to the transition temperature T_{c0}, i.e. for small values of $\delta\mu$ the homogeneous state $\mathbf{q} = 0$ is formed. In this temperature range, the critical mismatch follows a universal curve independent of the details of the system under consideration. With decreasing temperature a tricritical point appears at $T = 0.55T_{c0}$, below which the transition from the normal state into a modulated state with $\mathbf{q} \neq 0$ occurs. The critical value for the mismatch, assuming $\mathbf{q} = 0$, becomes the supercooling field below which the homogeneous superconductor is the thermodynamically stable phase.

The influence of dimensionality on the $\delta\mu$–T phase diagram is clearly evident from Fig. 4 (left panel) which displays the variation with tempera-

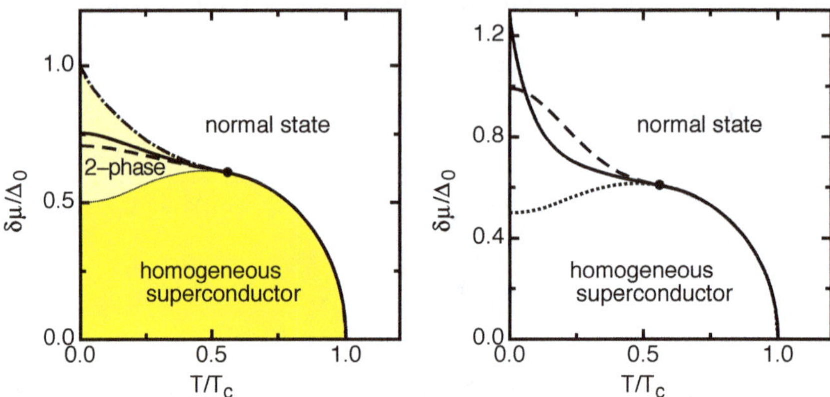

Fig. 4. Comparison of $\delta\mu$-T phase diagrams. Left panel: s-wave superconductor. For temperatures above the tricritical point (black dot) the normal state becomes unstable relative to a homogeneous superconducting state (for sufficiently low $\delta\mu$). Below the tricritical point, the homogeneous superconducting phase can coexist with the normal state in a $\delta\mu$ regime bound by the supercooling field (dotted line) and the superheating field (dash-dotted line). In a 3D system with spherical Fermi surface a second-order transition from the normal state to an FFLO state (solid line) is found for $\delta\mu$ slightly above the critical value for the first-order transition between the homogeneous phases (dashed line). For a 2D system with cylindrical Fermi surface the second-order transition coincides with the superheating field of the homogeneous phases. Right panel: d-wave superconductor. Second-order transition from the normal to the $d_{x^2-y^2}$-wave FFLO state in a 2D superconductor with cylindrical Fermi surface for modulation vector parallel to the nodal direction (dashed line) and anti-nodal direction (solid line). The supercooling field (dotted line) is included for comparison.

ture of the critical mismatch for s-wave superconductors with spherical and cylindrical Fermi surfaces as well as the two-phase regime for homogeneous superconductivity bounded by the superheating and the supercooling fields as well as the corresponding line of first-order transitions. In an anisotropic superconductor the variation with angle of the gap function selects the optimal orientation of the modulation vector $\mathbf{q}$. The latter may change discontinuously with temperature leading to kinks in the critical-mismatch curve. This effect was predicted for superconductors with a $d_{x^2-y^2}$-wave order parameter and a cylindrical Fermi surface by Maki and Won[23] as shown in Fig. 4 (right panel).

The second-order transition lines give good qualitative insight into the overall behavior of superconductors with population imbalance. The explicit evaluation of the full phase diagram including potential first-order transitions requires the calculation of the free energy for general inhomogeneous order parameters.

3. Properties of the Inhomogeneous States

At sufficiently low temperatures, the polarized normal state of an unbalanced Fermi liquid can become unstable against an inhomogeneous superconducting state. The modulation vector minimizing the free energy, however, is not unique. Usually, there are several equivalent modulation vectors $\mathbf{q}_m$, with $m = 1, \ldots, M$, in agreement with the point-group symmetry of the underlying crystal. In the vicinity of the second-order transition, the corresponding plane-wave order parameters $\sim e^{i\mathbf{q}_m \cdot \mathbf{R}}$ are equivalent. In idealized isotropic systems, all plane-wave states $\sim e^{i\mathbf{q} \cdot \mathbf{R}}$ with the same wavelength are degenerate at the transition. In the fully developed superfluid phase below the transition this degeneracy is lifted by nonlinear terms in the free-energy as shown by Larkin and Ovchinnikov.[2] Forming linear combinations of the equivalent plane-wave states results in an order parameter whose amplitude is periodically modulated which results in "crystalline" superconductivity.

Determining the optimal spatial structure of the periodic order parameter and predicting the properties of the corresponding superconducting phase is a problem which has been solved only partially. An adequate microscopic analysis of the nonuniform FFLO phases requires simultaneous selfconsistent calculations of the following properties which are intimately intertwined: (i) the inhomogeneous order parameter structure, (ii) the local excitation spectrum, and (iii) the quasiparticle selfenergy accounting for the renormalizations and Fermi-liquid interactions. The technical difficulties are associated with the fact that often there is a great number of different yet energetically almost degenerate states.

3.1. *Mean-field results: quasiclassical theory*

A suitable framework for a numerical study of the complete phase diagram of inhomogeneous superconductors is the quasiclassical theory of superconductivity developed by Eilenberger, Larkin and Ovchinnikov and Eliashberg (ELOE).[24–26] The quasiclassical theory of superconductivity is valid on the coarse-grained length and time scales of the superconducting phenomena set by the coherence length $\xi_0 = \hbar v_F / \pi \Delta_0$ and the inverse gap frequency Δ^{-1}. At the outset, properties determined by the Fermi wavelength k_F^{-1} and the Fermi energy ϵ_F are explicitly eliminated from the theory. We would like to emphasize that, since the traditional BCS theory is restricted in accuracy to terms of order T_c/ϵ_F an advantage of the quasiclassical method is that it acknowledges this fact and eliminates as many intermediate steps as possible. For the special case of a superconductor close to the transition point a quasiclassical description was derived by deGennes.[27]

The quasiclassical theory of ELOE is based on formal many-body perturbation theory, expressed in terms of the thermodynamic (imaginary time) Green's function $\mathbf{G}(\mathbf{x}, \mathbf{x}'; i\omega_n)$. Here $\mathbf{G}$ is a 4×4 Nambu–Gorkov matrix Green's function[28,29] which contains the "anomalous" Green's functions in the off-diagonal quadrants. In what follows, only the static limit will be important, and, consequently, only one frequency, the energy variable, appears as an argument. In the first step, center-of-mass and relative variables are introduced and a Fourier transformation with respect to the latter is performed $\mathbf{G}(\mathbf{R} + \frac{1}{2}\mathbf{r}, \mathbf{R} - \frac{1}{2}\mathbf{r}; i\omega_n) \longrightarrow \mathbf{G}(\mathbf{k}, \mathbf{R}; i\omega_n)$. The central quantity of the theory is the "quasiclassical" or "ξ-integrated" Green's function $\mathbf{g}(\mathbf{k}_F, \mathbf{R}; i\omega_n)$ which is derived from the full propagator $\mathbf{G}(\mathbf{k}, \mathbf{R}; i\omega_n)$ by averaging over the kinetic energy of the quasiparticles. The equations determining $\mathbf{g}(\mathbf{k}_F, \mathbf{R}; i\omega_n)$ take the form of transport-like equations which are supplemented by a normalization condition. For a comprehensive review, see Ref. 30.

The quasiclassical method was successfully used to study the evolution of FFLO states characterized by one-dimensional stripe patterns for the order parameter. For quasi-2D s-wave superconductors, Burkhardt and Rainer[31] examined selfconsistently the FFLO state including Fermi-liquid interactions. An important result is that the transition from the inhomogeneous FFLO state to the uniform superconducting phase should be of second order. The authors showed that both the critical value of the chemical-potential difference as well as the order of the normal-to-superconducting phase transitions sensitively depend upon the isotropic exchange interaction among the quasiparticles. Variation with temperature of the order of the normal-to-superconducting transition was also found for isotropic 3D s-wave superconductors by Matsuo *et al.*[32] and confirmed by Combescot and Mora.[22] The full phase diagram for a $d_{x^2-y^2}$-wave order parameter in an isotropic quasi-2D superconductor was studied by Vorontsov *et al.*[33] In unconventional superconductors, the orientation of the stripe pattern relative to the nodal direction of the gap can vary discontinuously with temperature.[23,34–36] The change in the modulation vector $\mathbf{q}$ is found below $T^* \simeq 0.06T_c$. Concerning the identification of FFLO phases, Vorontsov *et al.*[33] suggest to measure the Andreev resonance spectrum for which they predict characteristic features.

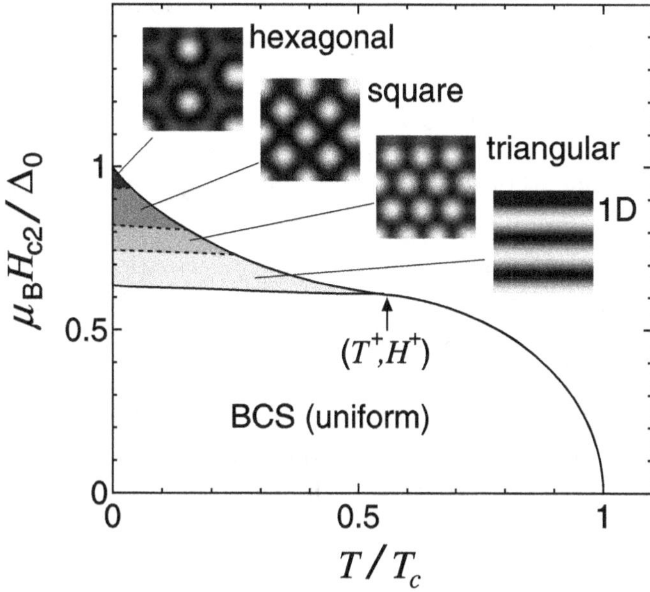

Fig. 5. Phase diagram of an s-wave superconductor with cylindrical Fermi surface.[3] The chemical-potential mismatch $\delta\mu = g\mu_B H$ is due to the Zeeman splitting of the quasiparticle states in a magnetic field applied perpendicular to the axis of the Fermi-surface cylinder. With decreasing temperature several types of periodic structures are predicted in the FFLO state.

3.2. *Mean-field results: Ginzburg–Landau approach*

While the solution of the full quasiclassical equations in the full $\delta\mu$ and T range could be solved only for stripe textures, more insight can be gained in restricted regions of the phase diagram. In the vicinity of the second-order transition from the normal to the superconducting phase a (generalized) Ginzburg–Landau expansion of the free energy in terms of the superconducting order parameter were derived from the microscopic theory. The central focus of these studies is "pattern formation" associated with the higher-order terms contributions. The latter lift the degeneracy of the primary plane-wave order parameters $\sim e^{i\mathbf{q}_m \cdot \mathbf{R}}$ and induce higher harmonics. Neglecting the admixture of higher harmonics, Larkin and Ovchinnikov[2] compared three cases, i.e. the single-vector state discussed in the previous section, the striped phase, and the simple cubic crystal and concluded that among these candidates the striped phase has the lowest energy. Shimahara predicts a series of phase transitions between different periodic structures as shown in Fig. 5.

Recently, a detailed study was presented by Bowers and Rajagopal[37] who analyzed 23 different crystalline structures. The coefficients of the free-energy functional derived from the microscopic Gorkov equations[29] allow for a determination of the stable structures within weak-coupling theory. The name of the multiple-wave configurations refers to the polyhedron inscribed in a sphere of radius $|\mathbf{q}|$ whose vertices correspond to the wave vectors $\mathbf{q}_m$. Among the structures which are stable within sixth-order expansion the octahedron gives the lowest free energy. Special attention should be paid to the cube structure for which the coefficients of the quadratic, quartic and sextic terms in the GL expansion are negative. In real space, $\Delta(\mathbf{R})$ exhibits a periodic face-centered cubic structure. These candidate states are displayed schematically in Fig. 6. To actually discuss the free-energy one has to go to the eighth order. Bowers and Rajagopal argue that this structure might correspond to a rather low value of the free energy.[37]

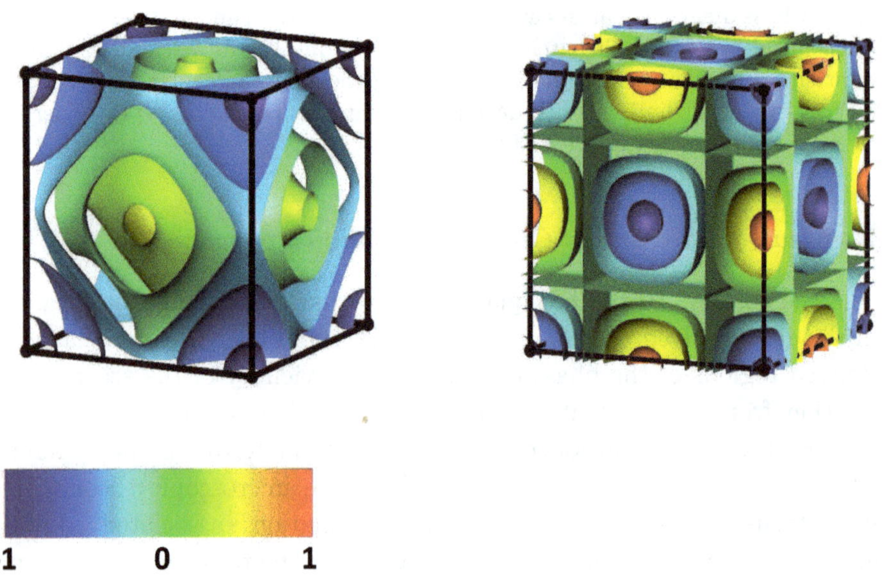

Fig. 6. Unit cell of candidate crystalline FFLO states in isotropic 3D superconductors. The octahedron state (left panel) gives the lowest energy within sixth-order Ginzburg–Landau expansion. Qualitative considerations suggest that the cube structure (right panel) should be the most favorable structure. The contours correspond to $\Delta(\mathbf{R})/\Delta_{\max} = -0.9, -0.5, -0.3, 0, 0.3, 0.5, 0.9$. The lengths of the cubes are $2\pi/|\mathbf{q}|$ and $2\pi/|\mathbf{q}|\sqrt{3}$ for the octahedron and the cube structure, respectively.

3.3. *Fluctuations*

Having discussed the mean-field predictions, we next turn to the superconducting fluctuations. In this context we have to distinguish between the thermal fluctuations at finite temperatures $T > 0$ and the $T = 0$ quantum fluctuations. While the former are predominantly classical, the dynamic nature of the latter is essential.

Thermal fluctuations in superconductors have been studied for a long time (for a review, see Ref. 38). The role of phase fluctuations of the modulated order parameter is most prominent in isotropic systems where the FFLO states are infinitely degenerate. The situation is different in the so-called "generic" case where the degeneracy manifold is reduced to isolated points in momentum space either by anisotropies in the normal-state quasiparticle properties or the node structure of the superconducting order parameter.

Quantum fluctuations at $T = 0$ have received little attention. Samokhin and Marenko[39] calculated the corrections to the spin susceptibility and the quasiparticle-decay rate of fermionic quasiparticles for both the isotropic and the generic case. The results reflect the singular behavior of the mean-field ground-state energy displayed in Fig. 3. Diener *et al.*[40] found Fermi-liquid interaction corrections to thermodynamic properties as a consequence of quantum fluctuations.

3.4. *Microscopic studies*

Microscopic model studies beyond the mean-field approximation have been performed for quasi-1D systems. The exact ground states of two-component Fermi liquids with attractive contact interaction and unequal densities are known from Bethe ansatz. Three phases appear at $T = 0$: the paired superfluid state, the polarized normal Fermi liquid and an intermediate partially polarized phase which exhibits FFLO-type correlations. These findings were recently confirmed by accurate Density Matrix Renormalization Group (DMRG) and Quantum Monte Carlo (QMC) calculations (see Refs. 41–44 and references therein).

In the weak-coupling limit the phase diagram is well described by mean-field theory.

4. Search for Experimental Realization

Shortly after the theoretical predictions for the possible existence of an inhomogeneous superconducting state at high magnetic fields and low temperatures, first experimental searches started. It took, however, more than 30 years until first claims for the observation of the FFLO state appeared. Yet, those proved to be not well founded. Only in the last decade, better justified evidences have been reported for homogeneous materials (heavy-fermion and organic superconductors) as well as for heterogeneous materials (artificial superlattices).

4.1. *Homogeneous materials*

In real superconducting materials two main conditions have to be fulfilled in order to allow the FFLO state to occur. First, the orbital critical field, $H_{\rm orb}$, should be sufficiently larger than the Pauli-paramagnetic limit, $H_P = \Delta_0/(\sqrt{2}\mu_B)$, where Δ_0 is the superconducting energy gap at $T = 0$. Or more precise, the Maki parameter,[45] $\alpha = \sqrt{2}H_{\rm orb}/H_P$, should be larger than 1.8.[46] Second, the superconductor needs to be in the clean limit with a mean-free path ℓ much larger than the coherence length ξ. These are rather rigorous conditions that are not satisfied by many superconductors. Likely candidates for a possible realization of the FFLO state are heavy-fermion or layered superconductors. In heavy-fermion materials the high effective mass, $m_{\rm eff}$, results in a small Fermi velocity, v_F, which in turn leads to a short coherence length ($\xi \propto v_F$) and, consequently, a large orbital critical field. With ℓ in high-quality heavy-fermion crystals much larger than ξ, the FFLO state may well be expected. Layered, or quasi-two-dimensional superconductors, such as some cuprates, many organic superconductors, or artificial superlattices may be even better suited for exhibiting the FFLO state.[3,6] For such materials, the orbital pair breaking can be greatly suppressed when applying the magnetic field parallel to the superconducting layers. At the same time, samples of very high quality are available with sufficiently large mean-free paths.

Nevertheless, even up to now only very few condensed-matter realizations exist that are reported to show indications for an FFLO-like state. First reports appeared in the 1990s, in which evidence for such an inhomogeneous state was suggested to exist in the heavy-fermion compounds UPd_2Al_3,[47] $CeRu_2$,[48] and UBe_{13}.[49] These claims, however, later had to be revised or are inconclusive. See Refs. 3, 6 and 50 for more details.

In the first decade of the 21st century, new experimental work strongly suggested the existence of an FFLO state in novel superconducting materials. These are the heavy-fermion compound $CeCoIn_5$[51,52] and the quasi-two-dimensional organic superconductors. Whereas for $CeCoIn_5$ some recent experimental results[53] cast some doubt on the genuine realization of the

FFLO state, the organic materials make a strong case in favor for such a state. In the following, the experimental status quo for these materials will be discussed in more detail.

4.1.1. *Heavy-fermion systems*

A recent prime candidate for the realization of an FFLO state is $CeCoIn_5$. It is a heavy-fermion superconductor with $T_c = 2.3$ K and tetragonal crystal structure.[54] It is a clean type-II superconductor with a large orbital critical field, a sufficiently large Maki parameter, and an unusual field-temperature phase diagram.[51,52,55] Therefore, all conditions to observe the FFLO state are fulfilled. And indeed, specific-heat data[51,52] as well as thermal-conductivity[56] and penetration-depth measurements using a self-resonant tank circuit[57] showed clear evidence for a phase transition within the superconducting state, as expected for the FLLO state.

The specific-heat data of Bianchi *et al.*[51] for magnetic fields aligned along the [110] direction is shown in the left panel of Fig. 7. First of all, above about 10 T and below about 1 K, the phase transition from the normal into the superconducting state changes from second to first order. This can be realized by the pronounced sharpening of the anomalies as well as by the occurrence of hysteresis in the specific-heat data collected by use of the relaxation method for rising and falling temperatures. Further to that, a

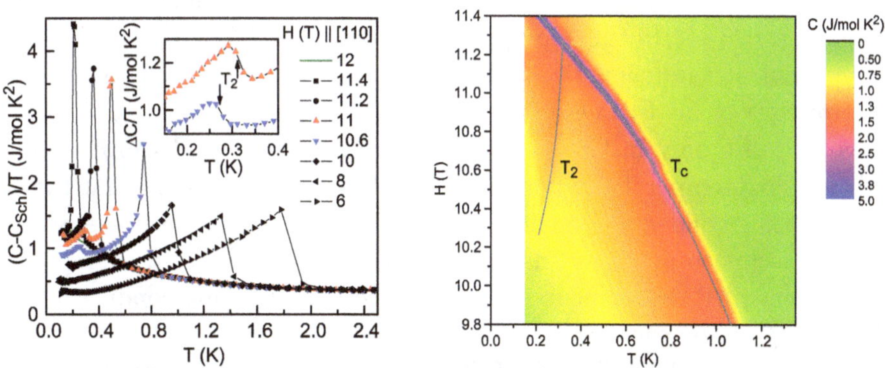

Fig. 7. (left panel) Specific heat of $CeCoIn_5$ in magnetic fields aligned parallel to the [110] direction. A Schottky-like nuclear contribution has been subtracted. The inset shows the anomalies, T_2 (originally called T_{FFLO}), at 10.6 and 11 T. (right panel) Contour plot of the electronic part of the specific heat divided by temperature for $H \,||\, [110]$. The grey lines denote T_c and T_2. After Ref. 51.

small but clear anomaly is resolved at lower temperatures around 300 mK, labeled T_2 (originally labeled $T_{\rm FFLO}$) in the inset of Fig. 7. This anomaly is most prominent for in-plane field orientations, but a smaller anomaly was found as well for the field aligned along the c axis.[51]

The phase diagram over a narrow region in field and temperature is shown in the right panel of Fig. 7 as a contour plot. The data were collected by use of the relaxation method.[51] Besides the phase transition from the normal to the superconducting state at T_c, a second phase transition at low temperatures and high fields within the superconducting region is clearly evident (T_2). Such kind of phase diagram, confirmed by other methods as well, is very suggestive for an FFLO state. There are, however, some features in contrast to the expected behavior. Especially, there is no saturation of the upper critical field, H_{c2}, due to the Pauli limiting observed and there is no upturn of H_{c2} at low temperatures at which the additional phase appears.

In order to get a definitive proof for the existence of an FFLO state, the spatial modulation of the superconducting order parameter should be measured. The method of choice for that purpose is small-angle neutron scattering. By that, the expected periodic oscillation of the magnetic-field strength along the applied field may in principle be detected. Indeed, in 2008 Kenzelmann *et al.* used high-field neutron diffraction to search for the magnetic response of $CeCoIn_5$ in the suggested FFLO phase.[53] Clear evidence of magnetic Bragg peaks with an incommensurate ordering vector $\mathbf{Q} = (q, q, 0.5)$ was found. This Bragg peaks are not present outside, what is now called, the Q phase. In the FFLO state, one would expect the Cooper pairs to carry a momentum that depends on the magnetic field via $|q| = 2\mu_B H/(\hbar v_F)$. In the experiment, however, $q = 0.442$ was found to be independent of the magnetic field.[53] This is in stark contradiction to the usual picture of an FFLO state. Anyway, the finding that antiferromagnetism appears only inside the superconducting state of $CeCoIn_5$ at high fields and low temperatures is remarkable. It strongly suggests that the superconducting condensate carries a field-independent momentum. A similar behavior seems to occur for magnetic fields applied parallel to the c direction. Recent μSR data hint to a field-induced coupled superconducting and spin-density-wave order for this field orientation.[58]

The findings of Kenzelmann *et al.*[53] have triggered numerous theoretical efforts trying to explain the nature of the highly unusual Q state. Recently, Mierzejewski *et al.* discussed the possibility of a mutual enhancement of an incommensurate magnetic order and FFLO superconductivity.[59] Before that, Yanase and Sigrist suggested that the observed antiferromagnetism might be stabilized by the FFLO state.[60] On the basis of Bogoliubov–de Gennes equations this was argued to be caused by the appearance of Andreev bound states localized around the zeros of the FFLO order parameter. In other words, the antiferromagnetic order should occur in regions where the superconducting order parameter vanishes, as opposed to Ref. 59 where the two orders mutually enhance each other. Anyway, in both cases the antiferromagnetic order might occur in the FFLO state even when it is neither stable in the normal state nor in the BCS state. In agreement with this prediction, recent nuclear magnetic resonance (NMR) experiments indicated that the incommensurate magnetic order in $CeCoIn_5$ is likely to coexist with a modulated superconducting phase, that may well be the FFLO state.[61] The authors further found some indication for a new region below the Q phase in the phase diagram, which could be an FFLO phase without magnetic order. It is evident that further work is necessary to settle all the open issues in this highly fascinating material.

4.1.2. *Organic superconductors*

Besides the heavy-fermion materials, the quasi-two-dimensional (2D) organic superconductors are prime candidates for exhibiting the FFLO state. When applying the magnetic field exactly parallel to the highly conducting planes, the orbital critical field will be greatly enhanced. In a number of different experiments, some indications for an FFLO state have been reported. In most of the experiments, however, no thermodynamic proof was given. Only recently, clear thermodynamic evidence for a narrow intermediate superconducting state was presented.[62]

One of the first reports claiming evidence for an FFLO state in a 2D organic superconductor was presented for κ-(BEDT-TTF)$_2$Cu(NCS)$_2$, where BEDT-TTF is bisethylenedithio-tetrathiafulvalene. This conclusion was based on measurements sensitive to the loss in vortex stiffness.[63] The observed rather broad features in the signal and the derived phase diagram are, however, largely at odd with the thermodynamic evidences for this material presented below.

For another 2D organic superconductor, λ-(BETS)$_2$GaCl$_4$, where BETS is bisethylenedithio-tetraselenafulvalene, thermal-conductivity, κ, data indicate a possible FFLO state.[64] Tanatar *et al.* investigated samples of different quality, which was determined by the resistivity ratio or by the relative thermal-conductivity maximum. For low-quality samples (2 and 5 in Fig. 8), the rather sharp increase of κ/T at H_{c2} might indicate a broadened first-order transition. For high-quality samples (1 and 6 in Fig. 8), this

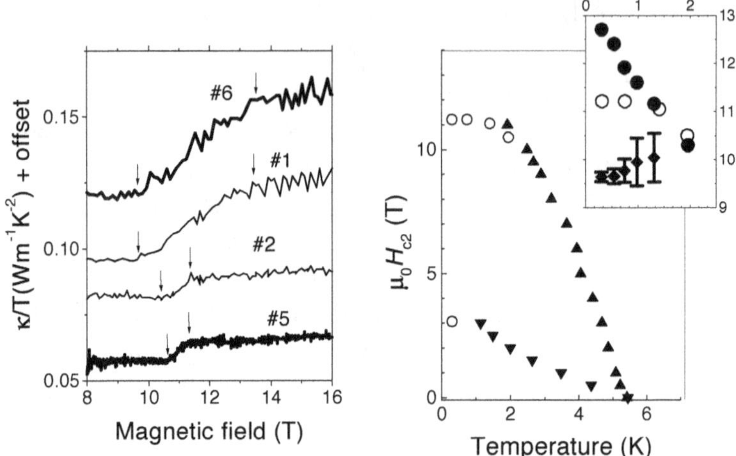

Fig. 8. (left) Field dependence of the thermal conductivity of four different λ-$(BETS)_2GaCl_4$ samples at 0.3 K. The field was aligned parallel to the conducting layers. The curves for sample 6 and 5 are shifted up and down by 0.04 $Wm^{-1}K^{-2}$, respectively. (right) Phase diagram for sample 2. Up and down triangles are from the resistive transitions for parallel and perpendicular field orientations, respectively. Open circles are from thermal conductivity. The inset shows the low-temperature part of the phase diagram for sample 1 (closed circles) and sample 2 (open circles), respectively. The diamonds are from the low-field slope change in κ/T (left arrows in the left panel) for sample 1. After Ref. 64.

feature is replaced by two distinct slope changes that are much farther apart than in the low-quality samples. The magnetic phase diagram for samples 1 and 2 are shown in the right panel of Fig. 8. For sample 2, a clear saturation of H_{c2} is observed below about 1 K, probably due to the Pauli-limiting effect. For the high-quality sample 1, the H_{c2} line continues to grow even below 1 K (inset of Fig. 8). When interpreting the low-field slope change as another phase transition (diamonds in the inset of Fig. 8), a phase diagram as expected for a superconductor with an FFLO state evolves. Although these experimental results give no definite proof for such an inhomogeneous state in this organic superconductor, all observed features are in accord with such a conclusion.

Another isostructural material, λ-$(BETS)_2FeCl_4$, where just gallium is replaced by magnetic iron, has become famous because of its field-induced superconducting state.[65,66] Thereby, after removing a zero-field antiferromagnetic state by the application of an in-plane magnetic field, superconductivity appears in the field range between 17 and 45 T. This behavior is well understood by the Jaccarino–Peter compensation mechanism.[67] Thereby,

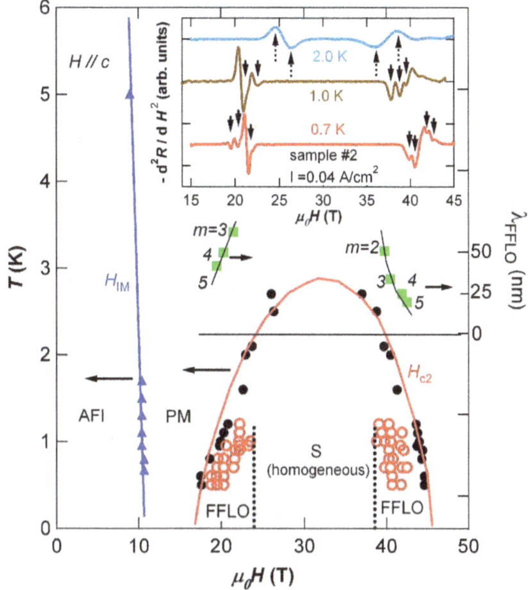

Fig. 9. Phase diagram of λ-$(BETS)_2FeCl_4$ for a magnetic field aligned parallel to the conducting layers ($H||c$). The triangles depict the transition from the antiferromagnetic-insulating (AFI) to the paramagnetic (PM) state, solid circles correspond to the onset of superconductivity, H_{c2}, and open circles are extracted from the dip positions in the second derivative of the resistance (solid arrows in the inset). The squares show estimated values of the order-parameter wavelength in the assumed FFLO states at $T = 0.7$ K (see text). The solid lines are guides to the eye. The dotted arrows in the inset correspond to broader features at H_{c2}. After Ref. 68.

the internal exchange field of the iron atoms acting on the π electrons of the BETS molecules becomes compensated in the applied external field.

Since λ-$(BETS)_2FeCl_4$ is a clean-limit superconductor, as evidenced by the observation of magnetic quantum oscillations, and the Pauli limit is estimated at 13 T,[68] the FFLO state may be expected. When Uji *et al.* measured the interplane resistance for in-plane fields, anomalous dip structures were observed.[68] These features appear much clearer in the second derivative of the field-dependent resistance curves (inset of Fig. 9) which allowed to construct the phase diagram shown in Fig. 9. The Jaccarino–Peter compensation field is about 32 T. Therefore, critical fields appear both below as well as above this value with two regions for possible FFLO states. The multiple dip structures observed are suggested to be caused by a commensurability effect.[68] According to Bulaevskii *et al.*[69] a possible Josephson-vortex lattice (with lattice constant l) may get pinned at the oscillating FFLO order parameter (with wavelength λ_{FFLO}). This effect may lead to the observed dip structures whenever $m = l/\lambda_{\mathrm{FFLO}}$ becomes an integer. Whether this assumption is correct requires further studies.

Another recent report suggested a possible FFLO state in the 2D organic superconductor β''-(BEDT-TTF)$_2$SF$_5$CH$_2$CF$_2$SO$_3$.[70] This was based on in-plane penetration-depth measurements by using a tunnel-diode-oscillator technique. Some features in the second derivative of the observed frequency changes were attributed to transitions into the FFLO phase and to H_{c2}. Below about 1.23 K, H_{c2} exceeds the Pauli-limiting field and the H_{c2} line becomes steeper. At the same time the second feature at lower fields suggests an emerging FFLO state.[70] The resulting phase diagram needs, however, further support.

Further to that, there has recently been a report suggesting an FFLO state in the quasi-one-dimensional organic superconductor (TMTSF)$_2$ClO$_4$, where TMTSF stands for tetramethyltetraselenafulvalene.[71] Constructing the magnetic phase diagrams from resistivity measurements, one low-quality sample revealed a saturation of H_{c2} reminiscent of Pauli limiting, whereas a better-quality sample showed a clear upturn in H_{c2} at lowest temperatures. The resulting H_{c2} data are very similar to the phase diagrams shown in Fig. 8 here for λ-(BETS)$_2$GaCl$_4$.[64]

In all of the above cases no thermodynamic proof for an FFLO state in organic superconductors was given. Only for κ-(BEDT-TTF)$_2$Cu(NCS)$_2$ such evidence has been presented recently.[62] For magnetic fields aligned exactly parallel to the 2D layers spike-like anomalies in the specific heat are observed for fields above 21 T and temperatures below 3 K [Fig. 10(a)]. These features can be much better resolved after subtraction of the phonon contribution to the specific heat. For that, data in the normal state measured at 14 T applied perpendicular to the layers have been used.[62] The resulting electronic contribution to the specific heat, C_e, divided by temperature is shown in Fig. 10(b). At about 21 T, a sharpening of the specific-heat anomaly and at 21.5 T a spike in C_e occurs. This is clear evidence for a first-order nature of the phase transition from the normal into the superconducting state. For higher fields, a second very sharp anomaly appears within the superconducting state. This proves the existence of an additional thermodynamic phase. The transition at lower temperatures shows a well-resolvable hysteresis, while the main superconducting transition reveals a smaller hysteresis [inset of Fig. 10(b)]. Above 23 T the transition left the temperature window accessible in the experiment.[62]

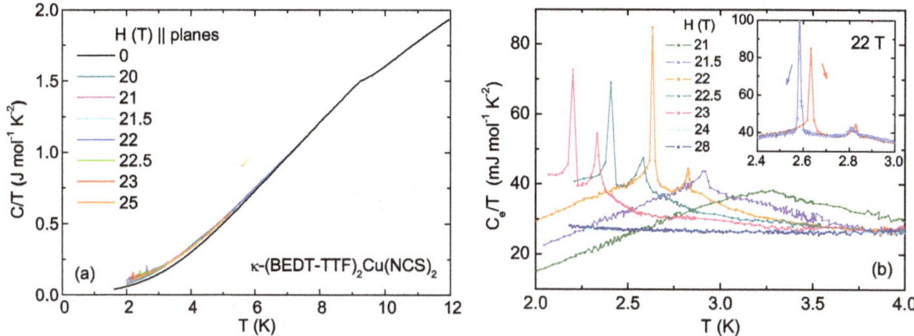

Fig. 10. (a) Temperature dependence of the specific heat divided by temperature, C/T, of κ-(BEDT-TTF)$_2$Cu(NCS)$_2$ for magnetic fields applied parallel to the superconducting layers. (b) Low-temperature part of the electronic part of the specific heat, C_e/T. Data taken in 14 T applied perpendicular to the layers were used to separate the phonon contribution. See Ref. 62 for more details.

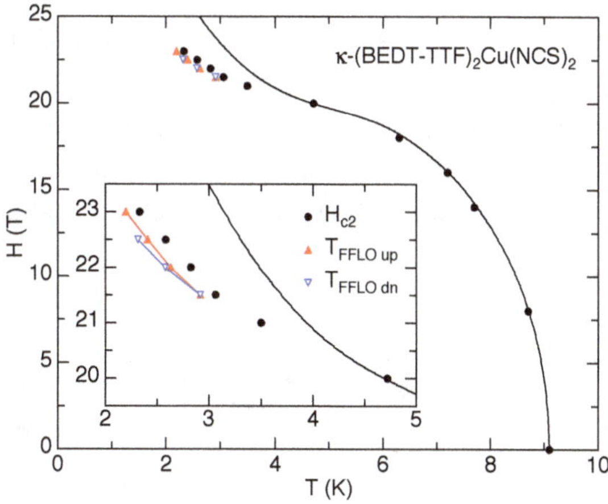

Fig. 11. Phase diagram of κ-(BEDT-TTF)$_2$Cu(NCS)$_2$ for fields applied parallel to the superconducting layers. The solid line represents the calculated H_{c2} using the known band-structure parameters. After Ref. 62.

From these data the magnetic phase diagram shown in Fig. 11 was extracted. From the very steep initial increase of H_{c2} an orbital critical field of roughly 130 T can be estimated. This large value results from the strongly reduced orbital currents for the field orientation parallel to the layers. Towards higher fields, however, H_{c2} levels off towards saturation at lower temperatures. This clearly signals the crossover to Pauli-paramagnetic limitation. The Pauli limit, H_P, can be determined quite accurately from the known gap value, Δ_0, extracted in earlier specific-heat studies.[72,73] Using $\Delta_0/k_BT_c = 2.4$,[73] leads to $H_P = 23$ T which fits nicely with the field of ~ 21 T where the sharp first-order phase transition emerges [Fig. 10(b)].

Above this field, the curvature of the H_{c2} line clearly changes. Together with an increased slope of this line, the second phase transition appears. This is favorably in line with the expectations for the existence of an FFLO state. A comparison with the calculated phase diagram is shown by the solid line in Fig. 11. For that, the well-known electronic band-structure parameters of κ-(BEDT-TTF)$_2$Cu(NCS)$_2$ have been considered and s-wave superconductivity has been assumed.[72–74] The latter, however, is not an essential ingredient. The calculated line and the experimental data agree very well except at lower temperatures. This as well as the unexpected small area of the FFLO phase may be caused by a small misalignment of the sample in the experiment. Also, the theoretical description might be improved when taking more detailed information on the electronic parameters into account.

For the 2D organic superconductors based on BEDT-TTF further evidence for the existence of the FFLO state has been gathered. For κ-(BEDT-TTF)$_2$Cu(NCS)$_2$, the results from magnetic-torque studies are shown in Fig. 12.[75] With decreasing temperature the magnetic torque, which is proportional to the magnetization, shows an increasingly sharp transition at

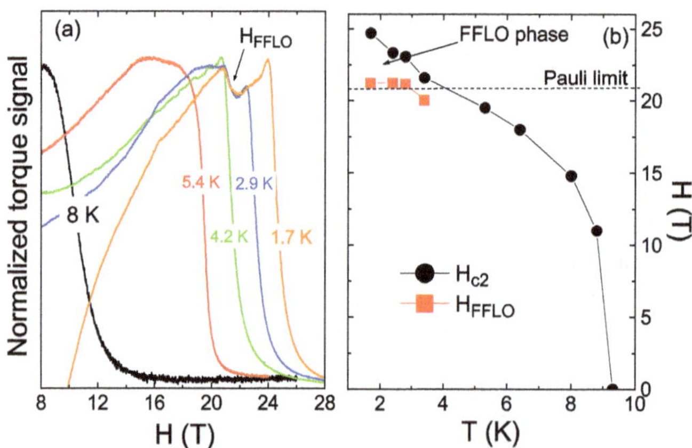

Fig. 12. (a) Field dependence of the magnetic-torque signal of κ-(BEDT-TTF)$_2$Cu(NCS)$_2$ for in-plane magnetic fields at different temperatures. (b) Magnetic phase diagram constructed from the data. After Ref. 75.

H_{c2}. This indicates that the Pauli limit, H_P, is reached. Nevertheless, superconductivity survives even beyond this field for lower temperatures. In addition, a second sharp feature appears in the torque signal below H_{c2}, signaling the evolution of the FFLO phase. The magnetic phase diagram constructed from the torque data is shown in the right panel of Fig. 12.[75] These results nicely confirm the specific-heat data for the H_{c2} line. There is, however, a discrepancy in the extension of the FFLO phase. This might be explained by the much better in-plane alignment of the magnetic field in the torque measurements which was made possible by an *in situ* rotation of the torque platform. The alignment in the specific-heat experiment is estimated to be not better than about 1°.

4.2. *Heterogenous systems*

Besides the still partially controversial evidences for the FFLO state in the homogeneous systems, a quasi-one-dimensional FFLO-like state may exist in artificially layered superconducting/ferromagnetic superlattices.[76] Due to the proximity effect, Cooper pairs from the superconductor can enter the ferromagnetic region. In the spin-split bands of the ferromagnet, however, the wavevectors of the spin-up and spin-down electrons must be largely different resulting in a Cooper pair with finite momentum.[77] Consequently, the real-space pairing function oscillates spatially inside the ferromagnet. Of course, on top of that oscillation the pairing function decays rapidly inside the ferromagnet. However, if the ferromagnetic layer is thin enough, the pairing-function wave may be reflected at the opposite ferromagnetic-layer surface, similar as in a Fabry–Pérot interferometer. Upon changing the ferromagnetic layer thickness, d_F, the pairing functions running oppositely may interfere either constructively or destructively. This eventually may lead to an oscillatory proximity effect, i.e. T_c might oscillate as function of d_F.[78,79] For optimum parameters, this could even result in a reentrant superconductivity.

A first report on the observation of reentrant superconductivity in Fe/V/Fe trilayers as a function of the iron thickness was given in Ref. 80. The T_c oscillation length is given by the magnetic coherence length $\xi_{FO} = \hbar v_F / E_{\mathrm{ex}}$, where E_{ex} is the exchange-splitting energy. For the strong ferromagnet iron, this length scale is very short (0.7–1.0 nm). Indeed, superconductivity in the Fe/V/Fe system vanished at this iron-layer thickness, but the number of experimental points is low and shows a large scatter.[80]

A much clearer evidence for such a reentrant behavior has been given by Zdravkov *et al.* for the bilayer system $Nb/Cu_{1-x}Ni_x$.[81] Alloying the strong ferromagnet nickel by copper reduces E_{ex} strongly which led to $\xi_{FO} = 8.6$ nm for the chosen $Cu_{41}Ni_{59}$ layer. This largely reduces the influence of surface roughness on the interference effects. The thicknesses of the superconducting niobium layers were chosen so that T_c was in the right range to be strongly susceptible on the thickness of the $Cu_{41}Ni_{59}$ layer, d_{CuNi} (inset of Fig. 13). In order to produce samples with different d_{CuNi}, but exactly the same niobium thickness, a wedge-shape ferromagnetic $Cu_{41}Ni_{59}$ layer was deposited on top of the niobium layer (right panel of Fig. 13). After evaporation of a silicon cap layer, samples of equal width were cut perpendicular to the wedge.[81] The resistively determined T_cs as a function of d_{CuNi} for two different niobium layer thicknesses are shown in the left panel of Fig. 13. For the thicker niobium layer a clear minimum in T_c is resolved. For the 7.3 nm thick niobium layer, T_c vanishes at $d_{CuNi} \approx 5$ nm and reappears at about 12 nm. The data can be well described using a model for a clean ferromagnet, i.e. a ferromagnet where the electronic mean-free path is larger than ξ_{FO} (solid lines in Fig. 13).[82,83] Consequently, superconducting/ferromagnetic superlattices seem to be a good realization for studying FFLO-like behavior.

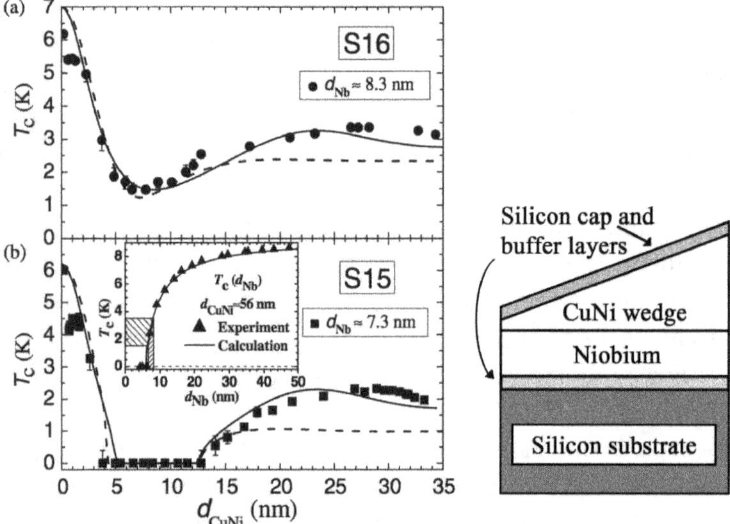

Fig. 13. (left) Critical temperature, T_c, of an (a) 8.3 and (b) 7.3 nm thick niobium layer as a function of the thickness of a ferromagnetic $Cu_{41}Ni_{59}$ layer. The solid and dashed lines are calculations for the clean and dirty cases, respectively. The inset shows T_c as a function of niobium-layer thickness for a 56 nm thick $Cu_{41}Ni_{59}$ layer with the solid line calculated for the clean case. (right) Sketch of the investigated superconductor-ferromagnet bilayer. After Ref. 81.

5. Summary and Outlook

The central focus of the present work are the Fulde–Ferrell–Larkin–Ovchinnikov states which, in addition to off-diagonal long-range order break spatial symmetries. These superfluid phases are suggested by mean-field analyses to occur at low temperatures in multi-component fermionic superfluids with population imbalance. The theoretical considerations apply to a wide variety of systems including metals, ultra-cold atoms, nuclear matter, and dense quark systems. The discussion presented here refers to the weakly interacting BCS regime where the systems mentioned are all described by essentially the same microscopic theory. This fact implies that the knowledge and insights gained by studying superconducting metals in high magnetic fields or ultra-cold atoms can provide explanations for phenomena observed in nuclear or astrophysical contexts.

There are indications that FFLO phases are realized in some highly anisotropic superconductors. Especially, for the low-dimensional organic superconductors, a number of experimental facts point to the existence of such an inhomogeneous state, whereas the situation is much more controversial for heavy-fermion materials. The main criteria employed to identify perspective candidate systems have been the upturn of the Pauli-limited critical field and the appearance of a first-order transition from the normal to the superconducting state at low temperatures as well as the existence of transitions within the superconducting phase. Besides these homogeneous systems, FFLO-like behavior may exist in artificial superconducting/ferromagnetic superlattices that show an oscillating or even reentrant superconducting phase transition as a function of the ferromagnetic layer thickness.

An unambiguous identification of FFLO states requires a better qualitative understanding of superfluids with periodically textured order parameters. The investigations should concentrate on characteristic features which are tied to the quantum nature of the order parameter and reflect its periodicity. Prominent candidates are Andreev bound-states which should form bands in a superconductor with a periodically modulated order parameter. The spatial variation of the latter should give rise to characteristic van Hove-type structures in the corresponding spectral density. The latter can be observed by tunneling spectroscopy. In addition, it should affect transport properties.

The interpretation of experimental data needs detailed quantitative reference calculations which account for realistic quasiparticle dispersion and include quasiparticle interactions. An appropriate theoretical treatment will require new effective calculational tools. In particular, it is highly desirable to have methods which allow for direct "unbiased" minimization of the free energy. Finally, the role of (quantum) fluctuations should be considered.

We hope that this article will stimulate further work on the exciting subject of fermionic superfluids with population imbalance and the FFLO phases.

Acknowledgments

We have benefitted from discussions with many colleagues, especially we would like to thank P. Fulde, R. Lortz, A. D. Bianchi, B. Bergk, G. Eilenberger, M. Sigrist, L. N. Bulaevskii, M. R. Beasley and D. J. Scalapino for stimulating discussions. G. Zwicknagl acknowledges the hospitality of the Kavli Institute for Theoretical Physics where part of the manuscript was completed. This work was supported in part by the National Science Foundation under Grant No. PHY05-51164 and by EuroMagNET (EU contract No. 228043).

References

1. P. Fulde and R. A. Ferrell, *Phys. Rev.* **135**, A550 (1964).
2. A. I. Larkin and Y. N. Ovchinnikov, *JETP (U.S.S.R.)* **47**, 1136 (1964).
3. Y. Matsuda and H. Shimahara, *J. Phys. Soc. Jpn.* **76**, 051005 (2007).
4. A. I. Buzdin, *Rev. Mod. Phys.* **77**(3), 935 (2005).
5. D. E. Sheehy and L. Radzihovsky, *Ann. Phys.* **322**, 1790 (2007).
6. R. Casalbuoni and G. Nardulli, *Rev. Mod. Phys.* **76**, 263 (2004).
7. B. S. Chandrasekhar, *Appl. Phys. Lett.* **1**, 7 (1962).
8. A. M. Clogston, *Phys. Rev. Lett.* **9**, 266 (1962).
9. G. Sarma, *J. Phys. Chem. Solids.* **24**, 1029 (1963).
10. D. Saint-James, D. Sarma and E. J. Thomas, *Type II Superconductivity* (Pergamon, New York, 1969).
11. E. Gubankova, W. V. Liu and F. Wilczek, *Phys. Rev. Lett.* **91**, 032001 (2003).
12. A. A. Abrikosov, *Zh. Eksp. Teor. Fiz.* **32**, 1442 (1957).
13. K. Maki, *Phys. Rev. B* **148**, 362 (1966).
14. P. Fulde, *Adv. Phys.* **22**, 667 (1973).
15. V. P. Mineev and K. V. Samokhin, *Introduction to Unconventional Superconductivity* (Gordon and Breach Science Publishers, 1999).
16. G. Eilenberger, in *Supraleitung und verwandte Quantenphänomene*, Kernforschungsanlage Jülich, Ferienkurs (1988), ed. W. Schmitz.
17. J. Bardeen, L. N. Cooper and J. R. Schrieffer, *Phys. Rev.* **108**, 1175 (1957).
18. J. Bardeen, *Rev. Mod. Phys.* **34**, 667 (1962).
19. G. Zwicknagl and P. Fulde, *Z. Phys.* **43**, 23 (1981).
20. P. Fulde and G. Zwicknagl, *J. Appl. Phys.* **53**, 8064 (1982).

21. W. Baltensperger and S. Straessler, *Phys. Kondens. Materie* **1**, 20 (1963).
22. R. Combescot and C. Mora, *Phys. Rev. B* **71**, 144517 (2005).
23. K. Maki and H. Won, *Czech. J. Phys.* **46**, 1035 (1996).
24. G. Eilenberger, *Z. Phys.* **214**, 195 (1968).
25. A. I. Larkin and Y. N. Ovchinnikov, *Zh. Eksp. Teor. Fiz.* **55**, 2262 (1968).
26. G. M. Eliashberg, *Soviet Physics JETP* **34**, 668 (1972).
27. P. G. deGennes, *Rev. Mod. Phys.* **36**, 225 (1964).
28. Y. Nambu, *Phys. Rev.* **117**, 648 (1960).
29. L. P. Gorkov, *Zh. Eksp. Teor. Fiz.* **36**, 1918 (1959).
30. J. W. Serene and D. Rainer, *Phys. Rep.* **101**, 221 (1983).
31. H. Burkhardt and D. Rainer, *Ann. Phys.* **3**, 181 (1994).
32. S. Matsuo, S. Higashitani, Y. Nagato and K. Nagai, *J. Phys. Soc. Jpn.* **67**, 280 (1998).
33. A. B. Vorontsov, J. A. Sauls and M. J. Graf, *Phys. Rev. B* **72**, 184501 (2005).
34. H. Shimahara, *J. Phys. Soc. Jpn.* **67**, 1872 (1998).
35. H. Shimahara and D. Rainer, *J. Phys. Soc. Jpn.* **66**, 3591 (1997).
36. K. Yang and S. L. Sondhi, *Phys. Rev. B* **57**, 8566 (1998).
37. J. A. Bowers and K. Rajagopal, *Phys. Rev. D* **66**, 065002 (2002).
38. A. I. Larkin and A. A. Varlamov, in *Superconductivity*, eds. K. H. Bennemann and J. B. Ketterson, Vol. I (Springer, Berlin, 2008), p. 369.
39. K. V. Samokhin and M. S. Mar'enko, *Phys. Rev. B* **73**, 144502 (2006).
40. R. B. Diener, R. Sensarma and M. Randeria, *Phys. Rev. A* **77**, 023626 (2008).
41. A. Feiguin and F. Heidrich-Meisner, *Phys. Rev. B* **76**, 220508(R) (2007).
42. A. E. Feiguin and D. A. Huse, *Phys. Rev. B* **79**, 100507(R) (2009).
43. G. G. Batrouni, M. H. Huntley, V. G. Rousseau and R. T. Scalettar, *Phys. Rev. Lett.* **100**, 116405 (2008).
44. M. Casula, D. M. Ceperley and E. J. Mueller, *Phys. Rev. A* **78**, 033607 (2008).
45. K. Maki and T. Tsuneto, *Prog. Theor. Phys.* **31**, 945 (1964).
46. L. W. Gruenberg and L. Gunther, *Phys. Rev. Lett.* **16**, 996 (1966).
47. K. Gloos, R. Modler, H. Schimanski, C. D. Bredl, C. Geibel, F. Steglich, A. I. Buzdin, N. Sato and T. Komatsubara, *Phys. Rev. Lett.* **70**, 501 (1993).
48. A. D. Huxley, C. Paulsen, O. Laborde, J. L. Tholence, D. Sanchez, A. Junod and R. Calemczuk, *J. Phys.: Condens. Matter* **5**, 7709 (1993).
49. F. Thomas, B. Wand, T. Lühmann, P. Gegenwart, G. R. Stewart, F. Steglich, J. P. Brison, A. Buzdin, L. Glémot and J. Flouquet, *J. Low Temp. Phys.* **102**, 117 (1996).
50. A. Bianchi, R. Movshovich, N. Oeschler, P. Gegenwart, F. Steglich, J. D. Thompson, P. Pagliuso and J. L. Sarrao, *Phys. Rev. Lett.* **89**, 137002 (2002).
51. A. Bianchi, R. Movshovich, C. Capan, P. Pagliuso and J. Sarrao, *Phys. Rev. Lett.* **91**, 187004 (2003).
52. H. A. Radovan, N. A. Fortune, T. P. Murphy, S. T. Hannahs, E. C. Palm, S. W. Tozer and D. Hall, *Nature* **425**, 51 (2003).
53. M. Kenzelmann, T. Strässle, C. Niedermayer, M. Sigrist, B. Padmanabhan, M. Zolliker, A. D. Bianchi, R. Movshovich, E. D. Bauer, J. L. Sarrao and J. D. Thompson, *Science* **321**, 1652 (2008).
54. C. Petrovic, P. G. Pagliuso, M. F. Hundley, R. Movshovich, J. L. Sarrao, J. D.

Thompson, Z. Fisk and P. Monthoux, *J. Phys.: Condens. Matter* **13**, L337 (2001).
55. A. D. Bianchi, M. Kenzelmann, L. DeBeer-Schmitt, J. S. White, E. M. Forgan, J. Mesot, M. Zolliker, J. Kohlbrecher, R. Movshovich, E. D. Bauer, J. L. Sarrao, Z. Fisk, C. Petrović and M. R. Eskildsen, *Science* **319**, 177 (2008).
56. C. Capan, A. Bianchi, R. Movshovich, A. D. Christianson, A. Malinowski, M. F. Hundley, A. Lacerda, P. G. Pagliuso and J. L. Sarrao, *Phys. Rev. B* **70**, 134513 (2004).
57. C. Martin, C. C. Agosta, S. W. Tozer, H. A. Radovan, E. C. Palm, T. P. Murphy and J. L. Sarrao, *Phys. Rev. B* **71**, 020503 (2005).
58. J. Spehling, R. H. Heffner, J. E. Sonier, N. Curro, C. H. Wang, B. Hitti, G. Morris, E. D. Bauer, J. L. Sarrao, F. J. Litterst and H.-H. Klauss, *Phys. Rev. Lett.* **103**, 237003 (2009).
59. M. Mierzejewski, A. Ptok and M. M. Maśka, *Phys. Rev. B* **80**, 174525 (2009).
60. Y. Yanase and M. Sigrist, *J. Phys. Soc. Jpn.* **78**, 114715 (2009).
61. G. Koutroulakis, J. M. D. Stewart, V. F. Mitrović, M. Horvatić, C. Berthier, G. Lapertot and J. Flouquet, *Phys. Rev. Lett.* **104**, 087001 (2010).
62. R. Lortz, Y. Wang, A. Demuer, P. H. M. Böttger, B. Bergk, G. Zwicknagl, Y. Nakazawa and J. Wosnitza, *Phys. Rev. Lett.* **99**, 187002 (2007).
63. J. Singleton, J. Symington, M. Nam, A. Ardavan, M. Kurmoo and P. Day, *J. Phys.: Condens. Matter* **12**, L641 (2000).
64. M. Tanatar, T. Ishiguro, H. Tanaka and H. Kobayashi, *Phys. Rev. B* **66**, 134503 (2002).
65. S. Uji, H. Shinagawa, T. Terashima, T. Yakabe, Y. Terai, M. Tokumoto, A. Kobayashi, H. Tanaka and H. Kobayashi, *Nature* **410**, 908 (2001).
66. L. Balicas, J. S. Brooks, K. Storr, S. Uji, M. Tokumoto, H. Tanaka, H. Kobayashi, A. Kobayash, V. Barzykin and L. P. Gor'kov, *Phys. Rev. Lett.* **87**, 067002 (2001).
67. V. Jaccarino and M. Peter, *Phys. Rev. Lett.* **9**, 290 (1962).
68. S. Uji, T. Terashima, M. Nishimura, Y. Takahide, T. Konoike, K. Enomoto, H. Cui, H. Kobayashi, A. Kobayashi, H. Tanaka, M. Tokumoto, E. S. Choi, T. Tokumoto, D. Graf and J. S. Brooks, *Phys. Rev. Lett.* **97**, 157001 (2006).
69. L. Bulaevskii, A. Buzdin and M. Maley, *Phys. Rev. Lett.* **90**, 067003 (2003).
70. K. Cho, B. E. Smith, W. A. Coniglio, L. E. Winter, C. C. Agosta and J. Schlueter, *Phys. Rev. B* **79**, 220507(R) (2009).
71. S. Yonezawa, S. Kusuba, Y. Maeno, P. Auban-Senzier, C. Pasquier and D. Jérome, *J. Phys. Soc. Jpn.* **77**, 054712 (2008).
72. J. Müller, M. Lang, R. Helfrich, F. Steglich and T. Sakai, *Phys. Rev. B* **65**, 140509 (2002).
73. J. Wosnitza, S. Wanka, J. Hagel, M. Reibelt, D. Schweitzer and J. Schlueter, *Synth. Met.* **133–134**, 201 (2003).
74. H. Elsinger, J. Wosnitza, S. Wanka, J. Hagel, D. Schweitzer and W. Strunz, *Phys. Rev. Lett.* **84**, 6098 (2000).
75. B. Bergk, A. Demuer, I. Sheikin, Y. Wang, J. Wosnitza, Y. Nakazawa and R. Lortz, *Physica C*, in press, arXiv:0907.3769 (2010).
76. C. L. Chien and D. H. Reich, *J. Magn. Magn. Mat.* **200**, 83 (1999).

77. E. A. Demler, G. Arnold and M. R. Beasley, *Phys. Rev. B* **55**, 15174 (1997).
78. J. S. Jiang, D. Davidović, D. H. Reich and C. L. Chien, *Phys. Rev. Lett.* **74**, 314 (1995).
79. Z. Radović, M. Ledvij, L. Dobrosavljević-Grujić, A. I. Buzdin and J. R. Clem, *Phys. Rev. B* **44**, 759 (1991).
80. I. A. Garifullin, D. A. Tikhonov, N. N. Garif'yanov, L. Lazar, Y. V. Goryunov, S. Y. Khlebnikov, L. R. Tagirov, K. Westerholt and H. Zabel, *Phys. Rev. B* **66**(2), 020505 (2002).
81. V. Zdravkov, A. Sidorenko, G. Obermeier, S. Gsell, M. Schreck, C. Müller, S. Horn, R. Tidecks and L. R. Tagirov, *Phys. Rev. Lett.* **97**, 057004 (2006).
82. A. S. Sidorenko, V. I. Zdravkov, A. Prepelitsa, C. Helbig, Y. Luo, S. Gsell, M. Schreck, S. Klimm, S. Horn, L. R. Tagirov and R. Tidecks, *Ann. Phys.* **12**, 37 (2003).
83. L. R. Tagirov, *Physica C* **307**, 145 (1998).

III. New Superconductors

PREDICTING AND EXPLAINING T_c AND OTHER PROPERTIES OF BCS SUPERCONDUCTORS

Marvin L. Cohen
Department of Physics, University of California,
and
Materials Sciences Division,
Lawrence Berkeley National Laboratory,
Berkeley, CA 94720, USA
mlcohen@berkeley.edu

After providing some history and background material regarding the evolution of research on superconductivity, a description of the use of the BCS theory for the development of approaches for calculations and predictions of superconducting properties will be given. The emphasis is on estimates of T_c raising T_c, and predicting new superconductors. The basic analysis will be based on the BCS formalism with modern extensions. The focus here is on phonon mediated pairing of electrons as described in the original BCS paper augmented by current modifications and the use of modern calculational approaches.

1. Introduction

The famous photograph (Fig. 1) of the attendees at the 1911 Solvay Conference was taken at the Metropole Hotel in Brussels in the fall of that year. Heike Kamerlingh Onnes is shown standing next to Albert Einstein. One can speculate that the smile on Onnes' face reflects a very happy period in his research career since superconductivity was discovered[1] in his laboratory in the spring of 1911, and one can also wonder about what went on during the conversations between Onnes and Einstein on superconductivity at that time. It has been reported that Einstein expressed doubt that a general theory explaining Onnes' observations could be developed. It took 46 years. The BCS paper[2] was published in 1957, which was unfortunately after Einstein's death in 1955.

The theory has been extremely successful. It explained all the properties of *conventional* superconductors and provided concepts and formalisms that

Fig. 1. Photograph of the attendees of the 1911 Solvay Conference taken in the Metropole Hotel. Kamerlingh Onnes (third from right) stands next to Einstein (second from right).

eventually found their way into many areas of physics such as nuclear structure and the physics of Bose–Einstein condensates. The concept of pairing introduced by Leon Cooper[3] was an essential ingredient of the BCS theory as was the mechanism for the use of lattice vibrations to provide the attractive pairing interaction. Measurements of the isotope effects on T_c provided the stimulus for considering lattice vibrations, and theoretical studies such as the one by John Bardeen and David Pines[4] explicitly demonstrated the attractive phonon-induced pairing interaction and its form. The many-electron features of the BCS theory evolved from the many-body wave function suggested by J. Robert Schrieffer.[5] The details have been well documented elsewhere, and the BCS theory has joined the ranks of great theories and conceptual breakthroughs of twentieth century physics.

Here the focus will be on the electron pairing interaction, the so-called *mechanism* of superconductivity, and on the use of BCS theory to solve for the material properties of superconductivity and, in particular, T_c.

2. Materials and T_c

After Onnes' discovery of superconductivity in Hg, many materials were tested. Empirical rules were established to help find more superconductors. It was soon found that only metals exhibited superconductivity. Another example of a rule of thumb was that bad conductors were better candidates for superconductivity than good conductors. After BCS, this latter trend was attributed to the argument that the stronger electron-phonon scattering that gave rise to higher resistivities would be helpful for inducing the pairing interaction. Some rules have been dropped, like the one that suggested avoiding oxides. The earliest exception for the oxide rule was the discovery of superconductivity in $SrTiO_3$.[6] This rule was later found to be particularly misguided since copper oxide superconductors[7] have been found to yield the highest T_c materials found thus far. Another rule was to avoid *magnetism related* elements such as Fe. Again, this rule has been broken with the new results found for Fe compounds.[8]

Despite such shortcomings of the empirical approaches of experimentalists searching for new superconductors (sometimes referred to as *superconductivity alchemists*), they were the ones who did find the new superconductors. The highest T_c's went up at a steady average rate of 1 K every three years during the 75-year time period from 1911 to 1986. Despite the theoretical advances by the Londons[9] in 1935, Ginzburg and Landau[10] in 1950, and BCS in 1957, no changes in the time dependence of the maximum T_c or the frequency of discoveries of new superconductors occurred at those times. Theory did not lead in this area.

To discuss the role that theory did have, it is convenient to divide superconducting materials into classes 1 and 2, as shown in Fig. 2. Class 1 contains *conventional BCS superconductors* like Al, Sn, Pb, Hg, etc. For these materials, there are many experimental tests showing in detail that the BCS theory works. Similar tests and arguments can be made for the other members of this class, such as C_{60}-based[11] superconductors, some organic superconductors, degenerate semiconductors,[12] and MgB_2[13] which is the highest transition temperature member of class 1 known at this time. A theoretical description of their superconducting properties based on electron pairing induced by electron-phonon coupling works extremely well for class 1, and it is possible to use the theory to predict new superconductors.

For class 2, the high T_c copper oxides are the most studied. At this time, the highest T_c's achieved for these systems[14,15] are approximately 135 K at atmospheric pressure and 165 K at high pressures. Heavy fermion metals[16] with electronic masses comparable to that of muons have unusual superconducting properties but relatively low T_c's, some organics appear to be in this class, and the interesting, recently discovered Fe-based materials with T_c's up to around 57 K at this time are usually included in class 2. Although some features of class 1 superconductors, such as electron pairing of some

CLASSES OF SUPERCONDUCTORS

1 BCS: conventional metals, C60, some organics, doped semiconductors, MgB2, ...

----

2 EXOTIC: copper oxides, heavy fermion metals, some organics, (Fe,Ni) oxypnictides , FeSe, ...

Fig. 2. Approximate division of superconducting materials into two classes.

kind, is believed to be responsible for superconductivity in this class of materials, the general view is that augmentations or fundamental changes in the BCS theory may be necessary to explain class 2 superconductivity. The search for these modifications has been going on for more than 20 years, however at this time, there is not yet consensus among researchers in this field.

The current situation is summarized in Fig. 3. For class 1, all experimental data have been explained. In some areas of research, such as calculations of T_c, the precision needed has not been achieved. This shortcoming is often attributed to our inability to calculate the needed input to the BCS theory, i.e. the normal state properties, with sufficient precision. However, there is little concern at present with the applicability of the BCS theory for this class of materials.

Although there is lack of consensus for an overall theory for class 2, there has been some very interesting and imaginative theoretical research in this area. The search for a general theory, particularly for the cuprates, has led to a deeper understanding of spin related properties and the physics of non-Fermi liquid systems.[17] Although some researchers have ruled out the electron-phonon induced superconducting mechanism for the cuprates, others have not.[18] One early claim often heard was that this mechanism generally could not account for such high T_c's. For the range of T_c's found thus far, this is likely to be an incorrect argument. However, the current status is that most researchers in this area believe phonon induced pairing in the cuprates is not the primary mechanism.

CURRENT SITUATION

CLASS 1--- ELECTRON PAIRS
PAIRING INDUCED BY ELECTRON-PHONON INTERACTIONS.
ALL EXPERIMENTS EXPLAINED.

CLASS 2---ELECTRON PAIRS
MANY THEORIES, BUT NO CONSENSUS ON THE MECHANISM

Fig. 3. Synopsis of the current status of theory for the two classes of superconductors.

What about the highest T_c's? The current upper limit record of 165 K is found in a pressurized cuprate. This has been the record for about 15 years. For class 1 materials, the fullerene systems are pushing toward 40 K, and T_c is around 40 K for MgB_2. At this point, it is difficult to predict whether the next record breaking T_c will be for a material in class 1 or class 2. An important consideration when posing the question of higher T_c's is that there are promising possible high T_c systems within class 1, and the theory is well established and to a great extent understood for those materials. So why has it been so difficult for theory to predict new superconductors within this class and to raise T_c? As mentioned above, the answer to this question is primarily related to the fact that a precise calculation of the pairing interaction requires a precise description of the normal state. Success in this area began with the use of experimentally determined parameters as input. The extension to more *ab initio* approaches came later when the theory of the normal state became more reliable.[19] At that point, successful predictions were made because structural phases and electron-phonon couplings could be evaluated from first principles. By the time superconductivity was discovered in MgB_2, many theorists had access to highly developed methods to calculate normal-state properties, so the microscopic origins of the superconducting properties could be considered.[20] This research had added benefits. Because of the particular nature of the phonon-induced pairing in MgB_2, new perspectives arose and a great deal was learned. Similar situations[22] occurred related to the research on fullerene-based superconductors.

3. *Before and After the Fact* Calculations

The Yogi Berra statement that "predicting is hard especially about the future" is apropos when it comes to new superconductors. Hence calculations after an experimental discovery are easier since it is very helpful to know the possible range of parameters involved and to know the answer. When a new candidate system is chosen, there are often many unknown variables to consider. This is true even when working with class 1 materials.

If we focus on estimates of T_c and the question of whether superconductivity will exist in potential candidates for new class 1 superconductors, we are led to examine success stories in the past. Just as the alchemists developed useful rules and relationships for making useful materials, the so-called superconductivity alchemists had useful rules related to the electron to atom ratio and similar material properties. The successes of these approaches were later examined in the light of the BCS theory. A common conclusion was that the successful searches for higher T_c materials involved finding systems with high densities of electron states at the Fermi level, $N(0)$, and materials which were near structural instabilities. These re-evaluations in the *light of theory* were based primarily on the solution for T_c using a simplified model to solve the BCS gap equation. This model was similar to that used by Cooper in his study of a single electron pair. The important point is that the net attractive pairing interaction parameter in this model is directly proportional to $N(0)$. The reason for lattice instability considerations is because the attractive phonon induced pairing interaction parameter often results in a softening of phonons, and when phonon frequencies become negative, crystals have structural instabilities. The formula for T_c discussed above based on the *BCS model* assumed a weak coupling pairing interaction $N(0)V$ where V is the net attractive pairing potential. As shown in Fig. 4, the exponential dependence of T_c on $N(0)$ is the reason for the strong focus on this quantity. The interaction V is composed of an attractive phonon component and the repulsive Coulomb potential. These compete as seen in the formula for the transition temperature T_c based on the BCS model,

$$T_c \sim \mathrm{E(phonon)} \exp[-1/(\lambda - \mu^*)] \,. \tag{1}$$

Here E(phonon) represents an average energy of the phonons such as the Debye temperature, and the net interaction in the exponent is the difference between the dimensionless attractive phonon pairing interaction λ and the modified repulsive electron-electron interaction μ^*.

These are the two components of $N(0)V$, i.e. $N(0)V$ can be viewed as $\lambda - \mu^*$. Superconductivity is expected when this difference is positive. The

T_c formulas

$$T_c = 1.14\omega_{ph} \exp\left(\frac{-1}{N(0)V}\right)$$ BCS, 1957

$$T_c = \frac{T_D}{1.45} \exp\left(-\frac{1.04(1+\lambda)}{\lambda - \mu^*(1+0.62\lambda)}\right)$$ McMillan, 1968

$$T_c = \frac{\langle\omega\rangle_{\log}}{1.20} \exp\left(-\frac{1.04(1+\lambda)}{\lambda - \mu^*(1+0.62\lambda)}\right) \quad \text{for } \lambda < 1.5$$

$$T_c = 0.183\sqrt{\lambda\langle\omega^2\rangle} \quad \text{for } \lambda > 10,\ \mu^* = 0$$ Allen and Dynes, 1975

Or the Eliashberg equations can be solved numerically once the Spectral Function has been calculated:

$$\lambda \equiv 2\int_0^{\omega_{\max}} \alpha^2 F(\omega)\omega^{-1} d\omega$$

Can include anisotropic electrons, anharmonic phonons, etc...

Fig. 4. Some of the formulas discussed in the text for calculating the superconducting transition temperature.

Coulomb interaction is modified (hence the asterisk superscript) because of the difference in energy scales between the two pairing potentials.[23]

To include the effects of renormalization and of stronger coupling, McMillan[24] extended the coupling range to be applicable when the electron-phonon parameter λ became larger and renormalization of this parameter is included.

The McMillan equation (Fig. 4) can be used for values of λ up to about 1.25. The extensions of the McMillan equation beyond the above limit led to incorrect estimates of the maximum transition temperature, $T_c^{\max}$. This gave a pessimistic view since the resulting expression for $T_c(\lambda)$ saturates. Later, Allen and Dynes[25] solved the Eliashberg[26] equations for several model phonon spectra and found that, for large λ, T_c was proportional to $\sqrt{\lambda}$ with no saturation.

Modern calculations of T_c are usually done numerically where the modified BCS theory, as embodied in the Eliashberg equations, gives precise estimates when precise normal state properties are used as input. There are model solutions that result in simple formulas (for $\mu = 0$) such as the Kresin–Barbee–Cohen[27,28] model based on a flat phonon spectrum

$$T_c = 0.25\sqrt{\langle\omega^2\rangle}(e^{2/\lambda} - 1)^{-1/2} \tag{2}$$

where $\langle\omega^2\rangle$ is an appropriately averaged square of the phonon frequency. This expression gives the correct $\sqrt{\lambda}$ limit at high λ and the exponential BCS model solution for small λ.

The shift in searching for superconductors and higher T_c superconductors, in particular, by replacing the search for large $N(0)$ materials to looking for large λ's was next. The separation suggested by the NV parameter is not appropriate for most materials. Most systems are not modeled well by the BCS model solution. A possible exception is the C_{60} superconductors[29] where it can be argued that N and V are for the most part independent of each other. The phonons responsible for the pairing are associated with the intramolecular vibrations of C_{60}. The N factor comes in through the electrons that originate from the alkali metals used for doping. Experimentally, increasing the lattice constant to increase N results in an increase in T_c in an expected way. Increasing N by exploring metal components with d-electrons has also been suggested.

Another aspect of the fullerene systems is the proposal to increase V by considering carbon molecules with higher curvature such as the case for C_{36}. Higher curvature is expected to yield stronger coupling between the intramolecular vibrations and the itinerant electrons. A consequence of this argument is that it suggests a trend in which this coupling would be weakest in graphite, stronger for carbon nanotubes, and even stronger for C_{60} and C_{36}.[29] This is consistent with the limited measurements obtained up to this point. The T_c of graphite is about 0.5 K, carbon nanotubes exhibit superconducting fluctuations around 15 K,[30,31] A_3C_{60} (A represents an alkali metal) shows superconductivity above 30 K. No data is available yet for C_{36} systems.

The measurements on carbon nanotubes are currently being refined. At this point, it appears that the theoretical prediction[30] of an inverse relationship between T_c and the tube diameter is borne out. The data also show some evidence that correlations between tubes are important. This area could yield new insights into the nature of superconductivity for lower dimensional systems.

To appreciate the thinking behind many of the proposals based on BCS theory for achieving higher T_c's, it can be argued that for this theory, most expressions for the transition temperature have the form, $T_c \sim \omega_0 f(\lambda)$, where ω_0 is a frequency related to the energy scale of the boson causing electron pairing and $f(\lambda)$ is a function that increases with λ. This suggests two paths to higher T_c's, increasing ω_0 or increasing λ. For the first path, one wants high frequency phonons which points to exploring compounds with atoms having light masses. Electronic collective excitations have also been suggested because of their high energy. For the second path, the search has been for large λ's. Hence, when considering the shift from looking for large $N(0)$ to large λ's, an important caveat to add is that strong electron-phonon interactions often lead to structural instabilities as described earlier. Therefore, the desired material may not be stable at atmospheric pressure.

4. *Ab Initio* Calculations for Real Material

In the last section, an overview of the approaches used by theorists to explain and predict properties of class 1 superconductors was given. The assumption was that the BCS theory with phonon mediated electron pairing was operative, and the focus was on T_c calculations using approximate solutions of the BCS gap equation in its original form. In early work, sometimes the T_c formulas described in Fig. 4 and Eqs. (1) and (2) were used,[12] and sometimes multiple square well solutions to the BCS gap equation were employed.[32] The input for the normal state parameters such as effective masses and phonon couplings were taken from experiment. The result was several successful predictions[6,12] of new superconductors such as some degenerate semiconductors, GeTe, SnTe, and $SrTiO_3$, where the normal state data used were available because of transport and optical experiments. These successful predictions were surprising because the carrier densities in these systems are so low[6,12,32] relative to standard metals. The fact that the carrier densities could be varied and the availability of reliable normal state data supplied important information made these estimates of T_c possible. Another interesting case[33] is the transition metal compound Nb_3Nb which is Nb in the A15 structure. Here again the availability of data and normal state calculations for Nb and Nb compounds in the A15 structure provided input for the semi-empirical calculation of T_c.

The successes using *ab initio* calculations came with the development of the Eliashberg theory[26] and the advances made in calculating normal state properties.[19] If we restrict ourselves to phonon mediated pairing and Eliashberg theory, then an expression for calculating λ is given in Fig. 4 in terms of the Eliashberg function,

$$\alpha^2 F(\mathbf{k}, \mathbf{k}', \omega) \equiv N(\varepsilon_F) \sum_j |g^j_{\mathbf{kk}'}|^2 \delta(\omega - \omega_{jq}), \tag{3}$$

where $N(\varepsilon_F)$ is the density of states per spin at the Fermi level, $g_{kk'}$ is the electron-phonon matrix element, and ω_{jq} is the frequency of the phonon in jth branch with $q = k - k'$. The expression in Fig. 4 for the Eliashberg function has been averaged over wavevectors k and k' and hence is a function of frequency only, whereas the expression given in Eq. (3) includes the initial and final electron wavevectors.

It is often convenient to solve the Eliashberg equations in the Matsubara representation[34] where

$$\lambda(\mathbf{k},\mathbf{k}',n) \equiv \int_0^\infty d\omega\alpha^2F(\mathbf{k},\mathbf{k}',\omega)\frac{2\omega}{\omega^2+(2n\pi T)^2}\,. \tag{4}$$

Here, the conventional λ is represented by the average

$$\lambda = \langle\lambda(\mathbf{k},\mathbf{k}',0)\rangle\,. \tag{5}$$

The first *ab initio* calculations dealt with the averaged values for the input to the Eliashberg equations, and it was based on a method that was developed for calculating electron-phonon couplings.[35] The application to high pressure Si was a dramatic example of the power of this approach. Using only the atomic number and atomic mass for Si the total energy[36] for candidate crystal structures were compared at various pressures. Several new forms of Si were successfully predicted.[37] Two of these, simple (or primitive) hexagonal, sh, and hcp Si were studied in detail. The existence of these phases, their stability, lattice constants, mechanical properties, electronic structures, vibrational properties and electron-phonon coupling were then determined from first-principles calculations. Using the densities of states at the Fermi energy as a scaling parameter, the Coulomb potential μ^* could be estimated. The result was a prediction[38] of superconductivity for both phases and an estimate of T_c and its pressure dependence. Considering the limited input and successful predictions, it could be said that high pressure Si became the best-understood superconductor.

As was pointed out earlier, MgB_2 was not predicted to be superconducting even though the theoretical tools were available. However, this material proved to be an interesting prototype system to study. For metals with fairly isotropic Fermi surfaces, λ can be related to the normal state mass enhancement which in turn can be extracted from heat capacity data. In contrast, for MgB_2 the anisotropic properties of the Fermi surface have to be considered. This is possible using the theory associated with Eqs. (3) and (4), and the result was an almost *ab initio* calculation of the properties of MgB_2,[20,21,39,40] including the transition temperature, isotope effects, tunneling gaps, specific heat, temperature dependence of the gaps, and other features. The agreement with experiment was excellent, and the calculations demonstrated that specific parts of the Fermi surface played a special role which could be demonstrated by the existence of two superconducting energy gaps. It was shown that a particular phonon mode strongly couples to specific electronic states resulting in a relatively high T_c.

5. Increasing T_c

The quest for higher superconducting transition temperatures began in 1911, and although our theoretical understanding of superconductivity has advanced considerably and successful predictions of new superconductors have been made, especially for class 1 superconductors, we are still far from having a well-defined path for increasing T_c. For strong electron-phonon coupling, the Allen–Dynes equation is more optimistic than the McMillan equation which saturates as the coupling gets stronger. As discussed earlier, strong coupling can lead to lattice instabilities, and it was also shown[41] that for some cases, other issues arise such as the role of the frequency differences between the Coulomb and phonon interactions.

The search for larger electron-phonon couplings also depended on the development of methods capable of calculating the couplings with high precision. The early approach[35] was developed to work with a plane-wave basis set, and calculations at many points in reciprocal space were needed to obtain a good estimate of λ. A recent development[42] which has had, and likely will have, a big impact in this area is the use of Wannier functions to evaluate electron phonon couplings. This real space approach is illustrated schematically in Fig. 5 where an electron in one cell is coupled to an electron in another cell via a vibrating atom in a third cell. This approach allows a detailed calculation of the electron and phonon self-energies illustrated by the diagrams in Fig. 6. The calculations using this approach are equivalent to plane wave calculations at millions of points in reciprocal space. At this point, applications have been made to explore superconductivity in doped diamond[43] and SiC,[44] the relationship between phonon softening and superconductivity,[45] and to conduct tests of the strength of electron-phonon couplings in cuprates[46] and Fe based superconductors.[47] The latter two are important tests since there is still considerable interest in the role of phonons for pairing in these materials. In particular, the *kink* observed[48] in angular resolved photoemission experiments led to speculations that the size of the λ extracted assuming the kink was caused by electron-phonon effects could lead to the observed high T_c. The calculations, however, did not support this view when assuming standard theory for the electron-phonon interactions.

As described earlier, the quest for higher T_c materials has become a search for conditions that will lead to large λ's, either via large densities of states or

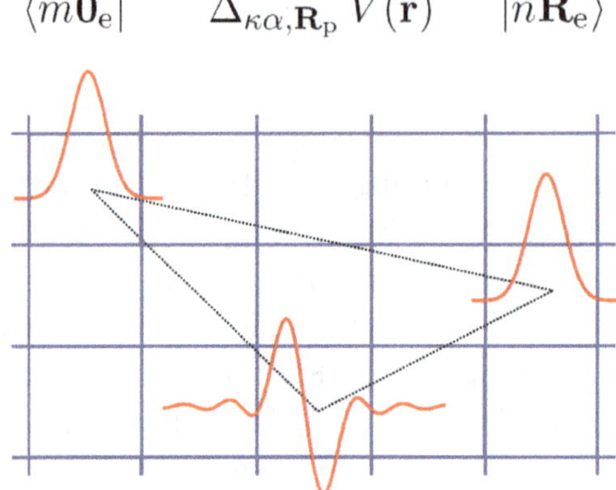

Fig. 5. Schematic diagram illustrating the interaction of two electrons in different crystal cells coupled by a vibrating atom in another cell.

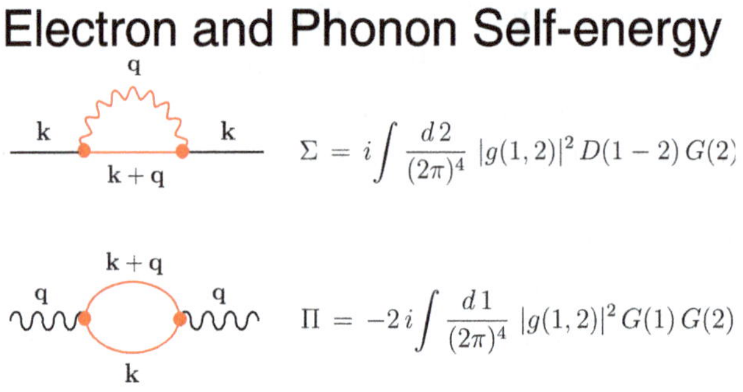

Electron and Phonon Self-energy

q
k k
k + q

$$\Sigma = i \int \frac{d2}{(2\pi)^4} |g(1,2)|^2 D(1-2)\, G(2)$$

k + q
q q
k

$$\Pi = -2i \int \frac{d1}{(2\pi)^4} |g(1,2)|^2 G(1)\, G(2)$$

Fig. 6. Feynman diagrams for electron and phonon self-energies that can be calculated using the Wannier function approach.

large electron-phonon pairing potentials. An interesting approach is to look at λ as a dimensionless ratio of electronic and vibrational spring constants

$$\lambda \sim \frac{N(E_F)\langle I^2\rangle}{M\langle\omega^2\rangle}, \tag{6}$$

where $\langle I^2\rangle$ is the square of the displacement induced electron scattering matrix element averaged over the Fermi surface, $\langle\omega^2\rangle$ is an average of the square of the phonon frequencies, and $N(E_F)$ is the electronic density of states at the Fermi energy.[49] For harmonic systems, the $M\langle\omega^2\rangle$ and λ are roughly independent of M, however perhaps a better view is to look at T_c

for large coupling, where

$$T_c \sim (\lambda \langle \omega^2 \rangle)^{1/2} \sim \left(\frac{\eta}{M}\right)^{1/2} . \tag{7}$$

Equation (7) implies that what is needed for higher T_c's is stronger electronic spring constants η and light masses. This can be extended to explore compounds with multiple masses in a unit cell, and estimates can be made using information from molecular fragments.[50] Roughly speaking, strong electronic spring constants are expected for strong bonds such as the covalent or partially covalent bonds found in diamond and SiC. The problem here is getting enough mobile carriers,[51] but progress is being made.[52]

6. Conclusions

It is clear that for class 1 materials the BCS theory with modern field theoretic extensions is extremely robust and predictive. There are no *counter experiments* this author is aware of. The theory explains the superconducting properties and gives reliable predictions. Although extensions using bosons other than phonons appear promising, there are no experimental demonstrations that superconductivity has been achieved in this manner even for class 2 superconductors. Tests of the phonon mechanism for class 2 materials using standard electron-phonon theory are not encouraging. If phonon induced pairings are important, modifications of the current theory are needed. At this point, it appears that only the electron pairing aspects of the original BCS theory has received the blessing of the superconductivity community for this class. No consensus exists at this time for a correct underlying theory (or combination of theories) appropriate for the members of class 2 superconductors despite the interesting candidates available.

Regarding the quest for higher T_c superconductors, it is not clear whether class 1 or class 2 type materials are more promising. The known materials of class 2 have the higher T_c's, but without a descriptive theory, it is difficult to make quantitative predictions. For class 1, at present MgB_2 and materials based on C_{60} have the highest T_c's. Modifications of MgB_2 suggest[53] modest increases in T_c. For C_{60} systems, recent verifications[54] of theoretical predictions[55] are encouraging. There appear to be no barriers to rule out the notion that class 1 materials can rival class 2 materials for the highest T_c's.

Acknowledgments

This work was supported by the NSF grant DMR07-05941 and Director, Office of Science, Basic Energy Sciences, Materials Sciences and Engineering Division of the U.S. Department of Energy under contract No. DE-AC02-05CH11231.

References

1. H. K. Omnes, *Leiden Comm.*, 120b, 122b, 124c (1911).
2. J. Bardeen, L. N. Cooper and J. R. Schrieffer, *Phys. Rev.* **108**, 1175 (1957).
3. L. N. Cooper, *Phys. Rev.* **104**, 1189 (1956).
4. J. Bardeen and D. Pines, *Phys. Rev.* **99**, 1140 (1955).
5. J. R. Schrieffer, *Theory of Superconductivity* (Benjamin, New York, 1964).
6. J. F. Schooley, W. R. Hosler and M. C. Cohen, *Phys. Rev. Lett.* **12**, 474 (1964).
7. J. G. Bednorz and K. A. Müller, *Z. Phys. B* **64**, 189 (1986).
8. Y. Kamihara, T. Watanabe, M. Hirano and H. Hosono, *JACS* **130**, 3296 (2008).
9. F. London and H. London, *Proc. Roy. Soc. A* **149**, 71 (1935).
10. V. L. Ginsburg and L. D. Landau, *JETP* **20**, 1064 (1950).
11. A. F. Hebard, M. J. Rosseinsky, R. C. Haddon, D. W. Murphy, S. H. Glarum, T. T. M. Palstra, A. P. Ramirez and A. R. Kortan, *Nature* **350**, 549 (2000).
12. M. L. Cohen, in *Superconductivity*, ed. R. D. Parks (Marcel Dekker, Inc., New York, 1969), p. 615.
13. J. Nagamatsu, N. Nakagawa, T. Muranaka, Y. Zenitani and J. Akimitsu, *Nature* **430**, 63 (2001).
14. C. W. Chu, *Nature* **365**, 323 (1993).
15. L. Gao, Y. Xue, F. Chen, Q. Xiong, R. L. Meng, D. Ramirez and C. W. Chu, *Phys. Rev. B* **50**, 4260 (1994).
16. F. Steglich, *Physica Scripta T* **29**, 15 (1989).
17. P. W. Anderson, *Theory of Superconductivity in the High T_c Cuprates* (Princeton Press, Princeton, NJ, 1997).
18. J. C. Phillips, *Phys. Rev. B* **75**, 214503 (2007).
19. M. L. Cohen, *Physica Scripta T* **1**, 5 (1982).
20. H. J. Choi, D. Roundy, H. Sun, M. L. Cohen and S. G. Louie, *Phys. Rev. B* **66**, 020513 (2002).
21. H. J. Choi, D. Roundy, H. Sun, M. L. Cohen and S. G. Louie, *Nature* **418**, 758 (2002).
22. M. L. Cohen and V. H. Crespi, in *Buckminsterfullerenes*, eds. W. E. Billups and M. A. Ciufolini (VCH Publishers, New York, 1993), p. 197.
23. N. N. Bogoliubov, V. V. Tolmachev and D. V. Shirkov, *A New Method in the Theory of Superconductivity* (Consultants Bureau, Inc., New York, 1959).
24. W. L. McMillan, *Phys. Rev.* **167**, 331 (1968).
25. P. B. Allen and R. C. Dynes, *Phys. Rev. B* **12**, 905 (1975).
26. G. M. Eliashberg, *JETP* **11**, 696 (1960).
27. V. Z. Kresin, *Bull. Am. Phys. Soc.* **32**, 796 (1987).

28. L. C. Bourne, A. Zettl, T. W. Barbee III and M. L. Cohen, *Phys. Rev. B* **36**, 3990 (1987).
29. M. L. Cohen, in *High Temperature Superconductivity*, eds. S. E. Barnes, J. Askenazi, J. L. Cohn and F. Zuo (Am. Inst. Phys., New York, 1999), p. 359.
30. L. X. Benedict, V. H. Crespi, S. G. Louie and M. L. Cohen, *Phys. Rev. B* **52**, 14935 (1995).
31. Z. K. Tang, L. Zhang, N. Wang, X. X. Zhang, G. H. Wen, G. D. Li, J. N. Wang, C. T. Chan and P. Sheng, *Science* **292**, 2462 (2001).
32. M. L. Cohen, *Phys. Rev.* **134**, A511 (1964).
33. W. E. Pickett, K. M. Ho and M. L. Cohen, *Phys. Rev. B* **19**, 1734 (1979).
34. P. B. Allen and B. Mikovic, in *Solid State Physics*, eds. H. Ehrenreich, F. Seitz and D. Turnbull, Vol. 32 (Academic Press, New York, 1982), p. 1.
35. M. M. Dacorogna, M. L. Cohen and P. K. Lam, *Phys. Rev. Lett.* **55**, 837 (1985).
36. J. Ihm, A. Zunger and M. L. Cohen, *J. Phys. C* **12**, 4409 (1979). [Erratum: *J. Phys. C* **13**, 3095 (1980)].
37. M. T. Yin and M. L. Cohen, *Phys. Rev. Lett.* **45**, 1004 (1980).
38. K. J. Chang, M. M. Dacorogna, M. L. Cohen, J. M. Mignot, G. Chouteau and G. Martinez, *Phys. Rev. Lett.* **54**, 2375 (1985).
39. J. M. An and W. E. Pickett, *Phys. Rev. Lett.* **86**, 4366 (2001).
40. A. Liu, I. I. Mazin and J. Kortis, *Phys. Rev. Lett.* **87**, 087005 (2002).
41. M. L. Cohen and P. W. Anderson, in *Superconductivity in d- and f-band Metals*, ed. D. H. Douglass (Am. Inst. Phys., New York, 1972), p. 17.
42. F. Giustino, M. L. Cohen and S. G. Louie, *Phys. Rev. B* **76**, 165108 (2007).
43. F. Giustino, J. R. Yates, I. Souza, M. L. Cohen and S. G. Louie, *Phys. Rev. Lett.* **98**, 047005 (2007).
44. J. Noffsinger, F. Giustino, S. G. Louie and M. L. Cohen, *Phys. Rev B* **79**, 104511 (2009).
45. J. Noffsinger, F. Giustino, S. G. Louie and M. L. Cohen, *Phys. Rev. B* **77**, 180507 (2008).
46. F. Giustino, M. L. Cohen and S. G. Louie, *Nature* **452**, 975 (2008).
47. J. Noffsinger, F. Giustino, S. G. Louie and M. L. Cohen, *Phys. Rev. Lett.* **102**, 147003 (2009).
48. A. Lanzara, P. V. Bogdanov, X. J. Zhou, S. A. Kellar, D. L. Feng, E. D. Lu, T. Yoshida, H. Elsaki, A. Fujimori, K. Kishio, J.-I. Shimoyama, T. Noda, S. Uchida, Z. Hussain and Z.-W. Shen, *Nature* **412**, 510 (2001).
49. J. J. Hopfield, *Phys. Rev. B* **186**, 443 (1969).
50. J. E. Moussa and M. L. Cohen, *Phys. Rev. B* **78**, 064502 (2008).
51. J. E. Moussa and M. L. Cohen, *Phys. Rev. B* **77**, 064518 (2008).
52. E. A. Ekimov, V. A. Sidorov, E. D. Bauer, N. N. Mel'nik, N. J. Curro, J. D. Thompson and S. M. Stishov, *Nature* **428**, 542 (2004).
53. H. J. Choi, S. G. Louie and M. L. Cohen, *Phys. Rev. B* **80**, 064503 (2009).
54. A. Y. Ganin, Y. Takabayashi, Y. Z. Khimyak, S. Margadonna, A. Tamai, M. J. Rosseinsky and D. Prassides, *Nature Materials* **7**, 367 (2008).
55. S. Saito, K. Umemoto, S. G. Louie and M. L. Cohen, *Solid State Comm.* **130**, 335 (2004).

THE EVOLUTION OF HTS: T_c-EXPERIMENT PERSPECTIVES

C. W. Chu

University of Houston,
Lawrence Berkeley National Laboratory
**Hong Kong University of Science and Technology*

Dedicated to John Bardeen, Leon Cooper and Bob Schrieffer, the architects of superconductivity.

The rise of the superconducting transition temperature T_c has been reviewed in three major superconducting systems: the cuprate, the Fe-pnictide and the heavy fermion. While the first two systems display high T_cs, heavy fermion superconductors show low T_c but embody many crucial features found in the others. The prospect of future superconductors with higher T_c, preferably close to room temperature, is also discussed. Those interested in the detailed physics of high temperature superconductivity are referred to the article by E. Abrahams in the next chapter of this book and reviews published elsewhere.

1. Introduction

The search for superconductors with a higher transition temperature (T_c) has been one of the major driving forces in the long-sustained research effort on superconductivity ever since its discovery in 1911.[1] The effort has led to the rise of T_c and to the discoveries of novel compounds, both superconducting and non-superconducting; new physics, such as non-Fermi liquid behavior in cuprates superconductors; and new phenomena, such as heavy electrons in lanthanide compounds.

Traditionally, high temperature superconductor (HTS) is a relative term referring to superconducting compounds with a T_c comparable to the existing T_c-record, which rises with time. Before 1953, any superconductors with a T_c comparable to 9 K of Nb were considered to be HTSs. However, after the discovery of the A15 inter-metallic superconductors, only those with a

*Until September 1, 2009.

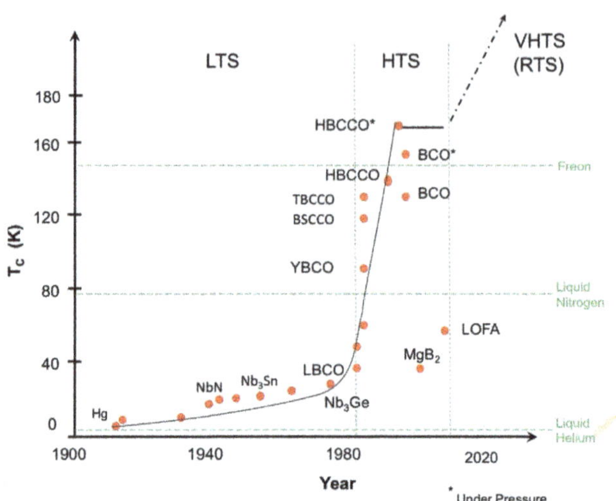

Fig. 1. The evolution of T_c with time and the three periods, i.e. low temperature superconductivity (LTS), high temperature superconductivity (HTS) and very high temperature superconductivity (VHTS) or room temperature superconductivity (RTS).

T_c close to 23 K of Nb3Ge were treated as HTSs. Following the discovery of cuprate superconductors in 1986, HTSs have been more or less referred to superconductors with a T_c above 23 K and sometimes only to those with a T_c above 77 K, the liquid nitrogen boiling temperature.

The rise of T_c with time is summarized in Fig. 1. In accordance with the materials, the evolution of T_c with time may be roughly divided into three periods: low temperature superconductivity (LTS) before 1986 that deals mainly with inter-metallic compounds or alloys with the highest T_c at 23 K; high temperature superconductivity (HTS) between 1986 and 2009 (or later) that concerns mostly cuprates with the highest T_c at the current records of 134 K at ambient and 164 K under high pressure and the recently discovered iron-pnictides with a highest T_c at 57 K; and very high temperature superconductivity (VHTS) after 2009 (or later when a new T_c-record or a new material system is found) in the hope of achieving a T_c higher than the current record and/or new insight to HTS, not to mention uncovering possible new physics.

2. The LTS-Period: 1911–1986

Before 1986, extensive efforts to search for higher T_c had been carried out mainly on elements, alloys and inter-metallic compounds. In 1955, Bernd

Matthias[2] proposed the empirical rule according to which T_c generally peaks at the valence-electron-per-atom ratios (e/a) of $\sim$4.75 and 6.4. In the ensuing 33 years, mainly due to Matthias's effort, a generation of superconducting inter-metallic alloys and compounds was born, giving rise to the then-record T_c of 23 K observed in Nb_3Ge with an $e/a = 4.75$ in 1974,[3] attainable in a liquid hydrogen environment. The results had also demonstrated the possible correlation of instabilities with superconductivity of high T_c, as evidenced by the simultaneous occurrence of lattice instabilities and high temperature superconductivity in different material families with a relatively high T_c within their respective families.

In 1957, John Bardeen, Leon Cooper and Bob Schrieffer developed the comprehensive microscopic theory of superconductivity, the well-known BCS theory.[4] According to the theory, the electrons, in the presence of a phonon-mediated effective attraction V between them, form pairs and undergo a phase-coherent superconducting transition at a temperature

$$T_c = 1.14\Theta_D \exp[-1/N(E_F)V]\,, \tag{1}$$

where Θ_D is the Debye temperature and $N(E_F)$ the electron density of states at the Fermi surface E_F. According to this relationship, a higher Θ_D, $N(E_F)$ and/or V is expected to give rise to a higher T_c. The descriptive power of the theory has been amply demonstrated by the accounts of many successful experiments in the ensuing decades. For instance, the Matthias empirical rule can be understood in terms of the variation of $N(E_F)$ with e/a and the simultaneous occurrence of lattice instabilities and high T_c in terms of the large $N(E_F)V$. Unfortunately, BCS theory fails to provide a definitive guide in the search for superconductors with high T_c other than the simplistic suggestion mentioned above. The failure can be attributed to: (1) the small superconducting energy scale in comparison with the large energy scales associated with the many other excitations in solids and (2) the coupling between the three parameters, Θ_D, $N(E_F)$ and V, in the BCS T_c-formula. The mismatch of energy scales makes an accurate T_c-calculation difficult and the coupling between Θ_D, $N(E_F)$ and V renders the simplistic approach to raise T_c mentioned above impractical.

The correlation between high temperature superconductivity and lattice instabilities had therefore become one of the most investigated problems in superconductivity in the 1970s and 1980s. The goal was to determine if lattice instabilities would be detrimental to or helpful for superconductivity at high temperature. This is because in almost all superconductors with a relatively high T_c, such as A15 superconductors, lattice instabilities were detected. They reveal themselves as a martensitic transformation with a phonon softening indicative of the large electron-phonon coupling favorable to superconductivity. However, phonon stiffening returns upon the completion of the martensitic transformation, suggestive of a reduction in electron-phonon interaction. The correlation had thus been studied extensively by adjusting the lattice instabilities and T_c via high pressure without introducing any chemical or physical complexity. It was found that lattice instabilities do affect the T_c of the A15 superconductors, but only slightly by no more than a few tenths of one degree.[5] The observation led us to conclude that the incipient instabilities associated with large N and/or V do not present an insurmountable obstacle to higher T_c, provided a catastrophic instability is avoided. This gave my group the confidence that superconductivity at higher T_c might be achievable.

Unfortunately, T_c stagnated at 23 K after 1973, which was reinforced by the T_c-limit of 30's K predicted theoretically.[6] Indeed, one of the most difficult challenges superconductivity researchers faced then was the crisis in confidence. The field was put in a very depressed state in 1986 as the government and industry largely withdrew support for superconductivity study, bringing the search for higher T_c almost to a halt. However, the discovery of cuprate HTSs in 1986 drastically changed the situation.

3. The HTS-Period: 1986–2009

In the past 23 years, many HTSs have been discovered in different material systems and many studies carried out. However, in this paper I shall review only the T_c-evolution of three systems, i.e. the cuprate system that is the largest and possesses the highest T_c, the newly discovered iron-pnictide system that may provide important insight into the inner working of HTS and the low T_c-heavy fermion system that is rich in physics and bears great similarities to the cuprate and iron-pnictide HTSs. Over the years, studies have revealed exciting physics and challenges in all three compound systems that have been covered by numerous review articles. Those whose interest goes beyond T_c may refer to the article by E. Abrahams in the next chapter of this book and review articles cited later.

3.1. *The cuprate system*

Disappointed by the slow progress in raising T_c in the inter-metallic compounds over the years, Alex Mueller and Georg Bednorz of IBM Zurich Research Lab decided to leave the beaten path of inter-metallics and dive into

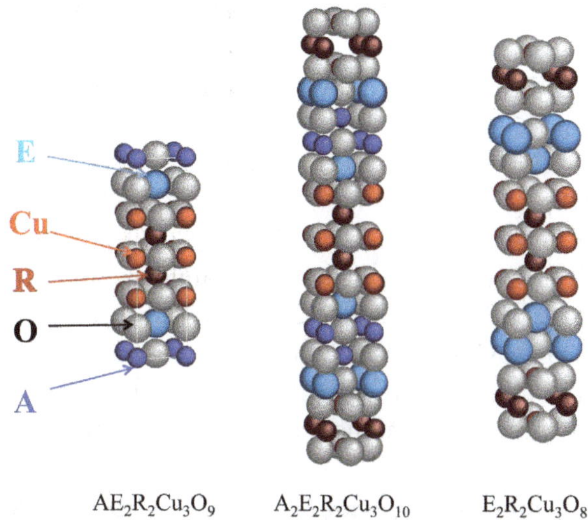

Fig. 2. The layer structures of the three major cuprate HTS families, $AE_2R_{n-1}Cu_nO_{2n+3}$, $A_2E_2R_{n-1}Cu_nO_{2n+4}$ and $E_2R_{n-1}Cu_nO_{2n+2}$, as exemplified by members with $n = 3$.

the woods of oxides for higher T_c. They were encouraged by the relatively high T_c of ~ 14 K detected in the perovskite-like oxides of $Li_{1+x}Ti_{2-x}O_4$ and $Ba(Pb_{1-x}Bi_x)O_3$ and wanted to raise the T_c of oxide superconductors further by increasing the electron-phonon interaction and the carrier concentration. Their seminal work on La-Ba-Cu-O in 1986 heralded the new era of HTS of cuprates. Many cuprate HTSs have since been found. Despite the differences in details, all cuprate HTSs discovered exhibit a layer structure as shown in Fig. 2 and can be represented by a generic formula $A_mE_2R_{n-1}Cu_nO_{2n+m+2}$, where A = Bi, Tl, Pb, Hg or Cu; E = Ca, Sr or Ba; R = Ca, Y or rare-earth; $m = 0$, 1 or 2 and $n = 1, 2, \ldots$. The generic formula can be rewritten as $A_mE_2R_{n-1}Cu_nO_{2n+m+2} = [(EO)(AO)_m(EO)] + [(CuO_2)R_{n-1}(CuO_2)_{n-1}]$, which consists of two substructures: the active block of $[(CuO_2)R_{n-1}(CuO_2)_{n-1}]$ and the charge reservoir block of $[(EO)(AO)_m(EO)]$. The space group for compounds with $m = 2$ or 0 is I4/mmm but changes to P4/mmm when $m = 1$. The active block comprises *n-square-planar*-(CuO_2)-layers interleaved by $(n-1)$-R-layers and the charge reservoir block contains m-(AO)-layers bracketed by 2-(EO)-layers which are relatively inert chemically. Superconducting current flows mainly in the (CuO_2)-layers and doping takes place in the charge reservoir block, which transfers charges without introducing defects into the (CuO_2)-layers (similar to modulation-doping in semiconducting superlattices). All cuprate

HTSs can simply be designated as Am2(n-1)n or just 02(n-1)n-E when the (AO)-layers are absent.

We shall recall briefly the discovery of the main cuprate HTS families to which the following HTSs belong: (1) the first 35 K HTS, the Ba-doped La_2CuO_4 [214 or La2001]; (2) the first 93 K HTS, $YBa_2Cu_3O_7$ [123, YBCO or Cu1212]; (3) the first 110 K HTS, $Bi_2Sr_2Ca_2Cu_3O_{10}$ [BSCCO or Bi2223]; (4) the first 120 K HTS, $Tl_2Ba_2Ca_2Cu_3O_{10}$ [TBCCO or Tl2223]; (5) the current T_c-record holder of 134 K at ambient and 164 K at 30 GPa, $HgBa_2Ca_2Cu_3O_{9-\delta}$ [HBCCO or Hg1223]; (6) the base of HTS without a toxic element, $Ba_2Ca_2Cu_3O_x$ [BCCO or 02(n-1)n-Ba]; and (7) the simplest infinite layer HTS family, $RCuO_2$ [RCO or 0011-R]. Reviews on the experimental results of these HTS families can be found in Ref. 7.

3.1.1. *The first cuprate HTS family with a T_c up to $\sim$35 K: doped R_2CuO_4, where R = La, rare-earth [R214 or R2001]*

The seminal discovery[8] of the high temperature superconductors, Ba-doped La_2CuO_4 [La214 or La2001] with a T_c of $\sim$ 35 K, in mid April 1986 by Alex Mueller and Georg Bednorz of Zurich IBM Research Lab marked the start of the new era of HTS of cuprates. It showed that HTS is possible in oxides.

A comprehensive account of the discovery of the Ba-doped 214, $(La_{1-x}Ba_x)_2CuO_4$ by Mueller and Bedorz has been published elsewhere.[9] In their search for superconductivity with a high T_c in materials with a strong electron-phonon coupling and higher carrier concentration, Mueller and Bednorz started in 1983 by examining the perovskite oxides known to display the Jahn–Teller effect, which promotes electron-phonon interaction. In January 1986, they detected resistively superconductivity up to 35 K in their multiphase La-Ba-Cu-O samples, setting a new T_c-record as shown in Fig. 3. The resistive result was published in the September 1986 issue of *Zeitschrieft für Physik* in an article entitled "Possible High T_c Superconductivity in the Ba-La-Cu-O System."[8]

Initially, the resistive data did not attract too much attention except by a few, since oxides are often insulators and not even metallic, let alone superconducting at high temperature. Their magnetic data showing the decisive Meissner effect did not reach the U.S. until late January 1987. Little action was taken after the news broke except by a lucky few. Among these few, Shoji Tanaka and Koichi Kitazawa's group at the University of Tokyo and my group at the University of Houston quickly reproduced the Zurich resistive results, and so reported on December 4, 1986, at the Materials Research Society (MRS) Fall Meeting in Boston. The rapid reproduction

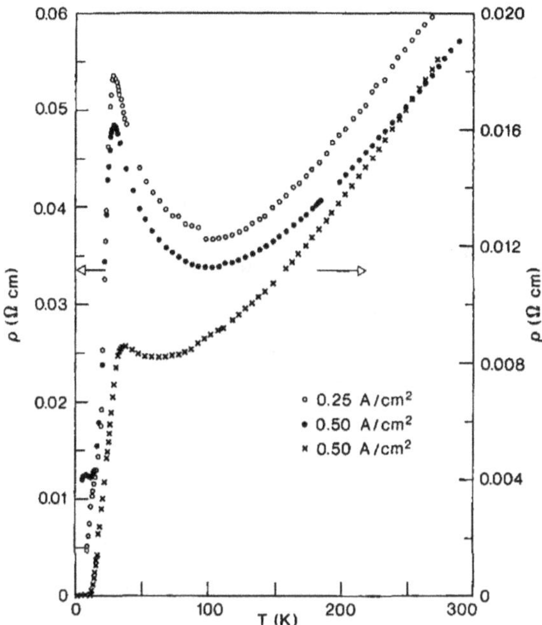

Fig. 3. The $\rho(T)$ of La-Ba-Cu-O shows a T_c of 35 K (La214) in 1986 by Bednorz and Müller [J. G. Bednorz and K. A. Müller, *Z. Phys. B* **64**, 189 (1986)].

by both groups was helped by their prior experience studying the oxide superconductors $Ba(Pb_{1-x}Bi_x)O_3$ and $Li_{1+x}Ti_{2-x}O_4$. The Meissner effect was reproduced by the Tokyo group quickly. The same group also identified the superconducting phase in the mixed-phase sample to be the Ba-doped La_2CuO_4 [La214 or La2001], as announced at the 1986 MRS Fall Meeting. This compound $(La_{1-x}Ba_x)_2CuO_4$ has the La_2CuO_4 layer structure with a layer stacking sequence $La_2CuO_4 = (LaO)(LaO)(CuO_2)$.

After reproducing the results of Mueller and Bednorz, the Houston group applied pressures to the La-Ba-Cu-O mixed-phase samples, focusing only on the superconducting phase in an attempt to understand the cause for such an unexpectedly high T_c. The T_c was quickly raised to 40 K and then to 52 K[10] at a rate more than 10 times that of the inter-metallic superconductors. The observation of a T_c higher than 40 K shattered the then-theoretically predicted T_c-limit of 30s K.[6] The unusually large positive pressure effect on T_c observed also suggested that cuprate superconductors may form a class of their own and warrant further study, although some still suggested at the time that the 35 K cuprate superconductor was nothing more than a conventional BCS-superconductor because its T_c still fell within what theory predicted.[11]

It was later observed that superconductivity with a similar T_c can also be induced in La214 by Sr-doping or excess oxygen. The structure of these La214 compounds displays the so-called 1T-phase and the carriers exhibit hole-characteristics. A similar layer structure with a different stacking known as the 1T'-phase was stabilized in R124 with R = smaller rare-earth such as Nd, Sm, Eu, Gd or Tb and can be electron-doped by partial Ce-replacement for 1T'-Nd214 to induce superconductivity.[12] A 1T*-phase, which is a mixture of the T- and T'-phases, each filling half of the unit cell, was also synthesized by properly adjusting the matching between the various atomic radii and synthesis conditions. These new 214-phases display a lower T_c but offer the opportunity to examine the symmetry between the roles of electron- and hole-dopings in HTSs.

3.1.2. *The first liquid nitrogen cuprate HTS family with a* $T_c \sim 93$–100 *K,* $RBa_2Cu_3O_7$*, where* R = *rare-earth* [*R*123, *RBCO or Cu*1212]

The discovery[13] of $YBa_2Cu_3O_7$ [Y123, YBCO or Cu1212] with a $T_c \sim 93$ K in late January 1987 by Paul C. W. Chu, Maw-Kuen Wu and colleagues in their respective groups at the University of Houston and the University of Alabama at Hunstville ushered in the modern era of HTS. It brought down the liquid nitrogen temperature barrier of 77 K, representing a giant advancement in modern science and technology and drastically changing the psyche of superconductivity research. It is the most desirable HTS compound for device applications to date.

As mentioned earlier, the Houston group was among the very few non-skeptics who took seriously the initial report from the Zurich IBM Research Lab.[8] This was because, since the mid-70s, they had actively investigated the unstable superconductors, including the perovskite-type oxides, such as $BaPb_xBi_{1-x}O_3$ and $Li_{1+x}Ti_{2-x}O_4$. Not only did they learn that superconductivity at higher temperature is possible, but they also mastered the oxide synthesis skills crucial for later cuprate HTS studies. The observations of an onset $T_c \sim 40$–52 K in the mixed-phase La-Ba-Cu-O samples under pressure,[10] above the then-theoretically predicted T_c-limit of 30s K, enhanced their confidence in achieving higher T_c and raised doubts about the old theoretical models. The unusually large positive pressure effect on T_c observed further suggested to me that higher T_c might be obtained through chemical pressures by replacing elements in the compound with those of smaller ionic radii of the same valences, such as Ba by Sr or Ca and La by the non-magnetic Y or Lu.[14] The Ba–Sr replacement to raise T_c was quickly confirmed, but the Ba-Ca substitution was unfortunately found to lower the T_c.

In many scientific discoveries, "a kick of luck," as Mueller put it, often plays a role. At the 1986 Fall MRS Meeting in Boston on December 4, Kitazawa from Tokyo announced the identification of the superconducting La214-phase in the La-Ba-Cu-O mixed-phase samples, after my presentation on $BaPb_xBi_{1-x}O_3$ and disclosure of duplicating the observation by Mueller and Bednorz. It was natural for people, including ourselves, to make pure 214-phase samples, preferably single crystals, and to examine the origin of the 35 K-T_c before contemplating how to raise T_c. We tried but failed to grow La214 single crystals, following the destruction of two of our three expensive crystal-growing Pt-crucibles. I made a crucial decision to turn our attention to stabilizing the high temperature resistivity anomalous drops, indicative but not yet a proof of superconductivity, that we detected sporadically in the multiphase La-Ba-Cu-O samples but not in the pure 214 ones, by changing the stoichiometry of La:Ba:Cu and/or replacing La with Y and Lu.[14] It should be noted that the first sign of superconductivity evidenced by a resistivity drop at a temperature of ~ 75 K was detected on November 25, 1986, in one of our La-Ba-Cu-O mixed phase samples, although it was too fleeting to make a definitive characterization due to the unstable nature of the samples. However, I showed the preliminary data to M. K. Wu at the Fall MRS meeting in Boston and successfully convinced him to join the search. On January 12, 1987, we observed a large diamagnetic shift or Meissner signal up to ~ 96 K in one of our mixed-phase La-Ba-Cu-O samples, representing the first definitive superconducting signal detected above the liquid nitrogen temperature of 77 K, as displayed in Fig. 4. Unfortunately, the sample degraded and the diamagnetic signal disappeared the following day. No effort of ours in the ensuing two weeks succeeded to reproduce and stabilize this high temperature superconducting signal. I decided to report details of the observation and let other groups stabilize and identify the high temperature superconducting phase. No sooner than half of the paper was drafted, M. K. Wu called with great excitement from Alabama in the afternoon of January 29, 1987, and said that a resistive drop indicative of a stable superconducting transition above 77 K was detected in the mixed-phase Y-Ba-Cu-O samples. On January 30, we observed the Meissner effect in their sample. Stable superconductivity at ~ 93 K was finally achieved (Fig. 5), nearly tripling the T_c of La214. The 93 K superconductivity was attributed to a non-214 phase, based on the high pressure studies carried out immediately afterward. The results appeared in the March 2, 1987, issue of *Physical Review Letters* in an article entitled "Superconductivity at 93 K in a New Mixed-Phase Y-Ba-Cu-O Compound System at Ambient Pressure."[13]

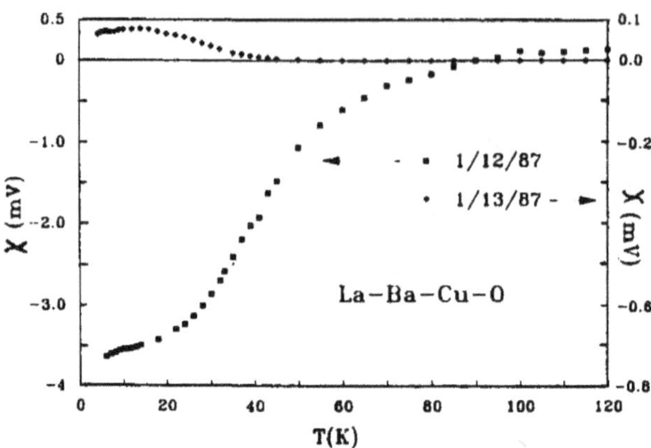

Fig. 4. $\chi(T)$ of La-Ba-Cu-O shows the first Meissner signal above 77 K (La123) (but unstable) by Chu *et al.* in 1987 [C. W. Chu, *AIP Conf. Proc.* **169**, 220 (1988)].

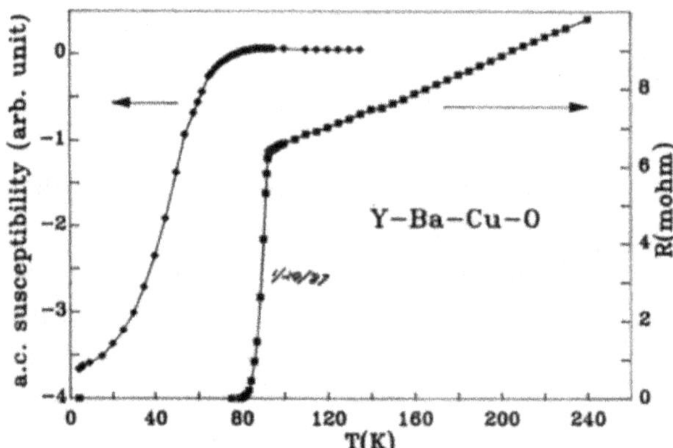

Fig. 5. $\rho(T)$ and $\chi(T)$ of Y-Ba-Cu-O shows the T_c above 93 K (Y123) in 1987 by Wu/Chu *et al.* [M. K. Wu *et al.*, *Phys. Rev. Lett.* **58**, 908 (1987)].

Contrary to the earlier R214-case, labs all over the world reproduced our results soon after the news broke out. In about a month, the superconducting phase $YBa_2Cu_3O_7$ (known as Y123, YBCO or Cu1212) was identified and its structure resolved in collaboration with Bob Hazen *et al.* from the Carnegie Geophysical Laboratory in Washington, D.C. The Y123-phase has the layer structure with layers stacking in the sequence: $YBa_2Cu_3O_7 = (BaO)(CuO)(BaO)(CuO_2)(Y)(CuO_2)$. It was interesting to note that the La123-phase with a high T_c had already been made in the La-Ba-Cu-O sample on January 12, 1987, when the X-ray diffraction taken then was compared to the Y123 pattern later. With the structure information in hand, we quickly found through partial magnetic substitution of Y that Y in YBCO is electronically isolated from the superconducting component of the compound and serves only as a stabilizer to hold the crystal structure together. The information led us to the subsequent discovery of the whole series of $RBa_2Cu_3O_7$ (RBCO, R123 or Cu1212) with a T_c varying between 93 and 100 K where $R = Y$ and all rare-earth elements except Ce and Pr.[15] Two related structures were later found by others, i.e. $YBa_2Cu_4O_8$ (Y124) and $Y_2Ba_4Cu_7O_{15}$ (Y247), with a T_c around 80 K. The layer-stacking sequence for $YBa_2Cu_4O_8$ is (BaO)(CuO)(CuO)(BaO)(CuO_2)(Y)(CuO_2). $Y_2Ba_4Cu_7O_{15}$ can be considered as an intergrowth of Y123 and Y124.

The discovery broke the liquid nitrogen temperature barrier of 77 K for superconductivity. It brings superconductivity technology a giant step closer to applications by using liquid-nitrogen as the low-cost, plentiful, more efficient and easier-to-handle coolant and at the same time poses serious challenges to physicists concerning the origin of HTS. The excitement at the time was amply demonstrated in the Special Panel Discussion on Novel Materials and High Temperature Superconductivity initiated by me and hastily organized by the American Physical Society (APS) on March 18, 1987, in the Sutton Complex at the New York Hilton and the July 1987 Federal Conference at Washington, D.C., attended by President Reagan and his cabinet members. The APS Panel Discussion started at 7:30 p.m. with short presentations by five panelists: Alex Mueller of IBM Zurich Lab; Shoji Tanaka of the University of Tokyo; Paul Chu of the University of Houston; Bertrum Batlogg of Bell Labs; and Zhongxian Zhao of the Physics Institute of the Chinese Academy of Sciences, followed by short contributions and discussions that lasted until the wee hours of the next morning. The 1,200-seat meeting room was packed with more than 2,000 people. Many more who could not get into the room watched on TV screens outside the room to witness this exciting event, called the Woodstock of Physics by the late Bell Labs physicist Mike Schluter, referring to the legendary 1969 Woodstock Music and Art Festival in upstate New York. In spite of later discoveries of many other cuprate HTS families, some of which have higher T_c, YBCO remains to be the most desirable material for HTS science and technology due to its superior sample quality, current carrying capacity in the presence of high magnetic fields and physical robustness in thin-film form. A YBCO puck together with a note was selected as an entry for the White House's National Millennium Time Capsule in 2000 (Fig. 6). The capsule was created in the spirit of "honor the past – imagine the future" to contain

Fig. 6. Closing Ceremony for the White House Millennium Time Capsule "Honor the Past — Imagine the Future" that included a YBCO puck, held at the National Archives in Washington, D.C., December 6, 2000.

discoveries and achievements in all areas by Americans over the last 100 years considered to be important to the present and/or the future. It will be opened in 2100 to communicate to the future generations about the accomplishments and visions realized in the U.S. in the 20th Century.

3.1.3. *The first cuprate HTS family without rare-earth with a T_c up to 110 K: $Bi_2Sr_2Ca_{n-1}Cu_nO_{2n+4}$, where $n = 1, 2, 3, \ldots$ [BSCCO or Bi22(n-1)n]*

The discovery[16] of the $Bi_2Sr_2Ca_{n-1}Cu_nO_{2n+4}$ [BSCCO or Bi22(n-1)n] with a T_c up to 110 K in January 1988 by Hirosh Maeda *et al.* of the National Research Institute for Metal suggested that HTS had a broader material-base than originally thought and that more HTS materials with a higher T_c might be discovered. The graphitic-like behavior of BSCCO due to its weak inter-layer coupling makes the compound a good material for spectroscopy studies and for the first-generation HTS-wires, in spite of the weak flux pinning associated with the compound family.

Euphoria permeated the field following the announcement of superconductivity above 90 K in YBCO and RBCO, and the sky seemed to be the only limit to T_c. As 1987 was drawing to an end, an impatient industrial physicist told the Wall Street Journal that the accumulated man-hours devoted to cuprate-HTSs in 1987 exceeded all previous effort expended on LTSs in the preceding 75 years since its discovery and that any cuprate with a T_c above 90s K should have been found. He went on to propose

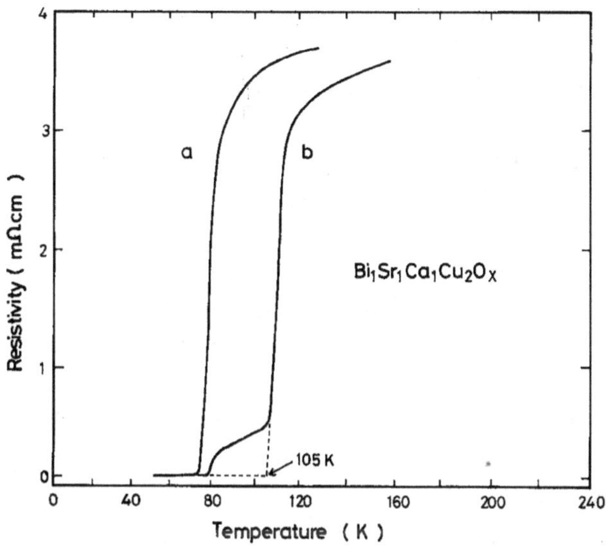

Fig. 7. $\rho(T)$ of Bi-Si-Ca-Cu-O (Bi1223) shows a T_c up to 105 K in 1988 by Maeda *et al.* [H. Maeda *et al.*, *Jpn. J. Appl. Phys.* **27**, L209 (1988)].

searching outside cuprates for superconductors with higher T_c. The fallacy of prophecy based on statistics immediately faced the truth in the first week of January 1988 when Maeda *et al.* of the National Research Institute for Metal reported superconductivity above 100 K in Bi-Sr-Cu-O that later appeared in the February 20 issue of *Japanese Journal of Applied Physics* in an article entitled "A New High-T_c Oxide Superconductor without a Rare Earth Element."[16]

In their attempt to broaden the material base for HTSs, Maeda *et al.* examined elements in the VB group of the periodic table, such as Bi, that are trivalent and of similar ionic radii to those of the rare-earth elements. They discovered superconductivity in samples of Bi-Sr-Ca-Cu-O above 105 K as shown in Fig. 7. The resistivity data suggested that the sample might consist of multiple phases. Indeed, shortly afterward, three members of the Bi22(n-1)n family with $n = 1$, 2 and 3 were isolated and the structures determined by several groups, including the Carnegie Institute Geophysical Lab and our own. The layer stacking sequence is, e.g. $(SrO)(BiO)(BiO)(SrO)(CuO_2)(Ca)(CuO_2)(Ca)(CuO_2)$ for $Bi_2Sr_2Ca_2Cu_3O_{10}$ with $n = 3$. Layer structural modulation appears in the (BiO) double layers for all n. The maximum T_cs for members of the homologous series are $\sim$22, $\sim$80 and $\sim$110 K for increasing $n = 1$, 2 and 3, respectively. However, for $n > 3$, T_c starts to drop and this drop is attributed to the combined influence of electrostatic shielding and proximity effects of the CuO_2-layers. Aided by the experience on R123, it took less than two days for Hazen *et al.* of Carnegie Geophysical Lab and Veblen *et al.* of Johns Hopkins to crack the structure of BSCCO after receiving the samples from us, in contrast to the more than 10 days required for Hazen *et al.* to do the same for Y123 in 1987. From their structures, it was immediately evident that the weak van der Waal force between the (BiO)-double-layers in BSCCO is responsible for the graphitic-behavior of the compound, making cleaving the sample in vacuum easy for spectroscopy studies and mechanical rolling effective for alignment of the (CuO_2)-layers for the first generation-HTS-wire processing.

It is interesting to note that in the summer of 1987 Akimutsu *et al.* of Aoyama-Gakuin University and Raveau *et al.* of University of Caen reported an 8 K superconducting transition in the Bi-Sr-Cu-O, which was later found to be associated with the $n = 1$ member of the BSCCO-family. This important piece of information was lost at the time in the mad rush for superconductors with a higher T_c than YBCO. I even marked their preprints with "exciting, more study needed!" in August 1987, but took no immediate action. With greater attention to these early results, the discovery of the 110 K superconducting BSCCO-family without rare-earth could have been advanced by half a year. A lesson from this episode is never to get trapped in the turbulence of excitement presented by fashionable ideas.

3.1.4. *The second cuprate HTS family without rare-earth with a* T_c *up to* 125 *K*: $Tl_2Ba_2Ca_{n-1}Cu_nO_{2n+4}$, *where* $n = 1, 2, 3, \ldots$ [*TBCCO or Tl*22(*n*-1)*n*]

The discovery[17] of $Tl_2Ba_2Ca_{n-1}Cu_nO_{2n+4}$ [TBCCO or Tl22(n-1)n] with a T_c up to 125 K in early February 1988 by Zhengzhi Sheng and Allen Hermann of the University of Arkansas appeared to justify the optimism suggested following the discovery of YBCO and BSCCO that more superconductors with higher T_c might be discovered soon. TBCCO was considered by some to be a preferred material for HTS applications, especially for thin film devices, due to their higher T_c, easier formation and higher stability when compared with BSCCO. Unfortunately, the softness of TBCCO makes it less desirable for thin film device applications, although the experience gained from its processing proved useful for the later processing of HTS with toxic elements.

After examining the role of R in R123, Sheng and Hermann of the University of Arkansas started to replace the trivalent R in R123 with the trivalent Tl with an ionic radius similar to R. They found superconductivity up to

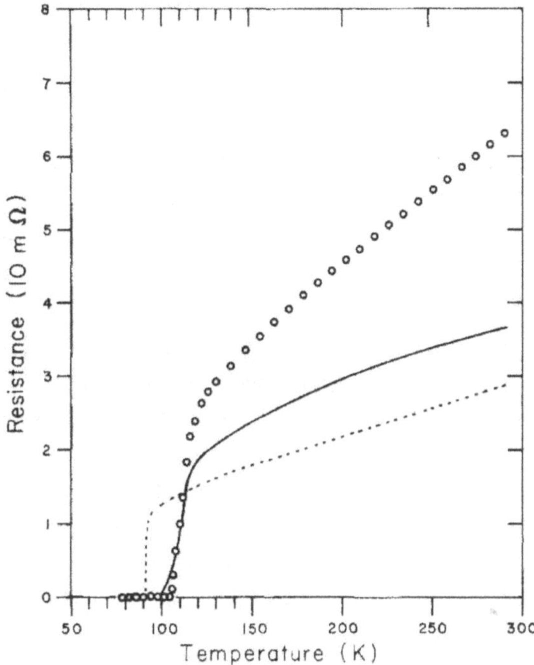

Fig. 8. $\rho(T)$ of Tl-Ba-Ca-Cu-O shows a T_c up to 120 K (Tl2223) in 1988 by Sheng and Hermann [Z. Z. Sheng and A. M. Hermann, *Nature* **332**, 138 (1988)].

90 K in their mixed phase samples of Tl-Ba-Cu-O with a nominal composition of $Tl_2Ba_2Cu_3O_{8+x}$ in early January 1988, about the same time as Maeda *et al.* announced their finding of superconductivity up to 110 K in Bi22(n-1)n. Sheng and Hermann soon partially replaced Ba with Ca and observed superconductivity above 120 K in their multiphase samples during the second week of February 1988 as shown in Fig. 8. The results appeared in the March 10, 1988, issue of *Nature* in an article entitled "Bulk superconductivity at 120 K in the Tl-Ca/Ba-Cu-O system."[17] The $n = 2$ and 3 members of the Tl22(n-1)n homologous series were quickly identified in these samples by Hazen *et al.* of the Carnegie Geophysical Lab and Torardi *et al.* of Du Pont Research Lab only two days after the announcement of the 120 K superconductivity. Parkin *et al.* of the IBM Almaden Lab obtained pure-phase samples of Tl2223 and achieved a T_c of 125 K, which was the record-T_c at ambient pressure until April 1993. The T_c of Tl2223 was raised to 131 K by Berkeley *et al.* at the Naval Research Lab by pressures up to 7 GPa in September 1992.

The homologous series Tl22(n-1)n display the layered stacking sequence, e.g. (BaO)(TlO)(TlO)(BaO)(CuO_2)(Ca)(CuO_2)(Ca)(CuO_2) for

$Tl_2Ba_2Ca_2Cu_3O_{10}$ with $n = 3$. Similar to Bi22(n-1)n, the T_c of Tl22(n-1)n increases with n up to 3 and decreases when $n > 3$. The maximum T_cs are 90, 110 and 125 K, respectively, for $n = 1$, 2 and 3. Depending on the oxygen content, the T_c can be varied by more than 10 K. The structure of Tl22(n-1)n is similar to that of Bi22(n-1)n. In spite of the gross similarity between Tl22(n-1)n and Bi22(n-1)n, there exists no structural modulation along the double (TlO)-layers in contrast to that along the double (BiO)-layers, suggesting that such a structural anomaly does not play a significant role in their high T_c as was initially thought.

The single (TlO)-layer TBCCO homologous series Tl12(n-1)n was later discovered. The layer stacking sequence is similar to that for Tl22(n-1)n except that the double (TlO)-layers in Tl22(n-1)n are replaced by the single (TlO)-layer. The T_c of members of this series is, in general, lower than that for the corresponding members of Tl22(n-1)n. It was found to increase continuously with n up to $n = 4$ before it decreases, i.e. $T_c = 50$, 82, 110 and 120 K for $n = 1$, 2, 3 and 4, respectively. The reasons for the lower T_cs and for the continuous increase of T_c to $n = 4$ could be interesting for understanding the inner atomic structure influence on T_c, but unfortunately remain unknown.

3.1.5. *The cuprate HTS family with the highest T_c of* 134 *K at ambient and* 164 *K at* 30 *GPa: $HgBa_2Ca_{n-1}Cu_nO_{2n+3-\delta}$, where $n = 1, 2, 3, \ldots$ [HBCCO or Hg12(n-1)n]*

The discovery of $HgBa_2Ca_{n-1}Cu_nO_{2n+3-\delta}$ [HBCCO or Hg12(n-1)n] with a record-T_c up to 134 K in mid April 1933 by A. Schilling *et al.* of ETH at ambient[18] and to 164 K at 30 GPa in Fall 1993 by C. W. Chu *et al.* of Houston and H. K. Mao *et al.* of Carnegie Geophysical Lab[19] provides new challenges as well as opportunities for HTS science and technology. The T_cs of the optimally doped samples increase in parallel by about 30 K for all members of Hg12(n-1)n up to $\sim$ 30 GPa in contrast to other HTSs. A T_c of 134 K is above the temperature in the cargo bay of the Space Shuttle when orbiting in Earth's shadow and above the boiling point of liquid natural gas (LNG) at $\sim$ 110 K. The former makes HBCCO a possible material for HTS devices operable on the Space Shuttle without cryogen and the latter may enable the development of the combined HTS/LNG system for the efficient delivery of electrical and chemical energies simultaneously. In addition, superconductivity at 164 K in HBCCO under pressure can be achieved by the use of the common air-conditioning technology.

In late 1989, the T_c of cuprates appeared to have stagnated at 125 K since the Spring 1988. A prominent chemist speculated that the T_c of

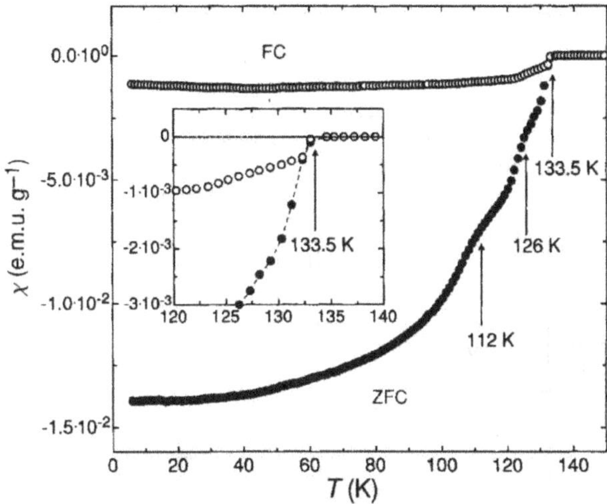

Fig. 9. $\rho(T)$ of Hg-Ba-Ca-Cu-O shows a T_c up to 134 K (Hg1223) by Schilling *et al.* in 1993 [A. Schilling *et al.*, *Nature* **363**, 56 (1993)].

cuprates could not exceed 160 K based on some physical chemistry arguments. However, $HgBa_2Ca_2Cu_3O_{7-\delta}$ was discovered in 1993 to display a superconducting transition at $\sim$133 K by Schilling *et al.* of ETH at Zurich as shown in Fig. 9 and the results appeared in the May 6, 1993, issue of *Nature* in an article entitled "Superconductivity above 130 K in the Hg-Ba-Ca-Cu-O system."[18] The homologous series of HBCCO, Hg12(n-1)n, was quickly identified to have the layer stacking sequence, e.g. $(BaO)(HgO_{1-\delta})(BaO)(CuO_2)(Ca)(CuO_2)(Ca)(CuO_2)$ for $HgBa_2Ca_2Cu_3O_{9-\delta}$ with $n = 3$. When Hg1223 is subjected to high pressures, its T_c rises to 164 K at $\sim$30 GPa before exhibiting signs of saturation as shown in Fig. 10, clearly contradicting the predicted 160 K-limit made in 1989 for cuprate HTSs. A $T_c = 134$ K and 164 K remain the records to date at ambient and high pressure, respectively.

It is interesting to note that attempts were made as early as in 1991 to substitute the linearly coordinated Hg^{+2} for the similarly coordinated Cu^{+2} in the (CuO)-chain-layer of R123. Compounds of $HgBa_2EuCu_2O_x$ with a structure similar to Tl1212 were made but found not superconducting. A small superconducting signal due to an impurity phase at 94 K was detected in 1991 in one of our $HgBa_2EuCu_2O_x$ samples without recognizing that the impurity phase was Hg1201. The whole story of HBCCO did not start to unfold until September 1992 when Antipov of Moscow State University and Marezio of CNRS at Grenoble succeeded in synthesizing Hg1201 with a $T_c =$ 94 K. Knowing that increasing the number of (CuO_2)-layers per cell will lead

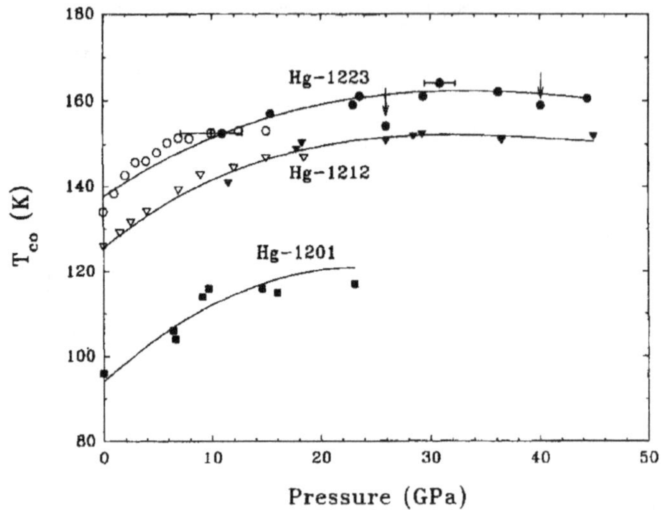

Fig. 10. $\rho(T)$ of Hg12(n-1)n for $n = 1, 2, 3$ under pressures up to 45 GPa shows a T_c up to 164 K in 1993 by Gao *et al.* [L. Gao *et al.*, *Phys. Rev. B* **50**, 4260(R) (1994)].

to an increase of T_c, A. Schilling *et al.* of ETH added Ca to the compound to achieve $n = 2$ and 3 for Hg12(n-1)n and reported an enhanced T_c up to ~133 K. After overcoming issues with the complex chemistry of HBCCO, we and several other groups isolated the different phases and showed the maximum T_c's to be 97, 128 and 134 K, for $n = 1$, 2 and 3, respectively.

While Hg12(n-1)n has a layer structure similar to Tl12(n-1)n, there exists a subtle difference, presumably arising from the linear oxygen coordination of Hg^{+2}-ions in HBCCO as reflected in the relatively short Hg-O bond length along the *c*-axis and the large number of voids in the $HgO_{1-\delta}$-layer. We therefore proposed that HBCCO could be different from other cuprate HTSs in that Hg might contribute to the density of states near the Fermi surface and that the O-doping range might be large. Band calculations also showed that large density of states associated with a van Hove singularity existed near the Fermi surface. Higher T_c is therefore expected in HBCCO under pressures. Experiments on optimally doped pure Hg1201, 1212 and 1223 were carried out under pressures. We quickly found that the T_c grows with pressure without saturation up to 18 GPa. Later, in October 1993, we extended the pressure to 45 GPA in collaboration with Mao *et al.* of the Carnegie Geophysical Lab. The T_c was found to peak at ~118 K in Hg1201 at ~24 GPa, at ~154 K in Hg1212 at ~29 GPa and at 164 K in Hg1223 at ~30 GPa.[19] The present day record T_c of 164 K was thus obtained. The pressure-dependent T_c is shown in Fig. 10. The T_cs of the three members of the Hg12(n-1)n series appear to vary with pressure in a parallel fashion independent of n and to show no sign of pressure-induced structure change up to 45 GPa. The observation strongly suggests that superconductivity of members of the Hg12(n-1)n homologous series arises from a common origin. However, the unusually large pressure-induced T_c-enhancement, even in their optimally doped states, cannot be accounted for by models commonly used to explain the pressure effect on T_c for other HTSs. A modified rigid band model has been proposed for the observations.

3.1.6. *The interstitially doped HTS family that forms the base of TBCCO and HBCCO without a volatile toxic element with a T_c of 126 K at ambient: $Ba_2Ca_{n-1}Cu_nO_x$ [BCCO or 02(n-1)n]*

The discovery[20] of $Ba_2Ca_{n-1}Cu_nO_x$ [BCCO or 02(n-1)n] with a T_c up to 126 K at ambient by C. W. Chu *et al.* of the University of Houston in 1997 demonstrates that a formal multi-substructure in a HTS serving as a charge reservoir may not be essential, provided that alternative means, such as interstitial doping outside the active block, can be found.

It has been demonstrated that the layer cuprate HTSs consist of two substructures, namely, the active block of $[(CuO_2)R_{n-1}(CuO_2)_{n-1}]$ and the charge reservoir block of $[(EO)(AO)_m(EO)]$. The active block comprises *n-square-planar*-(CuO_2)-layers interleaved by $(n-1)$-R-layers and the charge reservoir block contains m-(AO)-layers bracketed by two *chemically inert* (EO)-layers. It is known that processing of cuprate HTS materials into practical forms is a challenge due to the chemical and physical complexity of the materials. We therefore tried to synthesize layer cuprate HTSs of simpler structure, e.g. without the toxic elements such as Tl and Hg in the charge reservoir block. We attempted to deactivate the charge reservoir block by removing the m-(AO)-layers and allowing interstitial doping to take place in the two (EO)-layers that usually bracket the m-(AO)-layers. The compound does not form at ambient. However, through high pressure synthesis processing in a C- and O-free environment, we succeeded in stabilizing the $Ba_2Ca_{n-1}Cu_nO_x$ [BCCO or 02(n-1)n-Ba]. The XRD shows a layer lattice following the I4/mmm space group symmetry. It superconducts with a T_c of 126 K for the $n = 4$ member and the T_c increases to $\sim$150 K under pressure. The results (Fig. 11) were published in the August 22, 1997, issue of *Science* in an article entitled "Superconductivity up to 126 Kelvin in Interstitially Doped $Ba_2Ca_{n-1}Cu_nO_x$ [02(n-1)n-Ba]."[20]

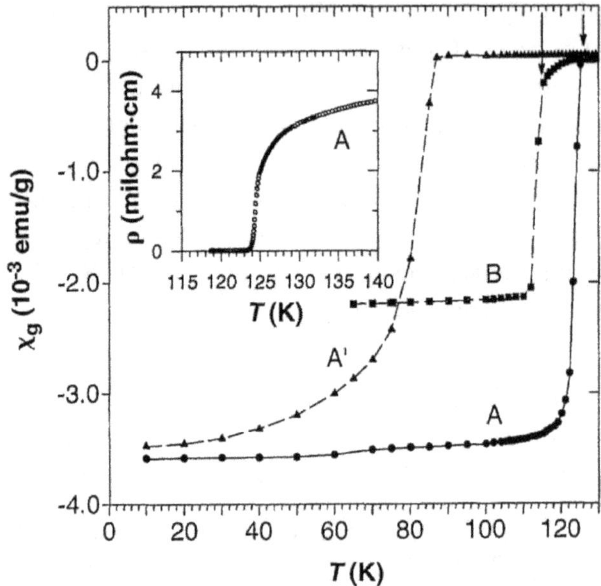

Fig. 11. $\chi(T)$ of $Ba_2Ca_{n-1}Cu_nO_x$ shows a T_c up to 126 K in 1997 by Chu *et al.* [C. W. Chu *et al.*, *Science* **277**, 1081 (1997)].

Although $Ba_2Ca_{n-1}Cu_nO_x$ [BCCO or 02(n-1)n] meet our original goal to possess a simpler structure without toxic elements and are interesting from chemistry and physics points of view, the compounds are not very stable and degrade in the presence of humid air, and thus of no application value. Its homologues $Sr_2Ca_{n-1}Cu_nO_x$(SCCO) were found also to exist with high T_c.

3.1.7. *The simplest infinite layer HTS family defected $RCuO_2$ where $R = Ca$ with a $T_c \sim 40$–110 K [RCO or 0011-R]*

The discovery[21,22] of superconductivity in defected $CaCuO_2$, the building block of cuprate HTSs, with a $T_c \sim 40$–110 K by John Goodenough *et al.* of the University of Texas and Takano *et al.* of Kyoto University in 1991–1992 could provide the basis to unravel the mysteries of HTS in cuprates if the microstructure of the defected $CaCuO_2$ is determined to ascertain that the superconductivity observed is caused by the infinite-layer phase.

After the discovery of R123, it was suggested that a higher T_c might be achieved by increasing the number of (CuO_2)-layers per unit cell (n) in cuprates. Indeed, T_c was later found to increase with n, although only up to 3 or 4. It seemed to be interesting to synthesize a cuprate with large n, preferably for $n = \infty$, the building block of cuprate HTSs, i.e. $RCuO_2$ [RCO or 0011-R]. In 1988, Siegrst *et al.* succeeded in stabilizing $(Ca_{0,85}Sr_{0.15})CuO_2$ [0011-Ca]. Unfortunately, it was not superconducting and changing the Ca/Sr-ratio away from 0.85/0.15 immediately triggered the collapse of the crystal. Three years later, Goodenough *et al.* used the high pressure synthesis technique and stabilized $Sr_{1-x}Nd_xCuO_2$ with a $T_c = 40$ K at ambient.[21] Later, Takano *et al.*[22] demonstrated superconductivity up to $\sim$110 K in $(Sr_{1-x}Ca_x)_{0.9}CuO_2$ after it was prepared at 6 GPa. Unfortunately, high resolution electron micrographs revealed that the superconducting sample was loaded with defects of Ca- and Sr-vacancies, which may be one of the superconducting members of the Sr- deficient series of $Sr_{n-1}Cu_{n+1}O_{2n}$ with $n = 3, 5, \ldots$, and not the infinite layer cuprate. As a result, the exact nature of superconductivity in the defected CCO or 0011-Ca remains unclear.

3.2. *The iron pnictides and chalcogenide superconductors*

In the last 23 years, almost all major HTS systems were discovered as a result of a conscientious search for novel superconductors with higher T_c. One exception is the discovery of F-doped LaFeAsO with a T_c of 26 K by Hideo Hosono *et al.*[23] of Tokyo Institute of Technology in early 2008 that inaugurated the exciting iron-pnictide and chalcogenide HTS period. The discovery was an outgrowth of the effort of Hosono *et al.* on the functionality of novel layer transparent oxide semiconductors, such as LaCuOCh where Ch = chalcogen. LaCuOCh consists of two subdtructures: one a narrow-gap semiconducting slab (CuCh) and the other a wide-band insulating slab (LaO), reminiscent of the multi-substructure of cuprate HTSs. Similar to the cuprate HTS, modulation doping can be carried out by introducing charge carriers to semiconducting slabs without defects through chemical doping into the insulating slabs.

The announcement of the LaFeAsO discovery in early 2008 could not have come at a better time. It has rejuvenated the field of HTS, boosted the morale of scientists in the field and hence effectively issued a reprieve of the "death sentence" proposed for HTS in 2006. While scientists continue to work on HTS, T_c stopped rising after 1994 and the theory remains unsettled. A feeling of pessimism began to creep in. The situation seems to have been epitomized in the 2006 article by Andreas Barth of FIZ Karlsruhe and Werner Marx of MPI Stuttgart, Germany titled "Mapping High Temperature Superconductors: A Scientometric Approach."[24] Using scientometric analysis of the time-dependence of the overall numbers of articles and patents, and the time-variation of publications related to specific compound subsets and subject categories, they sentenced HTS to die sometime

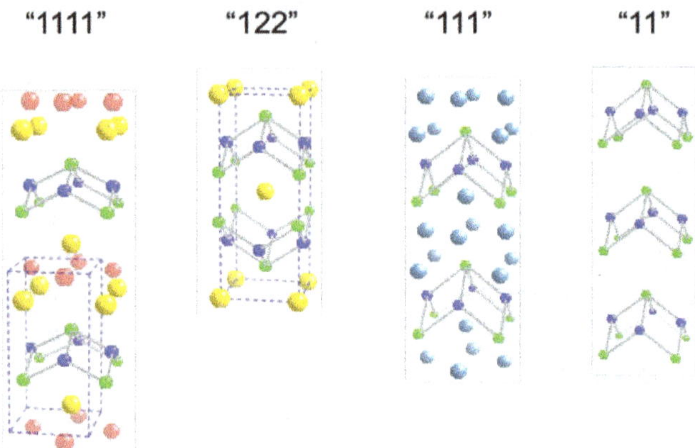

Fig. 12. The structures of Fe-pnictides and Fe-chalcogenides [1111], [122], [111] and [11].

between 2010 and 2015 by linear extrapolation. While experts in database collection and research consider the analysis accurate and sound, many scientists point out the pitfall of statistics in predicting the future of science. Scientific breakthroughs go beyond statistics. Many must remember that superconductivity had also been sentenced to die in the late 1980s.

Because of the unusually rich Fe-content and the high T_c of 26 K, the discovery of F-doped LaFeAsO by Hosono *et al.* in early 2008 has attracted worldwide attention. The research intensity on this and related materials was unmatched except by that associated with YBCO. In the ensuing months, four families of Fe-pnictide and chalcogenide superconductors with their respective derivatives were discovered with a T_c as high as 57 K (Fig. 12). They share many similarities with the cuprate HTSs but differ in electronic structures, e.g. single band for cuprates whereas multibands for Fe-pnictides. The four families are doped FeAsRO [1111], where R = rare-earth; doped $AeFe_2As_2$ [122], where Ae = alkaline earth = Ca, Sr, Ba; undoped AFeAs [111], with A = alkaline; and Fe(Se,Te) [11], as shown in Fig. 12. Distinct gaps exist among the T_cs of the four families with their maxima at 57, 38, 20 and 10 K, respectively. All Fe-pnictide and chalcogenide superconductors display a layer structure that consists of the $(FeAs)_2$-slabs, where Fe-ions form a square planar sheet sandwiched between two As-layers. The presence or absence of the different layers separating the $(FeAs)_2$-slabs in the compounds differentiate them from one another.

I shall address the discovery of each of the four families of Fe-pnictide and chalcognide systems. Summaries of other aspects of Fe-pnictide and chalcogneide superconductors can be found in several review articles.[25]

3.2.1. *The first Fe-pnictide superconductor family with a T_c up to 57 K: doped RFeAsO, where R = rare-earth* [R1111]

The discovery[23] of F-doped LaFeAsO [La1111] with a T_c = 26 K in early January 2008 by Hideo Hosono *et al.* of Tokyo Institute of Technology has inaugurated the new Fe-pnictide superconductor era, demonstrating that superconductivity with a relatively high T_c can occur in a Fe-rich environment, contrary to conventional wisdom, providing a new opportunity to explore the relationship between HTS and magnetism and offering a possible new avenue for high T_c.

Hosono and colleagues had engaged extensively in modifying the functionality of transparent novel oxide semiconductors with 2D layer structures and with two separate or loosely coupled substructures, and gained extensive knowledge on the chemistry, structure and electron-band manipulation of inorganic solids. Realizing the importance of layer structure and magnetic interaction in HTS cuprates, they investigated LaTPO and LaTAsO, where T is a transition metal element. They detected superconductivity in compounds LaFePO and LaNiPO when the number of d-electrons in T is even, and magnetism in LaMnPO and LaCoPO when the number of d-electrons in T is odd. The T_cs observed in 2006 of LaNiPO and LaFePO were 3 K and 5–12 K, respectively. Unfortunately, the results did not attract much attention by the HTS community due to the low T_c. The situation changed when Hosono *et al.* observed superconductivity up to 26 K in F-doped LaFeAsO, or $La(O_{1-x}F_x)FeAs$, and the results (Fig. 13) appeared in the March 19, 2008, issue of Journal of American Chemical Society in an article entitled "Iron-Based Layered Superconductor $La[O_{1-x}F_x]FeAs$ ($x = 0.05$–0.12) with $T_c = 26$ K."[23]

The parent compound LaOFeAs [La1111] is a member of the well-known equiatomic quaternary layered compounds ROTPn where R = La, Nd, Sm, Gd; T = Mn, Fe, Co, Ni, Cu; and Pn = P, As, Sb. They exhibit a tetragonal layer structure of the ZrCuSiAs type with a space group of P4/nmm. LaOFeAs consists of (Fe_2As_2)-layers sandwiched by the (LaO)-layers. Each (Fe_2As_2)-layer contains a squared-planar Fe-sheet sandwiched by two As-sheets; and each (LaO)-layer comprises an O-sheet sandwiched between two La-sheets. Similar to the cuprates, the (Fe_2As_2)-layers form the active block where the charge carriers flow, while the (LaO)-layers constitute the charge-reservoir block that inject charge carriers into the former without degrading the integrity of the (Fe_2As_2)-layers. But different from the cuprates: the divalent Fe is tetrahedrally coordinated with four As-ions in LaOFeAs, whereas the divalent Cu forms the fourfold square plane in the

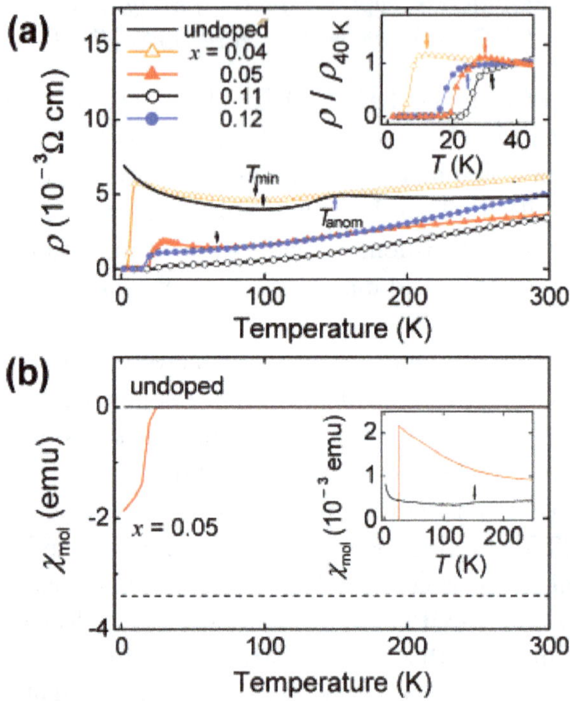

Fig. 13. $\rho(T)$ and $\chi(T)$ of $\mathrm{LaFeAs(O_{1-x}F_x)}$ [1111] shows a $T_c = 26$ K in 2008 by Hosono *et al.* [Y. Kamihara *et al.*, *J. Am. Chem. Soc.* **130**, 3296 (2008)].

cuprate HTSs; LaOFeAs is a semimetal and undergoes a crystallographic transition at ~155 K from tetragonal to orthorhombic symmetry followed by an antiferromagnetic transition [or a spin-density-wave (SDW) transition] at $T_N \sim 145$ K. Accompanying these transitions, anomalies in resistivity, magnetic susceptibility, specific heat, thermoelectric power, Moessbauer, μSR and NMR measurements have been observed. LaOFeAs becomes superconducting when charge carriers are introduced via electron-doping by partial substitution of O by F or by O-reduction to suppress the T_N.

The electron-doped La(O,F)FeAs [La1111] displays a T_c as high as 26 K.[23] The discovery has generated great excitement because 1111 is a new material compound system with a large concentration of magnetic Fe that may lead to higher temperature and provide a basis for unraveling the origin of HTS. In the ensuing few weeks after the report, the T_c was raised by replacing La with rare-earth elements of smaller radii. Pressure was also later found to raise the T_c of La(O,F)FeAs to 43 K. Development parallel to that for the cuprates seemed to be taking place with some thinking that another

1987 "cuprate miracle" might be in the offing. Euphoria permeated the community and once again, the sky was considered the only limit to T_c.

Soon after the discovery of the 26 K superconductor of the doped La1111, the doped $Sm(O_{1-x}F_x)FeAs$ [Sm1111] was found to have a T_c of 43 K, then highest among the non-cuprate superconductors. We examined the pressure effect on the T_c of Sm1111 with different x to determine whether the T_c of Sm1111 in particular and R1111 in general can be further enhanced by pressure or doping. The results show that pressure can either enhance or suppress the T_c, depending on the doping x, similar to the cuprate R123. The observation of the same sign of dT_c/dP and dT_c/dx suggests that both pressure and doping have a similar effect on superconductivity and therefore on the SDW-state. Given the gross similarity of the R1111 to the cuprate R123, we therefore proposed that members of R1111 are expected to have a similar maximum T_c of $\sim$ 50s K,[26] independent of R, as was confirmed by later theoretical calculations. Unfortunately, the prediction remains correct to date.

Following the observation of superconductivity in $La(O_{1-x}F_x)FeAs$, the T_c of $R(O_{1-x}F_x)FeAs$ has been raised from 26 K to 41 K, 52 K, 52 K, 55 K, 36 K, 46 K and 45 K for R = Ce, Pr, Nd, Sm, Gd, Tb and Dy, respectively. Electron doping via O-reduction in $RO_{1-x}FeAs$ results in similar T_c as through F-doping. Partial replacement of elements in other sites of R1111 can also induce superconductivity with high T_c, for example, in $(La_{1-x}Th_x)OFeAs$ ($T_c \leq 30.3$ K), $(La_{1-x}Sr_x)OFeAs$ ($T_c \leq 25$ K), $LaO(Fe_{1-x}Co_x)As$ ($T_c \leq 22$ K), and $(A_{1-x}R_x)FeAsF$ where A = Ca, Sr ($T_c \leq 50$ K), accompanied by suppression of the SDW-state. It is evident that doping the O-site gives a higher T_c in R1111. High pressure studies up to 30 GPa on the T_c of many of these compounds, superconducting or not, have been carried out.[27] The T_c shows an initial rapid rise to a maximum < 57 K followed by a slow decrease at higher pressure, but for some, their T_c does not display the initial increase, depending on the detailed doping state and pressure homogeneity. For those non-superconducting R1111, pressure turns them superconducting while suppressing the SDW-state. It is clearly evident that the suppression of the SDW-state by doping and pressure is critical to the appearance of superconductivity, although some experiments do show the coexistence of the superconducting and the SDW-states in the intermediate region. We therefore propose that defects externally introduced such as by irradiation should suppress the SDW and induce superconductivity in R1111. In spite of the great progress made on R1111, no T_c higher than 60 K has been realized.

3.2.2. *The second Fe-pnictide oxygen-free superconducting family with a T_c up to 38 K: doped AFe_2As_2, where A = Ba, Sr, Ca [A122]*

The discovery[28] of K-doped $BaFe_2As_2$ [Ba122] with a T_c up to 38 K in late May 2008 by Marianne Rotter *et al.* of Ludwig-Maximilians-University of Muenchen demonstrated that two (Fe_2As_2)-layers per unit cell is achievable, a simple elemental A-layer can serve as the charge reservoir and oxygen is not critical for superconductivity in Fe-pnictides, giving hope that higher T_c in Fe-pnictides may be possible due to their material flexibility.

After the discovery of the 26 K-superconductivity in R1111, it was soon recognized that the (Fe_2As_2)-layer is crucial for the high T_c in Fe-pnictides. In the hope of raising T_c by increasing the (Fe_2As_2)-layers per unit cell, it is rather natural to examine the well-known AFe_2As_2 [A122], which contains two (Fe_2As_2)-layers per unit cell. The compound family was first synthesized in the early 1980s, with A = the alkaline earth Ba, Sr, or Ca. The undoped A122 crystallizes in the tetragonal layer structure of the $ThCr_2Si_2$-type of the I4/mmm space group with the (Fe_2As_2)-(A)-(Fe_2As_2)-(A) layer-stacking. They are semi-metallic. Upon cooling, they undergo a tetragonal to orthorhombic structural transition at T_o = 122 K, 205 K and 170 K for A = Ba, Sr and Ca, respectively, similar to R1111. These structural transitions coincide with the antiferromagnetic transitions at T_N, indicative of the onset of the SDW-state in contrast to the R1111 where $T_o > T_N$. From valence consideration, A122 and R1111 are similar. Doping and pressure are therefore expected to induce superconductivity in undoped A122.

Realizing the existence of the iso-structure compound KFe_2As_2, Rotter *et al.*[28] hole-doped $BaFe_2As_2$ to induce superconductivity in Ba122 by suppressing the SDW-state. They succeeded in achieving a T_c of 38 K in $(Ba_{1-x}K_x)Fe_2As_2$ [$(Ba_{1-x}K_x)$122] when $x = 0.4$ in late May 2008 and published the results in the September 5, 2008, issue of *Physical Review Letters* in an article entitled "Superconductivity at 38 K in the iron arsenide $(Ba_{1-x}K_x)Fe_2As_2$" (Fig. 14). At about the same time, the iso-electronic and iso-structural $(Sr_{1-x}K_x)Fe_2As_2$ [$(Sr_{1-x}K_x)$122] was independently found by us to be superconducting with a maximum T_c = 38 K at x = 0.4–0.5, similar to $(Ba_{1-x}K_x)$122. At the same time, we found that undoped $A'Fe_2As_2$ with A' = K, Rb and Cs are superconducting at 3.7 K, 2.8 K and 2.6 K, respectively, as expected based on the valence count of the elements in A'122. A complete T_c-x phase diagram of $(Sr_{1-x}K_x)Fe_2As_2$, the first for Fe-pnictide superconductors, was therefore constructed and a systematic study was made. It displays the possible coexistence of superconductivity

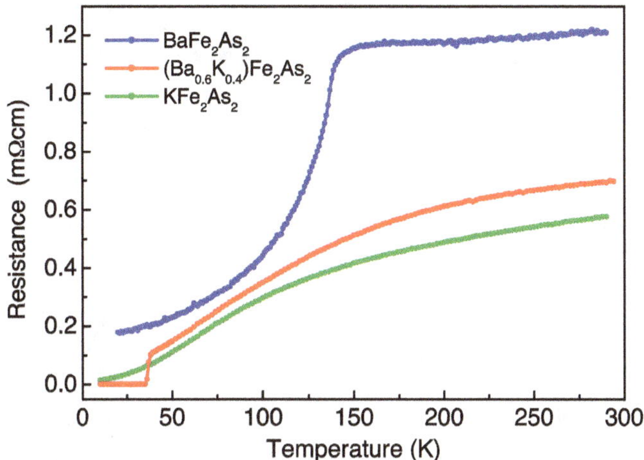

Fig. 14. $\rho(T)$ of $(Ba_{0.6}K_{0.4})Fe_2As_2$ [122] shows a T_c up to 38 K in 2008 by Rotter *et al.* [M. Rotter *et al.*, *Phys. Rev. Lett.* **101**, 107006 (2008)].

and SDW at $40.15 < x < 0.35$. It also showed that dT_c/dP and dT_c/dx have the same sign, similar to cuprate HTSs, lending support to the previously suggested T_c-limit of $\sim$50s K for Sm1111 and R1111.

Various doping and high pressure experiments similar to those made on R1111 were carried out on A122 and were observed to generate grossly similar, although qualitatively different, results, leading to the same conclusion that superconductivity competes against the SDW and becomes a robust ground state when SDW is suppressed either completely or partially by doping or pressure. Unfortunately, no effort using high pressure to date has been able to raise the T_c of A122 to above 38 K, the highest attained by doping. The lower T_c of A122 than that of R1111 may signal the need to look for a new avenue to higher T_c rather than increasing the (Fe_2As_2)-layers per unit cell. This leaves a distinct T_c-gap between R1111 and A122 for reasons that remain unknown.

3.2.3. *The third family of Fe-pnictide superconductors of the simplest structure with a $T_c \sim 23$ K without doping and 33 K with doping: undoped A′FeAs where A′ = Li, Na [A′111]*

The discovery[29] of superconductivity in LiFeAs [Li111] without doping (or in the absence of SDW) with a T_c up to 18 K by Joshua Tapp *et al.* of the University of Houston and independently by X. C. Wang *et al.* of the Physics Institute in Beijing in July 2008 showed that superconductivity occurs in the

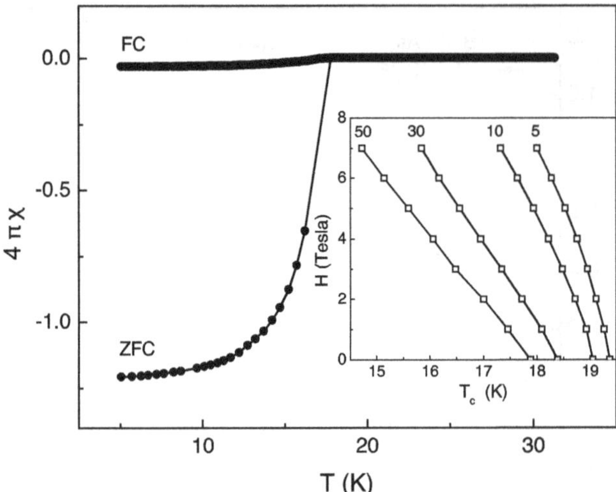

Fig. 15. $\chi(T)$ of LiFeAs [111] shows a $T_c \sim 18$ K in 2008 by Tapp *et al.* [J. H. Tapp *et al.*, *Phys. Rev. B* **78**, 060505(R) (2008); and X. C. Wang *et al.*, *Solid State Comm.* **148**, 538 (2008)].

simplest structure of Fe-pnictides with only the (Fe_2As_2)-layers separated by the nominal double layers of Li, demonstrating further the important role of the (Fe_2As_2)-layers in superconductivity of Fe-pnictides but at the same time raising questions concerning the role of magnetism in superconductivity in Li111.

The reason is not known for the lower T_c of A122 than of R1111 which has two (Fe_2As_2)-layers per unit cell compared with that of R1111 which possesses only one (Fe_2As_2)-layer per unit cell, in contrast to cuprate HTSs where T_c increases with the number of (CuO_2)-layers per unit cell. A Fe-pnictide superconductor with a simpler structure might help reveal the underlying reason. Li111 crystallizes in a tetragonal structure of the Cu_2Sb type of the P4/nmm space group. It has a structure similar to R1111 after replacing the (LaO)-layer in R1111 by the simpler (Li)-layer. Li111 was found[29] superconducting with a T_c of 18 K and the results appeared in the August 18, 2008, issue of *Physical Review B* in an article entitled "LiFeAs: An Intrinsic FeAs-based superconductor with $T_c = 18$ K" (Fig. 15).

The Li111 examined was found to be stoichiometric with Li:Fe:As = 1:1:1 within the resolutions of XRD and neutron analyses. It is rather unusual to find that Li111 becomes superconducting at a relatively high temperature but without external doping or any sign of SDW as in undoped R1111 and A122, in spite of their similarities in structure and valence count. Unlike $A'122$ with A' = alkaline, Li111 should not be a metal let alone a superconductor based on a valence consideration. The negative Seebeck coefficient detected seems to suggest that the Li111 samples investigated are effectively electron-doped, implying that the samples may be Li-rich, in contrast to XRD and neutron analyses. An alternative scenario is that the superconductivity in stoichiometric Li111 is caused by chemical pressure associated with the very small ionic radius of Li, which promotes the electron population in the bands of the compound and suppresses the SDW. This seems to be consistent with the large negative pressure effect on the T_c of Li111, putting the undoped compound in its overdoped region.

Later, the isostructural and isoelectronic NaFeAs [Na111] were successfully prepared and studied. Similar to Li111, the undoped Na111 was found superconducting with a $T_c \sim 23$ K. While the stoichiometric issue remains, its T_c increases rapidly to 31 K with pressure up to 3 GPa. Chemical pressure induced by partial replacement of As by P with a smaller ionic radius was found to enhance the T_c to 33 K.[30] It is very likely that the T_c of *P*-doped Na111 may be further raised by pressure, similar to Hg12(n-1)n.

3.2.4. *The first layer Fe-chalcogenide without intervening layers with a T_c of 12 K at ambient and 27 K at 1.5 GPa: FeSe [11]*

The discovery[31] of superconductivity in FeSe [11] with a T_c of 12 K by Maw-Keun Wu *et al.* of the Physics Institute in Taipei in 2008 and the subsequent rise of its T_c to 27 K under pressure by Yoshikazu Mizuguchi *et al.* of the National Institute for Material Science at Tsukuba in 2008 demonstrated that superconductivity with a relatively high T_c in Fe-chalcogenides is associated with the (Fe_2Se_2)-layer, analogous to Fe-pnictides with the (Fe_2As_2)-layer and that As is important but not indispensable for superconductivity at high temperature. It offers a simple material system for model calculations to identify the crucial parameters in the occurrence of superconductivity in both the Fe-pnictides and Fe-chaogenides.

Because of the toxicity issue of As and the significance of (Fe_2As_2)-layers in superconductivity in Fe-pnictide superconductors, Wu *et al.* of the Institute of Physics at Taipei decided to investigate the α-phase of FeSe which possesses the tetragonal PbO-type structure of the P4/nmm space group and consists of the (Fe_2Se_2)-layers. They[31] synthesized and discovered in June 2008 the Se-deficient α-$FeSe_{1-x}$ to be superconducting although at 12 K, the first layer *Fe-chalcogenide superconductor.* The results appeared in the September 23, 2008, issue of *Proceedings of the National Academy of Sciences* in an article entitled "Superconductivity in the PbO-type Structure α-FeSe" (Fig. 16).

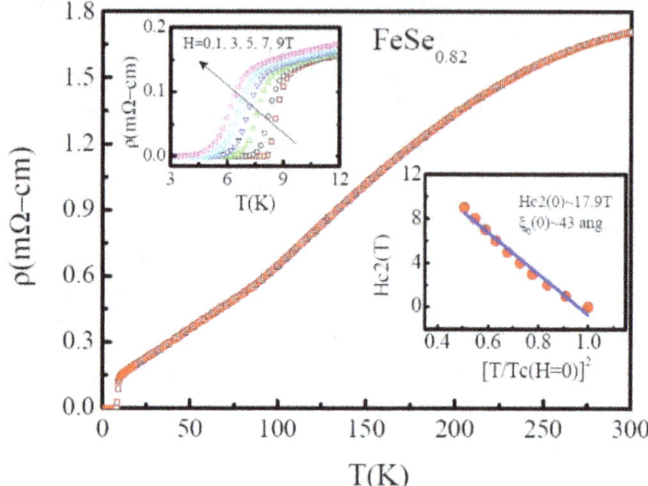

Fig. 16. $\rho(T)$ of $FeSe_{0.82}$ [111] shows a $T_c \sim 9$ K in 2008 by Wu *et al.* [F. C. Hsu *et al.*, *Proc. Natl. Acad. Sci. USA* **105**, 14262 (2008)].

Soon afterward, high pressure was found to raise the T_c to 27 K at 1.48 GPa,[32] exceeding the ambient T_c of La1111. Partial replacement of Se with Te or S makes the synthesis easier but enhances the T_c at ambient only slightly.

3.3. *Heavy fermion superconductors*

For a long time, physicists have been intrigued by the interaction of magnetism with superconductivity. Matthias introduced the rare-earth elements to a superconductor to probe the coupling between superconductivity and magnetism and to explore the possible coexistence of the two. This demonstrated the importance of the exchange interaction between the 4f electrons in the suppression of superconductivity with Ce as an exception and, by extrapolation, proposed the possible coexistence of ferromagnetism and superconductivity in the small doping region. The exceptionally large T_c-suppression by Ce was later attributed to the Kondo effect of Ce-ions. The interaction results from the strong hybridization of its magnetic 4f-electrons with the conduction electrons and thus suppresses superconductivity. However, at temperatures below the characteristic Kondo temperature, the depairing effect is reduced by Kondo-screening, which is responsible for the observation of e.g. reentrant superconductivity in $Ce_{1-x}La_xAl_2$. The question concerning the possible coexistence of ferromagnetism and supercondutivity could not be unambiguously resolved at the time because of the potential segregation of the magnetic dopant from the superconducting host. The impasse was alleviated later when rare-earth (R) containing ternary compounds were discovered, such as the rare-earth containing rhodium borides series RRh_4B_4 and Chevrel series RMo_6S_8. Superconducting, antiferromagnetic and ferromagnetic orderings were found in these compounds, clearly showing the coexistence of superconductivity with antiferromagnetism in bulk form and with ferromagnetism in a real-space modulated state known as the FFLO state. Antiferromagnetism was also shown to augment superconductivity as evidenced by the enhanced H_{c2} of some magnetic superconductors. While magnetism and superconductivity exist and interact in the chemically ordered lattice of these compounds, they occur in two loosely coupled sublattices, e.g. magnetism in the R-ion sublattice and superconductivity in the Mo_6S_8-cubes of RMo_6S_8.

In the 1970s, intermediate valence compounds, with an enhanced Sommerfeld coefficient $\gamma = Ce/T$ of the electron specific heat C_e and a quadratic temperature dependent resistivity $\rho = \rho_o + AT^2$ at low temperature, had attracted great attention. Since γ is proportional to the effective electron mass m^*, the enhanced value observed indicates that the electrons are strongly correlated and heavily renormalized. Some of these compounds display a γ that is orders of magnitude greater than that of the ordinary metallic compounds. This was clearly demonstrated by experiments on $CeAl_3$,[33] where the conduction electrons near the Fermi surface are strongly renormalized with a $\gamma = 1.620$ mJ/(mole.K^2), an exact T^2-term with $A = 35$ $\mu\Omega$ cm/K^2 and the development of coherence among the lowest lying virtual bound states below 0.2 K. Heavy fermion systems were born, arising from the close proximity of the $4f$ electrons to the Fermi energy.

The discovery of the heavy fermion superconductor $CeCu_2Si_2$ by Frank Steglich *et al.*[34] of the Technical University of Darmstadt in 1979 demonstrated for the first time that highly renormalized electrons with heavy mass can condense into an otherwise itinerant superconducting state. This represents a quantum advancement to unravel the intricate interplay of magnetism with superconductivity, to uncover the nature of the superconducting state and to reveal the fascinating normal state properties of rare-earth intermetallic compounds. It leads to the development of the exciting field of strongly correlated electron systems to which heavy fermion superconductors belong. In fact, the large γ associated with $CeCu_2Si_2$ also raised some interest in the early days to search for superconductors with high T_c in large γ materials. The effort was quickly abandoned, since the T_c of heavy fermion superconductors has never exceeded a few degrees, although finally reached 18.5 K years later in 2002.

I shall review the history of the discovery of the heavy fermion superconductors. Much of the exciting physics revealed by experiments over the last three decades on this class of materials has been detailed elsewhere.[35]

3.3.1. *Ce-based heavy fermion superconductors: CeM_2X_2 with M = Cu, Pd, Rh or Ni and X = Si or Ge; Ce_nMIn_{3n+2} where M = Co or Rh, Ir and n = 1 or 2; noncentrosymmetric superconductors*

3.3.1.1. CeM_2X_2 with M = Cu, Pd, Rh or Ni and X = Si or Ge

In the process of exploring how superconductivity or magnetism interacts with a "third kind" of collective phenomenon, i.e. the Kondo or intermediate-valence phenomenon, Frank Steglich *et al.* of the Technical University of Darmstadt discovered[34] the first heavy fermion superconductor, $CeCu_2Si_2$, characterized by the typical signatures of a bulk superconductor, i.e. a zero resistivity below $T_c \sim 0.5$ K, a Meissner effect below $\sim$0.6 K and a large specific heat jump ΔC at T_c (Fig. 17). What seems most unusual of all are its large $\gamma = 1$ J/mole-K^2, its large Pauli susceptibility and its $\Delta C/\gamma T_c \sim 1.4$ close to the BCS value. The first two show that the conduction electrons near the Fermi level in the normal state of $CeCu_2Si_2$ are highly renormalized heavy electrons due to their hybridization with the $4f$ magnetic electrons and the last demonstrates that the same heavy electrons participate in the superconducting process. The observations show that magnetism plays a crucial role in both the normal and superconducting states of $CeCu_2Si_2$. In contrast to previous magnetic superconductors, $CeCu_2Si_2$ embodies superconductivity and magnetism in the same electron system. The exciting results changed and broadened Kondo physics research. They were published in the December 17, 1979, issue of *Physical Review Letters* in an article entitled "Superconductivity in the Presence of Strong Pauli Paramagnetism: $CeCu_2Si_2$" (Fig. 17).

$CeCu_2Si_2$ crystallizes in the tetragonal $ThCr_2Si_2$-type layer structure of the I4/mmm space group similar to AFe_2As_2, the A122 Fe-pnictide superconductors. Under pressure, the T_c rises from 0.5 K rapidly up to 2.25 K at 2.5 GPa[36] and decreases at higher pressures. Despite the significance of the discovery of $CeCu_2Si_2$, the material complexity encountered and the initial skepticism associated hampered studies of the compound for many years. For example, a slight deviation from stoichiometry, e.g. a few % Cu-deficient, drastically changes its ground state from superconducting to antiferromag-

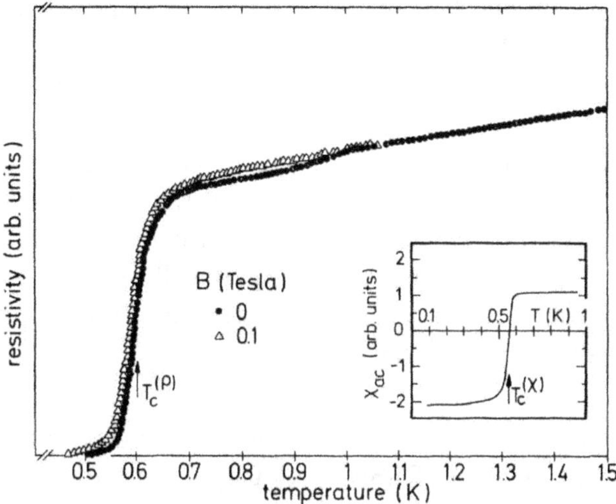

Fig. 17. $\rho(T)$ of $CeCu_2Si_2$ shows a $T_c \sim 0.6$ K in 1979 by Steglich *et al.* [F. Steglich *et al.*, *Phys. Rev. Lett.* **43**, 1892 (1979)].

netic. The isostructural series CeM_2X_2 with M = Cu, Pd, Rh or Ni and X = Si or Ge were successful synthesized only after 1993. They were found to be heavy fermion compounds and become superconducting under pressures with a T_c below 1 K, except in the case of $CeNi_2Ge_2$, which is superconducting below 0.64 K at ambient. The ensuing extensive investigation on pure and mixed compounds of the CeM_2X_2 series shows that they are strongly correlated electron systems and that the superconductivity in $CeCu_2Si_2$ is not conventional of the BCS-type but arises from electron-electron interaction that is magnetic in nature. Clearly, the CeM_2X_2 compounds sit on the border between magnetism and superconductivity, exhibiting competition and/or enforcement between the two interactions and also display instabilities associated with the ambivalent nature of Ce. The CeM_2X_2 superconductors and cuprate HTSs bear many resemblances, although with a huge gap between their T_cs.

3.3.1.2. Ce_mMIn_{3m+2} where M = Co or Rh, Ir and $m = 1, 2, \ldots, \infty$

Ce_mMIn_{3m+2} crystallize in the tetragonal layer structure of space group of P4/mmm, which consists of layers of $CeIn_3$ interleaved by MIn_2-layers. They also sit close to the border between antiferromagntism and superconductivity. While $CeCoIn_5$ and $CeIrIn_5$ are superconducting at ambient at the edge of an antiferromagnetic quantum critical point, $CeIn_3$, $CeRhIn_5$ and Ce_2RhIn_8 are antiferromagnetic at ambient but become superconducting under pressures. Aside from detailed and some subtle differences, they exhibit the complex phase diagrams that are similar to CeM_2X_2 and bear some resemblance to cuprate HTSs. For instance, $CeIn_3$, which can be considered the parent compound of the series, i.e. for $m = \infty$, orders antiferromagnetically at a $T_N = 10.2$ K and becomes superconducting at a $T_c = 0.18$ K when T_N is suppressed to < 0.18 K by a pressure of 2.6 GPa. Its $P(T)$ phase diagram is similar to that for the cuprate HTSs and its superconductivity has been shown to be magnetically mediated. It should be noted that $CeCoIn_5$ has the highest T_c among the Ce-based heavy fermion superconductors, but only at 2.3 K.[37]

3.3.1.3. The noncentrosymmetric heavy fermion superconductors: $CePt_3Si$, $CeIrSi_3$, $CeRhSi_3$, $CeCoGe_3$

The study of the effect of crystal symmetry on T_c has a long history in superconductivity research, reaching back to Matthias' days. For instance, it was thought that cubic compounds favored high T_c. The absence of centrosymmetry in thin-film and amorphous superconductors has long been recognized but without detailed research. The relationship between the spatial symmetry and the electron-pairing symmetry of a superconductor is of great academic and device interest but was not actively pursued until the discoveries of heavy fermion superconductors and HTSs. The strong spin fluctuations in strongly correlated electron systems have been shown to suppress the conventional s-wave superconductivity but to allow the spin-triplet pairing, provided that time reversal symmetry and inversion symmetry are preserved. In other words, spin-triplet pairing cannot exist in a heavy fermion superconductor without centrosymmetry. The recent discovery of noncentrosymmetric heavy fermion superconductors $CePt_3Si$, $CeIrSi_3$, $CeRhSi_3$ and $CeCoGe_3$ was surprising.[38] They all undergo an antiferromagnetic transition at different T_Ns < 21 K and enter a superconducting state under pressures at lower temperatures. Strong spin-orbit coupling has been proposed to alleviate the impasse in these noncentrosymmetric heavy fermion superconductors. Extensive experiments were carried out and indeed found spin-orbit coupling in these compounds although of different strength and isotropy.

3.3.2. *U-based heavy fermion superconductors: UBe_{13}, UPt_3, URu_2Si_2, UPd_2Al_3, UNi_2Al_3 UGe_2, URhGe, UCoGe*

In 1975, Bucher *et al.* of Bell Labs investigated the magnetic and crystal-field properties of the easily formed MBe_{13} series with M being a rare earth, Th or U, in their search for better compounds for nuclear cooling and nuclear ordering. In contrast to expectation, they did not detect the temperature-independent Van Vleck susceptibility or any magnetic ordering down to 0.65 K in UB_{13}. Instead, they observed a sharp superconducting transition at 0.97 K with a very high critical field and a specific heat that increases upon cooling below 4 K giving an estimated $\gamma \sim 1$ J/mole-K^2 at ~ 1 K. At the time, a very high critical field was known to occur only in superconductors in filamentary form. The observed superconductivity in UBe_{13} was therefore attributed to possible filamentary U impurity. The discovery of heavy fermion superconductivity was unfortunately missed. Eight years later in 1983, Hans Ott and colleagues at ETH Zurich and LANL believed that UBe_{13} might be the U-version of the $CeCu_2Si_2$ heavy fermion superconductor and decided to revisit it. They did systematic measurements on the magnetic and thermal properties of both the poly- and single-crystalline UBe_{13} and demonstrated unambiguously that UBe_{13} is a bulk superconductor as evidenced by the zero resistivity at 0.85 K accompanied by a specific heat jump (Fig. 18). Similar to Bucher *et al.*, they observed a large $\gamma = 1.1$ J/mole-K^2, a large Pauli magnetic susceptibility and a high critical field and concluded that UBe_{13} is indeed a heavy fermion superconductor. The discovery of the second heavy fermion superconductor and the first containing the $5f$-electrons of actinides was published in the May 16, 1983, issue of *Physical Review Letters* in an article entitled "UBe_{13}: An Unconventional Actinide Superconductor."[39]

UBe_{13} crystallizes in the cubic $NaZn_{13}$-type structure with the Fm3c space group. It is the only compound formed congruently in the U-Be system. Quality UBe_{13} samples can thus be obtained readily once the Be-toxicity and the U-supply issues are resolved. The unambiguous UBe_{13} results in comparison with $CeCu_2Si_2$ in its early stage accelerated the research on heavy fermion superconductivity. The T_c of UBe_{13} was found to decrease with pressure while the normal state resistivity returns more to the conventional coherent Kondo-lattice-like. The Th-doped samples $U_{1-x}Th_xBe_3$ exhibit a complex phase diagram with two superconducting phases for $2\% < x < 4\%$. Many studies on UBe_{13} in the ensuing years have seemed to show that it is rather different from other heavy fermion superconductors and still leave some questions about its magnetic and superconducting properties unanswered.

Soon after the discovery of UBe_{13}, UPt_3 was discovered[40] as the third heavy fermion superconductor with a $T_c \sim 0.54$ K and a large $\gamma \sim 1$ J/mole-K^2 in 1984. It is perhaps the most extensively studied heavy fermion superconductor with an unambiguously established unconventional

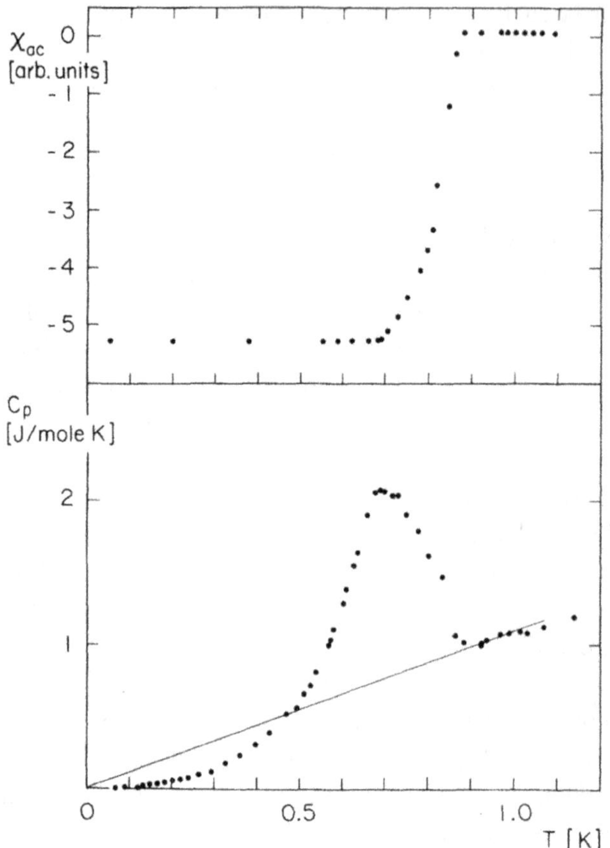

Fig. 18. $\chi(T)$ and $C_p(T)$ of UBe_{13} shows a $T_c \sim 0.85$ K in 1983 by Ott *et al.* [H. R. Ott *et al.*, *Phys. Rev. Lett.* **50**, 1595 (1983)].

superconducting state with a spin-triplet f-wave pairing. Later specific heat measurements showed a second superconducting transition at 0.48 K at zero field. In the presence of a magnetic field, $H_c(T)$s for the first and second superconducting phases cross at 0.44 K and 0.6 T and evolve into a third superconducting phase above 0.6 T between the two $H_c(T)$s. An unusual tetracritical point therefore forms. Seven more U-based heavy fermion superconductors were discovered after 1984. Some (UPt_3, URu_2Si_2, UPd_2Al_3, UNi_2Al_3) order antiferromagnetically before entering their superconducting state on cooling, while others (UGe_2, URhGe, UCoGe) become ferromagnetic before undergoing a superconducting transition (for UGe_2, under pressure, but for URhGe and UCoGe, at ambient). Similar to the Ce-based heavy fermion superconductors, their superconducting states take place deep in the magnetically ordered states, be it ferromagnetic or antiferromagnetic.

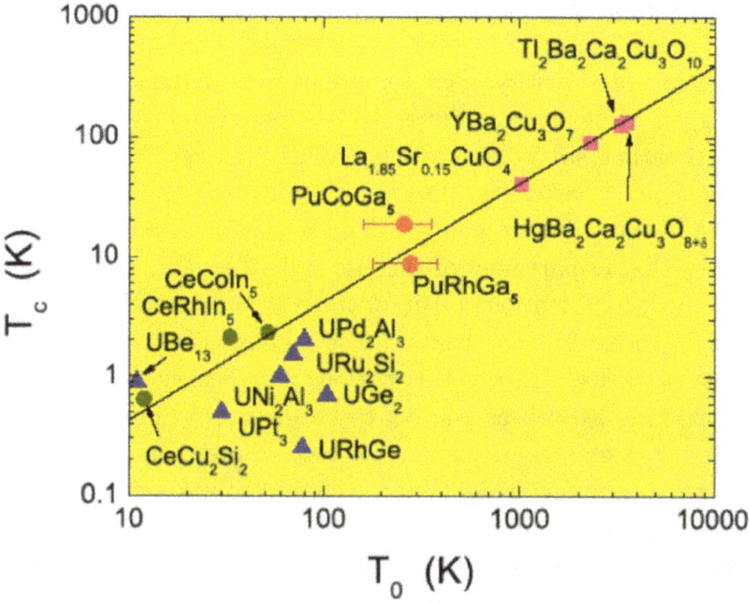

Fig. 19. T_c–T^* where T^* measures the strength of spin-fluctuations [C. Pfleiderer, *Rev. Mod. Phys.* **81**, 1551 (2009); plot from N. Curro *et al.*, *Nature* **434**, 622 (2005) as shown in J. L. Sarrao and J. D. Thompson, *J. Phys. Soc. Jpn.* **76**, 051013 (2007)].

3.3.3. *Actinide-based heavy fermion superconductors: $PuCoGa_5$, $PuRhGa_5$, $NpPd_5Al_2$*

The actinide-based $PuCoGa_5$, $PuRhGa_5$ and $NpPd_5Al_2$ are all unconventional superconductors at ambient with the highest T_c among all heavy fermion superconductors at 18.5 K, 8.7 K and 4.9 K, respectively.[41] The high T_cs have been attributed to the $5f$-electrons, which have been shown to sit at the border between itinerancy and localization, signaling some kind of instabilities, and to have a wider band-width and stronger spin-orbit coupling compared with the $4f$-electrons. These unusual characteristics place this group of compounds between the 3d-HTSs and the $4f$-heavy fermion superconductors and become the bridge between the two, as clearly demonstrated by the T_c–T^* relation where T^* is the characteristic temperature of the spin-fluctuations in the compounds (Fig. 19). Such a T_c–T^* seems to suggest that the study of heavy fermion superconductors may provide insight into the underlying working mechanism of HTS and may even lead to higher T_c.

4. The VHTS (or RTS)-Period: 2009–

The designation of the very high temperature superconductivity (VHTS) or, more aggressively, room temperature superconductivity (RTS) period after 2009 is rather subjective and speculative in terms of materials and commencing time. It reflects in part the hope and vision of the practitioners in the field and the public in general. The year 2009 also coincides with the year when the U.S. Department of Defense and other federal funding agencies started making a conscientious push for the search for novel superconductors with higher T_c to harness the fruits of the accumulative knowledge on HTS over the last 23 years. During this period, it is anticipated that the focus will be on existing as well as the yet-to-be discovered materials to achieve VHTS with a T_c higher than the current record, preferably RTS, and at the same time, to obtain new insight into HTS and to uncover new physics.

In the last 20 years, many theoretical models have been advanced to account for numerous unusual observations in HTSs and many HTS prototype devices constructed and demonstrated successfully with superior performance to their non-superconducting counterparts. To take advantage of the full prowess of superconductivity for our daily lives, RTSs will be the ultimate goal, although lofty. VHTSs will reduce the burden of cooling and RTSs will eliminate completely the inconvenience of cooling. Even before the discovery of superconductivity, people were already fascinated by the concept of perpetual motion machines in our lives and RTS is considered the closest thing to such a machine. Therefore, RTS found its way into popular culture from science-fiction to cinema before it became serious science. The most recent example is the "Unobtainium" in the highest grossing movie *Avatar*. RTS, if achieved, could profoundly change the world scientifically and technologically.

4.1. *A practical room temperature superconductor*

As a graduate student in 1967, I asked my thesis advisor, the late Professor Bernd T. Matthias, whether there existed a RTS and if yes, where. His answer was succinct and direct: "Yes, just go to the edge of the universe." At the time, he had just discovered a $T_c \sim 21$ K in a pseudoternary intermetallic compound, $Nb_3(Al,Ge)$ and the ambient temperature at the edge of the universe is 3 K, residue of the Big Bang. A 21 K superconductor is thus a RTS at the edge of the universe. However, the edge of the universe is more than 13 billion light years away from us, an astronomical distance that indeed can be reached by us only in our dreams. Therefore, strictly speaking, RTS is a relative term, depending on its operating environment temperature. In principle, one can achieve RTS either by raising the T_c of a superconductor or by lowering the ambient temperature of the operating environment so that the two temperatures can cross. In this discussion, our goal is to raise the T_c.

Over the years, different target-T_cs have been set at different times, e.g. 77 K (liquid nitrogen boiling point before YBCO), 100 K (inside the cargo bay of the Space Shuttle or on the moon's surface opposite the sun before TBCCO), 120 K (liquid natural gas boiling point before HBCCO), 148 K (liquid Freon boiling point before HBCCO under pressure), 198 K (dry ice temperature) and 300 K (temperature of our living environment). With the superconductors we have, many of these target temperatures have been reached. Unfortunately, they are still not practical for the ubiquitous applications of superconducting devices in our daily life. For instance, $HgBa_2Ca_2Cu_3O_9$ with the current record T_c of 164 K under high pressure can be considered a RTS in a liquid Freon environment, achievable by an air conditioner. However, the high pressure required renders it impractical not to mention the undesirable effect of Freon on the protective ozone layer in our upper atmosphere. As a result, the practical and desirable RTS that we want today is one that has a T_c of 300 K, high enough so that its superconducting state can be achieved in our living environment without any cryogenic cooling. On the other hand, in order to take advantage of 90% of the maximum current-carrying capacity of a superconductor, the operating temperature usually should be kept at $\sim$70% of its T_c or lower. For an operating temperature of 300 K of our living environment, the T_c required will therefore be $\sim$430 K. However, we shall be very happy to settle for a T_c of 300 K. The discovery of such a superconductor will have an all-encompassing impact on our lives and a new industrial revolution will follow. According to our current theoretical understanding and experimental evidence, there exists no reason why RTS should be impossible.

4.2. *Examples of interesting claims*

The long and tortuous path in the search for superconductors with higher T_c has been dotted with triumphs of success and agonies of failure, including extravagant claims. Being an optimist who believes that whatever is not prohibited by the basic laws of physics will happen, I do not dismiss claims outright until they are proven false by reasoning or by testing them experimentally to the best of my ability. Consequently, I have been contacted by many such claimers. Unfortunately, so far, to find them not superconducting is the norm, let alone superconducting at room temperature. Examples are abundant, but let me just cite just a few for amusement. A few years ago, the head of a California based company contacted me and asked me to test samples of an alleged RTS polymer material developed in the former Soviet Union with published references. I did but found them to be insulating. In another example, the anchor lady of a reputable TV program asked me to test a piece of material that was allegedly left by an extraterrestrial vehicle in an Arizona desert and determined by a few reputable labs to be superconducting at room temperature in the presence of a strong microwave. I found it to be a rather ordinary metal, containing elements such as Fe, Mn, In, etc., and not superconducting. Yet another example was a call several years ago from a Croatian physicist who asked me to sign a disclosure agreement before faxing me information about the RTS he discovered. After seeing his faxed message, I had some doubts but still repeated what was described in the message. When told of our negative results, he attributed them to our alleged poor sample quality. Still giving him the benefit of doubt, I asked for a piece of his sample to test, but he wanted me to purchase it after he mass-produced it and put it on the market before Easter that year. Unfortunately, I have yet to hear from him since that Easter.

While many of the claims can be ignored outright due to experimental artifacts or dismissed after cursory examinations, others are interesting and may be worth further examination. Revisiting these may lead to new understanding of HTS and new mechanisms for HTS, in addition to possible higher T_c. I shall briefly describe three such examples below:

(i) In 1946, Ogg reported[42] a large drop in resistance and the presence of persistent current in their sodium-ammonia solution at temperatures as high as 180 K upon rapid cooling but not on slow cooling. He attributed the observations to superconductivity arising from possible Bose–Einstein condensation of electron-pairs. Conflicting experimental reports followed immediately and interest died off rapidly. Interest in this material system was renewed in the early 1970s, mostly in the former Soviet Union, but waned in the late 1970s. However, the topic has been picked up by some recently in the West. As the study of HTS progresses, similarities between cuprates and the Na-ammonia can be found, such as phase-separation, possible occurrence of phase coherence and pairing at different temperatures, etc. It may be time to bring the vast knowledge and sophisticated diagnostic tools developed for HTS in the last 23 years to bear on unraveling the mystery of this interesting system.

(ii) In 1978, Brandt *et al.*[43] claimed to observe superconductivity up to 170 K in CuCl under pressures based on the large resistive and magnetic anomalies detected, and attributed the observation to electron pairing via the exchange of excitons. It may be interesting to note that ac susceptibility and resistivity anomalies above 90 K over a temperature range of 10–20 K in the thermally non-equilibrium state during rapid warming were also detected, corresponding to a possible paramagnetic-diamagnetic-paramagnetic transition. No definitive confirmation or refutation has been reported to date. It has been suggested that a more complex material than CuCl, due to its disproportionation in rapid cooling under pressure, may be responsible for the observation. Given the role of Cu in HTS and certain similarities appearing in HTS and CuCl under pressure and rapid cooling, another visit may be warranted.

(iii) Since the discovery of YBCO, there have been numerous claims of T_c in cuprates close to 300 K.[44] None satisfied the four criteria I set for a genuine superconductor: i.e. zero resistivity, large diamagnetic shift decisively showing the Meissner effect, high stability long enough for a definitive diagnosis and good reproducibility from sample to sample and from lab to lab. These reports may at best be called Unidentified Superconducting Objects (USOs). Some of the USOs have been shown to arise from experimental artifacts or from misinterpretation of the data. However, the high frequency of sighting USOs in the similar high temperature range ($\sim$240–300 K) and in some reputable labs makes these USOs too tantalizing to ignore. In view of the extremely unstable chemical nature of the cuprates, and the small coherence length for HTS with very high T_c, the fleeting nature of USOs reported may not be totally unexpected.

4.3. *Examples of visionary predictions*

Many visionary predictions have been made to reach the RTS-wonderland. A few examples proposed in 1960s and 1970s that may still provide valuable inspiration are given below:

- In 1964, Little[45] examined the question originally posed by London — whether superconductivity occurs in organic macromolecules within the framework of the BCS theory — and concluded that superconductivity above room temperature is not only possible but also expected in organic macromolecules of a special design. The proposed model macromolecule consists of a long spine and a series of side chains. By choosing the molecules in the side chains with proper oscillation of charges, the electrons in the spine can be polarized to form pairs via the exchange of excitons with the side chain molecules. According to Little's estimation, a superconductivity transition should occur at temperatures well above room temperature. One challenge in realizing Little's vision is in synthesizing and characterizing the macromolecules with the proposed design. With the great advancement in material syntheses and diagnostic techniques for nanomaterials made in recent years, an attempt to produce and characterize macromolecules including the model macromolecule proposed by Little may be worthwhile and timely.

- In 1964, Ginzburg[46] noticed the possible drawbacks of the one-dimensional organic macromolecule superconductor proposed by Little, such as fluctuations, instabilities and lack of Coulomb screening, and therefore focused instead on two-dimensional materials. He proposed a novel mechanism to enhance T_c within the weak limit of BCS, i.e. to increase T_c to as high as 105 K by taking advantage of the high characteristic temperature Θ_e of the electron-electron interaction via the exchange of excitons. In order to facilitate the so-called exciton mechanism, he conjectured that one should consider the following material systems: metallic thin films, metal surfaces covered by dielectrics, metal sandwiches with dielectrics in between and two-dimensional layer compounds. Many experiments were done in the ensuing years but did not yield clear confirmation. It may not be surprising to find a connection between the layered materials Ginzburg proposed and the layer cuprate and Fe-pnictide HTSs. New understanding and synthesis-knowledge of layer HTS materials in the last two decades will provide new means to test the proposal.
- In 1973, Allender, Bray and Bardeen[47] carried out a more rigorous analysis on a simple model system to explore the existence of the exciton mechanism and its impact on T_c, if it exists.[48] Their model material consists of a metal thin film of Fermi energy E_F on a semiconductor of a narrow gap of E_g. They concluded that T_c can indeed be enhanced by adding the exciton mechanism to the phonon mechanism when the material conditions are optimized and that the exciton mechanism is a promising vehicle for higher T_c. They also pointed out the difficulty in creating the model material system with the proper metal/semiconductor interfaces that satisfy the stringent requirements. Several experimental attempts had been made before the late 80s but with no success reported. With the recent advancement in multi-thin-film synthesis, one should be able to artificially prepare samples satisfying the stringent criteria specified for the exciton

mechanism. Naturally occurring metal/semiconductor layered compounds may be a good alternative if the E_F is adjusted to be located in the middle of E_g.

4.4. *Common features of superconductivity with high T_c*

At the moment, we do not yet know enough about a RTS other than that it has a $T_c > 300$ K, as we have chosen, to ask intelligent questions with answers that will lead us to the promised land of RTS. It is also not unlikely that a RTS to be discovered may turn out to be a very different material from the HTSs to which we are accustomed. However, being a superconductor, VHTS and/or RTS has to have the basic superconducting characteristics and very likely has to share at least some features with the HTSs we have. A summary of these features is considered most helpful and can serve as a launching pad for our attempt to look for VHTS and/or RTS.

4.4.1. *Electron-pairing and phase-coherence*

According to the BCS theory, electron-pairing and phase-coherence are the very basic features of a superconducting state that exhibits zero resistivity ($\rho = 0$) and perfect diamagnetism (magnetic induction $B = 0$). It has long been accepted that electron-pairing and phase-coherence occur at the same temperature, T_c, for LTSs. However, suggestions have been made that electron-pairing may precede phase-coherence at a temperature higher than T_c in the cuprate HTSs. If true, the feature provides a new degree of freedom in the search for VHTSs.

4.4.2. *Strongly correlated electron systems with multi-interactions*

The current HTSs belong to the class of materials of transition metal oxides and pnictides where strongly correlated electrons exist. Strongly correlated electron systems are rich in electronically induced phase transitions over a wide temperature range and therefore may provide various avenues to HTS. Figure 20 shows schematically the linkages between magnetism and superconductivity and between magnetism and ferroelectricity in strongly correlated electron systems. In contrast to the original thinking, we now find that, under the proper conditions, magnetism and superconductivity can enforce each other; similarly, magnetism and ferroelectricity can help each other. I believe that if the missing link between ferroelectricity and superconductivity is found, higher T_c can be achieved. Optimization of multi-interactions will be the key.

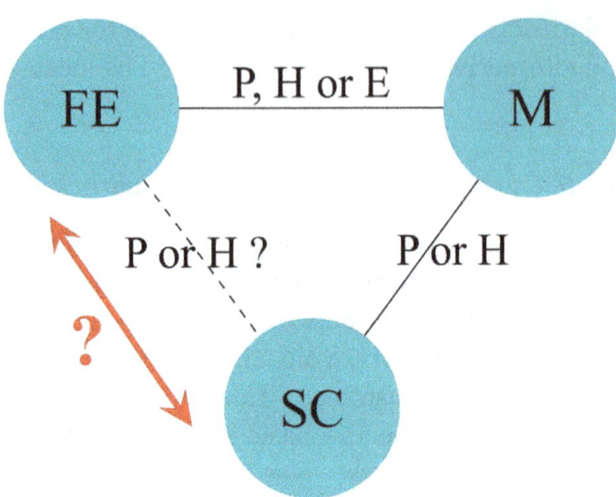

Fig. 20. The interplay between the superconducting-magnetic and ferroelectric-magnetic interactions in strongly correlated electron systems, where P is pressure, H is magnetic field and E is electric field.

4.4.3. *Instabilities and fluctuations*

In the strongly correlated electron systems to which HTSs belong, there exist various electronic, phononic and electric charge instabilities, e.g. superconducting, structural, antiferromagnetic, ferromagnetic, ferroelectric, antiferroelectric, metal-insulator, charge-density-waves, spin-density-waves, charge-order, orbital-order, etc., some of which can induce transitions at very high temperatures. The associated fluctuations may become a source of electron pairing for superconductivity. The effect becomes most effective at temperatures near the phase transition, which is tunable by both physical and chemical means.

4.4.4. *Layered structure with multi-substructures*

Reduced dimensionality has been shown to facilitate high T_c, generally due to the relatively large density of states and the stronger electron-electron interaction associated with a 2D electron system. Compounds with two distinct substructures have an extra advantage: the active block for superconductivity to take place and the charge reservoir block to provide charge carriers for inducement of superconductivity and suppression of unwanted ordering through doping and/or physical means, while keeping the integrity of the layers and minimizing defects in the layer substructure for superconductivity.

4.4.5. *Covalency, mixed valence and low carrier density*

Both cuprate and Fe-pnictide HTSs appear to originate from a magnetic insulating parent state through doping or pressure by changing the carrier density. Mixed valence states appear to be indispensable for the parent compounds to evolve into a conducting state with strong covalency to achieve high T_c. Conducting compounds of this kind will have low carrier density.

4.5. *A holistic, multidisciplinary, enlightened empirical approach*

In philosophy, scholars have long debated "Empiricism versus Rationalism." The difference between the two lies mainly in the different relationships between reason and knowledge and between experience and knowledge: empiricists consider experience to be the source of knowledge and independent of reason, while rationalists claim reason to be the source of knowledge and independent of experience. In the long search for superconductors of higher T_c, Matthias and Mueller reminded us at different times of the effectiveness of the empirical approach.

The study of HTS in the past shows that HTS and future RTS are both all-encompassingly complex in terms of materials, physics and chemistry. It appears that a holistic, multidisciplinary, enlightened empirical approach that embodies both experience and reason and engages researchers, experimental and theoretical, from different disciplines with complementary expertise appears to be most effective to lead us to achieve VHTSs and/or RTSs. The historical review given above can guide us enlightened-empirically to the most fruitful future path to VHTSs and/or RTSs. It includes also some of the interesting claims that may warrant further tests, some visionary predictions that may be pursued and the common features of superconductors with high T_c that may form the general basis for the search for novel superconductors. The worldwide extensive studies on HTS cuprates, Fe-pnictides and heavy fermion superconductors over the decades have given us unusual physical insight into materials, powerful computational capabilities for materials, sensitive material characterization techniques and new materials synthesis tools. The time is ripe to bring all these skills to bear in the search for novel superconductors with higher T_c, preferably at room temperature. If history is any guide, I strongly believe that, in the process, in addition to higher T_cs, new physics will be discovered and novel materials found with both scientific and technological significance.

5. Conclusion

High-temperature superconductivity never ceases to excite scientists, engineers and the general public because of the continuing challenges it poses to scientists and the promises it holds for technologists. The search for HTS commenced immediately after the discovery of superconductivity by Kamerlingh Onnes in 1911. As described in Sec. 1, HTS is a relative term because T_c varies with time and the whole period of superconductivity is divided into three periods for the convenience of presentation in this paper, namely, LTS, HTS and VHTS and/or RTS. Brief recounts of the discoveries of the major HTS material systems were given.

The era of modern HTS was heralded by the discovery of cuprate superconductors in 1986 and the tumbling down of the 77 K liquid-nitrogen temperature barrier in 1987. In the ensuing years, impressive progress has been made in HTS following the extensive world-wide effort: T_c has risen continuously to the current record of 134 K at ambient and 164 K at 30 GPa, many new compounds discovered, new physics uncovered, numerous theoretical models proposed, various theoretical and experimental techniques developed, and many prototypes constructed and demonstrated. However, the record-T_c has stagnated since 1993, a commonly accepted microscopic theory has yet to be found and the full technological prowess of HTS is yet to be developed.

Although HTS reduces the burden of cryogenic cooling and a VHTS will further reduce the burden, making superconductivity applications more practical, RTS will eliminate completely the need for cooling and applications will become ubiquitous. Unprecedented scientific excitement and challenges associated with VHTS and RTS are expected. There exists no theoretical or experimental reason why VHTS or RTS is impossible. In addition, I believe that whatever physics laws do not say cannot happen will happen eventually. It appears that a holistic comprehensive enlightened empirical approach may be most effective in the search for VHTS and/or RTS and some specific steps have been proposed by optimizing the multi-interactions, controlling the instabilities, and effectively introducing charge carriers to materials with characteristics as described in Sec. 4.4.

Acknowledgments

I would like to thank my colleagues Yuyi Xue and Bernd Lorenz for discussions and in the preparation of figures. Partial support for the work in

Houston by the U.S. Air Force Office of Scientific Research, the U.S. Department of Energy through Oak Ridge National Laboratory, the T.L.L. Temple Foundation, the John J. and Rebecca Moores Endowment, and the State of Texas through the Texas Center for Superconductivity at the University of Houston; and at Lawrence Berkeley Laboratory by the Director, Office of Science, Office of Basic Energy Sciences, Division of Materials Sciences and Engineering of the U.S. Department of Energy is greatly appreciated.

References

1. H. K. Onnes, *Leiden Comm.* 120b (1911); *ibid.* 122b (1911); *ibid.* 124c (1912).
2. B. T. Matthias, *Phys. Rev.* **97**, 74 (1955).
3. J. R. Gavaler *et al.*, *J. Appl. Phys.* **45**, 3009 (1974); L. R. Testardi *et al.*, *Solid State Comm.* **15**, 1 (1974).
4. J. Bardeen, L. N. Cooper and J. R. Schrieffer, *Phys. Rev.* **106**, 162 (1957).
5. C. W. Chu and L. R. Testardi, *Phys. Rev. Lett.* **32**, 766 (1974); C. W. Chu, *Phys. Rev. Lett.* **33**, 1283 (1974); C. W. Chu and V. Diatschenko, *Phys. Rev. Lett.* **41**, 572 (1978).
6. W. L. McMillan, *Phys. Rev.* **167**, 331 (1968).
7. See, for example, J. Orenstein and A. J. Millis, *Science* **288**, 468 (2000) and references therein.
8. J. G. Bednorz and K. A. Müller, *Z. Phys. B* **64**, 189 (1986).
9. J. G. Bednorz and K. A. Müller, *Nobel Lecture*, December 8, 1987.
10. C. W. Chu *et al.*, *Phys. Rev. Lett.* **58**, 405 (1987); C. W. Chu *et al.*, *Science* **235**, 567 (1987).
11. R. J. Cava *et al.*, *Phys. Rev. Lett.* **58**, 408 (1987).
12. Y. Tokura *et al.*, *Nature* **337**, 345 (1989).
13. M. K. Wu *et al.*, *Phys. Rev. Lett.* **58**, 908 (1987).
14. C. W. Chu, Patent application, January 12, 1987.
15. P. H. Hor *et al.*, *Phys. Rev. Lett.* **58**, 1891 (1987).
16. H. Maeda *et al.*, *Jpn. J. Appl. Phys.* **27**, L209 (1988).
17. Z. Z. Sheng and A. M. Hermann, *Nature* **332**, 138 (1988).
18. A. Schilling *et al.*, *Nature* **363**, 56 (1993).
19. L. Gao *et al.*, *Phys. Rev. B* **50**, 4260(R) (1994) and references therein.
20. C. W. Chu *et al.*, *Science* **277**, 1081 (1997).
21. M. G. Smith *et al.*, *Nature* **351**, 549 (1991).
22. M. Azuma *et al.*, *Nature* **356**, 775 (1992).
23. Y. Kamihara *et al.*, *J. Am. Chem. Soc.* **130**, 3296 (2008).
24. A. Barth and W. Marx, arXiv:cond-mat/0609114v2 [cond-mat.supr-con] (2006).
25. H. Hosono, *J. Phys. Soc. Jpn.* **77**, Suppl. C, 1 (2008); S. Uchida, *ibid.* **77**, Suppl. C, 9 (2008); K. Ishida *et al.*, *ibid.* **78**, 062001 (2009).
26. B. Lorenz *et al.*, *Phys. Rev. B* **78**, 012505 (2008).

27. See, for example C. W. Chu and B. Lorenz, *Physica C* **469**, 326 (2009) and references therein.
28. M. Rotter *et al.*, *Phys. Rev. Lett.* **101**, 107006 (2008).
29. J. H. Tapp *et al.*, *Phys. Rev. B* **78**, 060505(R) (2008); X. C. Wang *et al.*, *Solid State Comm.* **148**, 538 (2008).
30. T. L. Xia *et al.*, arXiv:1001.3311v2 [cond-mat.supr-con] (2010).
31. F. C. Hsu *et al.*, *Proc. Natl. Acad. Sci. USA* **105**, 14262 (2008).
32. Y. Mizuguchi *et al.*, *Appl. Phys. Lett.* **93**, 152505 (2008).
33. K. Andres *et al.*, *Phys. Rev. Lett.* **35**, 1779 (1975).
34. F. Steglich *et al.*, *Phys. Rev. Lett.* **43**, 1892 (1979).
35. C. Pfleiderer, *Rev. Mod. Phys.* **81**, 1551 (2009) and references therein.
36. B. Bellarbi *et al.*, *Phys. Rev. B* **30**, 1182 (1984).
37. C. Petrovic *et al.*, *J. Phys.: Condens. Matter* **13**, L337 (2001).
38. R. Settai *et al.*, *J. Phys. Soc. Jpn.* **77**, 073705 (2008) and references therein.
39. H. R. Ott *et al.*, *Phys. Rev. Lett.* **50**, 1595 (1983).
40. G. R. Stewart *et al.*, *Phys. Rev. Lett.* **52**, 679 (1984).
41. J. Sarrao *et al.*, *Nature* **420**, 297 (2002); F. Wastin *et al.*, *J. Phys.: Condens. Matter* **15**, S2279 (2003); D. Aoki *et al.*, *J. Phys. Soc. Jpn.* **76**, 063701 (2007).
42. R. A. Ogg, *J. Am. Chem. Soc.* **68**, 155 (1946).
43. N. B. Brandt *et al.*, *JETP Lett.* **27**, 33 (1978).
44. B. G. Levi, *Physics Today* **47**(2), 17 (1994); R. F. Service, *Science* **265**, 2014 (1994).
45. W. A. Little, *Phys. Rev.* **134**, A1416 (1964).
46. V. L. Ginzburg, *JETP* **47**, 2318 (1964).
47. D. Allender *et al.*, *Phys. Rev. B* **7**, 1020 (1973).
48. C. W. Chu, *AAPPS Bulletin* **18**, 9 (2008).

THE EVOLUTION OF HIGH-TEMPERATURE SUPERCONDUCTIVITY: THEORY PERSPECTIVE

Elihu Abrahams

Center for Materials Theory, Serin Physics Laboratory, Rutgers University, Piscataway, NJ 08854, USA
and
**Department of Physics and Astronomy, University of California, Los Angeles, 405 Hilgard Avenue, Los Angeles, CA 90095, USA*

Theoretical developments concerning the high transition temperature cuprate superconductors are reviewed.

1. Historical Remarks: Strongly-correlated Electrons

It is often pointed out that the discovery in 1986, by J. G. Bednorz and K. A. Müller, of superconductivity at 30 K in lanthanum barium copper oxide[1] generated a historically unprecedented volume of research papers. They ranged from materials preparation to refined experimental technique to novel theoretical proposals and tools. In the domain of theory, no paper has been more influential than that of P. W. Anderson on the resonating valence bond ("RVB") state and superconductivity[2] in 1987. This paper, although concentrating on RVB as a consequence of strong electron–electron correlations, did set a particular framework for addressing the physics of the cuprates, as well as emphasizing the problem of the physics of situations in which the largest energy scale is determined by electron–electron interactions, rather than by the kinetic energy (in contrast to the circumstances of the familiar Landau Fermi-liquid approach[3]).

Such situations are the domain of the modern many-body problem in condensed matter physics. The theoretical challenge is to find methods to treat systems in which strong interactions dominate the physics: when perturbative methods fail and/or when fluctuations are so strong that

---

*Current address

mean-field treatments are inadequate. The explosion of activity in this field of strongly-correlated electrons is due in part to the aforementioned seminal article of Anderson.

Of course, high-temperature superconductivity is not the first case of strongly-correlated electron physics in condensed matter. Other examples, which attracted much attention earlier and are still important problems include the Kondo problem, the so-called heavy-fermion metals, quantum antiferromagnetism and other magnetic states, metal-insulator transitions, non-Fermi liquid normal metallic states, quasi-one-dimensional systems ... Nevertheless, it is the case that much of the intense activity in recent years to construct frameworks beyond the Fermi liquid that are suitable for the strong-correlation situation has been generated by the mysteries of high-temperature superconductivity.

The Bardeen-Cooper-Schrieffer (BCS) theory of superconductivity,[4] which this volume commemorates, and its strong-coupling extensions,[5] brilliantly accounted for "conventional" superconductivity. It is also a beautiful exemplar of the fundamentals of the practice of theoretical condensed matter physics. Starting from the complicated many-body Hamiltonian for a solid, containing the dynamics and interactions of electrons and ions, one needs to systematically remove the irrelevant degrees of freedom. In general, this results in a *reduced Hamiltonian* that contains everything necessary to capture the low-energy physics of the system. Thus, in the BCS case, even the electron–phonon interaction on which the theory is based is not explicitly treated. Instead, the reduced Hamiltonian contains only the kinetic energy of the Fermi liquid quasiparticles of the normal state and a short-range attractive interaction between them. Remarkably, this is sufficient to account for all the essential physics of most conventional "low-temperature" superconductors.

2. Phase Diagram and Effective Hamiltonians

In contrast to conventional superconductors, the cuprates are complex materials with many unique properties determined by strong repulsion between the mobile carriers. It is worthwhile to illustrate the complexity of the physics of the cuprates by showing a sketch of a typical phase diagram, Fig. 1. In this phase diagram we can see a number of themes that dominate current research on strongly-correlated materials, in particular competing phases and quantum critical points. For the cuprates, three of these phases have received the most attention. The mystery of the superconducting phase

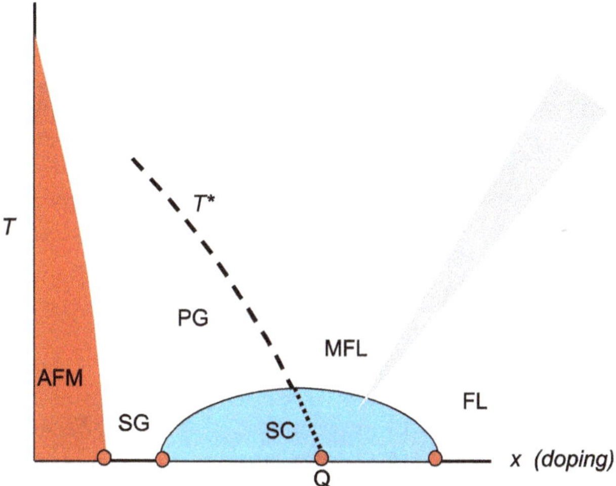

Fig. 1. Typical phase diagram of a hole doped high-T_c cuprate superconductor. AFM = antiferromagnet, SG = spin glass, SC = superconductor, PG = pseudo-gapped metal, MFL = marginal Fermi liquid, FL = Fermi liquid, gray wedge = crossover region, red dots = quantum critical points.

lies in the question "What is the mechanism?" Even before the symmetry of the superconducting order parameter was determined not to be the conventional s-wave, it was understood[6] that pairing by exchange of phonons as in traditional superconductors was not the answer. Clues were and are being sought in the behavior of the two unusual normal states: a "pseudogap" phase in the low-doping region and a strange metal, or non-Fermi liquid, phase in the region above the maximum of the superconducting dome. This region of doping is called "optimal." For reasons to be discussed below, the strange metal is labeled MFL, standing for "marginal Fermi liquid," in the figure. In the pseudogap region, a suppression of the density of states near the Fermi surface is observed. Whether all these are true thermodynamic phases is still an open question.[7] In any case, in spite of more than two decades of theoretical activity, there is as yet no comprehensive theory that gives all aspects of the unusual behaviors seen in the various regions of the phase diagram.

The morphology of all the superconducting cuprate compounds involves CuO_2 two-dimensional (2D) layers, as depicted in Fig. 2(a). The Cu configuration is $3d^9$ (Cu^{2+}) and each oxygen is $2p^6$ (O^{2-}). Simple tight-binding considerations involving the relevant Cu $3d_{x^2-y^2}$ and O $2p_x, 2p_y$ orbitals, as depicted in the figure, show that these lead to three planar bands. Since

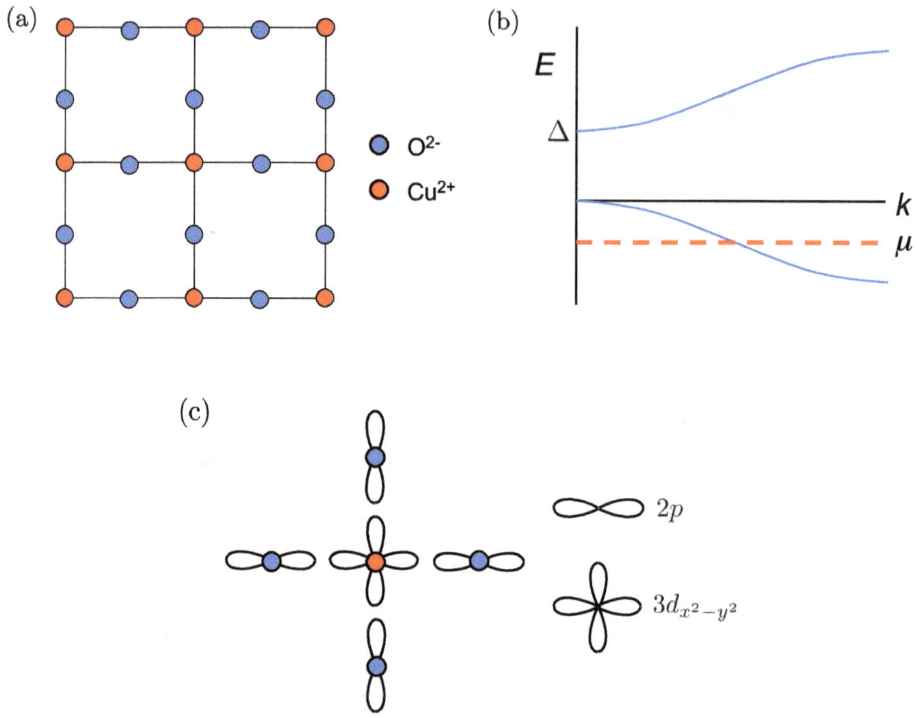

Fig. 2. Electronic structure: (a) Structure of CuO_2 planes. (b) Characteristic tight-binding band structure for holes. The chemical potential μ is shown for the half-filling (undoped) case. (c) Cu and O orbitals giving rise to the bands in (b).

each Cu has one hole in an otherwise full $d_{x^2-y^2}$ band and in the hole-doped superconductors, extra holes are added chemically, it is convenient to describe the energy bands in terms of hole states — thus, an antibonding band dispersing downward, a non-dispersive non-bonding band, and a bonding band dispersing upward, see Fig. 2(b).

In the stoichiometric (undoped, $x = 0$ in Fig. 1) compounds, the band arising from the Cu^{2+} ions is half full *i.e.* one hole per site, Fig. 2(b). In view of the fact that these are insulating antiferromagnets, it is natural to describe them as Mott insulators.[8] This is contrary to the conclusion one would get from band-structure calculations, which all predict the undoped compounds to be metallic; this is strong evidence that these materials are in the class of strongly-correlated systems. When divalent strontium is doped onto trivalent lanthanum sites in lanthanum cuprate (La_2CuO_4), for example, to give $La_{2-x}Sr_xCuO_4$ (LSCO), x extra holes away from half filling transform the compound into a metal and, for moderate x, below a transition temperature

T_c, a superconductor. Thus the superconducting compounds may briefly be described as two-dimensional doped Mott insulators.

Indeed, in single crystals, all properties are highly anisotropic. For example, the typical cuprate tetragonal crystal structure has a c/a ratio of almost five (in lanthanum cuprate) and in the normal state of the superconducting compounds the measured resistivities are in the ratio ρ_c/ρ_a of at least 10 and usually with opposite temperature dependences. Thus the physics, hence the observed magnetism and superconductivity, is dominated by the CuO_2 planes common to all the cuprates.

The phase diagram of Fig. 1 depicts the situation when extra holes are doped into the CuO_2 planes. There is also the class of electron-doped cuprates, such as $Nd_{2-x}Ce_xCuO_4$ (NCCO). The phase diagram for electron-doped materials usually shows the same main phases as the hole doped superconductors, but with a number of differences, principally a broader antiferromagnetic region. Research on the properties of electron doped compounds, while extensive,[9] is not yet as comprehensive as that on the hole doped cuprates. Taking the point of view that the experimental differences between the two classes do not signify that there is a basic divergence in the underlying physics, we concentrate in what follows on the hole-doped case.

There are three main families of the hole-doped high-T_c superconductors. They differ in the number per unit cell of CuO_2 planes, the manner of doping away from a parent antiferromagnetic insulator and the presence or absence of chains between the planes. There is the single-plane lanthanum cuprate family (LSCO) mentioned above; the two-plane compounds based on $YbBa_2Cu_3O_{7-x}$ (YBCO) where holes are introduced by oxygen deficiency; and the compounds with different number of planes composed of bismuth, strontium, copper, oxygen and with or without calcium (BSCCO).

This combination of low dimensionality and strong-correlation Mott physics led Anderson[2] and others to suggest that fundamentally new physical ideas are required to explain the behavior of the cuprates, which is sketched in Fig. 1. Probably the "simplest" effective Hamiltonian for a strongly-correlated system is the one-band Hubbard model (see Eq. (2)). But what is the essential reduced Hamiltonian for the cuprates as doped Mott insulators? If we restrict our attention to the CuO_2 planes, then only the Cu^{2+} $3d_{x^2-y^2}$ and the O^{2-} $2p_x, 2p_y$ orbitals are relevant, as shown in Fig. 2(c). An extended Hubbard model can capture both the insulating antiferromagnet of the undoped compound and the metallic character when extra holes are doped into the planes. Such a Hamiltonian reads (in electron notation)[10]

$$\mathcal{H}_{VSA} = -\sum_{\langle ij\rangle\sigma} t_{dp}(d^{+}_{i\sigma}p_{j\sigma} + \text{h.c.}) + U\sum_j n_{j\uparrow}n_{j\downarrow}$$
$$+V\sum_{\langle ij\rangle} n_i m_j + \Delta\sum_j (n_j - m_j), \tag{1}$$

where d^{+}_i creates an electron in a Cu $d_{x^2-y^2}$ orbital on Cu site i, p_j destroys an electron in a p_x or p_y orbital on an O site j and n or m counts the number of electrons in the relevant Cu or O orbitals. The first term is the kinetic energy due to electrons hopping between nearest neighbor Cu and O ions, the second is the short-range Hubbard repulsion on Cu, the third is the Coulomb repulsion between electrons on nearest neighbor sites and the last is the local ionic energy difference ($\Delta = E_p - E_d$) between O and Cu (shown in Fig. 2(b)). The phases of this extended Hubbard model were discussed by Zaanen, Sawatzky and Allen (ZSA).[11] They showed that when U is very large and Δ sufficiently exceeds t_{dp}, a hole in the Cu d^9 orbital becomes localized and an insulating state is formed with Δ playing the role that U plays in the simple one-band Hubbard model; then the charge gap in this case is the energy to transfer the hole from Cu to O, rather than U.

There is still no consensus about how much of this Hamiltonian is essential for the cuprates. Further reduction would lead to a simple one-band Hubbard model on a square lattice

$$\mathcal{H}_H = -\sum_{\langle ij\rangle\sigma} t(d^{+}_{i\sigma}d_{j\sigma} + \text{h.c.}) + U\sum_j n_{j\uparrow}n_{j\downarrow} \tag{2}$$

This is the effective Hamiltonian proposed by Anderson[2] and it must be said that it has been adopted in the majority of theoretical works on cuprate superconductivity. What happened to the oxygens? There are at least two points of view. One is that in view of the analysis of ZSA, the only essential thing for the behavior of the cuprates is holes hopping on a square lattice with strong on-site repulsion. The other is that a doped hole (beyond the undoped half-filled case) forms a mobile spin singlet complex that involves the four oxygens neighboring a copper and the original copper hole, the so-called Zhang-Rice singlet.[12] This leads, if the energy scales are appropriate, to an effective one-band model. The applicability of the Zhang-Rice picture depends on $\Delta > t_{dp}$, which may or may not be true, depending in part on what renormalizations are to be included in the definition of Δ. For example, the early density-functional calculations,[13] which included *e.g.* all the static renormalizations due to U and V give a Δ close to zero. On the other hand, the experimental fact is that the undoped compounds are antiferromagnetic insulators; according to ZSA this means that $\Delta > t_{dp}$, apparently justifying the Zhang-Rice picture.

The analysis of Zhang and Rice[12] leads finally to an effective one-band Hamiltonian that involves only d electrons:

$$\mathcal{H}_{t-J} = -\sum_{\langle ij\rangle\sigma} t(1-n_{i,-\sigma})d^{+}_{i\sigma}d_{j\sigma}(1-n_{j,-\sigma}) + J\sum_{\langle ij\rangle}[\mathbf{S}_i \cdot \mathbf{S}_j - n_i n_j/4], \quad (3)$$

where $J = (2V^4/\Delta^3)[1+2\Delta/U]$. Here the kinetic energy term contains projections that prevent double occupancy. This is identical to the $t-J$ Hamiltonian[14] derived in the large U limit from the one-band Hubbard model $\mathcal{H}_H$ in 2D. In that case, $J = 4t^2/U$. One sees immediately from Eq. (3) that at half-filling, where $n_i = 1$, one is left with the 2D Heisenberg antiferromagnet whose ground state is now known to be the usual two-sublattice Néel antiferromagnet. The values of the parameters in the $t-J$ model that are appropriate for a typical cuprate[15] are $t \sim 0.4$ eV, $J \sim 0.13$ eV. This corresponds to $U \sim 12t \sim 4.8$ eV for the original one-band Hubbard model.

As shown in Fig. 1, three main phases of the typical cuprate superconductor are tuned by doping and temperature. They are the normal (*i.e.* non-superconducting) strange-metal (region of optimal doping) and the pseudogap ("underdoped" region) states and the superconducting state itself. In a complete theory of the high-T_c materials, all of the regions of the phase diagram are to be understood, as well as the crossovers between them. This is a stronger demand on a theoretical framework than that faced by Bardeen, Cooper and Schrieffer and their reduced Hamiltonian. The conventional superconductors of 1957 are much less complex than the cuprates; only two phases were at play. For the high-T_c materials, it will be much more difficult to replicate the fantastic success of BCS theory, which not only accounted for all the observed properties of superconductors, but also predicted others.

There have been a number of reviews of theoretical treatments of high-temperature superconductivity.[16–19] A list of some of the theory ideas and techniques that have been developed in connection with high-T_c makes clear how difficult it might be to give a comprehensive survey:

> Spin fluctuations, anisotropic phonons, excitons, charge fluctuations, plasmons, circulating currents, bipolarons, resonating valence bonds, stripes, interlayer tunneling, spin bags, spin liquids, flux phases, BCS-BEC crossover,

> marginal Fermi liquid, van Hove singularities, quantum criticality, anyons, time reversal violation, dynamical mean field theory, slave bosons, gauge theory, *d*-density waves, gossamer superconductivity, SO(5), ...

Rather than a duplication of some earlier overviews, what follows here consists instead of short descriptions of some of the ideas that are important for historical background and/or are representative of current research.

3. RVB and Gauge Theories

Perhaps the earliest systematic attempts to describe the phases of the cuprates were carried out using the Hubbard model Eq. (2) in the limit of infinite U, that is to say with the constraint that double occupancy of any site is forbidden. This may be realized in a variety of ways. Since most of this topic has been extensively reviewed by Lee, Nagaosa and Wen,[17] only a brief summary is given here.

A variational method can be built with the use of a Gutzwiller projection,[20] $P_G = \prod(1 - n_{i\uparrow}n_{i\downarrow})$ operating on a trial wave function, for example the BCS wave function, which does not contain the effects of strong repulsion. This was applied by Anderson[2] in his formulation of the resonating valence bond ("RVB") description of the 2D lightly-doped Hubbard model (near one-electron per site). The importance and consequences of the projection have been continually emphasized by him and the method has been used frequently since Ref. 21. Anderson's RVB state arises because of the effective exchange interaction (*e.g.* from the $t - J$ model) that can give antiferromagnetism in the Mott insulating state at half-filling. At half-filling, the Cu spins-1/2 are localized and the antiferromagnetic exchange interactions between them can generate a spin-liquid state of fluctuating singlet pairs (RVB), which because of the effect of quantum fluctuations on the long-range ordered two-dimensional spin-1/2 Néel state can be nearby in energy to the latter. This RVB spin-liquid state has only short-range antiferromagnetic correlations and it has spin 1/2 fermionic excitations called spinons. This is in contrast to the antiferromagnetic state, which has $S = 1$ excitations. The RVB state is thought to be stabilized by the addition of holes, which destroys the Néel state. This picture has attractive features that we will mention below in connection with the properties of the underdoped (pseudogap) region of the phase diagram.

As an alternative to the Gutzwiller projection, some subsequent developments have been based on the recognition that the leading effect of very strong short-range repulsion (as in the Hubbard model) can be captured by a *constraint* which keeps electrons of antiparallel spin apart. Thus, no more than one electron per site ("no-double-occupancy"). The approach then requires a treatment of the constraint with weak residual interactions arising from deviations from the strong coupling limit. In leading order, this is the content of the $t-J$ model [Eq. (3)].

At half filling, one electron per site, $n_i = 1$, the kinetic energy drops out of the $t-J$ Hamiltonian and one is left with a (localized) spin only Heisenberg antiferromagnet. The metallic phases of the $t-J$ model occur when it is doped away from half-filling, *e.g.* by adding x holes, as in the hole-doped cuprate superconductors. The motion of holes under the constraint of no double occupancy leads to a long-range retarded interaction which is mediated by a dynamical gauge field with substantial low-energy weight. This interaction, arising solely from the constraint, is ultimately responsible for the low-temperature anomalies and non-Fermi liquid behavior in the normal-state properties of the model.

To understand the origin of this gauge field, we begin by examining the constraint $n_i = \sum_\sigma d^+_{i\sigma} d_{i\sigma} \leq 1$. It is convenient to replace the inequality with an equality for each site. This is achieved by regarding each site as being occupied by either a spin-up fermion, a spin-down fermion (spinons, $f_{i\sigma}$) or by a spinless boson representing a hole (holon, b_i). This is done by expressing the original electrons as $d_{i\sigma} = f_{i\sigma} b^+_i$. The constraint is then $\sum_\sigma f^+_\sigma f_\sigma + b^+ b = 1$, which explicitly eliminates double occupancy of any site from the Hilbert space. In the case of the Hubbard model, it has been shown[23] that a mean field treatment using the slave boson method is explicitly connected to the Gutzwiller approach of projected wave functions. The approach invites non-Fermi liquid behavior at the outset via the replacement of the physical electrons by the spinon-holon composite objects and it is a mathematical realization of the spin-charge separation that is characteristic of excitations in RVB theory and in one-dimensional interacting electron systems. There is still the possibility that the spinons can behave as a Fermi liquid. This occurs, for example, when the holons undergo Bose condensation, their dynamics then drops out of the problem.

Typically, the $t-J$ Hamiltonian is rewritten in terms of these new operators and mean-field solutions to the problem are obtained as usual by decoupling the interaction terms.

$$\mathcal{H}_{t-J} \rightarrow -t \sum_{\langle ij\rangle\sigma} f^+_{i\sigma} f_{j\sigma} b_i b^+_j - \frac{J}{2} \sum_{\langle ij\rangle\sigma\sigma'} f^+_{i\sigma} f_{j\sigma} f^+_{j\sigma'} f_{i\sigma'} . \qquad (4)$$

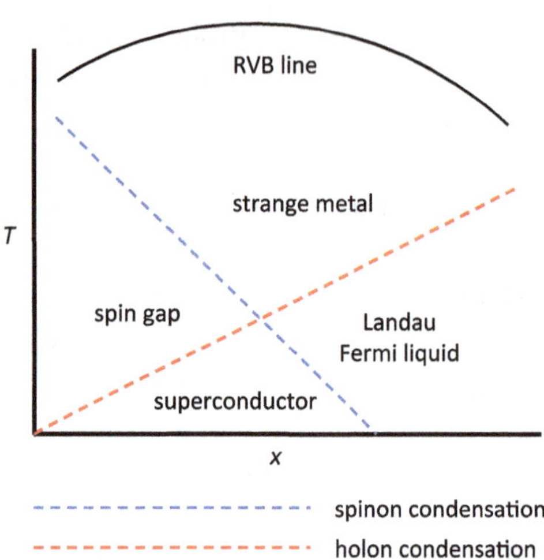

Fig. 3. Mean-field phase diagram of the slave-boson treatment of the $t-J$ model.

The constraints are generated with the use of Lagrange multipliers that modify the action, which defines the quantum statistical mechanical partition function. At the simplest mean-field level, these are taken to be all equal. That is, the constraint is taken globally rather than being satisfied at each site.

There were several successes of this simple slave-boson approach. Most prominently, although postulated one or two years earlier by a number of workers, "unconventional" d-wave superconductivity, later confirmed experimentally, was predicted in a slave boson calculation in 1988.[24] Although the antiferromagnet at half filling is not captured, the various mean-field treatments lead to a number of phases that are suggestive of the experimentally observed behavior. These are sketched in Fig. 3.[17,25] The reader should not fail to notice similarities between Figs. 1 and 3. The interpretation here is that below the "RVB line," the spinon bond operator $\sum_\sigma f_{i\sigma}^+ f_{j\sigma}$ acquires a nonzero expectation value $\chi_{ij} \neq 0$ and a uniform singlet RVB state begins to form, leading to a spin gap. The pseudogap then arises from the energy required to break singlets. This is reflected in the magnetic susceptibility, specific heat and, in particular, in the conductivity perpendicular to the CuO_2 planes. Other parts of the phase diagram are generated in mean field by holon Bose condensation $\langle b \rangle \neq 0$ (red line in the figure), and d-wave spinon pair condensation $\langle f_{i\uparrow} f_{j\downarrow} - f_{i\downarrow} f_{j\uparrow} \rangle \neq 0$ (blue line in the figure). When both order parameters are nonzero, long range order of d-wave superconductivity is obtained in a dome-shaped region of the phase diagram, just as in the actual materials.

It was soon recognized that these mean-field approximations inadequately represent the influence of the constraints, insofar as the latter are obeyed only on average. Consideration of fluctuations around the mean-field solutions leads to gauge theory. Specifically, the average version of the constraint is realized by the saddle point value of the Lagrange multiplier referred to above. Its fluctuations represent the time component of a U(1) gauge field whose spatial components are represented by the phase of the spinon bond field χ_{ij}. The spinons and holons couple to the gauge field and the result is in essence a strong-coupling problem. Further elaboration of the approach is described extensively in the review of Lee, Nagaosa and Wen.[17]

We have seen that the slave-boson approach to the $t-J$ model can serve as a microscopic basis for RVB states and spin-charge separation. A mean-field treatment gives a phase diagram that contains much of what is seen experimentally in the hole-doped cuprates. However, a number of difficulties presented themselves as the U(1) gauge theory was developed. One is that the predicted behavior of the superfluid density as the doping is decreased was at odds with the experimental results. This led to further elaboration of the gauge theory approach to include the SU(2) symmetry appropriate to the undoped situation. This is described in detail in Ref. 17. In spite of such difficulties and the mathematical complexities of further development, the general philosophy of RVB as a basis for the understanding of cuprate superconductors retains substantial adherents.[26]

It is important to note that the RVB and gauge theory developments described in this section are all based on enforcing the constraint of no double occupancy, that is to say the limit of infinite Hubbard repulsion U. The issue of the difference between infinite and finite U is not often discussed. The question is whether the physics of a moderate U system in which some double occupancy may occur, can be well-captured in the infinite U limit. In this connection, it should be mentioned that the proposal of gossamer superconductivity[27] rests on a *partial* Gutzwiller projection, which permits superconductivity (perhaps with extremely low superfluid density) to coexist with antiferromagnetism in the state that is insulating at infinite U. This shows that there can be profound differences between the two regimes.

4. Phenomenologies

The complexities of the cuprates and the rapid proliferation and refinement of experimental techniques have often determined the direction of theoretical activity. It is fair to say that absent a complete theory of the high-temperature superconductors, these efforts were essentially phenomenologically based. In the late 1980s, high transition temperatures, pairing mechanisms and symmetry of the superconducting order parameter were focus issues.

4.1. *Spin fluctuations*

In the first decade of the high-T_c era, spin fluctuation theories for the superconducting transition were prominent. The motivation was the proximity of the antiferromagnetism, evidence for antiferromagnetic spin fluctuations from neutron scattering and nuclear magnetic resonance studies, and the successes of a related approach for the understanding of unconventional superconductivity in the heavy fermion compounds. Superconductivity would arise because of pairing mediated by the spin fluctuations, analogous to phonon-mediated pairing in conventional BCS superconductivity. Unlike the heavy fermions, which have spin degrees of freedom that are not directly associated with the conduction electrons, spin fluctuations in the cuprates are made from the same electrons that provide the charge carriers that condense into the superconducting state. Thus a phenomenological construction of "spin-fermion" models, in which conduction electrons are coupled to spin fluctuations that are consistent with the frequency and wave number susceptibility observed in neutron scattering. The origin of such a coupling, is of course the Coulomb interaction, which is repulsive. However, the spatial structure of the antiferromagnetic spin fluctuations favors opposite spins on neighboring sites; this conspires to give an effective pair attraction provided the pair wave function is in the *d*-wave angular momentum channel. This was another prediction of *d*-wave superconductivity in addition to the one mentioned above.[24]

This phenomenological approach, while it lacks a microscopic derivation, has been used to calculate various properties of the superconducting and normal states, with mixed success. Reviews of the problems and successes of this phenomenology are available in several places.[19,28]

4.2. *Marginal Fermi liquid*

Besides the remarkable high transition temperatures of the copper oxide metals, it soon became apparent that the normal state properties were not those of ordinary Fermi-liquid-like metals. In particular, in the doping region of the maximum T_c, near the top of the superconducting dome seen in the phase diagram of Fig. 1, virtually every normal state thermodynamic and transport property is anomalous. To account for this, the marginal Fermi liquid phenomenology was created.[29] It was proposed that a single hypothesis about the particle-hole excitation spectrum can account for all the observed anomalies. This spectrum determines the charge and spin response, *i.e.* electronic polarizability and spin susceptibility, of the system to external probes. The analysis of Ref. 29 was based on the experimental observation of an unusual Raman scattering spectrum that indicated a polarizability $P(q,\omega)$, whose energy scale, rather than the Fermi energy as in the usual case, was the temperature. Explicitly,

$$\mathrm{Im}\, P(\mathbf{q},\omega,T) \propto \mathrm{sgn}(\omega) \begin{cases} |\omega|/T, & |\omega| \ll T \\ \text{const}, & T \ll |\omega| \ll \omega_c \end{cases} \tag{5}$$

up to a cutoff ω_c.

It is very important to note that this form is characteristic of scaling behavior of critical fluctuations near a quantum critical point ("QCP"), to which we shall return below. This was an early suggestion from theory that a QCP, marked with a "Q" in Fig. 1, could control the normal state behavior near optimal doping. Were superconductivity not to intervene below T_c, the QCP would govern a $T = 0$ transition between two phases. With the assumption that P is only weakly dependent on wave vector $\mathbf{q}$, the scattering of electrons from bosonic fluctuations of the above form leads to a linear-in-T resistivity, which is perhaps the most striking observed property of the strange metal phase near optimal doping.

The interaction between electrons and the bosons leads to an electron self energy whose real part has a logarithmic singularity at zero temperature and this produces a logarithmic singularity at the Fermi momentum in the single-particle momentum distribution. This is to be contrasted with an ordinary Fermi liquid that has a nonzero step discontinuity at the Fermi level and with a generic "non-Fermi liquid" that has a power law behavior. This logarithmic behavior is the origin of the term marginal Fermi liquid (MFL). In the language of Green's functions, the statement[29] is that quasiparticle renormalization factor Z at the Fermi surface goes to zero as $T \to 0$ as an inverse logarithm:

$$\mathcal{G}(\mathbf{k},\omega) = \frac{1}{\omega - \epsilon_{\mathbf{k}} - \Sigma(\mathbf{k},\omega)} \approx \frac{Z}{\omega - \tilde{\epsilon}_{\mathbf{k}} + i\Gamma}, \tag{6}$$

where the self energy is calculated from the interaction of the electrons with the fluctuations and is determined by a coupling constant λ and an ultraviolet cutoff ω_c

$$\Sigma_{\rm MFL}(\mathbf{k},\omega) \sim \lambda \left[\omega \ln \frac{x}{\omega_c} - i\frac{\pi}{2}x\right], \quad x = \max(|\omega|, T), \tag{7}$$

so that

$$1/Z = (1 - \partial\,{\rm Re}\Sigma/\partial\omega) \propto \ln(\omega_c/T), \qquad \Gamma \propto \max(\omega, T).$$

It is important to notice from Eq. (6) that the decay rate Γ from the inelastic scattering off the fluctuations is linear in the electron excitation energy, unlike the Fermi liquid where it is quadratic. In the MFL case, then, the quasiparticles are only marginally defined as the Fermi energy is approached.

This form of electron propagator can be used to calculate various normal state properties. Its imaginary part is the single-particle spectral function that is measured in angle resolved photoemission (ARPES) experiments. The marginal (*i.e.* logarithmic) behavior of the real part of the self energy results in much more spectral weight appearing in the tail of the energy distribution spectrum. The comparison to experiment is striking.[30] A calculation[31] of the optical response gives an enhancement, compared to the conventional Drude spectrum, of the spectral weight at high frequency, just as seen in experiment.

There is an excellent review of the MFL phenomenology and its origins in microscopic theory.[32] Elsewhere, there are several reviews[16,33] that give further discussion of the applicability of the MFL phenomenology in both the normal and superconducting states near optimal doping.

4.3. *Stripes*

The vast majority of theoretical attacks on the strong-correlation problem, of which, as we have seen, the cuprate superconductors are a prime example, have considered the physics of a *homogeneous* system. However, there is no *a priori* reason why charge and/or spin inhomogeneities should not be of fundamental importance. It is no secret that theoretical approaches to the problems of the cuprates have often been the subject of controversy. Perhaps none more so than the issue of the relevance of stripe inhomogeneity. This subject has been recently reviewed[34] comprehensively, with complete references to earlier overviews. Remarks here will be brief.

In the context of high-T_c superconductors, stripe phases were first proposed[35] within a mean-field treatment of the Hubbard model. Shortly thereafter, numerical evidence[36] for electronic phase separation in the doped

$t-J$ model was found. That is, the doped holes were energetically favored to segregate into a hole-rich region. It is clear why this may happen at low doping: isolated holes destroy four magnetic bonds in a square lattice, while clumped ones break fewer. Within the $t-J$ model, it can be appreciated that the doped holes might segregate in order to resolve the competition between kinetic energy and superexchange. Beyond the $t-J$ model, however, the long-range Coulomb repulsion between holes disfavors such a charge accumulation. The competition between the two effects could lead to the formation of an ordered array of stripes, in which the holes occupy charged lines in the lattice. Between the lines, there are no missing spins, hence it is expected that at low temperature, the spins will be ordered antiferromagnetically, as in the undoped situation. From the theoretical perspective, stripe physics is very attractive, as it involves essentially one-dimensional (1D) ribbons of charge. Interacting electrons in 1D is a thoroughly studied problem and the non-Fermi liquid behavior is well-known.[37]

What about experimental evidence for stripes in the cuprates? Static stripe order was first observed[38] in 1995 by neutron scattering on a particular version of lanthanum-strontium cuprate that is doped with neodymium. This discovery generated more theoretical activity on the picture described above. The Nd substitution stabilizes a low-temperature tetragonal structure instead of the usual orthorhombic phase of LSCO. Neutron scattering in the parent undoped compound shows commensurate magnetic peaks at the antiferromagnetic ordering vector $\mathbf{Q} = (\pi/a, \pi/a)$.[39] In the Nd-doped LSCO, four incommensurate magnetic peaks were observed displaced from $\mathbf{Q}$ by shifts of magnitude δ. Furthermore, new charge Bragg peaks appeared that were displaced from the usual lattice reflections by $\pm 2\delta$. These features are consistently accounted for by the formation of stripes along the Cu-Cu directions in which holes aggregate along domain walls separated by a distance π/δ. Between these, there is antiferromagnetic order of spins with phase shift π across them. The major probes for observation of stripe phases are neutron and x-ray scattering in the LSCO and YBCO compounds. In addition, scanning tunneling microscope studies have revealed stripe and checkerboard patterns in BSCCO. The latter technique, however, is strictly a surface probe. Observation and interpretation are made difficult by, among other problems, the extent of quenched disorder, surface conditions, size of single crystals.

Does stripe ordering in the cuprates have anything to do with superconductivity? While the physics of possible charge/spin stripe ordering may be of fundamental interest, many claim that it is a distraction in the context of high-temperature superconductors. If one believes that the mechanism for superconductivity in the electron-doped compounds,[9] such as NCCO, is the same as that in the hole-doped materials, then the absence of any experimental evidence for stripes in the former supports a view that they are either irrelevant or detrimental. Furthermore, it is only in the LSCO compounds where static stripe order has been seen and it is most prominent at 1/8 doping, where it coincides with a pronounced dip in the superconducting dome of Fig. 1. That is, it is accompanied by a large suppression of T_c. Nevertheless, there are a number of proposals[40] that invoke stripe fluctuations as a source of pairing or as an enhancement of T_c.

5. Pseudogap

The term pseudogap was invoked first by Jacques Friedel,[41] who sought to explain the lowering of T_c on the underdoped side of the dome by the development of a suppression of the density of states (pseudogap) as doping was decreased. No aspect of the cuprate phase diagram has received more theoretical attention in recent years than this pseudogap region, seen in Fig. 1. Below a temperature T^* that can be appreciably higher than T_c, a suppression of the density of states is seen in most cuprates in the underdoped region. This is observed in a variety of experiments, in nuclear magnetic resonance, Raman scattering, single-electron spectroscopy, transport, and in thermodynamics.[42] As shown in the figure, T^* decreases as doping is increased, opposite to the behavior of T_c, and it ultimately disappears into the superconducting dome near or above optimal doping.

Whether the T^* line represents a true thermodynamic phase transition, as is proposed in some theories of competing orders, or rather a crossover is still an open question. Distinct thermodynamic features have not (yet?) been found, but there are classes of phase transitions that do not exhibit the usual specific heat singularity. It has also been argued that the presence of disorder may round a transition sufficiently to make an underlying singularity difficult to observe.

ARPES experiments have been particularly informative about the physics of the cuprates since they reveal the momentum dependence of the Fermi surface, the superconducting gap and the pseudogap in the Brillouin zone. Significantly, the symmetry of the pseudogap in momentum space is d-wave-like, just as the superconducting gap. However, there is an important difference. Whereas the superconducting gap is zero only along lines in the $(\pm\pi, \pm\pi)$ ("nodal") directions in the Brillouin zone, the pseudogap vanishes

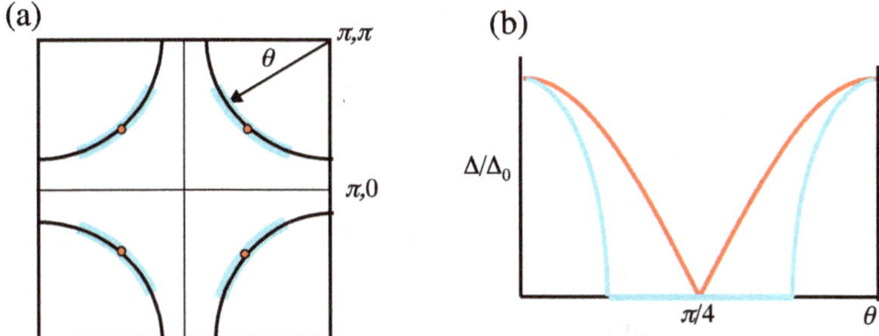

Fig. 4. (a) Typical 2D Brillouin zone for hole-doped cuprates. Black line: Normal state Fermi line. Red dot: Nodal point where superconducting gap is zero. Aqua line: Fermi arc where pseudogap vanishes. (π, π) = nodal direction. $(\pi, 0)$ = antinodal direction. (b) Sketch of angle dependence of gaps. Red: Superconducting gap $(\cos 2\theta)$. Aqua: Pseudogap and Fermi arc. Gaps are normalized to their respective maximum values.

along an arc of the Fermi surface, Fig. 4(a). The length of the arc is temperature dependent and extrapolates to zero at $T = 0$.[43] Thus, in the antinodal region a gap in the spectrum exists, but along the arcs there are coherent quasiparticle excitations. A complete analysis of the pseudogap physics must account as well for the origin of this "Fermi arc." The momentum dependence of the gaps is depicted in Fig. 4(b).

This striking phenomenon is unprecedented; it appears nowhere else in condensed matter physics and has been a source of enormous theoretical activity, ranging from speculation to microscopic calculation. In fact, almost all theoretical activity at the time of writing of this chapter appears to be focused on one or another aspect of the pseudogap phenomenon. Many scenarios for the pseudogap have been proposed. Most fall into one of four classes: preformed pairs, RVB, competing orders, onset of incoherence.

As is the case with most of the unusual experimental observations on these fascinating materials, several theoretical scenarios often seem to account for results in the pseudogap region. For example, it is observed that the in-plane conductivity is little affected by the formation of the pseudogap while the c-axis conductivity normal to the planes is significantly suppressed. Both the preformed pairs and the RVB pictures give simple explanations of this observation, see below.

5.1. *Preformed pairs*

It is a commonplace that fluctuations of an order parameter can have important effects on properties above a phase transition. In the cuprates, such fluctuations are likely to be important because of low dimensionality (quasi-two-dimensions) and the short coherence length in the superconducting state. Consequently, it was quite natural to propose that Cooper pairs formed at T^* but without the long-range phase coherence required for superconductivity, and that these would be in a singlet state with d-wave symmetry, thus a precursor to superconductivity. Of course, one could well argue, as some have,[44] that the pseudogap is instead a precursor for the Mott insulator at zero doping.

ARPES experiments have become very important for understanding the pseudogap, since they discriminate between the nodal and antinodal regions of the Fermi surface, whereas the other probes generally give the overall density of states. Perhaps the most important evidence for the preformed pairs scenario comes from relatively recent ARPES experiments[45] that reveal, in the pseudogap phase, a Bogoliubov quasiparticle spectrum, symmetric about the Fermi level, in the antinodal region and a normal quasiparticle spectrum in the region of the Fermi arc, where of course there is no pseudogap. In this description, the nodal region has an arc rather than a nodal point as in the superconducting state because in a nonzero region around the nodal point the pseudogap is small and is destroyed by quasiparticle damping.[46]

In this picture, the suppression of the c-axis charge transport at T^* is simply a consequence of band structure: the interplane hopping vanishes at the nodal points where the quasiparticles carry the in-plane current, while it is large in the antinodal region where the pseudogap opens, thus suppressing the interplane transport.

5.2. *RVB and the pseudogap*

We have already seen the RVB phase diagram in Fig. 3 that the RVB theory can give an account of crossovers, as a function of doping, from gapped state (pseudogap) to strange metal to Fermi liquid. Here, the pseudogap is essentially the energy required to break the spin singlets that constitute the RVB spin-liquid state below the blue dashed line in the figure.

On the issue of the c-axis conductivity versus the in-plane transport referred to above, the RVB picture is simple: the singlet pairing of the neutral spinons does not affect the in-plane charge transport. For the c-axis conductivity the transport between planes must be by the full electrons (or rather, holes) so that spinon and holon need to be recombined, which requires breaking spin singlets and paying the spin gap energy. In fact, all the single-carrier probes, such as photoemission, require the cost of the spinon spin gap energy.

This appealing picture derives from a mean-field treatment of the RVB state. However, low dimensionality and strong coupling suggest that fluctuation effects are important. Consequently, there have been many extensions and elaborations of the RVB theory, often in the language of gauge theory. These are comprehensively reviewed in Ref. 17, along with a discussion of many experimental results.

An RVB alternative to the gauge theory approach is based on the original Anderson idea[2] of the projected wave function, see Sec. 3. Analytic treatment of the projection is difficult so that much of what has been published is computational. Apart from computational approaches, a mean-field theory[47] based on the Gutzwiller projection and using the Gutzwiller approximation[20] of assigning weight factors to terms in the Hamiltonian was introduced very early and it is reviewed in Ref. 21. This has more recently been used to formulate an ansatz for the self energy[44] in the pseudogap regime. This gives an opportunity to make analytic analyses of pseudogap physics. A review of this theory with an application to ARPES experiments and pertinent references is given in Ref. 48.

5.3. *Competing orders*

A quite different scenario for the pseudogap phenomenon is that it is a consequence of a transition to an ordered phase that competes with superconductivity. The T^* line is then a thermodynamic phase transition line that would terminate in a $T = 0$ quantum critical point, were it not for the development of superconductivity at T_c. As in Friedel's original conjecture,[41] since the competition gets stronger the less the doping, this accounts for the suppression of the superconducting T_c as the doping is reduced into the underdoped regime. The competing orders that have been discussed include spin and charge density waves, including stripes and two types of circulating current phases. A brief description of the latter group follows.

5.3.1. *d-density wave*

The d-density wave (DDW) state is one in which there is staggered orbital current order across the plaquettes of a square lattice, see Fig. 5(a). First discussed in 1989[49] and 1991[50] and revived and renamed in 2000–2001,[51] and elaborated in a number of papers since by Chakravarty and coworkers.[52] In Ref. 51, the authors proposed a d-density wave phenomenology to characterize the pseudogap region of the cuprate phase diagram. As can be seen from the figure, the state breaks parity, time-reversal symmetry and the

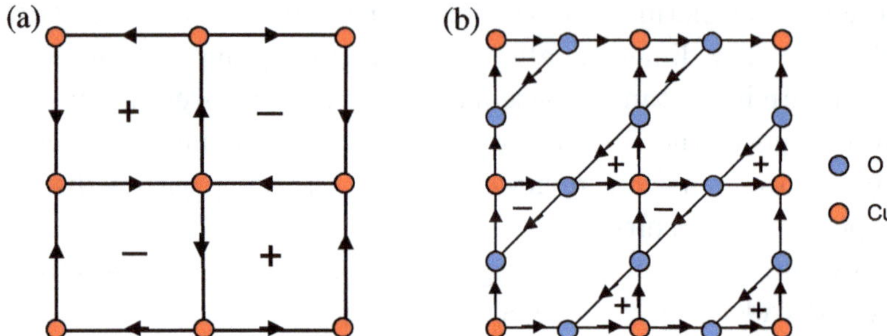

Fig. 5. Circulating current patters in (a) d-density wave order and (b) loop-current order. Plus (+) and minus (−) denote magnetic flux out or into the paper.

lattice translational symmetry, thereby doubling the unit cell. Therefore, it is to be expected that there will be a reconstruction of the original hole-like Fermi surface [Fig. 4(a)] and the new Fermi surface can have both electron and hole pockets.

The commensurate DDW order parameter is a particle-hole spin singlet, as it is in the familiar charge density wave (CDW) case, but with a d-wave form factor instead of s-wave.

$$\Psi_{\mathbf{k}} \propto \langle c^{+}_{\mathbf{k}+\mathbf{Q},\alpha} c_{\mathbf{k},\alpha} \rangle = iW_{\mathbf{k}} = iW_0(\cos k_x - \cos k_y), \qquad \mathbf{Q} = (\pi, \pi).$$

Where does the DDW state come from? It can arise from a mean-field treatment of some underlying Hamiltonian, which in mean field has the form

$$\mathcal{H} = \sum_{k,\alpha} \{(\epsilon_{\mathbf{k}} - \mu) c^{+}_{\mathbf{k},\alpha} c_{\mathbf{k},\alpha} + [iW_0(\cos k_x - \cos k_y) c^{+}_{\mathbf{k}+\mathbf{Q},\alpha} c_{\mathbf{k},\alpha} + \text{h.c.}]\}$$

Here, $\epsilon_{\mathbf{k}}$ is the electron band structure and $2W_0$ is the amplitude of the resulting DDW gap in the quasiparticle spectrum, which now has two branches:

$$E^{\pm}_{\mathbf{k}} = \frac{1}{2}\left[\epsilon_{\mathbf{k}} + \epsilon_{\mathbf{k}+\mathbf{Q}} \pm \sqrt{(\epsilon_{\mathbf{k}} - \epsilon_{\mathbf{k}+\mathbf{Q}})^2 + 4W^2_{\mathbf{k}}}\right] \tag{8}$$

As with most of the theoretical ideas about the cuprates, the DDW pseudogap scenario is controversial. It can account, perhaps not uniquely, for many observations in the underdoped region. This is summarized in Ref. 52. While the DDW phenomenology is concerned with the pseudogap and not with the mechanism of superconductivity, it should be appreciated that the competition of DDW and d-wave superconductivity is very direct and explicit: They compete for the same portions of the Fermi surface. This implies that as doping decreases and T^* goes up, the superconducting T_c goes down, as observed.

Perhaps the most compelling evidence for a Fermi surface reconstruction in the pseudogap region comes from recent quantum oscillation experiments[53] that measure Fermi surface areas. While there are still conflicting reports on the details, the simplest interpretation is that multiple Fermi pockets are observed. This implies breaking of the lattice translational symmetry. This would be consistent with a (possibly incommensurate) DDW phase. However, direct observation of DDW order via spin-polarized neutron scattering has so far returned mixed results.

5.3.2. *Loop currents in the CuO_2 lattice*

A different form of a circulating current phase was first proposed in 1997 by C.M. Varma, who has published a thorough review of this theory of the pseudogap phase.[54] The loop-current state is obtained in a mean-field treatment of an extended Hubbard model of the form of Eq. (1). Several requirements motivated the choice of effective Hamiltonian and the nature of the mean-field solution. The fact that charge fluctuations on the oxygens of the CuO_2 planes are not negligible leads one to consider a three-band [as in Fig. 2(b)] extended Hubbard model[10] [Eq. (1)] built from the p_x, p_y orbitals on oxygen and the $d_{x^2-y^2}$ orbital on copper. The success of the marginal Fermi liquid phenomenology (discussed in Sec. 4.2) makes it likely that the underlying fluctuations of Eq. (5) are associated with a quantum critical point underneath the superconducting dome in the region of optimal doping, "Q" in Fig. 1. The phase on the underdoped side of the QCP is responsible for the pseudogap. To extend the picture, one requires that the critical fluctuations P of Eq. (5), which exist in the quantum critical region near optimal doping, are those of the order parameter of the new phase; and to complete the picture, it is supposed that exchange of these same fluctuations provide the pairing interaction that produces d-wave superconductivity.

An important issue is whether the pseudogap phase breaks the lattice translation symmetry, as it does in the DDW scenario, above. Recent observations that bear on this issue are the magneto-oscillation experiments,[53] which can be interpreted by a Fermi surface reconstruction arising from a breaking of the translation symmetry. An alternate explanation[55] that in fact preserves the translation symmetry is based on a form of magnetic breakdown. If there is no broken lattice translation or rotation symmetry in this pseudogap phase, there remains the possibility that the new phase is characterized by broken time reversal and parity. For sufficiently large ionic Cu-O interaction, V in Eq. (1), a mean-field solution for this Hamiltonian gives stable states with circulating loop currents within a unit cell, such that lattice translation symmetry is preserved. The first experimental indication of the existence of such states came from ARPES.[57] The loop-current state of Fig. 5(b) was shown[54] to be consistent with the experimental data.

Since the first observation reported in Ref. 57, other experiments[58] on different classes of underdoped cuprate superconductors, using ARPES, polarized neutron scattering and magnetization measurements, have given evidence for a time-reversal breaking but translation preserving order below the pseudogap temperature T^*. The local orbital moments however appear not be perpendicular to the CuO_2 planes, which means that a current pattern may be more complicated, perhaps involving the oxygens out of the plane.[59]

The loop-current picture is then the following: guided by the success of the MFL phenomenology, its indication of the existence of a quantum critical point and various experimental constraints, a mean-field solution of the extended three-band Hubbard model gives a stable loop current phase. In order to demonstrate the gap in the electron excitation spectrum, it is necessary to derive the coupling of the electrons with the (bosonic) fluctuations of the loop currents. This is a very complicated procedure and is carried out in Ref. 54. The result is an anisotropic gap in the excitation spectrum at $T = 0$ of an unusual form:

$$E_{\mathbf{k}} = \left\{ \begin{array}{ll} \epsilon_{\mathbf{k}} + D(\mathbf{k}), & E_{\mathbf{k}} > \mu \\ \epsilon_{\mathbf{k}} - D(\mathbf{k}), & E_{\mathbf{k}} < \mu \end{array} \right\}, \qquad D(\mathbf{k}) \sim \cos^2(2\theta). \tag{9}$$

Here, θ is the angle on the Fermi surface as defined in Fig. 4. This pseudogap is tied to the Fermi level and has similarities to a d-wave gap ($\sim \cos 2\theta$) but lacks the latter's sign (phase) changes around the Fermi surface. At nonzero T, the point nodes of this gap broaden into arcs.[55]

5.4. *Antinodal incoherence*

A major difficulty of the strong-correlation many-body problem discussed in Sec. 1 is a shortage of comprehensive analytic calculational tools to analyze the relevant effective Hamiltonians. The dynamical mean field theory (DMFT)[60] and its cluster and cellular extensions[61] have been developed to treat situations in which dynamical correlations are dominant. In the context of the cuprate superconductors, DMFT has been applied primarily to the one-band Hubbard model and to the $t - J$ model. As a matter of fact, cluster DMFT methods applied to the one-band Hubbard model including next-neighbor hopping have given the essential features (antiferromagnetism, superconductivity, pseudogap) of the cuprate phase diagram as a function of temperature and doping.[62,63]

The cluster extensions of DMFT beyond the original single-site formulation permit an analysis that distinguishes different regions of momentum space, in particular, for understanding of the pseudogap, the nodal and the antinodal regions of the Brillouin zone. A cluster DMFT treatment[64] of the Hubbard model at large but finite on-site interaction U implies a scenario for the formation of the pseudogap and the Fermi arc in the underdoped regime. In the antinodal region, the strong short-range correlations generated by the effective exchange coupling (as in RVB and the $t-J$ model) result in the suppression of quasiparticle coherence in the antinodal region of the zone and the appearance of a pseudogap. At the same time, in the nodal regions, quasiparticle coherence is preserved and a Fermi arc remains over a nonzero region of the original Fermi surface, whose antinodal parts are destroyed by the pseudogap. All this is caused by the short-range part of the Coulomb interaction, since for smaller values of U, the entire Fermi surface is metallic. An extension[65] of this work to the behavior of the c-axis (interplane) transport in the underdoped regime gives support to this interpretation and gives good agreement with experimental results referred to above, just before Sec. 5.1.

5.5. *Some open questions on the pseudogap*

At the time of this writing, there are several aspects of the pseudogap phenomenon that have had conflicting experimental reports. Once sorted out, these features will impose important constraints on theories of the origin of the pseudogap and of the mechanism for superconductivity in the cuprates.

"One gap or two." This controversial issue refers to the nature of the gap in the superconducting state. Based on the observation that the pseudogap energy scale T^* and the superconducting energy scale T_c behave in an opposite fashion with doping concentration, it has been suggested that the gap in the antinodal region is the pseudogap persisting below T_c and that the nodal region has the true superconducting gap *i.e.* the latter appears on the Fermi arc below T_c. The one-gap scenario describes a single d-wave gap that persists outside the Fermi arc above T_c, as described briefly in Sec. 5.1. The preformed pairs scenario is a "one-gap" picture and most of the other theories give two gaps. Experiments give conflicting information on this issue.

Theories differ on whether the pseudogap is pinned to the Fermi level as it would be for preformed pairs. For example, as we have seen, the DDW gap can be centered above the Fermi level, while the loop currents gap is on the Fermi level. ARPES experiments have given conflicting results.

We have also seen that the shape of the pseudogap can be different for different theories. Hopefully, future high-resolution experiments will shed light on this question.

6. Quantum Criticality

The quantum critical phenomenon has become a major topic in condensed matter research in recent years, especially in the context of the heavy-fermion compounds.[66] What about the cuprates? An early theoretical suggestion of the existence of a putative quantum critical point ("QCP") under the superconducting dome ("Q" in Fig. 1) came from the marginal Fermi liquid theory.[29] On the experimental side, Loram, Tallon and coworkers[67] proposed the existence of a $T = 0$ critical point near or slightly above optimal doping.

As with almost everything else about the cuprates, the QCP issue is controversial. There is conflicting evidence about the position of the T^* line with respect to the superconducting dome. Does it enter the dome near optimal doping and terminate at a QCP as shown in Fig. 1 or does it merge with the overdoped side of the dome and introduce no new QCP. Further disagreement arises on the question of whether, if it exists at all, the QCP has anything to do with the superconductivity. Scenarios of competing orders involve a QCP that separates the pseudogap phase from a Fermi liquid, while for theories based on RVB ideas, *i.e.* proximity to the Mott insulator, quantum criticality is only a distraction: there is no QCP in the RVB phase diagram, Fig. 3.

As mentioned earlier, a QCP is an essential ingredient of the loop-current theory that leads to marginal Fermi liquid. The loop-current transition can be mapped[54] onto an quantum Ashkin-Teller model, whose correlator of quantum fluctuations has the form, Eq. (5), of the fluctuation susceptibility originally hypothesized for the marginal Fermi liquid phenomenology.[29] This gives the MFL picture a microscopic basis and makes clear the role of the quantum critical point. The loop-current theory then is based both on the existence of a QCP near optimal doping and on the three-band extended Hubbard model. However, it is not certain that this route to MFL is unique. A calculation[68] using the dynamic cluster approximation extension of dynamical mean field theory to treat the one-band Hubbard model with moderate $U/t = 6$ and 8 gives evidence of a QCP. A MFL region is found that appears to emanate from a QCP at about 15% doping. The region is bounded on the low doping side by a T^* line determining the onset of a pseudogap and by a crossover to a Fermi liquid on the high doping side, just as in Fig. 1.

In addition to the possible QCP near optimal doping, there are, of course, other QCPs as shown in the phase diagram, Fig. 1. A system that is tuned away from a Mott insulator by doping will exhibit a QCP that determines the metal-insulator transition and the pseudogap region could be interpreted[17] as the quantum critical region associated with this QCP. Such a QCP, interpreted as an "antiferromagnetic quantum critical point" (AFQP) has been discussed in the context of the spin-fluctuation phenomenology in Ref. 28b and also within a phenomenological two-fluid description.[69] In this latter work, an examination of various experimentally determined magnetic quantities leads the authors to propose a scaling behavior reflecting the AFQP that involves two components — a spin liquid of localized spins whose excitations couple to a Fermi liquid. This generates a gap that is identified with the experimentally observed pseudogap.

An in-depth discussion of cuprate quantum criticality can be found in Ref. 34, Sec. VI-B. As they are also an essential ingredient in the theory of fluctuating stripes, quantum critical fluctuations are reviewed in Ref. 37.

7. Superconductivity

The reader will have noticed that little of this review concerns the superconducting state of the cuprates. This is because a generalization of BCS theory to the *d*-wave symmetry of the order parameter has done very well to elucidate most of the physical properties of the superconducting state in the optimally doped and overdoped regions of the phase diagram, without of course revealing the mechanism of pairing. Indeed, as is reflected in this review, the most striking mysteries of the cuprates are found in the unusual normal state properties.

Theoretical ideas about the mechanism for superconductivity in the cuprates fall into two groups. One group deals with the pairing in the superconducting state as analogous to that of conventional superconductors; thus pairing mediated by some bosonic excitation, the "glue." The other group is based on the idea that the superconductivity arises out of the insulating Mott state upon doping and is not to be described by the pairing of coherent quasiparticles, although coherence may be restored below T_c. Much of the discussion here involves the question of whether the one-band Hubbard and $t-J$ models with realistic parameters contain the principal features of the cuprate phase diagram, in particular *d*-wave superconductivity.

There is now ample evidence, mostly from cluster dynamical mean field approaches, that the one-band Hubbard and $t-J$ models do describe the important properties of the cuprate superconductors, including d-wave superconductivity, the pseudogap at low doping below a T^* that decreases with doping, the phenomenon of Fermi arcs, the competition between superconductivity and antiferromagnetism, evidence for a quantum critical point beneath the superconducting dome.

DMFT methods rely heavily on numerical computation, but they have advantages over earlier standard computational techniques for strongly correlated systems, such as quantum Monte Carlo and exact diagonalization on small clusters. The DMFT methods, *embedding* as they do a finite size cluster in a self-consistent bath, are relevant for the thermodynamic limit. Furthermore, since the basic construction of DMFT is analytic and treats self energies explicitly, it gives an avenue to describing the underlying physics of the results obtained. For superconductivity in the cuprates, these results are described in Refs. 62 and 63 and can be summarized as follows.

At large values of the Hubbard on-site repulsion $U \geq 12t$,[62] the underdoped regime is characterized by a pseudogap in the one-particle spectrum and can be understood as holes moving in a sea of RVB singlets. To the right of the T^* line (Fig. 1), the optimally doped regime, there are no coherent quasiparticles as a result of a very large scattering rate, *i.e.* $\mathrm{Im}\,\Sigma$. Coherence is restored upon entering the superconducting state, at least in the nodal region. There is a crossover to a strongly-correlated Fermi liquid in the overdoped region. The main contribution to the superconducting condensation energy is the effective superexchange, that is, the equivalent of the J term in the $t-J$ model. The superexchange as the pairing mechanism was already anticipated in early RVB theory, where for example in Ref. 24, both the d-wave superconductivity and the pseudogap were predicted.

At moderate values of U,[63] of order $8t$, studies using the dynamical cluster approximation extension of DMFT have also reproduced the main features of the cuprate phase diagram. This method has also been used to determine the irreducible particle-particle scattering vertex Γ in the Hubbard model. Interestingly, the frequency dependence of the d-wave part of Γ reflects the frequency dependence of the spin susceptibility at the antiferromagnetic wave vector (π, π). This lends some support to phenomenologies based on spin-fluctuation mediated pairing.[28]

The bosons that have been invoked to be the pairing glue include phonons, stripe fluctuations, spin fluctuations, and loop-current fluctuations. Here, comments will be made on the last two of these. In both cases, unlike the phonon-mediated case, the bosonic variables are composites made up from the same degrees of freedom as the electrons that are being paired.

Then the calculations that are based on the familiar Eliashberg equations[70] used for conventional superconductors are actually semi-phenomenological in nature as they involve coherent quasiparticles interacting with the relevant independent bosonic degrees of freedom.[71] The case of spin fluctuation exchange pairing has already been discussed in Sec. 4.1 and mentioned in the previous paragraph.

As mentioned earlier in Secs. 5.3.2 and 6, the fluctuations of the loop-current order are given by a susceptibility of the form of P in Eq. (5). These fluctuations are then invoked and calculated as the bosonic pairing glue for superconductivity. It is necessary to determine the structure of the coupling of electrons to these fluctuations as input to the Eliashberg procedure. This has been carried out recently[72] and d-wave superconductivity is obtained. Using parameters obtained from the fit of MFL to normal state transport, the authors find reasonable values for T_c and the gap ratio Δ/T_c.

On the face of it, the strong-coupling and Eliashberg approaches stand in sharp contrast. In the former, the physics is determined by the competition between the superexchange, set by J and the kinetic energy of the doped holes, set by xt, where x is the hole concentration. The latter is essentially a weak-coupling formulation that phenomenologically treats the bosons and electrons as separate entities.

8. Conclusion

Why is the cuprate problem hard? The easy answer is that because of the simultaneous occurrence of strong coupling, low dimensionality and low spin, all conventional approaches such as perturbation theory and mean-field theory are inadequate. Perhaps it is a consequence of this confluence of difficulties that the cuprates are distinguished by having a number of different competing states close in energy, depending upon the level of doping. Furthermore the many unusual properties, *e.g.* pseudogap, strange metal, Fermi arcs, that have been discovered experimentally are quite unprecedented; this puts severe constraints on candidate theories for high-temperature superconductivity. Nevertheless, arguments can and have been made that the high-T_c problem is solved ("it depends on what you mean by 'solved'").

There seem to be two different fundamental answers to the question of what is it about the cuprates that is responsible for their high transition temperatures. One is that they are quasi-two-dimensional with spin 1/2 and they are proximate to a Mott insulator with its accompanying antiferromagnetic exchange interactions, which happen to be unusually strong. Then the conclusion is that the basic physics can be understood by studying the one-band two-dimensional Hubbard and $t - J$ models, which contain these elements. This approach of course originates in Anderson's seminal 1987 paper[2] and based on the ideas contained there, d-wave superconductivity and pseudogap were already predicted in 1988 in a slave boson mean field calculation.[24] It has been stated[73] that that already represents the solution. Since that early work, the same underlying physics has motivated further developments. These include gauge theories and cluster extensions of DMFT, which, as described in earlier sections, have elaborated many of the features of the cuprates. A more complete solution?

In sharp contrast is the view that what is important is the chemistry of the cuprates and the uniqueness of the energy scales associated with the two-dimensional CuO_2 lattice. Thus, a three-band model is considered essential,[54,69] in which the on-site Hubbard U may simply be treated in mean field. This is the approach taken, for example, in the loop-current theory,[54] which in its turn also describes many of the features of the cuprates. Another complete solution?

Setting aside the perhaps disagreeable situation of having too many solutions rather than none, one should recognize the important contributions of cuprate superconductivity to both experimental and theoretical condensed matter physics. Remarkable advances have been made in materials preparation and in experimental techniques. On the theory side, while a BCS d-wave theory may be serviceable in the superconducting state, its origins in the normal state remain controversial. In fact, the normal state, in its various manifestations as a function of doping (pseudogap, strange metal, Fermi liquid crossover), gives every indication of being the consequence of a new state of matter. This strong-correlation problem continues to generate entirely new theoretical frameworks, concepts and predictions that of necessity go beyond the Landau paradigms of coherent quasiparticles and ordinary phase transitions. In the years to come, this new physics will be applied to newly discovered correlated materials and will continue to dominate the frontier of condensed matter science.

References

1. J. G. Bednorz and K. A. Müller, *Z. Phys. B* **64**, 189 (1986).
2. P. W. Anderson, *Science* **235**, 1196 (1987).
3. See, for example, P. Nozières, *Theory of Interacting Fermi Systems* (Benjamin, New York, 1964).
4. J. Bardeen, L. N. Cooper and J. R. Schrieffer, *Phys. Rev.* **108**, 1175 (1957).

5. G. Eliashberg, *Sov. Phys. JETP* **11**, 696 (1960).
6. Ref. 2 and P. W. Anderson and E. Abrahams, *Nature* **327**, 363 (1987).
7. In this field, by "open question" it is not meant that no answers have been offered. It means that there is no agreement about which of the answers, if any, is correct.
8. We use the term Mott insulator to include both a phase of a single band correlated system and the "charge-transfer insulator" phase of a multiband system. See the text that follows.
9. A review: N. P. Armitage, P. Fournier and R. L. Greene, arXiv:0906.2931v1.
10. C. M. Varma, S. Schmitt-Rink and E. Abrahams, *Solid State Commun.* **62**, 681 (1987). In contrast to Fig. 2b, we are using electron notation here in $\mathcal{H}_{VSA}$.
11. J. Zaanen, G. Sawatzky and J. Allen, *Phys. Rev. Lett.* **55**, 418 (1985).
12. F. C. Zhang and T. M. Rice, *Phys. Rev. B* **37**, 3759 (1988).
13. L. F. Mattheiss, *Phys. Rev. Lett.* **58**, 1028 (1987).
14. P. W. Anderson, *Phys. Rev.* **115**, 2 (1959); D. J. Klein and W. A. Seitz, *Phys. Rev. B* **10**, 3217 (1974); K. A. Chao, J. Spałek and A. M. Oles, *J. Phys. C* **10**, L271 (1977); C. Gros, R. Joynt and T. M. Rice, *Phys. Rev. B* **36**, 381 (1987).
15. E. Pavarini, I. Dasgupta, T. Saha-Dasgupta, O. Jepson and O. Andersen, *Phys. Rev. Lett.* **87**, 047003 (2001).
16. M. R. Norman and C. Pépin, *Rep. Prog. Phys.* **66**, 1547 (2003).
17. P. A. Lee, N. Nagaosa and X.-G. Wen, *Rev. Mod. Phys.* **78**, 17 (2006).
18. J. Orenstein and A. J. Millis, *Science* **288**, 468 (2000).
19. M. R. Norman, *Handbook of Magnetism and Advanced Magnetic Materials*, Vol. 5, eds. H. Kronmuller and S. Parkin (Wiley, New York, 2007), p. 2671; arXiv:cond-mat/0609559.
20. M. Gutzwiller, *Phys. Rev. Lett.* **10**, 159 (1963).
21. See P. W. Anderson, P. A. Lee, M. Randeria, T. M. Rice, N. Trivedi and F.-C. Zhang, *J. Phys.: Condens. Matter* **16**, R755 (2004) and references therein.
22. S. E. Barnes, *J. Phys. F: Met. Phys.* **6**, 1375 (1976); P. Coleman, *Phys. Rev. B* **29**, 3035 (1984).
23. G. Kotliar and A. E. Ruckenstein, *Phys. Rev. Lett.* **57**, 1362 (1986).
24. G. Kotliar and J. Liu, *Phys. Rev. B* **38**, 3142 (1988).
25. P. A. Lee and N. Nagaosa, *Phys. Rev. B* **46**, 5621 (1992).
26. P. W. Anderson, P. A. Lee, M. Randeria, T. M. Rice, N. Trivedi and F. C. Zhang, *J. Phys.: Condens. Matter* **16**, R755 (2004).
27. R. B. Laughlin, *Phil. Mag.* **86**, 1165 (2006); B.A. Bernevig *et al.*, arXiv:cond-mat/0312573.
28. For example: **a**. D. J. Scalapino, *Phys. Rep.* **250**, 329 (1995); **b**. A. V. Chubukov, D. Pines and J. Schmalian, in *The Physics of Conventional and Unconventional Superconductors*. eds. K. H. Bennemann and J. B. Ketterson (Springer-Verlag, 2008).
29. C. M. Varma, P. B. Littlewood, S. Schmitt-Rink, E. Abrahams and A. E. Ruckenstein, *Phys. Rev. Lett.* **63**, 1996 (1989).
30. E. Abrahams and C. M. Varma, *Proc. Natl. Acad. Sci. USA* **97**, 5714 (2000); A. Kaminski *et al.*, *Phys. Rev. B* **71**, 014517 (2005); L. Zhu *et al.*, *Phys. Rev. Lett.* **100**, 057001 (2008).

31. E. Abrahams, *J. Phys. I France* **6**, 2191 (1996).
32. C. M. Varma, Z. Nussinov and W. van Saarloos, *Phys. Rep.* **361**, 267 (2002). See Sec. 7.
33. D. N. Basov and T. Timusk, *Rev. Mod. Phys.* **77**, 721 (2005).
34. M. Vojta, *Adv. Phys.* **58**, 699 (2009).
35. J. Zaanen and O. Gunnarsson, *Phys. Rev. B* **40**, 7391 (1989).
36. V. J. Emery, S. A. Kivelson and H. Q. Lin, *Phys. Rev. Lett.* **64**, 475 (1990).
37. S. A. Kivelson *et al.*, *Rev. Mod. Phys.* **75**, 1201 (2003) discusses the 1D physics, and reviews fluctuating stripes in the context of quantum criticality.
38. J. M. Tranquada *et al.*, *Nature* **375**, 561 (1995).
39. Here, a is the Cu-Cu lattice parameter.
40. Such proposals are reviewed in Ref. 34, Sec. VI D.
41. J. Friedel, *J. Phys. Condens. Matter* **1**, 7757 (1989).
42. A thorough review of the the evidence for pseudogap from early experiments is given in Ref. 17.
43. See, for example, A. Kanigel *et al.*, *Nature Physics* **2**, 447 (2006).
44. e.g. K.-Y. Yang, T. M. Rice and F.-C. Zhang, *Phys. Rev. B* **73**, 174501 (2006).
45. A. Kanigel *et al.*, *Phys. Rev. Lett.* **101**, 137002 (2008); H.-B. Yang *et al.*, *Nature* **456**, 77 (2008).
46. A. V. Chubukov, M. R. Norman, A. J. Millis and E. Abrahams, *Phys. Rev. B* **76**, 180501(R) (2007).
47. F. C. Zhang, C. Gros, T. M. Rice and H. Shiba, *Supercond. Sci. Technol.* **1**, 36 (1988).
48. K.-Y. Yang *et al.*, *EPL* **86**, 37002 (2009).
49. H. J. Schulz, *Phys. Rev. B* **39**, 2940 (1989).
50. T. C. Hsu, J. B. Marston and I. Affleck, *Phys. Rev. B* **43**, 2866 (1991).
51. C. Nayak, *Phys. Rev. B* **62**, 4880 (2000); S. Chakravarty, R. B. Laughlin, D. K. Morr and C. Nayak, *Phys. Rev. B* **63**, 094503 (2001).
52. See C. Zhang, S. Tewari and S. Chakravarty, arXiv:0910.1966v1, and references therein.
53. For example, A. Audouard *et al.*, *Phys. Rev. Lett.* **103**, 157003 (2009) and references therein.
54. C. M. Varma, *Phys. Rev. B* **73**, 155113 (2006).
55. C. M. Varma, *Phys. Rev. B* **79**, 085110 (2009).
56. Recent experimental evidence for Fermi surface reconstruction, hence broken lattice symmetry, are all in the presence of large magnetic fields, see the discussion in Sec. 5.3.1 and Ref. 53. Therefore, the possibility of field-induced translational symmetry breaking cannot be excluded.
57. A. Kaminski *et al.*, *Nature* **416**, 610 (2002).
58. For references to these experiments and discussion of the symmetry of loop-current order, see A. Shekhter and C. M. Varma, *Phys. Rev. B* **80**, 214501 (2009).
59. Y. Li *et al.*, *Nature* **455**, 372 (2008); C. Weber, A. Läuchi, F. Mila and T. Giamarchi, *Phys. Rev. Lett.* **102**, 017005 (2009).
60. G. Kotliar and D. Vollhardt, *Phys. Today* **57**(3), 53 (2004); A. Georges, G. Kotliar, W. Krauth and M. J. Rozenberg, *Rev. Mod. Phys.* **68**, 13 (1996).

61. G. Kotliar *et al.*, *Rev. Mod. Phys.* **78**, 865 (2006); T. Maier, M. Jarrell, T. Pruschke and M. H. Heitler, *Rev. Mod. Phys.* **77**, 1027 (2005).
62. K. Haule and G. Kotliar, *Phys. Rev. B* **76**, 092503 (2007).
63. T. A. Maier, M. S. Jarrell and D. J. Scalapino, *Physica C* **460–462**, 13 (2007).
64. M. Ferrero *et al.*, *Phys. Rev. B* **80**, 064501 (2009).
65. M. Ferrero *et al.*, arXiv:1001.5051v1.
66. A recent review: P. Gegenwart, Q. Si and F. Steglich, *Nat. Phys.* **4**, 186 (2008).
67. J. L. Tallon *et al.*, *Phys. Stat. Sol. B* **215**, 531 (1999).
68. N. S. Vidhyadhiraja, A. Macridin, C. Şen, M. Jarrell and M. Ma, *Phys. Rev. Lett.* **102**, 206407 (2009).
69. V. Barzykin and D. Pines, *Adv. Phys.* **58**, 1 (2009).
70. G. M. Eliashberg, *J. Exp. Theor. Phys.* **11**, 696 (1960); A. J. Millis, S. Sachdev and C. M. Varma, *Phys. Rev. B* **37**, 4975 (1988).
71. For a critique, see P. W. Anderson, *Science* **317**, 1705 (2007).
72. V. Aji, A. Shekhter and C. M. Varma, accepted in *Phys. Rev. B.*
73. P. W. Anderson, *Low Temp. Phys.* **32**, 282 (2006).

IV. BCS Beyond Superconductivity

THE SUPERFLUID PHASES OF LIQUID ^{3}He: BCS THEORY

A. J. Leggett
Department of Physics, University of Illinois at Urbana-Champaign, 1110 West Green Street, Urbana, IL 61801, USA
aleggett@illinois.edu

Following the success of the original BCS theory as applied to superconductivity in metals, it was suggested that the phenomenon of Cooper pairing might also occur in liquid 3-He, though unlike the metallic case the pairs would most likely form in an anisotropic state, and would then lead in this neutral system to superfluidity. However, what had not been anticipated was the richness of the phenomena which would be revealed by the experiments of 1972. In the first place, even in a zero magnetic field there is not one but two superfluid phases, and the explanation of this involves ideas concerning "spin fluctuation feedback" which have no obvious analog in metals. Secondly, the anisotropic nature of the pair wave function, which in the case of the B phase is quite subtle, and the fact that the orientation must be *the same* for all the pairs, leads to a number of qualitatively new effects, in particular to a spectacular amplification of ultra-weak interactions seen most dramatically in the NMR behavior. In this chapter I review the application of BCS theory to superfluid 3-He with emphasis on these novel features.

As is well known, the element helium is unique in remaining in the liquid state under its own vapor pressure down to (apparently) zero temperature, freezing only at a pressure of the order of 30 atm; this is true of both the stable isotopic species, ^{4}He and ^{3}He, although their freezing pressures are somewhat different. In the case of the heavy (and more common) isotope, ^{4}He, it has been known since 1938 that at a temperature of approximately 2.2 K (at saturated vapor pressure) the liquid makes a transition to an exotic phase (He-II) in which it shows behavior qualitatively different from that of a normal liquid, including most spectacularly the ability to flow through narrow capillaries with zero viscosity, the phenomenon that was christened by Kapitza "superfluidity"; thus the He-II phase is usually known informally as "superfluid ^{4}He". Following the original proposal of Fritz London, the modern understanding of the superfluid phase is based on the idea that this phase is characterized by the phenomenon of Bose–Einstein condensation (BEC), in which a nonzero fraction of all the atoms (which since $Z = 2$ and $A = 4$ are bosons) occupy a single one-particle state. This hypothesis leads to a very satisfactory qualitative and sometimes even quantitative picture of the properties of the superfluid phase of ^{4}He, see e.g. Ref. 1.

In the case of the light (and rare) isotope, ^{3}He, no evidence of superfluidity or any other exotic behavior is found above temperatures of the order of 3 mK; indeed, above this temperature liquid ^{3}He appears to behave qualitatively like any other liquid. However, quantitatively the behavior, particularly below ~ 100 mK, is rather unique among liquids; for example, the specific heat is approximately linear in temperature, and the spin susceptibility is temperature-independent. A qualitative understanding of this behavior may be obtained from the consideration that, since the ^{3}He atom contains two electrons with total spin zero and a nucleus with spin 1/2, it may be regarded as a composite spin-1/2 fermion and hence should obey Fermi statistics. If one considers a gas of noninteracting Fermi particles with the mass and density (at s.v.p.) of liquid ^{3}He, then the Fermi temperature should be of order 5 K; thus, at temperatures well below this, the particles should constitute a degenerate Fermi gas, similarly to the electrons in a normal metal, and thus should display the characteristic behavior of such a system. Of course, there are some important differences with the electrons in a normal metal: the ^{3}He atom is electrically neutral, so the interatomic forces are short-ranged rather than the r^{-1} Coulomb interaction, there is no background lattice and no immediate analog of the electron-phonon interaction or of impurity scattering. One result of this latter circumstance is that the transport coefficients in normal liquid ^{3}He are limited only by atom-atom collisions, the frequency of which is, because of the Pauli principle, proportional to T^2; thus the viscosity and spin diffusivity are proportional to T^{-2}, while the thermal conductivity (thermal diffusivity times specific heat) is proportional to T^{-1}. In addition, while the temperature-dependences of quantities like the specific heat and spin susceptibility of liquid ^{3}He are indeed those expected for a degenerate noninteracting Fermi gas, the absolute values are rather different. This is not surprising, given that the system is nothing like a noninteracting gas; in fact, the mean interatomic distance (~ 4 Å) is not much greater than the hard-core radius ($\sim 2 \cdot 6$ Å); what is at first sight surprising is that the Fermi-gas picture gives even qualitative agreement with the data.

Major insight into the reasons for this behavior came from the "Fermi-liquid" approach of Landau, published in 1956. Rather than starting from the noninteracting Fermi gas and trying to calculate the effects of the interactions quantitatively, Landau in some sense turned the problem around and asked: suppose we make the explicit assumption that perturbation theory in the interaction will eventually converge, then what conclusions can we draw about the groundstate and low-lying excitations of the strongly interacting system (the "Fermi liquid")? He showed that under that assumption there would be a one-one correspondence between the low-lying states of the noninteracting gas and the (fermionic) low-lying states of the Fermi liquid, so that one can parametrize the latter by the same quantum numbers as the former, namely the deviation $\delta n(\mathbf{p}, \sigma)$ of the number of particles with momentum p and spin σ from its value ($(\theta(p - p_F)$ where p_F is the Fermi momentum) in the groundstate. In the interacting system one may think intuitively of $\delta n(\mathbf{p}, \sigma)$ as the deviation from its groundstate value of the number of "quasiparticles", a quasiparticle being a real atom surrounded by a "dressing cloud" of other atoms. Landau was further able to show that when the Hamiltonian is expressed in terms of the quantities $\delta n(\mathbf{p}, \sigma)$ there are only two differences from that of the noninteracting gas: first, the dispersion relation $\epsilon(p)$ and hence the effective mass $m^* \equiv p_F/(d\epsilon/dp)_{p=p_F}$ would be modified from the mass m of a free ^{3}He atom, and secondly, there would arise various interaction terms that, thanks to the symmetry of the situation, can be expressed in the form of a set of "molecular fields," of which the simplest to visualize are the "Hartree" field arising from a variation in total density and the "Weiss" field generated by a spin polarization. In the Landau Fermi-liquid approach the strengths of the various molecular fields, like the effective mass renormalization, are not supposed to be calculated but have to be inferred from the experimental data. Consequently, in the first few years following Landau's 1956 paper, there was some scepticism among those working on liquid ^{3}He that this approach was anything more than a parametrization of the data; however, such doubts were eventually put to rest by the observation of a novel collective excitation ("zero sound") predicted by the theory, and of some anomalous effects in the spin-diffusion behavior in high magnetic fields, where one could account for a wide range of behavior at the cost of only one new fitted molecular-field parameter. By 1972 most of those working on liquid ^{3}He were convinced of the nontriviality and utility of the Fermi-liquid theory; indeed it had proved consistent with just about all the experimental behavior of this system. In the meantime, Silin and others had shown that a variant of the theory could be applied to the electrons in normal (nonsuperconducting) metals, again explaining the previously rather mysterious success of a simple Bloch–Sommerfeld model (a model that takes into account crystalline band-structure effects but not, except in a simple mean-field way, the effects of the strong Coulomb interaction) in accounting at least qualitatively for the experimental behavior.

Following the publication in 1957 of the BCS theory of superconductivity and its rapid acceptance by the community, people soon started to ask: if the normal state of liquid ^{3}He resembles so relatively closely that of the electrons in a normal metal, could the phenomenon of Cooper pairing not occur also in that system? Should this indeed occur, then since the ^{3}He atom is electrically neutral, the resulting state would clearly not be superconducting; however, it might be expected to show the analog of superconductivity in a neutral system, namely superfluidity (as in liquid ^{4}He); consequently, the possible Cooper-paired state of ^{3}He, at that time hypothetical, became informally known as "superfluid ^{3}He". In the years between 1958 and 1972 a substantial amount of theoretical work was done on the possible superfluid phase [*sic*!] of liquid ^{3}He, and a number of important results were established that have stood the test of time. First, it very soon became evident that the structure of any Cooper pairs formed in that system was likely to be rather different from that postulated in the original BCS theory of superconductivity. To discuss this question, let us define the two-particle order parameter

$$F(\mathbf{r}\mathbf{r}'\sigma\sigma') \equiv \langle \psi_\sigma^\dagger(\mathbf{r})\psi_{\sigma'}^\dagger(\mathbf{r}')\rangle \tag{1}$$

where the angular brackets have their usual significance in BCS theory of an "anomalous average," i.e. strictly speaking, if one insists on particle number conservation, the "average" should be understoood as a matrix element of the bracketed combination of operators between many-body states differing by 2 in the total particle number N. The quantity F is fundamental in BCS theory; its physical significance is that it is the unique eigenfunction of the two-particle reduced density matrix associated with a macroscopic $(O(N))$ eigenvalue, multiplied by the square root of that eigenvalue (N_0); so, apart from the factor of $\sqrt{N_0}$, it can be thought of as the "wave function of the Cooper pairs". (For more details, see e.g. Ref. 2, Sec. 5.4.) The more commonly encountered Ginzburg–Landau "order parameter" $\Psi(\boldsymbol{R})$ is, in the simple case considered by BCS, the special case of (1) corresponding to $\mathbf{r} = \mathbf{r}' \equiv \mathbf{R}$, $\sigma = -\sigma$. Let us for the moment assume that the center of mass of the pairs is at rest, so that F is a function only of the relative coordinate, which as there should be no danger of confusion we will simply call $\mathbf{r}$.

In the original BCS theory, the spin state of the pairs is a singlet and can be factorized out, so that (in an obvious notation)

$$F(\mathbf{r}, \sigma\sigma') = \frac{1}{\sqrt{2}}(\uparrow\downarrow - \downarrow\uparrow)\, F(\mathbf{r}) \tag{2}$$

Moreover, if with BCS we ignore the rather annoying but inessential complications due to the presence of a crystal lattice, $F(\mathbf{r})$ is taken to be simply a function of $r \equiv |\mathbf{r}|$ ("s-wave state"), so that the internal structure of the pairs is completely isotropic in both spin and orbital space. Since the form of the function $F(r)$ is fixed, at any given temperature and pressure, uniquely by the energetics, there are no degrees of freedom associated with this internal structure.

It was soon realized in thinking about the possibility of formation of pairs in liquid ^{3}He that their relative wave function is unlikely to be of a similar form. The reason is the following: the Fourier transform of $F(\mathbf{r})$, $F(\mathbf{k})$, is according to the BCS theory given by $\Delta/2E_k$, where Δ is the energy gap and $E_k \equiv \sqrt{(\epsilon_k - \mu)^2 + |\Delta|^2}$, with μ the chemical potential; Δ is proportional to the transition temperature T_c, which both in a typical (pre-1986) superconductor and in liquid ^{3}He is known from experiment to be at most of order 10^{-3} times the Fermi energy (which is essentially μ). Thus, the real-space pair wave function $F(\mathbf{r})$ is made up of a "packet" of (relative) plane waves with the magnitude of their $\mathbf{k}$-vectors approximately k_F ($\equiv p_F/\hbar$). If this packet is isotropic (s-wave) as in the BCS theory, then at short distances $F(r)$ is approximately proportional to $\sin(k_F r)/(k_F r)$, so that the probability to find the two fermions of the pair at a relative distance $\lesssim k_F^{-1}$($\sim$1 Å^{-1} for both metals and ^{3}He) is approximately some nonzero constant. For the case of a typical metallic superconductor this state of affairs is tolerable and indeed desirable, since at such short distances the phonon-induced indirect interaction is attractive and the direct Coulomb interaction, while repulsive, is too "soft" to make much difference. However, for the case of liquid ^{3}He there is a very strong hard-core interatomic repulsion, due primarily to the Pauli principle, at distances $\sim k_F^{-1}$, so that pairs formed in a relative s-state are likely to have a positive energy with respect to the normal state.

Thus, it was soon concluded that should Cooper pairs indeed form in liquid ^{3}He, they were likely to do so in a state of relative angular momentum higher than zero (it can be shown that in the limit $\Delta \ll E_F$ the relative angular momentum is approximately a good quantum number, i.e. the quantity $F(\mathbf{r})$ must be composed of spherical harmonics Y_{lm} of the same l-value (but possibly different m-values, see below). For the moment, we ignore the spin degree of freedom). In such a case, the probability density to be at relative distance r vanishes at the origin and has its first maximum near the point $r_l \equiv l/k_F$. Now the attractive part of the van der Waals potential between two ^{3}He atoms is maximum around 3 Å, and the closest r_l to this corresponds to $l = 2$, so most of the early papers concluded that this angular momentum was the most likely to occur ("d-wave pairing"). Evidently, by the Pauli principle, such a state would have to be a spin singlet, so the spin wave function factors out just as in the original BCS theory. In an important early paper, Anderson and Morel[3] gave a detailed analysis of this possibility, concluding that the pairs would most likely form not with a single value of m but in a combination that minimized the gap anisotropy and thus that of the pair wave function $F(\mathbf{r})$ (it turns out that in the case of $l \neq 0$ spin singlet pairing the angular anisotropy of the energy gap $\Delta(\mathbf{k})$ over the Fermi surface approximately tracks that of $F(\mathbf{k})$ and thus of the real-space function $F(\mathbf{r})$). Anderson and Morel also gave a brief discussion of the case of $l = 1$ (p-wave) spin triplet pairing; in this case, they chose to investigate one specific possible state, in which the "up" and "down" spins form pairs independently but with a common direction of relative angular momentum L. As we shall see below, this was a serendipitous choice.

A few other theoretical developments in the period before 1972 are worth mentioning. First, while early papers, including that of Anderson and Morel, had assumed that in the case of spin triplet pairing "up" spins paired only with up and "down" with down, it was pointed out by Vdovin[4] and independently by Balian and Werthamer[5] that it is possible to form a state in which, in addition, some up spins are paired with down (this is discussed in more detail below) and moreover that at least in the case of p-wave ($l = 1$) pairing the resulting state (which has become known in the literature as the BW state) would have, within a standard BCS-like approach, a lower energy than the kind of state considered by Anderson and Morel. Secondly, it was shown[6,7] that one could combine the ideas of BCS on pairing in a weakly interacting Fermi gas with those of Landau on a strongly interacting normal Fermi liquid to produce a theory of a "superfluid Fermi liquid"; interestingly, the molecular fields introduced in Landau's approach, which in the normal phase merely renormalize various constants, can in the superfluid state change the predicted temperature-dependences appreciably. Thirdly, it was pointed out[8] that in a Fermi system that is nearly ferromagnetic (as is, at least arguably, the case in liquid ^{3}He where the enhancement of the free-gas susceptibility by the relevant molecular field is about a factor of 4) there would be, in addition to any "bare" forces such as the van der Waals interaction, an extra interaction, analogous to the phonon-mediated interaction in a BCS superconductor, due to the exchange of virtual long-lived spin fluctuations, which would be repulsive in a spin singlet state but attractive in a triplet one.

Thus, in the spring of 1972 the general opinion in the community interested in liquid ^{3}He could probably have been summarized as follows:

1. It is quite likely that Cooper pairs will form at some temperature (only the braver souls among the theorists ventured to predict at what temperature, and their estimates ranged over 15 orders of magnitude);
2. If pairs form, they will probably form in a d-state but possibly (because of the effect of the spin-fluctuation-exchange interaction) in a spin triplet (p- or f-) state;
3. If the pair state is p-wave, it will be the BW state;
4. The most obvious diagnostic of the onset of Cooper pairing will be superfluidity, plus a BCS-like jump in the specific heat at the transition to the paired phase and a decrease in the spin susceptibility below the transition.

In the event, Nature had some surprises in store for us. In the late fall of 1971, Doug Osheroff, at the time, a graduate student in the group of David Lee at Cornell, was measuring the pressurization (P versus T) curve of a mixture of solid and liquid ^{3}He at temperatures below 3 mK. He observed two tiny but reproducible anomalies in the curve, which we now recognize as signaling, respectively, a second-order transition to one new phase (now known as the A phase) and a first-order transition between that phase and a *second* new phase (the B phase). Subsequently, yet a third new phase (the A_1 phase) was found to occur in high magnetic fields. Evidently, "superfluid ^{3}He" is a richer system than the theorists had anticipated! The phase diagram of liquid ^{3}He below 3 mK in zero magnetic field as we now know it is shown in Fig. 1.

What is the nature of these new phases? Although the earliest experiments did not give direct evidence for superfluidity (this came somewhat later) most of the community rapidly concluded that all three new phases were indeed Cooper-paired phases analogous to the superconducting phase of metals. But what was the symmetry (internal structure) of the pair wave function in each case (A, B, A_1)? To discuss this question, we need to introduce a little standard notation.

Let us consider, as above, the pair wave function $F(\mathbf{r}, \mathbf{r}', \sigma, \sigma')$ and for the moment set the dependence on the center-of-mass variable $\mathbf{R} \equiv (\mathbf{r}+\mathbf{r}')/2$ to

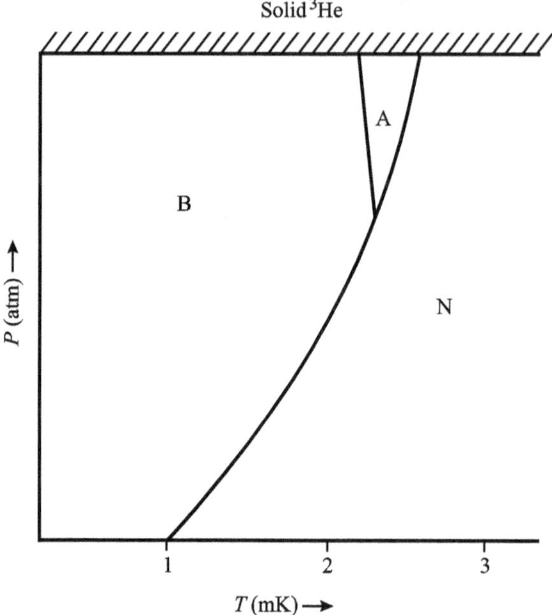

Fig. 1. Phase diagram of ^{3}He in zero magnetic field.

be a constant. Then, changing notation so that $\mathbf{r}$ now denotes the *relative* coordinate (formerly $\mathbf{r} - \mathbf{r}'$) we can write quite generally

$$F = F(\mathbf{r}, \sigma, \sigma') \tag{3}$$

According to the Pauli principle, if F is a singlet (triplet) in spin space it must be an even (odd) function of $\mathbf{r}$, and it turns out that except under rather pathological conditions it is energetically disadvantageous to mix these two possibilities, so we can discuss them separately. In the case of singlet pairing (which turns out not to be directly relevant to superfluid ^{3}He) the spin function factors out just as in the original BCS theory, i.e. we can write in an obvious notation

$$F = F(\mathbf{r}) \cdot \frac{1}{\sqrt{2}}(\uparrow\downarrow - \downarrow\uparrow) \tag{4}$$

What is the general structure of the orbital wave function $F(\mathbf{r})$? In the original BCS theory it is isotropic and as a function of $|\mathbf{r}|$ looks much like the wave function of two free atoms at the Fermi surface up to a distance of the order of a characteristic length $\xi \sim \hbar v_F / k_B T_c$ (the "pair radius," or within a factor of order unity, the "Pippard coherence length"), but beyond that point falls off as $\exp -|r|/\xi$, indicating that the two-particle state is bound as in Cooper's original calculation (for details see e.g. Ref. 2, Sec. 5.4). In the $l \neq 0$ case the behavior both of $F(\mathbf{r})$ and of the energy gap function $\Delta(\mathbf{k})$ is anisotropic, with $\Delta(\mathbf{k})$ having roughly the form of a combination of spherical harmonics $Y_{lm}(\hat{\mathbf{k}})$ (and, as in the s-wave case, little dependence on $|\mathbf{k}|$). Thus, while the spin properties of the system described by Eq. (3) should be isotropic as for the BCS case, orbital properties such as the (nonzero-T) superfluid density may display an experimentally measurable anisotropy (cf. Ref. 3). This would already be a significant difference from the BCS case.

However, the case of spin triplet (odd-l) pairing is even more interesting. Let us specialize for simplicity of presentation to the p-wave case (which as we shall see is generally believed to be that relevant to all three new phases of superfluid ^{3}He). We then have three possible Zeeman spin substates and three orbital ones, giving a nine-dimensional Hilbert space. Quite generally, we can resolve the pair wave function in terms of its amplitudes in the spin Zeeman substates: again using an obvious notation,

$$F(\mathbf{r}:\sigma\sigma') = F_{\uparrow\uparrow}(\mathbf{r})|\uparrow\uparrow\rangle + F_{\uparrow\downarrow}(\mathbf{r})\cdot\frac{1}{\sqrt{2}}(|\uparrow\downarrow + \downarrow\uparrow\rangle + F_{\downarrow\downarrow}(\mathbf{r})|\downarrow\downarrow\rangle \tag{5}$$

Since we are interested in the present context in the angular dependence of the relative wave function, it is convenient to take the Fourier transform $F(\mathbf{k})$ and sum it over the energy variable $|\mathbf{k}|$ (recall that in the BCS limit $T_c \ll E_F$, which should be a fairly good approximation for superfluid ^{3}He where $T_c/E_F \sim 10^{-3}$, the sum goes only over a range $\ll k_F$). As a function of angle $\hat{\boldsymbol{n}}$ on the Fermi surface the resulting (complex) expression is closely proportional to the (complex) energy gap $\Delta(\hat{\mathbf{n}})$, so it is conventional to discuss the latter, writing it in a form similar to (5):

$$\Delta(\mathbf{k}:\sigma\sigma') = \Delta_{\uparrow\uparrow}(\mathbf{k})|\uparrow\uparrow\rangle + \Delta_{\uparrow\downarrow}(\mathbf{k})\frac{1}{\sqrt{2}}\,|\uparrow\downarrow + \downarrow\uparrow\rangle + \Delta_{\downarrow\downarrow}(\mathbf{k})|\downarrow\downarrow\rangle \tag{6}$$

One more convenient piece of notation and we are through. If we for the moment put aside the A_1 phase, it turns out that all the p-wave Cooper-paired phases seriously considered as candidates for the experimentally observed A and B phases of superfluid ^{3}He have the following ("unitary") property: for each direction of $\hat{\mathbf{k}}$ on the Fermi surface separately, we can find a choice of spin axes (call it choice (a)) such that of the above three amplitudes, only $\Delta_{\uparrow\downarrow}(\hat{\mathbf{k}})$ is nonzero. (We emphasize that the choice need not be the same for all values of $\hat{\mathbf{k}}$.) Let us define, for any given direction $\hat{\mathbf{k}}$, a unit vector $\hat{\mathbf{d}}(\mathbf{k})$ as the z-axis in this basis; then a compact description of the symmetry of any given state, which is entirely equivalent to Eq. (6), is given by simply specifying the quantity $\mathbf{d}(\mathbf{k}) \equiv |\Delta_{\uparrow\downarrow}(\hat{\mathbf{k}})|\,\hat{\mathbf{d}}(\hat{\mathbf{k}})$ as a function of direction $\hat{\mathbf{k}}$ on the Fermi surface. The precise equivalence is given by the formulae (applicable for any **k**-*independent* choice of spin axes)

$$\begin{gathered}\Delta_{\uparrow\uparrow}(\mathbf{k}) = \big(d_x(\mathbf{k}) + id_y(\mathbf{k})\big), \quad \Delta_{\uparrow\downarrow}(\mathbf{k}) = d_z(\mathbf{k}), \\ \Delta_{\downarrow\downarrow}(\mathbf{k}) = \big(d_x(\mathbf{k}) - id_y(\mathbf{k})\big)\end{gathered} \tag{7}$$

The quantity $\Delta_0(\mathbf{k}) \equiv |\mathbf{d}(\mathbf{k})| = \left(|\Delta_{\uparrow\downarrow}(\mathbf{k})|^2 + |\Delta_{\uparrow\uparrow}(\mathbf{k})|^2\right)^{1/2}$ is the gap magnitude for direction $\hat{\mathbf{k}}$ on the Fermi surface, and plays the same role as the "energy gap" in the original BCS theory, i.e. quasiparticle energies are $E_k = \sqrt{(\epsilon_k - \mu)^2 + |\Delta_0(\mathbf{k})|^2}$, etc. Note that for a unitary state it is also always possible, for any given $\hat{\mathbf{k}}$, to find a choice of spin axes (different from the above one; call it choice (b)) such that the z-component vanishes; then (as indeed more generally) the magnitude of the "up-up" gap is identical to that of the "down-down" one though their phases may be different.

The forms of $\mathbf{d}(\hat{\mathbf{n}})$ appropriate to the experimentally observed A and B phases of superfluid ^{3}He have been inferred from various experimental anisotropies, in particular that of the NMR data, and are also consistent with theoretical predictions (see below); they are believed to be as follows. The A phase is identified with the "ABM" phase (the one originally studied by Anderson and Morel and later by Anderson and Brinkman); this is defined by the statement that

$$\mathbf{d}(\hat{\mathbf{n}}) = \mathbf{d}f(\hat{\mathbf{n}}) \tag{8}$$

where $\mathbf{d}$ is a fixed vector and with a suitable choice of orbital axes the function $f(\hat{\mathbf{n}})$ is

$$f(\hat{\mathbf{n}}) = \left(\frac{3}{2}\right)^{1/2} \Delta_0 \sin\theta \cdot \exp i\varphi \tag{9}$$

where θ and φ are the standard azimuthal angles and Δ_0 is the RMS gap over the Fermi surface. In other words, Eq. (8) tells us that the spin wave function factors out and with an appropriate choice of spin axes is simply the $S = 1$, $S_z = 0$ Zeeman state $\left(\frac{1}{\sqrt{2}}(\uparrow\downarrow + \downarrow\uparrow)\right)$; while Eq. (9) tells us that with an appropriate choice of orbital axes the pairs appear to have a relative angular momentum $L_z = 1$ around the z-axis. (On the question of the physical meaning of this observation, see below.) In a general reference frame the spin state is characterized by a single vector $\mathbf{d}$ and the orbital behavior by a single vector $\mathbf{l}$; our considerations up to now give no reason for any particular relative orientation of $\mathbf{d}$ and $\mathbf{l}$. An interesting point about the

ABM phase is that it is one of a class of so-called "equal-spin-pairing (ESP)" states that have the property that we can find a choice (b) of spin axes such that no matter where we are on the Fermi surface, up spins are paired only with up and down only with down, so that to a first approximation one can regard the spin populations as independent. It is then fairly obvious that application of a magnetic field along this axis will simply enlarge the up-spin Fermi surface relative to the down-spin one but will not otherwise interfere with the pairing process (at least at this level of discussion). Thus, assuming that the system is able to adjust the spin axes so that the "up" direction is indeed along the field, the susceptibility of the *ABM* phase should not be reduced from that of the normal phase; this indeed seems to be the case to a good approximation for the experimental *A* phase.

The *B* phase is usually identified with the "BW" phase predicted by Vdovin and by Balian and Werthamer. The form of $\mathbf{d}(\hat{\mathbf{n}})$ that correponds to the state explicitly considered by these authors is

$$\mathbf{d}(\hat{\mathbf{n}}) = \Delta_0 \hat{\mathbf{n}} \tag{10}$$

which when translated into the language of Zeeman components describes a state with each of the spin components paired with the opposite orbital component, so as to give $L = 1$, $S = 1$ but $\mathbf{J} = \mathbf{L} + \mathbf{S} = 0$; that is, what in the language of atomic physics would be denoted a $^3P_0$ state. By the Wigner–Eckhardt theorem such a state must be isotropic in all its properties, since it possesses no characteristic vector. Moreover, since with any choice of the spin axes one-third of the spins are paired "oppositely," the susceptibility is reduced from the normal-state value (actually by more than 1/3, because of molecular-field effects); such a reduction is indeed seen in the experimental *B* phase (and as a result the *B* phase is unstable with respect to the *A* phase at $T = 0$ in fields above ~ 0.55T). However, it turns out that the straightforward identification of the *B* phase with the simple $^3P_0$ state misses a subtlety; since the latter does not affect the discussion of the next few paragraphs, we postpone it for the moment.

At this point let us make a brief digression to discuss the A_1 phase. This phase, which occurs only in high magnetic fields, is non-unitary and thus description in terms of (a generalized) vector $\mathbf{d}(\mathbf{n})$, while possible, is not particularly convenient; a more compact description is that the A_1 phase is simply the *A* phase with only one spin population paired. This phase has some unique properties, particularly as regards the "two-fluid" description of its hydrodynamics (see e.g. Ref. 9), but there is no space to discuss them here.

Returning to the A and B phases, which take up the bulk of the phase diagram, we note that if the above identifications are correct the very existence of the A phase in zero magnetic field poses a puzzle, since according to the pre-1972 calculations, based on a straightforward generalization of BCS theory, the ABM phase, like all equal-spin pairing phases, should be unstable in zero field with respect to the BW phase, which is indeed, within this framework, the unique minimum of the free energy below T_c. This puzzle was brilliantly solved by Anderson and Brinkman,[10] who took up the earlier idea that a substantial part of the attraction responsible for forming the pairs might be due to the exchange of virtual spin fluctuations, and then pointed out a crucial difference with the situation in superconducting metals: in the latter case, the attraction comes from the exchange of virtual phonons, which are excitations of a different system (the ions) from the electrons which undergo the pairing; consequently, when pairs are formed the structure of the virtual phonons is effectively unchanged. By contrast, in the case of liquid ^{3}He the virtual spin fluctuations whose exchange provides the attraction are excitations of the very same system as is forming the pairs, and thus their structure can be modified by the onset of pairing. Now one might expect that the "strength" of the spin fluctuation exchange might be somehow proportional to the spin susceptibility of the liquid, and thus be reduced in the BW state, which would thereby be disadvantaged with respect to an ESP state. Although this argument as it stands is too crude, it turns out that a microscopic calculation of this effect[11] (which goes beyond standard BCS theory, in fact invokes terms of higher order in T_c/E_F than are usually kept in that theory) not only confirms the intuitive result that it favors any ESP state over the BW state, but, remarkably, shows that the specific ABM state is uniquely favored, and that its advantage is greatest in the high-T, high-pressure part of the phase diagram where the A phase is actually found to occur. This is probably the only respect in which a satisfactory microscopic theory of the superfluid phases of ^{3}He needs to invoke ideas that go qualitatively beyond the original BCS scheme or straightforward generalizations of it.

At the phenomenological level, on the other hand, there is a crucially important difference between superfluid ^{3}He and the superconductors so successfully described by BCS theory: as already remarked, in the latter case the internal structure of the Cooper-pair wave function is uniquely determined by the energetics, so that there are no degrees of freedom associated with it. In the case of superfluid ^{3}He, on the other hand, while the "shape" of the pair wave function in both the A and the B phases is determined, as sketched above, by the "gross" energies of the problem, i.e. the kinetic energy and interatomic (hard-core and van der Waals) potential energy, the *orientation* is not. As regards this orientataion, the gross energies in effect constrain all the pairs to behave in *the same* way, but do not determine in *which* way. The situation may be regarded as analogous to that in an isotropic Heisenberg ferromagnet, where the strong (exchange) forces constrain all the spins to lie parallel, but say nothing about the common direction in which they lie (see further below). In the case of ^{3}He this is most obvious in the A phase, because while the characteristic spin vector $\mathbf{d}$ and the characteristic orbital vector $\mathbf{l}$ must be the same for all pairs, nothing we have said so far in any way constrains their orientations, either absolute or relative to one another. However, it is also true in a more subtle way for the B phase, since nothing (at least nothing said so far) prevents us starting with the familiar $^3P_0$ state as above, and then rotating the spin coordinates by some arbitrary amount (i.e. by an arbitrary angle θ about an arbitary axis $\hat{\omega}$) relative to the orbital coordinates; since none of the energies considered so far couple the spin and orbital coordinates, the state so obtained must be degenerate with the original one. This property (which is shared by the A phase, in the sense that a relative rotation of $\mathbf{l}$ and $\mathbf{d}$ cannot change the energy) is sometimes called "spontaneously broken spin-orbit symmetry".

Before investigating some of the consequences of this state of affairs, let us digress for a moment to explore the physical significance of the orbital vector $\mathbf{l}$ in the A phase. Formally, it is the direction such that, if it is chosen as orbital z-axis and the azimuthal angle θ and polar angle φ defined in the conventional way, the relative wave function of the Cooper pairs has the angular dependence $\sin\theta \cdot \exp i\varphi$. If one were talking about a diatomic molecule (at rest), then such an angular dependence of the relative wave function would indicate clearly that the molecule has *total* angular momentum $L = L_z = \hbar$. Can we draw a similar conclusion for ^{3}He-A? That is, does the system possess a *total* angular momentum along $\mathbf{l}$, and if so what is its magnitude? Curiously, while just about everyone agrees that the answer to the first question is yes, the answer to the second is after 35 years still not universally agreed; the answers $N\hbar/2 \equiv L_0$, $\sim(\Delta/E_F)L_0$ and $\sim(\Delta/E_F)^2 L_0$ all find advocates in the literature. The question has proved very difficult to resolve experimentally; it is complicated by the fact that in real life the direction of $\mathbf{l}$ is likely to vary in space (see next paragraph). For what it is worth, the view of the present author is that the question may not even be well defined, so that even for an ideally uniform direction of $\mathbf{l}$ the total angular momentum may depend on the way in which the superfluid phase was reached. For further discussion, see Ref. 2, Appendix 7A.

The fact that the gross energies do not determine the "orientation" of the pair wave function means that this quantity may vary in space and/or in time, and this gives rise to some fascinating phenomena, in particular, the possibility of a variety of topological singularities which are a more sophisticated analog of the simple Abrikosov vortices which are found in (type-II) BCS superconductors. The study of the various possible singularities in the A and B phases with the aid of the branch of pure mathematics known as homotopy theory has been a major industry over the last 30 years, see e.g. Refs. 11 and 12. One very interesting output of these studies has been the conclusion that in the A phase, given that close to a wall of the sample cell the **l**-factor tends to orient itself normal to the wall, any simply connected sample must contain at least two topological singularities.

An even more fascinating consequence of the spontaneously broken spin-orbit symmetry is a property which I will call "superfluid amplification". To explain this, it is helpful to return to our ferromagnetic analogy. In a ferromagnetic material such as Fe, the electron spins may be subject to a weak magnetic field. If we are above the Curie temperature, the spins behave to a first approximation independently; the single-spin Zeeman energy E_Z is in competition with the much larger thermal energy k_BT, and the resulting polarization is of order $E_Z/k_BT \ll 1$. Below the Curie temperature the situation changes qualitatively: the "strong" (exchange) forces now constrain all or most of the N spins to lie parallel, and as a result the energy difference which competes with the thermal energy is no longer E_Z but rather NE_Z, which is much *larger* than k_BT. As a result the polarization becomes of order unity in even very weak external fields.

The analog of this behavior in liquid ^{3}He is that the "strong" terms in the Hamiltonian (kinetic energy, hard-core, van der Waals attraction, even the "effective" spin-spin interaction generated by exchange) are all invariant under a relative rotation of the spin and orbital coordinates; thus in the normal phase, where there is no strong correlation between the behavior of different *pairs* of atoms, these coordinates are effectively uncorrelated. However, there exists one term in the Hamiltonian which is *not* invariant under the relative spin-orbit rotation, namely the electromagnetic interaction of the nuclear dipole moments. Crudely speaking, this term has the effect that if two atoms have parallel spins they would prefer to orbit in a plane parallel to the spins rather than one perpendicular ("end-over-end" rather than "side-by-side," just as would two small bar magnets). However, the difference in energy between these two configurations (call it g_n) is very small indeed ($\sim 10^{-7}K$ even at the minimum interatomic distance), so in the normal phase it cannot compete with the thermal energy and, as already remarked, the spins and orbital coordinates of any given pair of atoms are effectively uncorrelated.

Now consider what happens in a superfluid (Cooper-paired) phase. The "strong" terms in the Hamiltonian now conspire to force all the pairs to behave in exactly *the same* way, but cannot determine in *which* way. Now the tiny nuclear dipole-dipole interaction comes into its own: since all pairs must now have the same relative orientation of their spins and orbital coordinates, the difference between the "good" and "bad" orientations is now not $\sim g_n$ but rather $\sim N_p g_n$, where $N_p \sim N$ is some intuitive measure of the "number of Cooper pairs," and this difference is now very much larger than the thermal energy. Consequently, the relative orientation of the spin and orbital coordinates will adjust itself so as to minimize the dipole energy. In the A phase the result is that the characteristic spin vector $\mathbf{d}$ orients itself parallel (or antiparallel) to the characteristic orbital vector $\mathbf{l}$, while in the B phase the spins of the pairs are rotated away from the "reference" $^3P_0$ configuration by the angle $\cos^{-1}(-1/4) \cong 104°$. Of course, these prescriptions still leave part of the original degeneracy (e.g. in the A phase, the common orientation of $\mathbf{l}$ and $\mathbf{d}$) unbroken, and to break this residual degeneracy we need to invoke other effects such as those of the cell walls; see e.g. Ref. 13, Sec. X.

These, then, are the *equilibrium* configurations in the A and B phases respectively. Now, what happens if we somehow tilt the orientation away from equilibrium? (As we shall see below, this can be done by applying an rf magnetic field, which acts on the spins but not on the orbital coordinates.) It turns out that this results in a torque on the *total* spin of the system (and an equal and opposite torque on the total orbital angular momentum), which adds to the effect of the external magnetic field and gives rise to anomalous NMR behavior. We can make this observation quantitative by exploiting the fact that the "microscopic" frequencies of the system, such as the "gap frequency" $\Delta/\hbar$, the quasiparticle collision rate, etc., are all high compared to characteristic NMR frequencies; as a consequence, it is a good zeroth-order approximation to describe the NMR behavior by a Born–Oppenheimer type of approximation, in which we express the Hamiltonian as a function only of the "slow" variables, which in this case are the total spin vector $\mathbf{S}$ of all the atoms of the system (irrespective of whether they are paired or not) and the *orientation* of the spins of the Cooper pairs. It is convenient to describe this latter variable, schematically, by the (vector) angle of rotation $\boldsymbol{\theta}$ relative to the equilibrium configuration; it is essential to appreciate that the variables $\boldsymbol{\theta}$ and $\mathbf{S}$ are *kinematically independent*; indeed, they may be shown to satisfy the commutation relations

$$[S_i,\ \theta_j] = i\delta_{ij} \tag{11}$$

The Hamiltonian can now be written in the form

$$H = H_0(\mathbf{S}, \boldsymbol{\mathcal{H}}) + H_d(\boldsymbol{\theta}) \tag{12}$$

where the first term contains the Zeeman and exchange terms, while the second expresses the dependence of the nuclear dipolar energy on θ. (It is implicitly assumed, here, that the orbital coordinates of the pairs are effectively frozen over times of the order of the NMR period.) Then the semiclassical equations of motion take the simple form

$$d\mathbf{S}/dt = \mathbf{N}_{\text{ext}} - \partial H_d/\partial\boldsymbol{\theta} \tag{13a}$$

$$d\boldsymbol{\theta}/dt = \partial H_0/\partial\mathbf{S} \tag{13b}$$

where $\mathbf{N}_{\text{ext}}$ is the usual torque due to the external magnetic field. These equations can be solved for a variety of interesting situations; in particular, it turns out that the linear NMR behavior can serve as a "fingerprint" to identify the structure of the order parameter. For an ESP phase with a single **d**-vector, such as the ABM phase, a "Pythagorean" shift of the resonance frequency is predicted:

$$\omega^2(T) = \omega_L^2 + \omega_0^2(T) \qquad (\omega_L \equiv \gamma\mathcal{H}) \tag{14}$$

in agreement with what is observed experimentally in the A phase. (For other types of ESP phase the resonance generally splits, a behavior for which there is no evidence in the experiments.)

The situation with regard to the linear NMR behavior in the BW phase is quite interesting. While the considerations given above about the dipole energy determine only the angle of the spin-orbit rotation, not its axis, there are both theoretical and experimental reasons to believe that there is a very tiny energy which in the bulk liquid tends to orient this axis (call it $\hat{\boldsymbol{\omega}}$) along the direction of the external magnetic field $\boldsymbol{\mathcal{H}}$. If now we do a standard NMR experiment in which the oscillating rf field is perpendicular to $\boldsymbol{\mathcal{H}}$, solution of Eq. (13b) shows that the effect of this field is, to lowest order, to change only the direction of $\hat{\boldsymbol{\omega}}$, not the angle of the rotation; thus the dipole energy is essentially unchanged and the second term in Eq. (13a) is zero, so no shift in the NMR frequency is expected. However, an oscillating rf field applied *parallel* to $\boldsymbol{\mathcal{H}}$ rotates according to Eq. (13b) the spin coordinates around

$\mathcal{H}$, changing the rotation angle and with it the dipole energy. We therefore expect a nonzero-frequency *longitudinal resonance* — exactly what is seen in ^{3}He-B. Thus the experimentally observed NMR behavior strongly supports the standard identification of the A phase of superfluid ^{3}He with the ABM phase and of the B phase with the BW phase.

The principle of amplification of ultra-weak effects, which as I have indicated is most spectacularly exemplified in the NMR behavior, may have other applications. For example, it is predicted[14] that the A phase may show a very weak but nevertheless "macroscopic" (i.e. extensive) electronic ferromagnetism, and this effect may have been observed.[15] Also, consider the possibility of an electric dipole moment $\mathbf{d}_{el}$ induced by the P-violating but T-conserving part (the dominant part) of the neutral-current terms in the weak interaction of particle physics. In any ordinary system, characterized by a single vector, the angular momentum $\mathbf{J}$, the Wigner–Eckhardt theorem states that the only possibility is

$$\mathbf{d}_{\mathrm{el}} = \mathrm{const} \cdot \boldsymbol{J} \tag{15}$$

which violates not only P but T; hence it is not a candidate. However, the Cooper pairs in the B phase of ^{3}He are unique in being characterized not by a total angular momentum $\boldsymbol{J}$ but rather by the T-invariant pseudo-vector $\langle \boldsymbol{L} \times \boldsymbol{S} \rangle = \mathrm{const} \cdot \hat{\boldsymbol{\omega}}$. Hence symmetry permits a macroscopic electric dipole moment of the form

$$\mathbf{d}_{\mathrm{el}} = \mathrm{const} \cdot \hat{\boldsymbol{\omega}} \tag{16}$$

which violates P but not T.

Should future experiments detect such a dipole moment (which calculations[16] suggest is likely to be very tiny but not necessarily hopelessly so), it would constitute the first ever example of a *macroscopic* violation of parity due to the weak interaction.

In conclusion, while most of our understanding of the superfluid phases of liquid ^{3}He is based firmly on a generalization of the original ideas of BCS, these systems provide a richness of behavior that goes far beyond that seen in the original application to superconductors. Indeed, with the possible exception of the subsequently discovered fractional quantum Hall effect, we may say that superfluid ^{3}He is the most sophisticated physical system of which we can currently claim not just a qualitative but a quantitative understanding.

Acknowledgment

This work was partially supported by the National Science Foundation under grant no. NSF DMR09-06921.

References

1. D. R. Tilley and J. Tilley, *Superfluidity and Superconductivity*, 3rd edition, Adam Hilger, Bristol (1990).
2. A. J. Leggett, *Quantum Liquids: Bose Condensation and Cooper Pairing in Condensed Matter Systems* (Oxford University Press, Oxford, 2006).
3. P. W. Anderson and P. Morel, *Phys. Rev.* **123**, 1911 (1961).
4. Yu. A. Vdovin, in *Application of Methods of Quantum Field Theory to Problems of Many Particles*, ed. A. I. Alekseeva (Moscow, Gosatomizdat, 1963) (in Russian).
5. R. Balian and N. R. Werthamer, *Phys. Rev.* **131**, 1553 (1963).
6. A. I. Larkin and A. B. Migdal, *Zhurn. Eksperim. i Teor. Fiz.* **44**, 1703 (1963); [translation: *Sov. Phys. JETP* **17**, 1146 (1963)].
7. A. J. Leggett, *Phys. Rev. A* **140**, 1865 (1965).
8. A. Layzer and D. Fay, *Int. J. Magnetism* **1**, 135 (1971).
9. M. Liu, *Phys. Rev. Lett.* **43**, 1740 (1979).
10. P. W. Anderson and W. F. Brinkman, *Phys. Rev. Lett.* **30**, 1108 (1973).
11. W. F. Brinkman, J. W. Serene and P. W. Anderson, *Phys. Rev. A* **10**, 2386 (1974).
12. N. D. Mermin, *Revs. Mod. Phys.* **51**, 591 (1979).
13. G. E. Volovik, *Exotic Properties of Superfluid* ^{3}*He* (World Scientific, Singapore, 1992).
14. A. J. Leggett, *Revs. Mod. Phys.* **47**, 331 (1975).
15. A. J. Leggett, *Nature* **270**, 585 (1977).
16. D. N. Paulson and J. C. Wheatley, *Phys. Rev. Lett.* **40**, 557 (1978).
17. A. J. Leggett, *Phys. Rev. Lett.* **39**, 587 (1977).

SUPERFLUIDITY IN A GAS OF STRONGLY INTERACTING FERMIONS

W. Ketterle, Y. Shin, A. Schirotzek and C. H. Schunk

MIT-Harvard Center for Ultracold Atoms,
Research Laboratory of Electronics,
Department of Physics, Massachusetts Institute of Technology,
Cambridge, MA 02139, USA

After an introduction into 100 years of research on superfluidity and the concept of the BCS-BEC crossover, we describe recent experimental studies of a spin-polarized Fermi gas with strong interactions. Tomographically resolving the spatial structure of an inhomogeneous trapped sample, we have mapped out the superfluid phases in the parameter space of temperature, spin polarization, and interaction strength. Phase separation between the superfluid and the normal component occurs at low temperatures, showing spatial discontinuities in the spin polarization. The critical polarization of the normal gas increases with stronger coupling. Beyond a critical interaction strength all minority atoms pair with majority atoms, and the system can be effectively described as a boson-fermion mixture. Pairing correlations have been studied by RF spectroscopy, determining the fermion pair size and the pairing gap energy in a resonantly interacting superfluid.

1. From 1908 to 2008

The field of low-temperature physics has a long history. Many people regard the liquefaction of helium in 1908 as the beginning of modern low-temperature physics. This long-standing tradition continues in the research on ultracold bosonic and fermionic atomic gases, and it is interesting to draw a few analogies between current research and what happened 100 years ago. Many cold fermion clouds are cooled by sympathetic cooling with a bosonic atom which is evaporatively cooled into or close to Bose condensation. Popular combinations are ^{6}Li and ^{23}Na (used in our work at MIT), and ^{40}K and ^{87}Rb. It is remarkable that the first fermionic superfluids were also cooled "sympathetically" by ultracold bosons (liquefied ^{4}He) when Onnes cooled down mercury in 1911, finding that the resistivity of the metal suddenly dropped to nonmeasurable values at $T_C = 4.2$ K, it became "superconducting". Tin (at $T_C = 3.8$ K) and lead (at $T_C = 6$ K) showed the same remarkable phenomenon. This was the discovery of superfluidity in an electron gas.

Although superfluidity of bosons was directly observed only in 1938,[1,2] a precursor was already observed earlier by Kamerlingh Onnes when he lowered the temperature of the liquefied ^{4}He below the the λ-point at $T_\lambda = 2.2$ K. In his Nobel lecture in 1913, he notes "that the density of the helium, which at first quickly drops with the temperature, reaches a maximum at 2.2 K approximately, and if one goes down further even drops again. Such an extreme could possibly be connected with the quantum theory."[3] But instead of studying, what we know now was the first indication of superfluidity of bosons, he first focused on the behavior of metals at low temperatures and observed superconductivity in 1911.

The fact that bosonic superfluidity and fermionic superfluidity were first observed at very similar temperatures, is due to purely technical reasons (because of the available cryogenic methods) and rather obscures the very different physics behind these two phenomena. Bosonic superfluidity occurs at the degeneracy temperature, i.e. the temperature T at which the spacing between particles $n^{-1/3}$ at density n becomes comparable to the thermal de Broglie wavelength $\lambda = \sqrt{\frac{2\pi\hbar^2}{mk_BT}}$, where m is the particle mass. The predicted transition temperature of $T_{\mathrm{BEC}} \sim \frac{2\pi\hbar^2}{m} n^{2/3} \approx 3$ K for liquid helium at a typical density of $n = 10^{22}$ cm^{-3} coincides with the observed lambda point. In contrast, the degeneracy temperature (equal to the Fermi temperature $T_F \equiv E_F/k_B$) for conduction electrons is higher by the mass ratio $m(^4\mathrm{He})/\mathrm{m_e}$, bringing it up to several ten-thousand degrees. Of course, we know now, from the work of Bardeen, Cooper and Schrieffer,[4] that the critical temperature for superfluidity is reduced from the degeneracy temperature to the Debye temperature T_D (since electron-phonon interactions lead to Cooper pairing) times an exponentially small prefactor, $e^{-1/\rho_F|V|}$, with the electron-electron interaction V, attractive for energies smaller than k_BT_D and the density of states at the Fermi energy, $\rho_F = m_e k_F/2\pi^2\hbar^2$. The Debye temperature is typically 100 times smaller than the Fermi energy, and the exponential factor suppresses the transition temperature by another factor of 100, with the result that typical values for T_c/T_F are 10^{-4}.

When the interactions between the electrons are parameterized by an s-wave scattering length a, the transition temperature is given by the expression

$$T_{C,\mathrm{BCS}} = \frac{e^{\gamma}}{\pi} \frac{8}{e^2} E_F e^{-\pi/2k_F|a|} \tag{1}$$

with Euler's constant γ, and $e^{\gamma} \approx 1.78$. Now, for resonantly interacting fermions (i.e. near a Feshbach resonance), the scattering length a becomes infinite. The above equation is no longer valid, but implies correctly that the transition temperature will approach the Fermi temperature T_F. The value of T_C for $a = \infty$ (i.e. at unitarity) has been calculated analytically,[5–8] via renormalization-group methods[9] and via Monte-Carlo simulations.[10,11] The result is $T_C = 0.15 - 0.16\ T_F$.[8,11] It is at the unitarity point that fermionic interactions are at their strongest. Further increase of attractive interactions will lead to the appearance of a bound state and turn fermion pairs into bosons. As a result, the highest transition temperatures for fermionic superfluidity are obtained around unitarity and are on the order of the degeneracy temperature. Finally, almost 100 years after Kamerlingh Onnes, it is not just an accidental coincidence anymore that bosonic and fermionic superfluidity occur at similar temperatures! It is in this regime that our experiments are conducted.

(Note that this paper is an updated version of an earlier review paper written on the occasion of the LT25 conference in Amsterdam[12]).

2. Ultralow-density Condensed Matter Physics

Many people regard the extremely low nanokelvin temperatures of ultracold atoms as their distinguishing feature. One can take the position that what matters more is their extreme diluteness, at number densities around 10^{13} or 10^{14} cm^{-3}, a million times more diluted than air. With interatomic distances of several 100 nm, the atoms are fairly isolated, and allow the application of all the methods for manipulation and detection developed in atomic physics, including RF spectroscopy, optical detection, preparation in different hyperfine states. Most importantly, these gases are ideal realizations of bosons and fermions with short-range interactions, idealized by a delta function potential and characterized by the s-wave scattering length. Therefore, their properties are *fully* described by simple Hamiltonians (such as the hard sphere Bose gas, or, when exposed to a periodic potential, by Hubbard models).

This gives cold atoms a new and important link between the materials of the real world with all their richness and complexity, and the simple models used for their description in many-body physics. Often, predictions of models cannot be rigorously tested, because available materials have more complexity (and impurities) than the models, or, with the case of high-T_C superconductors as an example, it is not even clear if the models capture the essential physics of the material. In contrast, using the tools of atomic physics, it is possible to exactly engineer Hamiltonians for ultracold atoms. In this regard, cold atom experiments are quantum simulations of Hamiltonians, but we prefer to say that they realize new forms of matter, which are described by these Hamiltonians.

Of special interest are Hamiltonians which cannot be solved, even numerically. In this case, cold atom experiments may become a tool to verify or falsify whether certain approximation schemes are adequate, i.e. they capture the essential physics either in a qualitative or quantitative way. One example is the fermionic Hubbard model with repulsive interactions[13] suggested as a toy model for high-T_C superconductors, but there is so far no rigorous proof that its ground state is a d-wave superfluid.

The growing list of condensed matter systems which have been realized and studied with cold atoms include the weakly interaction Bose gas,[14,15] the Bose-Hubbard model with the superfluid to Mott insulator transition,[16] several regimes of the hard sphere one-dimensional Bose gas (Yang-Yang thermodynamics,[17] Tonks-Giradeau gas[18]), fermions with "infinite" interaction strength (i.e. resonant interactions in the unitarity limit),[19] the BEC-BCS crossover,[19] the Fermi-Hubbard model with the crossover to the Mott insulator,[20,21] and Anderson localization of non-interacting matter waves.[22,23]

3. Realization of the BEC-BCS Crossover

When the theory of superconductivity was developed in the '50s, there were controversial discussions about the role of Bose-Einstein condensation. Schafroth, Blatt and Butler speculated that a Bose-Einstein condensate of electron pairs is responsible for superconductivity, but could formalize their ideas only for the case of localized pairs.[24] In contrast, Bardeen, Cooper and Schrieffer pointed out that, for typical conditions, there are around 10^6 electrons in a coherence volume, and therefore the BCS transition is not analogous to Bose-Einstein condensation.[4] We know now that BEC and BCS are the two-limiting cases of the BCS-BEC crossover which smoothly connects the so-called pairing in momentum space (BCS limit) with localized pairs (BEC limit), and the condensation of preformed pairs (the bosons in the BEC limit) with pairing occurring only at the phase transition (BCS).

Soon after the formulation of the BCS theory, Blatt and others showed (see e.g. Ref. 25 and references therein) that the BCS wavefunction

$$|\Psi_{\rm BCS}\rangle = \prod_k (u_k + v_k c^\dagger_{k\uparrow} c^\dagger_{-k\downarrow})\,|0\rangle \tag{2}$$

can be expressed as an anti-symmetrized wavefunction of $N/2$ fermion pairs:

$$|\Psi\rangle_N = {b^\dagger}^{N/2}\,|0\rangle \tag{3}$$

The BCS wavefunction given above mixes up the number of particles (in the spirit of a grand-canonical description), whereas the wavefunction with the product of pairs assumes a fixed number of fermions, but both approaches can be formulated for both fixed and fluctuating particle numbers.

Here the pair creation operator

$$b^\dagger = \sum_k \varphi_k c^\dagger_{k\uparrow} c^\dagger_{-k\downarrow} \tag{4}$$

is defined using the creation operator $c^\dagger_{k\uparrow}$ for particles with momentum k and the Fourier transform $\varphi(\mathbf{r}_1 - \mathbf{r}_2) = \sum_k \varphi_k \frac{e^{i\mathbf{k}\cdot(\mathbf{r}_1-\mathbf{r}_2)}}{\sqrt{\Omega}}$ of the pair wave function $\varphi(\mathbf{r}_1, \mathbf{r}_2)$. Ω is the volume of the system. To write the BCS wavefunction as a "condensate of pairs" is the essence of the BCS-BEC crossover, since one can now define pair wave functions $\varphi(\mathbf{r}_1, \mathbf{r}_2)$ for smaller and smaller pair size and approach the BEC limit of isolated bosons. The credit for having predicted the possibility of such a crossover usually goes to Eagles for an early suggestion[26] and to Leggett for a complete presentation of the concept.[27]

It is only in the limit of small pairs (i.e. pairs spread out in momentum space), that the pair operators $b^\dagger$ obey bosonic commutation relations. For the commutators, one obtains $[b^\dagger, b^\dagger] = 0$, $[b, b] = 0$ and $[b, b^\dagger] = \sum_k |\varphi_k|^2(1 - n_{k\uparrow} - n_{k\downarrow})$. The third commutator is equal to one only in the limit where the pairs are tightly bound and occupy a wide region in momentum space. In this case, the occupation numbers n_k of any momentum state k are very small and $[b, b^\dagger]_- \approx \sum_k |\varphi_k|^2 = \int \mathrm{d}^3 r_1 \int \mathrm{d}^3 r_2\, |\varphi(\mathbf{r}_1, \mathbf{r}_2)|^2 = 1$.

The realization of the BCS-BEC crossover requires a wide tunability of density[26] or of the attractive interactions between the fermions.[27] After decades of theoretical work, it was only in 2003, that the crossover region was experimentally accessed using ultracold atoms.[28–30] The tunability of the interactions was implemented using Feshbach resonances. By varying a magnetic field, a (highly vibrationally excited) molecular state is tuned into resonance with two colliding fermions, resulting in a scattering resonance. By tuning across the resonance, the pair size of the fermions could be varied from

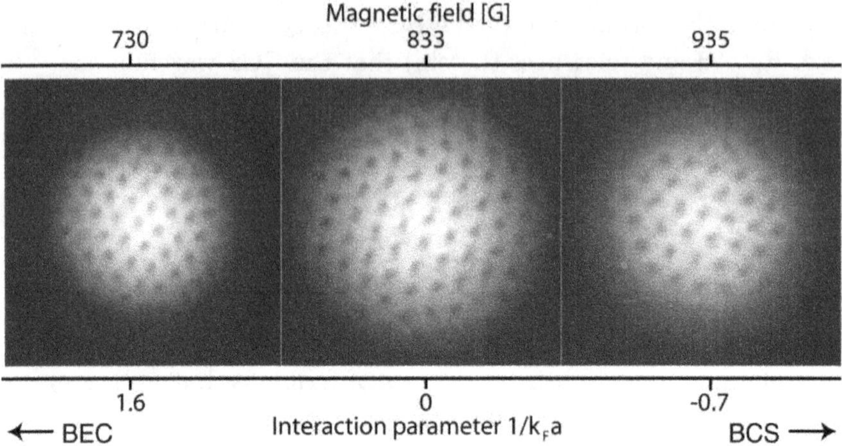

Fig. 1. Observation of vortices in a strongly interacting Fermi gas, below, at and above the Feshbach resonance. This establishes superfluidity and phase coherence in fermionic gases. After a vortex lattice was created at 812 G, a field favorable for generating vortices, the field was ramped in 100 ms to 792 G (BEC-side), 833 G (resonance) and 853 G (BCS-side), where the cloud was held for 50 ms. The field of view of each image is 880 μm $\times$ 880 μm. More recent version of Fig. 3 can be found in Ref. 34.

(somewhat) larger than the interparticle spacing (BCS side) to (somewhat) smaller (BEC side).

In most situations, the onset of superfluidity implies the formation of a pair condensate.[19,31] The BEC-BCS crossover was first characterized by observing such a pair condensate,[28–30,32,33] until superfluid flow was directly observed through quantized vortex lattices in rotating clouds[34] (Fig. 1). The field has been reviewed in the Varenna proceedings.[19]

4. Superfluidity with Population Imbalance

Once a superfluid (or superconducting) system is realized, one characterizes the stability of the superfluid phase by exploring all possible ways of destroying it, e.g. by raising the temperature, applying a critical magnetic field (which for neutral superfluids would correspond to a critical angular velocity), varying the strength of the interaction, and by imbalancing the population of the spin up and down components. Each way provides unique insight into the mechanism of pairing. In the BCS picture, pairing occurs preferably at the Fermi surface and therefore becomes energetically less favorably if the two Fermi surfaces do not overlap. Eventually superfluidity will break down when the difference in Fermi energies exceeds the energy gain Δ from pairing. This is the so-called Chandrasekhar-Clogston (CC) limit of superfluidity.[35,36] Pairing and superfluidity in an imbalanced Fermi mixture has been an intriguing topic for many decades, especially because of the possibility of new exotic ground states such as the Fulde-Ferrell-Larkin-Ovchinnikov (FFLO) state[37,38] in which either the phase or the density of the superfluid has a spatial periodic modulation.

Imbalanced Fermi systems can be realized with electron gases by applying a magnetic field. However, the situation in conventional superconductors is more complicated due to spin-orbit coupling, i.e. the field is shielded by the Meissner effect, unless the Meissner effect is suppressed, e.g. in thin films.[39] On the other hand, in atomic Fermi gases one can prepare a mixture with an arbitrary population ratio, since collisional relaxation processes are very slow. A few years ago, using population-imbalanced atomic Fermi gases, a behavior consistent with the CC limit has been observed,[40,41] i.e. a superfluid becomes more robust against imbalance with stronger coupling. The apparent absence of the CC limit in mesoscopic, highly elongated samples[42,43] is not yet understood, seems to depend on the aspect ratio of the cloud shape, and possibly reflects a non-equilibrium situation.[44]

In the remainder of the paper, we present recent results of the MIT group on the BEC-BCS crossover. One study addresses the phase diagram of a two-component Fermi gas of ^{6}Li atoms with strong interactions. We have identified and/or determined several important critical points including a tricritical point where the superfluid-to-normal phase transition changes from first-order to second-order, critical spin polarizations of a normal phase, and a critical interaction strength for a stable fermion pair in a Fermi sea of one of its constituents.[45–47] We also present recently measured RF spectra, where we have determined the fermion pair size and the superfluid gap energy in a resonantly interacting Fermi gas.[48,49]

5. Two-component Fermi Mixture in a Harmonic Potential

In our experiments, we prepared a two-component spin mixture of ^{6}Li atoms, using two states of the three lowest hyperfine states, around a Feshbach resonance. The population imbalance between the two components was controlled by a radio frequency (RF) sweep with an adjustable sweep rate. The atom cloud was confined in a three-dimensional harmonic trap with cylindrical symmetry, thus having an inhomogeneous density distribution. Due to the population imbalance, the chemical potential ratio of the majority (labeled as spin $\uparrow$) and the minority (spin $\downarrow$) components varies spatially over the trapped sample. Under the local density approximation (LDA), each sample represents a line in the phase diagram. Using spatially resolved measurements, we have mapped out the phase diagram of the system. The temperature was controlled by adjusting the trap depth, which determined the final temperature of evaporative cooling.

For typical conditions, the spatial size of our sample was $\sim 150\ \mu\text{m} \times 150\ \mu\text{m} \times 800\ \mu\text{m}$ with a total atom number of $\sim 10^7$ and a radial (axial) trap frequency of $f_r = 130$ Hz ($f_z = 23$ Hz). Our experiments benefit from the big size of the sample. Using a phase-contrast imaging technique, we obtained the *in situ* column density distributions of the two components $\tilde{n}_{\uparrow,\downarrow}(r)$, and the three-dimensional density profiles $n_{\uparrow,\downarrow}(r)$ were tomographically reconstructed from the averaged column density profiles (Fig. 2). The imaging resolution of our setup was $\sim 2\ \mu$m.

At low temperature, the outer part of the sample is occupied by only the majority component, forming a non-interacting Fermi gas. This part fulfills the definition of an ideal thermometer, namely a substance with exactly understood properties in contact with a target sample. We determined temperature from the *in situ* majority wing profiles. This *in situ* method provides a clean solution for the long-standing problem of measuring the temperature of a strongly interacting sample.

The parameter space of the system can be characterized by three dimensionless quantities: reduced temperature $T/T_{F\uparrow}$, interaction strength $1/k_{F\uparrow}a$ and spin polarization $\sigma = (n_\uparrow - n_\downarrow)/(n_\uparrow + n_\downarrow)$, where $T_{F\uparrow}$ and $k_{F\uparrow}$ are the Fermi temperature and wave number of the majority component, respectively, and a is the scattering length of the two components. The BCS-BEC crossover physics has been studied in the $\sigma = 0$, equal-mixture plane.

6. Phase Diagram at Unitarity

In the case of fixed particle numbers, it has been suggested that unpaired fermions are spatially separated from a BCS superfluid of equal densities due to the pairing gap energy in the superfluid region.[50–52] At low temperature, we have observed such a phase separation between a superfluid and a normal component in a trapped sample. A spatial discontinuity in the spin polarization clearly distinguishes two regions (Fig. 2). By correlating a non-zero condensate fraction[40] with the existence of the core region, we verified that the inner core is superfluid.[45] At the phase boundary two critical polarizations σ_s and σ_c are determined for a superfluid and normal phase,

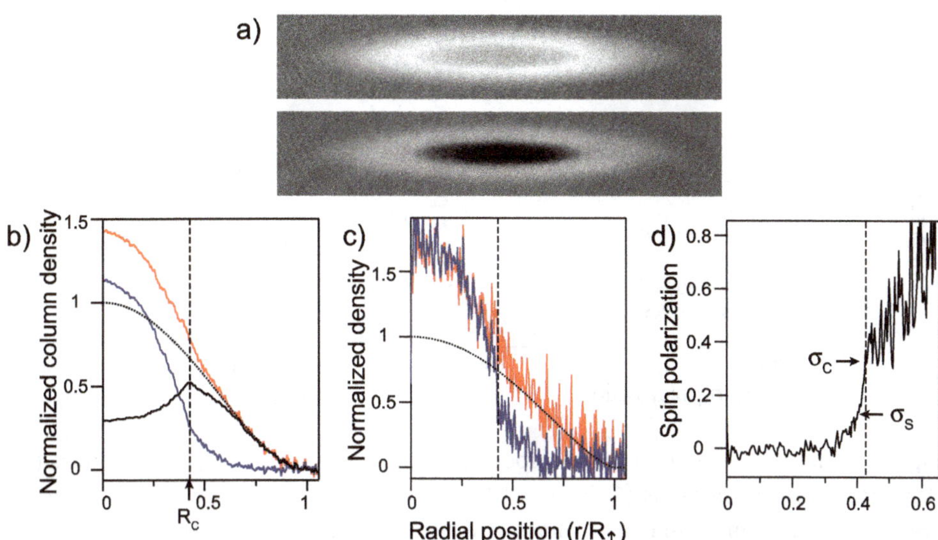

Fig. 2. Spatial structure of a trapped Fermi mixture with population imbalance. (a) The *in situ* column density distributions are obtained using a phase contrast imaging technique.[45] The probe frequencies of the imaging beam are different for two images so that the first image measures the density difference $n_\uparrow - n_\downarrow$ and the second image measures the weighted density difference $0.76n_\uparrow - 1.43n_\downarrow$. (b) The smooth column density profiles are obtained from the elliptical averaging of the images under the local density approximation (red: majority, blue: minority, black: difference). (c) The reconstructed three-dimensional density profiles. (d) The spin polarization profile shows a sharp increase, indicating the phase separation between a core superfluid and a outer normal region. The vertical dashed line marks the location of the phase boundary.

respectively. $\sigma_s \neq \sigma_c$ means that there is a thermodynamically unstable window, $\sigma_s < \sigma < \sigma_c$, leading to a first-order superfluid-to-normal phase transition. As the temperature increases, the discontinuity reduces with decreasing σ_c and increasing σ_s, and eventually disappears above a certain temperature. This is a tricritical point where the nature of the phase transition changes from first-order to second-order.[53] Above the tricritical point, the system shows smooth behavior across the superfluid-to-normal phase transition in density profiles and condensate fraction, which is characteristic of a second-order phase transition.

The phase diagram with resonant interactions ($1/k_{F\uparrow}a = 0$) is presented in Fig. 3(a),[46] characterized by three distinct points: the critical temperature T_{c0} for a balanced mixture, the critical polarization σ_{c0} of a normal phase

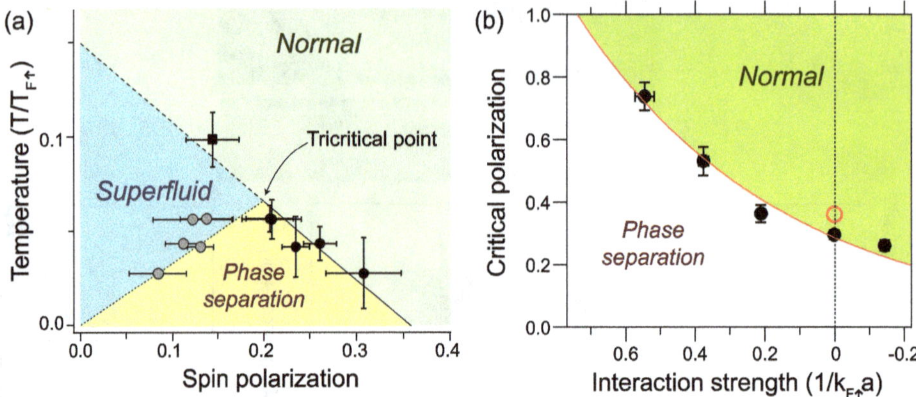

Fig. 3. Phase diagram of a two-component Fermi gas with strong interactions. (a) With resonant interactions ($1/k_{F\uparrow}a = 0$). At low temperature, the system shows a first-order superfluid-to-normal phase transition via phase separation, which disappears at a tricritical point where the nature of the phase transition changes from first-order to second-order. (b) The critical polarization σ_c of a partially-polarized normal phase increases with stronger interactions. Above a critical interaction strength ($1/k_{F\uparrow}a \approx 0.7$, $\sigma_c = 1$), all minority atoms can pair up to form a superfluid.

at zero temperature and the tricritical point (σ_{tc}, T_{tc}). From linear interpolation of the measured critical points, we have estimated $T_{c0}/T_{F\uparrow} \approx 0.15$, $\sigma_{c0} \approx 0.36$ and $(\sigma_{tc}, T_{tc}/T_{F\uparrow}) \approx (0.20, 0.07)$. The quantitative analysis of the *in situ* density profiles at the lowest temperature reveals the equation of state of a polarized Fermi gas,[54] showing that the critical chemical potential difference is $2h_c = 2 \times 0.95\mu$, where $\mu = (\mu_\uparrow + \mu_\downarrow)/2$. The pairing gap energy Δ of a superfluid has been measured to be $\Delta \geq \mu$,[49] and the observation of $h_c < \Delta$ excludes the existence of a polarized superfluid at zero temperature. A polarized superfluid at finite temperature results from thermal population of spin-polarized quasiparticles.[53]

7. Strongly Interaction Bose-Fermi Mixture

On the BEC side, two different fermions in free space have a stable bound state, forming a bosonic dimer which undergoes Bose-Einstein condensation at low temperature. Therefore, in the BEC limit a two-component Fermi gas with population imbalance will evolve into a binary mixture of bosonic dimers and unpaired excess fermions. Strong interactions and high degeneracy pressure can affect the structure of the composite boson and eventually quench the superfluid state. This is the reason why we have a partially-polarized normal phase near resonance even at zero temperature. With stronger coupling, the critical polarization σ_c of a partially-polarized normal phase increases, and becomes unity at a critical interaction strength of $1/k_{F\uparrow}a \approx 0.7$.[47] This means that beyond the critical coupling all minority atoms pair up with majority atoms and form a Bose condensate. This is the regime where a polarized two-component Fermi gas can be effectively described as a Bose-Fermi mixture.

In the limit of a BF mixture,[55] we have observed repulsive interactions between the fermion dimers and unpaired fermions. They are parameterized by an effective dimer-fermion scattering length of $a_{\mathrm{bf}} = 1.23(3)a$. This value is in reasonable agreement with the exact value $a_{\mathrm{bf}} = 1.18a$ which has been predicted over 50 years ago for the three fermion problem,[56] but has never been experimentally confirmed. Our finding excludes the mean-field prediction $a_{\mathrm{bf}} = (8/3)a$. The boson-boson interactions were found to be stronger than the mean-field prediction in agreement with the Lee-Huang-Yang prediction.[57] Including the LHY correction, the effective dimer-dimer scattering length was determined to be $a_{\mathrm{bb}} = 0.55(1)a$, which is close to the exact value for weakly bound dimers $a_{\mathrm{bb}} = 0.6a$.

8. Tomographic RF Spectroscopy with a New Superfluid

RF spectroscopy of a two-component Fermi gas measures a single-particle excitation spectrum by flipping the spin state of an atom to a third spin state. Since a fermion pair can be dissociated via spin flip, RF spectroscopy provides valuable information about the pair such as binding energy and size. In early experiments,[58,59] a spectral shift has been observed in a Fermi gas at low temperature and interpreted as a manifestation of pairing. However, it turned out that the spectral line shape is severely affected by the strong interactions of the third, final spin state and broadened due to the inhomogeneous density distribution of a trapped sample, preventing clear comparison of the experimental results to theory. Recently, we have developed several experimental techniques to overcome these problems. In order to minimize final state effects we have exploited a new spin mixture of states $|1\rangle$ and $|3\rangle$ of ^{6}Li atoms[48] (corresponding to $|F = 1/2, m_F = 1/2\rangle$ and $|F = 3/2, m_F = -3/2\rangle$ at low field), and using a tomographic technique, we have obtained local RF spectra from an inhomogeneous sample.[60]

Figure 4 shows the RF spectra of the various phases in a trapped sample with population imbalance. For a balanced superfluid, the majority and the

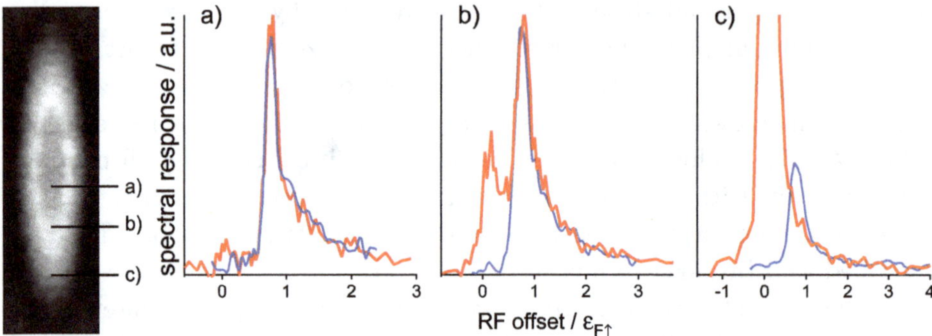

Fig. 4. Tomographic RF spectroscopy of strongly interacting Fermi mixtures. A trapped, inhomogeneous sample has various phases in spatially different regions. The spectra of each region (red: majority, blue: minority) reveals the nature of pairing correlation of the corresponding phase. (a) Balanced superfluid. (b) Polarized superfluid. The additional peak in the majority spectrum is the contribution of the excess fermions, which can be identified as fermionic quasiparticles in a superfluid. From the separation of the two peaks, the pairing gap energy of a resonantly interacting superfluid has been determined.[49] (c) Highly polarized normal gas. The minority peak no longer overlaps with the majority spectrum, indicating the transition to polaronic correlations.

minority spectra completely overlap, showing the characteristic behavior of pair dissociation, i.e. a sharp threshold and a slow high-energy tail. From the spectral width, we have determined the pair size to be $2.6(2)/k_F$ at unitarity, about 20% smaller than the interparticle spacing.[48] These are the smallest pairs so far observed in fermionic superfluids, highlighting the importance of small fermion pairs for superfluidity at high critical temperature.[61]

Excess fermions in a low-temperature superfluid constitute quasiparticles populating the minimum of the dispersion curve. The RF spectrum of a superfluid with such quasiparticles shows two peaks, which, in the BCS limit, would be split by the superfluid gap Δ. Therefore, RF spectroscopy of quasiparticles is a direct way to observe the superfluid gap in close analogy with tunneling experiments in superconductors. In a polarized superfluid near the phase boundary, we have obtained a local majority spectrum with a double-peak structure, from which the superfluid gap has been determined to be $\Delta = 0.44(3)E_{F\uparrow}$ at unitarity.[49] In addition, a Hartree term of $-0.43(3)E_{F\uparrow}$ is necessary to explain the observed spectral behavior. Evidence for a gap was also obtained using momentum resolved RF spectroscopy.[62]

The peak positions of the majority and the minority spectra become different in the partially-polarized normal phase, but still overlap in the high-energy tail. At large spin polarization, the limit of a single minority immersed in a majority Fermi sea is approached, where several theoretical studies suggest a polaron picture, associating the minority with weakly interacting quasiparticles in a normal Fermi liquid.[63–65] More recently, this picture has been experimentally confirmed and the binding energy[66] and the effective mass[67] of these polarons have been determined. Figure 4 shows that the different kinds of pairing correlations are smoothly connected across the superfluid-to-normal phase transition at finite temperature.

9. Summary and Discussion

In a series of experiments with population-imbalanced Fermi mixtures near Feshbach resonances, we have established the phase diagram of a two-component Fermi gas with strong interactions. This includes the identification of a tricritical point at which the critical lines for first-order and second-order phase transitions meet, and the verification of a zero-temperature quantum phase transition from a balanced superfluid to a partially-polarized normal gas at unitarity. The observed critical points such as the critical polarization of a normal phase and the critical interaction strength of a composite boson in a Fermi sea provide quantitative tests for theoretical treatments of strongly interacting fermions.

Our work can be summarized with the phase diagram of the system in a 3D parameter space (versus temperature, spin polarization and interaction strength) shown in Fig. 5. For a complete understanding, this macroscopic characterization of the different phases should be complemented by an investigation of their microscopic properties. At low temperature, one can interpret the observed polarized superfluid as a result of thermal population of spin-polarized quasiparticles. However, the behavior at higher temperature or/and in a stronger coupling regime is not yet completely understood and could include a gapless region ($h > \Delta$). Another open question is whether the Fermi liquid description is still valid for high minority concentrations, where Pauli blocking in the minority Fermi sea might play an important role. Furthermore, it has been speculated that exotic pairing states might exist in the partially-polarized phase.[68] So far, predicted exotic superfluid states such as the breached-pair state in a stronger coupling regime and the FFLO state in a weaker coupling regime have not been observed, but they may be hidden in thin layers near the superfluid-normal boundary. This discussion underlines that the population imbalanced Fermi system is the richest system realized thus far with ultracold gases and therefore nicely illustrates the

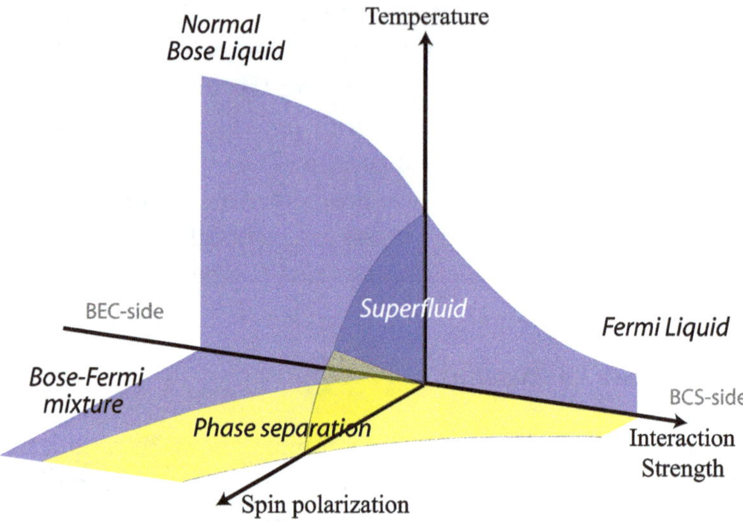

Fig. 5. Various phases of a two-component Fermi gas. The structure of the phase diagram is illustrated in the parameter space of temperature, interaction strength and spin polarization.

novel approach to engineer interesting many-body systems using the tools and methods of atomic physics.

Acknowledgments

This work was supported by NSF, ONR, MURI and ARO Award W911NF-07-1-0493 (DARPA OLE Program).

References

1. P. Kapitza, Viscosity of liquid helium below the λ-point, *Nature* **141**, 74 (1938).
2. J. F. Allen and A. D. Misener, Flow of liquid helium II, *Nature* **141**, 75 (1938).
3. K. Onnes, Investigations into the properties of substances at low temperatures, which have led, amongst other things, to the preparation of liquid helium. In *Nobel Lectures, Physics 1901–1921* (Elsevier Publishing Company, Amsterdam, 1967).
4. J. Bardeen, L. N. Cooper and J. R. Schrieffer, Theory of superconductivity, *Phys. Rev.* **108**, 1175 (1957).
5. P. Nozières and S. Schmitt-Rink, Bose condensation in an attractive fermion gas: From weak to strong coupling superconductivity, *J. Low Temp. Phys.* **59**, 195 (1985).

6. C. A. R. S. de Melo, M. Randeria and J. R. Engelbrecht, Crossover from BCS to Bose superconductivity: Transition temperature and time-dependent Ginzburg-Landau theory, *Phys. Rev. Lett.* **71**, 3202 (1993).
7. H. Hu, X.-J. Liu and P. D. Drummond, Temperature of a tarpped unitary Fermi gas at finite entropy, *Phys. Rev. A* **73**, 023617 (2006).
8. R. Haussmann, W. Rantner, S. Cerrito and W. Zwerger, Thermodynamics of the BCS-BEC crossover, *Phys. Rev. A* **75**, 023610 (2007).
9. Y. Nishida and D. T. Son, ϵ expansion for a Fermi gas at infinite scattering length, *Phys. Rev. Lett.* **97**, 050403 (2006).
10. A. Bulgac, J. E. Drut and P. Magierski, Spin 1/2 fermions in the unitary regime: A superfluid of a new type, *Phys. Rev. Lett.* **96**, 090404 (2006).
11. E. Burovski, N. Prokof'ev, B. Svistunov and M. Troyer, Critical temperature and thermodynamics of attractive fermions at unitarity, *Phys. Rev. Lett.* **96**, 160402 (2006).
12. W. Ketterle, Y. Shin, A. Schirotzek and C. H. Schunk, Superfluidity in a gas of strongly interacting fermions, *J. Phys.: Cond. Matt.* **21**, 164206 (2009).
13. D. J. Scalapino, The case for $dx2 - y2$ pairing in the cuprate superconductors, *Phys. Rep.* **250**, 329 (1995).
14. E. A. Cornell and C. E. Wieman, Nobel lecture: Bose-Einstein condensation in a dilute gas, the first 70 years and some recent experiments, *Rev. Mod. Phys.* **74**, 875 (2002).
15. W. Ketterle, Nobel lecture: When atoms behave as waves: Bose-Einstein condensation and the atom laser, *Rev. Mod. Phys.* **74**, 1131 (2002).
16. M. Greiner, O. Mandel, T. Esslinger, T. W. Hansch and I. Bloch, Quantum phase transition from a superfluid to a Mott insulator in a gas of ultracold atoms, *Nature* **415**, 39 (2002).
17. A. H. van Amerongen, J. J. van Es, P. Wicke, K. V. Kheruntsyan and N. J. van Druten, Yang-Yang thermodynamics on an atom chip, *Phys. Rev. Lett.* **100**, 090402 (2008).
18. T. Kinoshita, T. Wenger and D. S. Weiss, Observation of a one-dimensional Tonks-Girardeau gas, *Science* **305**, 1125 (2004).
19. M. Incuscio, W. Ketterle and C. Salomon (eds.), *Ultra-cold Fermi Gases*, Proceedings of the International School of Physics "Enrico Fermi" Course CLXIV (2007). IOS press.
20. R. Jördens, N. Strohmaier, K. Gunter, H. Moritz and T. Esslinger, A Mott insulator of fermionic atoms in an optical lattice, *Nature* **455**, 204 (2008).
21. U. Schneider, L. Hackermüller, S. Will, T. Best, I. Bloch, T. A. Costi, R. W. Helmes, D. Rasch and A. Rosch, Metallic and insulating phases of repulsively interacting fermions in a 3d optical lattice, *Science* **322**, 1520 (2008).
22. J. Billy, V. Josse, Z. Zuo, A. Bernard, B. Hambrecht, P. Lugan, D. Clement, L. Sanchez-Palencia, P. Bouyer and A. Aspect, Direct observation of Anderson localization of matter-waves in a controlled disorder, *Nature* **453**, 891 (2008).
23. G. Roati, C. D'Errico, L. Fallani, M. Fattori, C. Fort, M. Zaccanti, G. Modugno, M. Modugno and M. Incuscio, Anderson localization of a non-interacting Bose-Einstein condensate, *Nature* **453**, 895 (2008).
24. M. R. Schafroth, J. M. Blatt and S. T. Butler, *Helv. Phys. Acta.* **30**, 93 (1957).

25. M. Baranger, Reformulation of the theory of pairing correlations, *Phys. Rev.* **130**, 1244 (1963).
26. D. M. Eagles, Possible pairing without superconductivity at low carrier concentrations in bulk and thin-film superconducting semiconductors, *Phys. Rev.* **186**, 456 (1969).
27. A. J. Leggett, Diatomic molecules and Cooper pairs, *Modern Trends in the Theory of Condensed Matter* (Springer-Verlag Berlin, 1980), pp. 13–27.
28. M. Greiner, C. A. Regal and D. S. Jin, Emergence of a molecular Bose-Einstein condensate from a Fermi gas, *Nature* **426**, 537 (2003).
29. S. Jochim, M. Bartenstein, A. Altmeyer, G. Hendl, S. Riedl, C. Chin, J. H. Denschlag and R. Grimm, Bose-Einstein condensation of molecules, *Science* **302**, 2101 (2003).
30. M. W. Zwierlein, C. A. Stan, C. H. Schunck, S. M. F. Raupach, S. Gupta, Z. Hadzibabic and W. Ketterle, Observation of Bose-Einstein condensation of molecules, *Phys. Rev. Lett.* **91**, 250401 (2003).
31. I. Bloch, J. Dalibard and W. Zwerger, Many-body physics with ultracold gases, *Rev. Mod. Phys.* **80**, 885 (2008).
32. C. A. Regal, M. Greiner and D. S. Jin, Observation of resonance condensation of fermionic atom pairs, *Phys. Rev. Lett.* **92**, 040403 (2004).
33. M. W. Zwierlein, C. A. Stan, C. H. Schunck, S. M. F. Raupach, A. J. Kerman and W. Ketterle, Condensation of pairs of fermionic atoms near a Feshbach resonance, *Phys. Rev. Lett.* **92**, 120403 (2004).
34. M. W. Zwierlein, J. R. Abo-Shaeer, A. Schirotzek, C. H. Schunck and W. Ketterle, Vortices and superfluidity in a strongly interacting Fermi gas, *Nature* **435**, 1047 (June, 2005).
35. B. S. Chandrasekhar, A note on the maximum critical field of high-field superconductors, *Appl. Phys. Lett.* **1**, 7 (1962).
36. A. M. Clogston, Upper limit for the critical field in hard superconductors, *Phys. Rev. Lett.* **9**, 266 (1962).
37. P. Fulde and R. A. Ferrell, Superconductivity in a strong spin-exchange field, *Phys. Rev.* **135**, A550 (1964).
38. A. I. Larkin and Y. N. Ovchinnikov, Inhomogeneous state of superconductors, *Sov. Phys. JETP* **20**, 762 (1965).
39. M. Strongin and O. F. Kammerer, Effect of electron-spin paramagnetism on the critical field of thin Al films, *Phys. Rev. Lett.* **16**, 456 (1966).
40. M. W. Zwierlein, A. Schirotzek, C. H. Schunck and W. Ketterle, Fermionic superfluidity with imbalanced spin populations, *Science* **311**, 492 (2006).
41. M. W. Zwierlein, C. H. Schunck, A. Schirotzek and W. Ketterle, Direct observation of the superfluid phase transition in ultracold Fermi gases, *Nature* **442**, 54 (2006).
42. G. B. Partridge, W. Li, R. I. Karmar, Y. Liao and R. G. Hulet, Pairing and phase separation in a polarized fermi gas, *Science* **311**, 503 (2006).
43. G. B. Partridge, W. Li, R. I. Karmar, Y. Liao and R. G. Hulet, Deformation of a trapped Fermi gas with unequal spin populations, *Phys. Rev. Lett.* **97**, 190407 (2006).

44. M. M. Parish and D. A. Huse, Evaporative depolarization and spin transport in a unitary trapped Fermi gas, *Phys. Rev. A (Atomic, Molecular, and Optical Physics)* **80**, 063605 (2009).
45. Y. Shin, M. W. Zwierlein, C. H. Schunck, A. Schirotzek and W. Ketterle, Observation of phase separation in a strongly interacting imbalanced Fermi gas, *Phys. Rev. Lett.* **97**, 030401 (2006).
46. Y. Shin, C. H. Schunck, A. Schirotzek and W. Ketterle, Phase diagram of a two-component Fermi gas with resonant interactions, *Nature* **451**, 689 (2008).
47. Y. Shin, C. H. Schunck, A. Schirotzek and W. Ketterle, Realization of a strongly insteracting Bose-Fermi mixture with a two-component Fermi gas, *Phys. Rev. Lett.* **101**, 070404 (2008).
48. C. H. Schunck, Y. Shin, A. Schirotzek and W. Ketterle, Determination of the fermion pair size in a resonantly interacting superfluid, *Nature* **454**, 739 (2008).
49. A. Schirotzek, Y. Shin, C. H. Schunck and W. Ketterle, Determination of the superfluid gap in atomic Fermi gases by quasiparticle spectroscopy, *Phys. Rev. Lett.* **101**, 140403 (2008).
50. P. F. Bedaque, H. Caldas and G. Rupak, Phase separation in asymmetrical fermion superfluids, *Phys. Rev. Lett.* **91**, 247002 (2003).
51. J. Carlson and S. Reddy, Asymmetric two-component fermion systems in strong coupling, *Phys. Rev. Lett.* **95**, 060401 (2005).
52. D. E. Sheehy and L. Radzihovsky, BEC-BCS crossover in "magnetized" Feshbach-resonantly paired superfluids, *Phys. Rev. Lett.* **96**, 060401 (2006).
53. M. M. Parish, F. M. Marchetti, A. Lamacraft and B. D. Simons, Finite-temperature phase diagram of a polarized Fermi condensate, *Nature Phys.* **3**, 124 (2007).
54. Y. Shin, Determination of the equation of state of a polarized Fermi gas at unitarity, *Phys. Rev. A* **77**, 041603(R) (2008).
55. P. Pieri and G. C. Strinati, Trapped fermions with density imbalance in the Bose-Einstein condensate limit, *Phys. Rev. Lett.* **96**, 150404 (2006).
56. G. V. Skorniakov and K. A. Ter-Martirosian, *Sov. Phys. JETP* **4**, 648 (1957).
57. T. D. Lee, K. Huang and C. N. Yang, Eigenvalues and eigenfunctions of a Bose system of hard spheres and its low-temperature properties, *Phys. Rev.* **106**, 1135 (1957).
58. C. Chin, M. Bartenstein, A. Altmeyer, S. Riedl, S. Jochim, J. H. Denschlag and R. Grimm, Observation of the pairing gap in a strongly interacting Fermi gas, *Science* **305**, 1128 (2004).
59. C. H. Schunck, Y. Shin, A. Schirotzek, M. W. Zwierlein and W. Ketterle, Pairing without superfluidity: The ground state of an imbalanced Fermi mixture, *Science* **316**, 867 (2007).
60. Y. Shin, C. H. Schunck, A. Schirotzek and W. Ketterle, Tomographic RF spectroscopy of a trapped Fermi gas at unitarity, *Phys. Rev. Lett.* **99**, 090403 (2007).
61. F. Pistolesi and G. C. Strinati, Evolution of BCS superconductivity to Bose condensation: Role of the parameter $k_f\xi$, *Phys. Rev. B* **49**, 6356 (1994).
62. J. T. Stewart, J. P. Gaebler and D. S. Jin, Using photoemission spectroscopy to probe a strongly interacting Fermi gas, *Nature* **454**, 744 (2008).

63. C. Lobo, A. Recati, S. Giorgini and S. Stringari, Normal state of a polarized Fermi gas at unitarity, *Phys. Rev. Lett.* **97**, 200403 (2006).
64. R. Combescot, A. Recati, C. Lobo and F. Chevy, Normal state of highly polarized Fermi gases: Simple many-body approaches, *Phys. Rev. Lett.* **98**, 180402 (2007).
65. N. Prokof'ev and B. Svistunov, Fermi-polaron problem: Diagrammatic Monte Carlo for divergent sign-alternating series, *Phys. Rev. B* **77**, 020408(R) (2008).
66. A. Schirotzek, C.-H. Wu, A. Sommer and M. W. Zwierlein, Observation of Fermi polarons in a tunable Fermi liquid of ultracold atoms, *Phys. Rev. Lett.* **102**, 230402 (2009).
67. S. Nascimbène, N. Navon, K. J. Jiang, L. Tarruell, M. Teichmann, J. McKeever, F. Chévy and C. Salomon, Collective oscillations of an imbalanced Fermi gas: Axial compression modes and polaron effective mass, *Phys. Rev. Lett.* **103**, 170402 (2009).
68. A. Bulgac, M. M. Forbes and A. Schwenk, Induced p-wave superfluidity in asymmetric Fermi gases, *Phys. Rev. Lett.* **97**, 020402 (2006).

BCS FROM NUCLEI AND NEUTRON STARS TO QUARK MATTER AND COLD ATOMS

Gordon Baym

Department of Physics, University of Illinois,
1110 W. Green St., Urbana, IL 61801, USA

This article describes extensions of BCS theory to atomic nuclei and neutron stars, to high density quark matter and connections to pairing in ultracold clouds of atomic fermions.

1. Introduction

Starting very shortly after its initial discovery, BCS theory began to be applied to physical systems beyond laboratory superconductors, starting with pairing of nucleons in atomic nuclei,[1] and its extension soon after to neutron star matter.[2,3] Its impact on models of elementary particles and broken symmetry in high energy physics[4] has been profound. BCS formed the basis of predicting a superfluid state[5] of degenerate ^{3}He dissolved in superfluid ^{4}He, which would produce a novel mixture of a fermion superfluid dissolved in a Bose superfluid.[a] Application to the helium liquids reached fruition with the discovery[6] of the low temperature superfluid states of ^{3}He. Later BCS was applied to pairing of quarks in degenerate quark-gluon plasmas,[7] and more recently to ultracold atomic fermion clouds.

In view of the articles in this volume by Leggett on superfluid ^{3}He, Ketterle on cold atoms, and Weinberg and Wilcek on high energy applications, this article will focus on BCS in atomic nuclei, and its extensions to neutron star matter, and then high density quark matter, as well as touching on a few connections of quark matter physics to ultracold atomic physics.

---

*Research discussed here has been supported in part by the U.S. National Science Foundation over the years, most recently by NSF Grant PHY07-01611.

[a]Unfortunately the BCS transition temperature was in the microkelvin range, well beyond reach.

2. Application to Nuclei: Copenhagen, 1957

The first application of BCS, to pairing of even numbers of neutrons or protons outside closed shells in atomic nuclei, had its inception before the ink was even dry on the original BCS paper.[8] In the summer of 1957, David Pines brought news of the BCS theory to Niels Bohr's Institute in Copenhagen — the University Institute for Theoretical Physics — where he was asked by Aage Bohr and Ben Mottelson to lecture on the theory. The lecture, with Niels Bohr in the audience,[b] inspired discussion among A. Bohr, Mottelson, and Pines of possible extensions of BCS to nuclei; they soon realized that BCS would provide the solution to the outstanding puzzle of the much larger single particle excitation energies, of order in MeV, in nuclei that have an even number of neutrons and protons (even–even nuclei), compared with the excitation energies in nuclei in which the neutron number or proton number is odd (odd-A nuclei). The effect — seen in the mass range $130 < A < 190$ and $A > 228$, and strikingly in the high-resolution spectroscopic studies of ^{182}W versus ^{183}W — was as they realized the analog of the excitation gap in metallic superconductors; the ground states of even–even nuclei are effectively BCS paired with a gap $\sim 12/\sqrt{A}$ MeV. Pines gave a preliminary report of this first application of BCS to nuclei in September at the International Conference on Nuclear Structure at the Weizmann Institute,[11] and their paper, "Possible analogy between the excitation spectra of nuclei and those of the superconducting state," was submitted at the end of the year.[1]

In addition to pinning down the origin of the energy gap in nuclei, the paper makes the crucial observation that, "The correlations giving rise to the energy gap may also affect many other nuclear properties; thus, they appear to be responsible for the observed fact that the rotational moments of inertia are appreciably smaller than the values corresponding to rigid rotation."[12] The rotational spectrum has the simple semi-classical form $E = \hbar^2 I(I+1)/2\mathcal{I}$, where $\hbar I$ is the total angular momentum of the nucleus, and $\mathcal{I}$ is the rotational moment of inertia; more widely spaced levels indicate immediately that the moment of inertia is reduced from its classical value. This reduction of the moment of inertia is in fact the analogue in neutral

---

[b]Schrieffer, who was in Copenhagen in Fall 1957, recalls Niels Bohr commenting on the BCS theory, "It's an interesting idea but nature isn't that simple."[9] Bohr himself had struggled with superconductivity; his paper, "Zur frage der Superleitung," submitted to Naturwissenschaften in June 1932, was withdrawn in proof, in response to Felix Bloch's strong criticisms.[10]

systems of the Meissner effect in superconductors.[13] The detailed theory was soon worked out by Beliaev,[14] who came to Copenhagen for a year at the end of 1957, and by Migdal.[2] As Beliaev showed, the moment of inertia is given by

$$\mathcal{I} = 2\sum_{\nu,\nu'} \frac{|\langle \nu'|J_z|\nu\rangle|^2}{E_\nu + E_{\nu'}} |u_\nu v_{\nu'} - v_\nu u_{\nu'}|^2, \tag{1}$$

where the u and v, the usual BCS coherence factors associated with the single particle levels ν and ν', are responsible for the reduction of the moment of inertia from its classical value, similar to the reduction of the normal electron fraction in BCS. Explicit calculations by Migdal and others later gave good agreement with experiment.[15]

3. Neutron Stars

Migdal, in the introduction to his paper on superfluidity and moments of inertia of nuclei,[2] makes the prescient remark that "superfluidity of nuclear matter may lead to some interesting cosmological phenomena if stars exist which have neutron cores. A star of this type would be in a superfluid state with a transition temperature corresponding to 1 MeV." The theme of superfluidity of nuclear matter in neutron stars was later amplified by Ginzburg and Kirzhnits,[3] prior to the discovery of pulsars and their identification as rotating neutron stars, and by Ruderman,[16] and Baym, Pethick and Pines,[17] among others.

Figure 1 shows the cross-section of a neutron star. With increasing depth in the star, the matter becomes more and more neutron rich as the electron Fermi energy rises; electron captures, $e + p \to n + \nu_e$, convert protons into neutrons in the formation of the neutron star (the produced neutrino escapes). At the "neutron drip" point, the first solid line in the crust in Fig. 1, the matter become so neutron rich that the bound neutron states are filled and the continuum neutron states begin to fill; the still solid matter becomes permeated by a sea of free neutrons in addition to the sea of electrons. Above the neutron-drip point, the protons remain localized in nuclei in shells with Z = 40 or 50 as the matter continues to become more and more neutron rich, until the "pasta nuclei" regime, where the nuclear shapes becomes highly distorted. The neutron fluid is expected to be superfluid, with BCS pairing in the crust in $^1S_0$ and pairing in $^3P_2$ states in the liquid core. The proton fluid in the interior is expected to be a Type II superconductor paired in $^1S_0$ states.

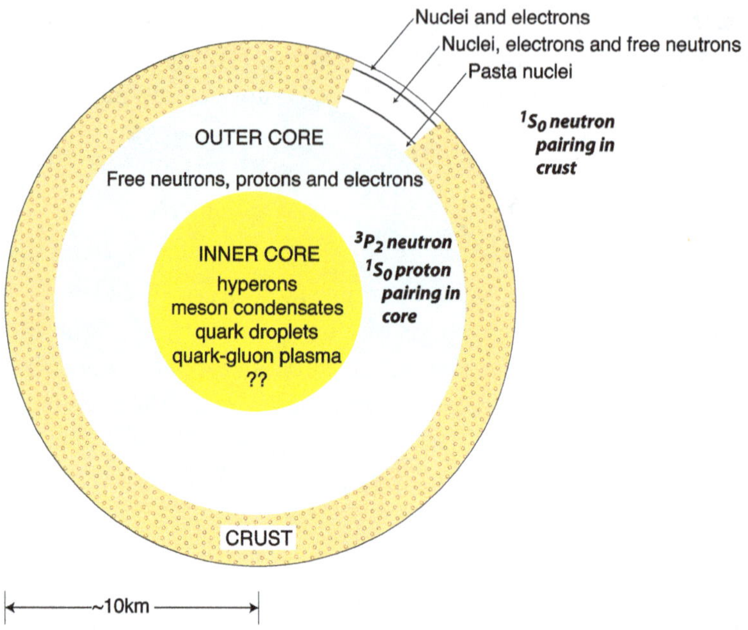

Fig. 1. Cross-section of a neutron star. The neutron superfluid in the crust is expected to be paired in $^1S_0$ states, as is the proton superfluid in the core. The neutron superfluid in the core is expected to be paired in $^3P_2$ states. The electrons are normal, however. A quark fluid in the core is also expected to be BCS paired in $^1S_0$ states.

The original estimates of pairing gaps were based on nucleon–nucleon scattering phase shifts, e.g. by Ref. 18 for neutrons, and by Ref. 19 for protons. The corresponding transition temperatures for pairing are illustrated in Fig. 2. Over the years, more and more sophisticated calculations of nucleonic pairing, taking into account dynamical screening of the effective interactions between nucleons, have been performed. More recently, numerical simulations of dense nuclear matter via Monte Carlo have been carried out to determine pairing gaps; various calculations are reviewed in Ref. 21. Figure 3 shows the results of a quantum Monte-Carlo (QMC) calculation by Gezerlis and Carlson for the zero temperature BCS pairing gap in low density pure neutron matter (as found in the crust of a neutron star); the calculation has been done for 66 to 68 particles in a box with an s-wave nucleon–nucleon potential with scattering length $a = -18.5$ fm and an effective range $r_e = 2.7$ fm.[20–21] For comparison, the plot also shows the BCS mean-field gap for the same s-wave scattering length.

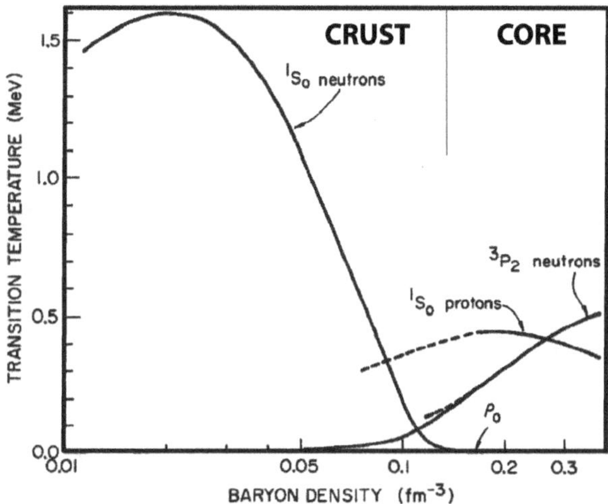

Fig. 2. Estimates of transition temperature for nucleonic pairing, based on nucleon–nucleon phase shifts.

The highly relativistic electrons, one should note, are too weakly interacting to undergo BCS pairing at ambient neutron star temperatures; the transition temperature for electron superconductivity would be $\sim \mu_e e^{-1/\alpha}$, where $\alpha = e^2/\hbar c$ is the electromagnetic fine structure constant and μ_e is the electron chemical potential.[17]

BCS pairing in neutron stars leads to macroscopic behavior familiar in laboratory superfluids and superconductors. In a rotating neutron starneutron star, the neutron superfluid rotates by forming vortex lines parallel to the rotation axis, around which the superfluid circulation is quantized in units of $\kappa = \pi\hbar/2m_n$, where m_n is the neutron mass. The proton fluid, on the other hand, is threaded by a triangular (Abrikosov) array of magnetic flux tubes each carrying a flux quantum, $\phi_0 = \pi\hbar c/e$, parallel to the magnetic field, as in a Type II superconductor. Even though ambient magnetic fields are below the critical magnetic field for flux expulsion, the enormous electrical conductivity of the normal state implies that the characteristic times for flux diffusion from macroscopic regions, of order $R^2/c^2\tau$, where R is the neutron star radius and τ is the microscopic scattering time, are typically comparable with the age of the universe.[23,24] Proton superconductivity forms in the magnetic field present at the birth of the neutron star, despite the Meissner effect.

While the vortex cores are of order 10 fm in radius, the characteristic spacing of neutron vortices $\simeq 0.01\sqrt{P(s)}$ cm (where $P(s)$ is the pulsar

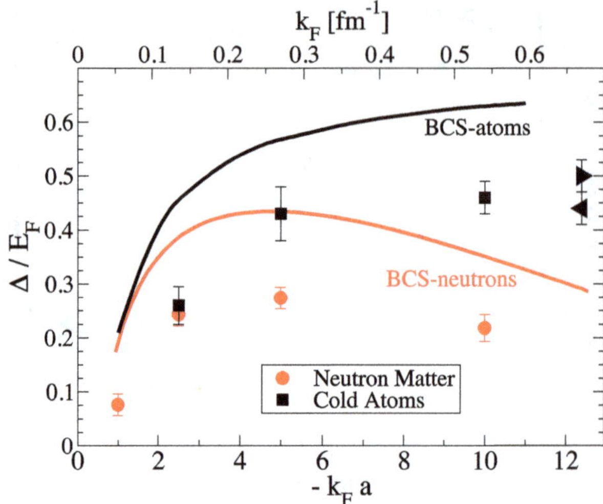

Fig. 3. Superfluid pairing gap for low density neutron matter, as a function of the neutron Fermi momentum, k_F calculated by QMC and compared with the mean-field BCS gap. The figure also shows the equivalent result for pairing of cold atomic fermions, where the effective range is negligible, and the unitarity limit of the QMC calculation (right arrow) compared with experiment (left arrow). Figure from Gezerlis's thesis.[20]

rotational period in seconds) is macroscopic. The proton vortices, on the other hand, are separated by sub-Ångstrom scales owing to the enormous magnetic fields ($B \sim 10^{12}$ G) in typical neutron stars.

The phenomenon of pulsar *glitches*, or sudden rotational speed-ups, provides an unusual opportunity to probe the superfluidity of neutron star matter. Over 90 glitches have been observed in over 30 pulsars. Since its discovery in 1968, the Vela pulsar PSR 0833-45, with a period 0.089 ms, has undergone some 15 such sudden increases of its rotation rate Ω, of size $\Delta\Omega/\Omega \sim 10^{-6}$, corresponding to an increase in rotational energy, $\Delta E_{\rm rot} = I\Omega\Delta\Omega \sim 10^{43}$ erg, where $I \sim 10^{45}$ g cm^2 is the stellar moment of inertia. The Crab pulsar, since its discovery in 1969, has undergone some 14 glitches of average size $\Delta\Omega/\Omega \sim 10^{-8} - 10^{-9}$. After a speed-up, the rotation relaxes back to its preglitch behavior over days to years.

A pulsar glitch, involving an energy comparable to the entire energy of the pulsar magnetosphere but with little change in the electromagnetic pulsations, appears to originate within the neutron star. On the one hand, relaxation processes within a normal star are microscopically rapid,[25] strongly suggesting that the long relaxation times observed in glitches are related to superfluidity. As reviewed in Ref. 26, the most likely mechanism involves pinning of superfluid neutron vortices to the nuclei in the inner crust and their slow outward creep through the nuclear lattice.[27] Since the angular momentum of a vortex, $\sim N\hbar(1 - r^2/R^2)$ (where N is the number of particles in the superfluid, r is the radial position of the vortex in polar coordinates, and R is the stellar radius) decreases as the vortex moves outwards, the spin-down of a neutron star is controlled by the rate at which vortices can move radially under dissipation. To the extent that vortices remain fixed to the crystal lattice, the superfluid angular velocity remains constant, allowing the difference between the angular velocities of the superfluid and the solid crust to grow as the crust slows. This differential motion can then act as the source of energy that powers glitches, with sudden transfers of angular momentum from the superfluid to the crust caused by catastrophic unpinning of many vortex lines, or by cracking of the crust to which the vortex lines are pinned. An unpinned line experiences a frictional force which transfers angular momentum to the crust on minute time scales. Following eventual repinning the vortices resume their slow outward creep, with the long-term observed post-glitch relaxation reflecting variations in the creep rate.

4. Pairing in Quark Matter

With increasing density in a neutron star, the nucleons, composed of three valence quarks, are squeezed together and the liquid becomes dominated by quark degrees of freedom, eventually becoming "quark matter." The components of quark matter are the very light up (u) and down (d) flavor quarks; at higher densities more massive strange (s) quarks enter as the chemical potential of d quarks exceeds the strange quark mass, ~ 100 MeV, plus interaction corrections. At non-zero temperature gluons are also present. Understanding the transition to quark matter remains an important theoretical challenge. The transition is most likely gradual; indeed once nucleons overlap considerably, the matter should percolate, opening the possibility of their quark constituents propagating throughout the system.

Quark matter at low temperatures becomes a color superconductor, driven (at very high densities) by exchange of gluons.[28,29] Two $^1S_0$ pairing states, with condensates antisymmetric in color (which takes on three values, nominally red, blue, and green) and flavor are most energetically favorable: a two-flavor color-antitriplet 2SC or isoscalar state, in which only u and d quarks are paired; and for massless u, d and s quarks, a color-flavor locked (CFL) state[28] that breaks both color and flavor symmetry, containing

18 condensates, e.g. pairing of a red u quark with a blue s quark, a blue s quark with a green d quark, and a green d quark with a red u quark, etc. The CFL state is the most stable for three flavors of massless quarks in the weak coupling limit, both at zero temperature, and near the critical temperature T_c.[30]

Quark pairing would play a role in the dynamics of neutron stars. The 2SC and CFL phases respond quite differently to magnetic fields and rotation.[24] The isoscalar state, if a Type II superconductor, would behave analogously to that of superconducting protons in a neutron star, forming magnetic vortices in response to ordinary magnetic field with flux quantum $6\pi\hbar c/\sqrt{3g^2+e^2}$, where g is the qcd coupling constant and e the electron charge. As in a rotating superconductor, this state responds to rotation by forming a very weak London magnetic field, $B \lesssim 1$ G (the field is actually a superposition of electromagnetic and color-gluon magnetic fields; the latter dominates).

On the other hand, the CFL phase forms U(1) vortices in response to rotation, as do superfluid neutrons, with a quantum of circulation, $3\pi\hbar c^2/\mu$, where μ is the baryon chemical potential.[24] Such vortices could play a role[31] in the pinning and depinning of rotational vortices that give rise to pulsar glitches.[26] Vortices involving only an electromagnetic U(1) phase of the gap are unstable[24]; however, the system does support stable magnetic vortices which involve the gradients of the full $SU(3)_c \times U(1)_{em}$ color structure.[32] As with superconducting protons in neutron stars, magnetic fields in color superconducting matter would be frozen in an intermediate state composed of alternating regions of normal and superconducting material if the system is a Type I superconductor, or in a lattice of vortices if a Type II superconductor.

5. Coupling of Pairing and Chiral Symmetry

The rich structure of quark matter opens new physics of superconductivity. A basic symmetry of the strong interaction is chiral symmetry, the conservation of the handedness or helicity (the spin of a particle along its direction of momentum) of massless quarks. Chiral symmetry is spontaneously broken in the normal vacuum, giving rise to a quark-antiquark condensate $\langle \bar{q}q \rangle$; such a condensate is an analog of magnetization in condensed matter systems. Figure 4 shows a possible schematic phase diagram of dense matter in the temperature — baryon chemical potential μ plane. At low temperatures and densities, matter is in the hadronic phase whose degrees of freedom are

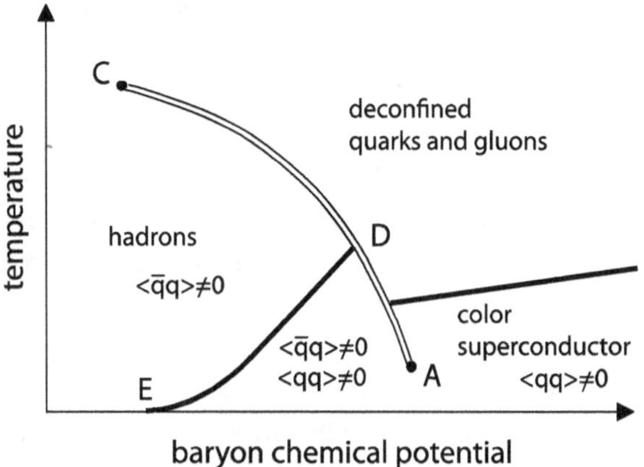

Fig. 4. Schematic phase diagram of high density quark matter. In the hadronic regime, chiral symmetry is spontaneously broken, with a non-zero order parameter $\langle \bar{q}q \rangle$. In the high temperature quark-gluon plasma phase chiral symmetry is restored, with the quarks unpaired. In the lower quark-gluon plasma phase to the right, quarks are BCS paired with an order parameter $\langle qq \rangle$, and chiral symmetry weakly broken. In the region below EDA quarks remain paired, likely as diquarks, and chiral symmetry is broken.

the familiar strongly interacting neutrons, protons, pi mesons, and other hadrons. This phase contains a non-zero chiral condensate. As the temperature or baryon density increases, matter enters the quark-gluon plasma (QGP) phase. Along the line CD the transition is likely first order, and at densities below that at C, the transition is second order for massless quarks. In the QGP, chiral symmetry is restored, $\langle \bar{q}q \rangle = 0$. Far to the right, at low temperatures, the system is color-paired, as described above.

Not all the symmetries of the unrenormalized QCD Lagrangian are realized in nature. In particular, owing to the axial anomaly, renormalization breaks the axial U(1) symmetry producing an effective attractive interaction between six quarks.[33] This coupling leads to an attractive third order coupling of the chiral condensate $\langle \bar{q}q \rangle \equiv \sigma$ to itself,[34] as well an effective attractive interaction between the chiral condensate and the quark pairing field[35] $\langle qq \rangle \equiv d$, schematically of the form $-\frac{1}{3}c\sigma^3 - \gamma d^2\sigma$, where c and d are positive constants. With this axial anomaly term, the Ginzburg–Landau interaction describing the coupled effective σ and d fields is schematically[35]

$$H_{\text{int}} = \frac{a}{2}\sigma^2 - \frac{c}{3}\sigma^3 + \frac{b}{4}\sigma^4 + \frac{f}{6}\sigma^6 + \frac{\alpha}{2}d^2 + \frac{\beta}{4}d^4 - \gamma d^2\sigma + \lambda d^2\sigma^2. \qquad (2)$$

The first four terms describe the chiral field self-interaction and the usual Ginzburg–Landau terms for BCS pairing of quarks, while the final two terms describe the coupling of the chiral and pairing fields. Owing to this latter coupling, the intermediate region around point A can develop a new critical point at finite baryon density and low temperature (always below the onset of pairing).[35] The symmetry group of color-flavor locked paired quark matter is in fact the same as that of chiral-symmetry-broken hadronic matter with equal mass u, d and s quarks,[28] SU(3) $\times$ Z_2, suggesting the possibility of a continuous transition below the critical point from the paired quark state to the hadronic phase. The solid line DE represents the transition from quark pairing to hadrons. As we discuss in a moment, the region below the critical point likely exhibits a BEC-BCS crossover, in which as the density is decreased large BCS pairs become small diquarks.

6. BEC-BCS Crossover and the Deconfinement Transition

Historically, the fields of superconductivity and Bose–Einstein condensation, which focused primarily on superfluid ^{4}He, developed along rather independent paths. Thus, it was not surprising that the connection between the two phenomena was not on the front burner. Indeed, the original BCS paper[37] comments on the newly proposed Schafroth, Butler and Blatt "quasichemical approach" to superconductivity in which small electron molecules are formed and Bose condense,[38] in the following footnote, "Our picture differs from that of Schafroth, Butler and Blatt, who suggest that pseudomolecules of pairs of electrons of opposite spin are formed. They show if the size of the pseudomolecules is less than the average distance between them, and if other conditions are fulfilled, the system has properties similar to that of a charged Bose–Einstein gas, including a Meissner effect and a critical temperature of condensation. Our pairs are not localized in this sense, and our transition is not analogous to 'a Bose–Einstein condensation.'" Shortly after the BCS paper was written, Bardeen sent a letter to Dyson, who was a firm believer in the equivalence of the two theories, emphasizing the difference of the two theories: "We believe that there is no relation between actual superconductors and the superconducting properties of a perfect Bose–Einstein gas. The key point in our theory is that the virtual pairs all have the same net momentum. The reason is not Bose–Einstein statistics, but comes from the exclusion principle ..."[39] We now understand more fully the connection of the two approaches thanks to the work of Eagles,[40] Leggett,[41] and Nozières and Schmitt-Rink,[42] who showed how BCS pairs continuously transform

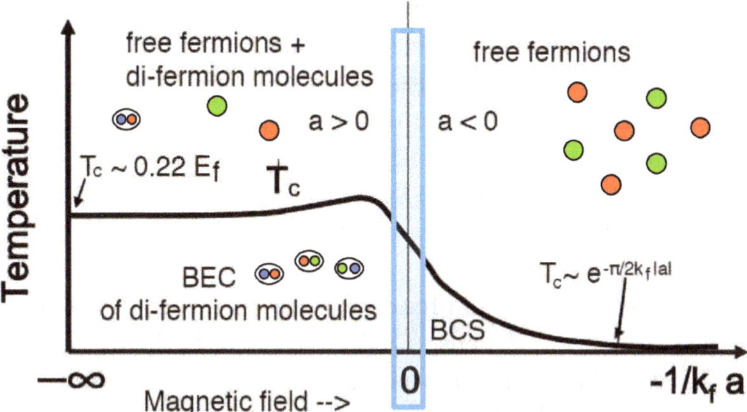

Fig. 5. Phase diagram of a gas of two equally populated hyperfine states of ultracold atomic fermions as a function of the negative of the inverse scattering length a, in units of the Fermi momentum. The continuous curve is the transition temperature for condensation, approaching the weakly interacting Bose–Einstein transition temperature to the left and the BCS transition temperature to the right; E_f is the free particle Fermi energy. The transition between the BEC and BCS regions is a smooth crossover.

into molecules with increasing interaction strength between the paired fermions.[c]

Figure 5 shows the finite temperature phase diagram of a gas of two equally populated hyperfine states of ultracold fermionic atoms. The horizontal axis is $-1/k_f a$, where a is the s-wave scattering length, and k_f is the Fermi momentum of the gas. At small positive a, to the left of the phase diagram, the system has strong bound states — di-fermion molecules — which as the temperature is lowered undergo Bose condensation, while at small negative a, to the right, the fermions become BCS paired at low temperatures. Remarkably, in the region between these two extremes the system undergoes a gradual crossover from a BEC superfluid of molecules to a BCS paired superfluid.[41,42] Nothing dramatic happens as one goes through "unitarity" ($|a| \to \infty$); rather the molecules continuously expand in size from tightly bound in the BEC regime to widely spaced pairs in the

[c]Leon Cooper believes that Bardeen did, in fact, understand the connection, but felt it wisest to focus on how the two pictures differed. (Private communication to G. Baym, March 2009). Leon's own view on the BEC-BCS crossover is discussed briefly in his chapter in this volume.

BCS regime. The system remains superfluid, as verified in a rotating system by the presence of quantized vortices from the BEC to BCS regimes.[43]

The transition temperature rises slightly with increasing $a > 0$ from the ideal Bose limit on the left, initially linear in a. Determining the effect of a repulsive interaction on the transition temperature of the ideal Bose gas was controversial for over four decades,[44] and was only resolved a decade ago.[45] The precise coefficient of the linear rise can only be calculated numerically.[46]

The correspondences of the phase diagram of ultracold atomic fermions with the phase diagram of dense QCD matter, Fig. 4, are apparent. At very large baryon chemical potential, μ, and low temperature the matter is BCS paired, as are the atomic systems at large $-1/k_f a$ and low temperature. As the temperature increases the atomic system becomes a fluid of normal unpaired fermions, while the QCD system becomes an unpaired plasma. Furthermore, with decreasing $-1/k_f a$, the atomic system continuously transforms into a Bose–Einstein condensate of di-fermion molecules; similarly QCD matter at low temperature goes from the color-paired states at large μ smoothly under the critical point, to the coexistence state in which both $\langle qq \rangle$ and $\langle \bar{q}q \rangle$ are non-zero.

Were the color symmetry group only SU(2), the analogy between the atomic and quark case would be strong — the baryons would be diquark pairs, bosons, which would go over into BCS pairs at higher densities, just as the atomic molecules slowly expand into BCS pairs with increasing magnetic field through a Feshbach resonance. However in the real world of SU(3)-color the baryons are three quark objects; understanding how the system crosses over from strong three-quark correlations at low densities to two-quark correlations at high densities remains an open issue. Since at lower densities two-flavor 2SC color pairing is favored, the atomic systems raise the possibility that at low temperatures three-quark correlations could enter through (2SC) BCS pairs shrinking into diquarks as the system enters the coexistence region, turning continuously into a strongly interacting diquark BEC. Then at line DE in Fig. 4, the diquarks would bind to the unpaired quarks to form baryons at lower density.[d]

[d]A simple simulation of this transition with ultracold atoms, via binding of bosonic atoms (analogs of diquarks) with fermionic atoms (analogs of the unpaired quarks) into molecules, the analog of the nucleon, is suggested in Ref. 47. More generally, atomic fermionic systems with three internal states, e.g. the three lowest hyperfine levels of ^{6}Li, offer the possibility of studying laboratory analogs of the QCD hadronization-deconfinement transition with pairing correlations.[48–50]

7. Concluding Remarks

To close on a personal note, I would like to dedicate this paper to the memory of John Bardeen, who was a warm colleague and caring mentor of me as a young faculty member in Urbana. It was a source of great pleasure to John to see how BCS expanded our understanding of modern physics beyond laboratory superconductors — from nuclear physics to astrophysics to high energy physics, a shaping that continues to this day, from cold atomic physics to quark matter.

References

1. A. Bohr, B. R. Mottelson and D. Pines, *Phys. Rev.* **110**, 936 (1958).
2. A. B. Migdal, *Nucl. Phys.* **13**, 655 (1959).
3. V. L. Ginzburg and D. A. Kirzhnits, *Zh. Eksp. Teor. Fiz.* **47**, 2006 (1964).
4. Y. Nambu and G. Jona-Lasinio, *Phys. Rev.* **122**, 345 (1961); **124**, 246 (1961).
5. G. Baym, J. Bardeen and D. Pines, *Phys. Rev. Lett.* **17**, 372 (1966); *Phys. Rev.* **156**, 207 (1967).
6. D. D. Osheroff, R. C. Richardson and D. M. Lee, *Phys. Rev. Lett.* **28**, 885 (1972); A. J. Leggett, *Rev. Mod. Phys.* **47**, 331 (1975).
7. B. C. Barrois, *Nucl. Phys.* **B129**, 390 (1977); D. Bailin and A. Love, *Phys. Rep.* **107**, 325 (1984).
8. J. Bardeen, L. N. Cooper and J. R. Schrieffer, *Phys. Rev.* **108**, 1175 (1957); submitted early July 1957.
9. Interview with J. R. Schrieffer by J. Warnow and R. M. Williams, 26 September 1974.
10. L. Hoddeson, G. Baym and M. Eckert, in *Out of the Crystal Maze: The History of Solid State Physics*, eds. L. H. Hoddeson, E. Braun, J. Teichmann and S. Weart, ch. 2 (Oxford Univ. Press, 1992), pp. 146–150.
11. *Proc. Rehovoth Conf. Nuclear Structure*, Weizmann Institute of Science, Rehovoth, September 1957, ed. H. J. Lipkin (North-Holland Pub. Co., Amsterdam, 1958).
12. A. Bohr and B. R. Mottelson, Kgl. Danske Videnskab. Selskab, Mat.-fys. Medd. **30**, 1 (1955).
13. G. Baym, in *Mathematical Methods in Solid State and Superfluid Theory*, eds. R. C. Clark and G. H. Derrick (Oliver and Boyd, Edinburgh, 1969), pp. 121–156.
14. S. Beliaev, *Mat.-fys. Medd.* **31**, 11 (1959).
15. A. Bohr, *Rev. Mod. Phys.* **48**, 365 (1976).
16. M. Ruderman, in *Proc. Fifth Eastern Theor. Phys. Conf.*, ed. D. Feldman (W. A. Benjamin, New York, 1967).
17. G. Baym, C. J. Pethick and D. Pines, *Nature* **224**, 673 (1969).
18. M. Hoffberg, A. E. Glassgold, R. W. Richardson and M. Ruderman, *Phys. Rev. Lett.* **24**, 775 (1970).

19. N.-C. Chao, J. W. Clark and C.-H. Yang, *Nucl. Phys. A* **179**, 320 (1972).
20. A. Gezerlis, Ph.D. dissertation, University of Illinois at Urbana-Champaign, Urbana, Illinois (2009).
21. A. Gezerlis and J. Carlson, *Phys. Rev. C* **81**, 025803 (2010).
22. A. Gezerlis and J. Carlson, *Phys. Rev. C* **77**, 032801 (2008).
23. G. Baym, C. J. Pethick and D. Pines, *Nature* **224**, 674 (1969).
24. K. Iida and G. Baym, *Phys. Rev. D* **66**, 014015 (2002).
25. J. A. Sauls, in *The Lives of Neutron Stars*, eds. M. A. Alpar, Ü. Kiziloğlu and J. van Paradijs (Kluwer, Dordrecht, 1995), p. 457.
26. G. Baym, R. Epstein and B. Link, *Physica* **B178**, 1 (1992).
27. P. W. Anderson and N. Itoh, *Nature* **256**, 25 (1975).
28. K. Rajagopal, *Nucl. Phys. A* **661**, 150c (1999); M. Alford, K. Rajagopal and F. Wilczek, *Nucl. Phys. B* **537**, 443 (1999); T. Schäfer and F. Wilczek, *Phys. Rev. Lett.* **82** (1999) 3956; *Phys. Rev. D* **60**, 114033 (1999).
29. M. G. Alford, A. Schmitt, K. Rajagopal and T. Schäfer, *Rev. Mod. Phys.* **80**, 1455 (2008).
30. K. Iida and G. Baym, *Phys. Rev. D* **63**, 074018 (2001).
31. M. Alford, J. A. Bowers and K. Rajagopal, *Phys. Rev. D* **63**, 074016 (2001).
32. I. Giannakis and H.-C. Ren, *Nucl. Phys. B* **669**, 462 (2003).
33. M. Kobayashi and T. Maskawa, *Prog. Theor. Phys.* **44**, 1422 (1970); G. 't Hooft, *Phys. Rev. Lett.* **37**, 8 (1976); *Phys. Rev. D* **14**, 3432 (1976).
34. R. D. Pisarski and F. Wilczek, *Phys. Rev. D* **29**, 338 (1984).
35. T. Hatsuda, M. Tachibana, N. Yamamoto and G. Baym, *Phys. Rev. Lett.* **97**, 122001 (2006); N. Yamamoto, M. Tachibana, T. Hatsuda and G. Baym, *Phys. Rev. D* **76**, 074001 (2007); G. Baym, T. Hatsuda, M. Tachibana and N. Yamamoto, *J. Phys. G: Nucl. Part. Phys.* **35**, 104021 (2008).
36. W. Ketterle, this volume.
37. Ref. 1, note on p. 1177.
38. M. R. Schafroth, S. T. Butler and J. M. Blatt, *Helv. Phys. Acta* **30**, 93 (1957).
39. J. Bardeen to F. Dyson, 23 July 1957, Bardeen papers.
40. D. M. Eagles, *Phys. Rev.* **18**, 456 (1969).
41. A. J. Leggett, *J. Phys.* (Paris) **C7**, 19 (1980).
42. P. Nozières and S. Schmitt-Rink, *J. Low Temp. Phys.* **59**, 195 (1985).
43. M. W. Zwierlein, J. R. Abo-Shaeer, A. Schirotzek, C. H. Schunck and W. Ketterle, *Nature* **435**, 1051 (2005).
44. T. D. Lee and C. N. Yang, *Phys. Rev.* **112**, 1419 (1957); K. Huang, in *Stud. Stat. Mech*, **II**, eds. J. de Boer and G. E. Uhlenbeck (North Holland Pub., Amsterdam, 1964), 1; H. T. C. Stoof, *Phys. Rev. A* **45**, 8398 (1992); M. Bijlsma and H. T. C. Stoof, *Phys. Rev. A* **54**, 5085 (1996); P. Grüter, D. Ceperley and F. Laloë, *Phys. Rev. Lett.* **79**, 3549 (1997); M. Holzmann and W. Krauth, *Phys. Rev. Lett.* **83**, 2687 (1999).
45. G. Baym, J.-P. Blaizot, M. Holzmann, F. Laloë and D. Vautherin, *Phys. Rev. Lett.* **83**, 1703 (1999); M. Holzmann, G. Baym, J.-P. Blaizot and F. Laloë, *Phys. Rev. Lett.* **87**, 120403 (2001).
46. V. A. Kashurnikov, N. V. Prokof'ev and B. V. Svistunov, *Phys. Rev. Lett.* **87**, 120402 (2001).

47. K. Maeda, G. Baym and T. Hatsuda, *Phys. Rev. Lett.* **103**, 085301 (2009).
48. A. Rapp, G. Zarand, C. Honerkamp and W. Hofstetter, *Phys. Rev. Lett.* **98**, 160405 (2007); A, Rapp, W. Hofstetter and G. Zarand, *Phys. Rev. B* **77**, 144520 (2008).
49. R. W. Cherng, G. Refael and E. Demler, *Phys. Rev. Lett.* **99**, 130406 (2007).
50. T. Ozawa and G. Baym, to be submitted.

ENERGY GAP, MASS GAP, AND SPONTANEOUS SYMMETRY BREAKING

Yoichiro Nambu

Enrico Fermi Institute, University of Chicago,
5640 South Ellis Ave., Chicago, IL 60637, USA

This article is based on a talk given at a Symposium at the University of Illinois on the occasion to commemorate the 50th anniversary of BCS — I gave a historical overview of how BCS theory has come to be transplanted to particle physics and has helped solve its problems.

In the early 1950s while I was at the Institute for Advanced Study, Professor Pines invited me to give a seminar at the University of Illinois as I was interested in the plasma theory of Bohm and Pines. That was before BCS, and before I joined the theorists Wentzel and Goldberger at the University of Chicago.

Now to provide a short story about my background, I studied physics at the University of Tokyo, graduating at the MS level according to the system of those days. I was attracted to particle physics because of three famous names, Y. Nishina, S. Tomonaga and H. Yukawa, who were the founders of particle physics in Japan. In fact, since this year and last there have been consecutive centennial celebrations of Tomonaga and Yukawa. But these people were at different institutions than mine. On the other hand, condensed matter physics was pretty good at Tokyo. R. Kubo, well known for his work in statistical mechanics, was my contemporary. I got into particle physics only when I came back to Tokyo as a kind of research associate after the war. Tomonaga had just started to develop the renormalization theory at his university nearby, and some of my room mates were working with him, as he also had served as a visiting professor at Tokyo after my graduation. The remarkable events of the Lamb shift and the pion-muon decay chain soon occurred, too. In hindsight, though, I must say that my early exposure to condensed matter physics has been quite beneficial to me.

I will now briefly review the history of particle physics to the extent that is necessary for the main theme of my talk. Particle physics is an outgrowth of nuclear physics which began in the early 1930s with the discovery of the neutron by J. Chadwick, the invention of the cyclotron by E. Lawrence, and the "invention" of meson theory by H. Yukawa. The appearance of an ever increasing array of new particles in the subsequent decades, and the advances in quantum field theory gradually led to our understanding of the basic laws of nature, culminating in the present Standard Model of particle physics. We have come to know that the elementary particles, or the fundamental fermions, consist of two kinds, leptons (so named by L. Rosenfeld) and quarks (so named by M. Gell-Mann). They are characterized by the internal quantum numbers, flavor and color. The forces or interactions among them (except for gravity) are described by the various gauge fields. A set of two leptons and six quarks make up a so-called generation, and there exist three generations, so far, with increasingly heavier masses. The electromagnetic and weak forces are due to the flavor gauge fields and are shared by all the fermions. The strong color forces are unique to the quarks, and lead to formation of permanent quark composites called hadrons (named by L. B. Okun), i.e. baryons and mesons.

On the theorists' side, faced with those new particles, their first attempts were to make sense out of them by finding some regularities in their properties. They invoked symmetry principles to classify them. A symmetry in physics leads to a conservation law. Some conservation laws are exact, like energy and electric charge, but these classification attempts were mostly based on approximate similarities of masses and interactions, but not exact symmetries.

Nevertheless, seeing similarities is a natural trait of the human mind, and in fact it has proved very useful. The near equality of proton and neutron masses and their interactions led to the concept of isospin SU(2) symmetry of N. Kemmer, Wentzel's former student. One could argue, or hope, that the differences were attributed to the electromagnetic interactions which had a different symmetry property from the strong interaction of the meson theory. When strange hadrons showed up, the search for an enlarged symmetry led to the SU(3) symmetry of M. Gell-Mann and Y. Ne'eman that covered all hadrons known up to the 1960s, although its origin was, and still is, not clear.

On the other hand, one could also go in the opposite direction, and elevate the symmetry to a general dynamical principle. Then symmetry will determine all physics, a most attractive possibility. Thus the beautiful properties of electromagnetism was extended to the SU(2) non-Abelian gauge field by C. N. Yang and R. L. Mills[1] as an ideal model for the strong interaction. But this is a huge oversimplification. Strong interactions are short range, and the mesons are all massive. Giving a mass to a gauge field destroys gauge invariance. Nevertheless, this gauge field analogy arose in the 1950s in two different contexts.

1. The existence of vector mesons ρ (770) (isospin 1) and ω (782) (isospin 0). Indeed the ρ (770) can give attractive forces to hadrons, and seemed to have a universal coupling strength $(g^2/4\pi^2)$ to isospin current, which led J. J. Sakurai,[2] my colleague, to promote a vector dominance theory of strong interactions.
2. The $V-A$ property of the weak interaction as was found by R. Feynman and M. Gell-Mann,[3] and by R. Marshak and G. Sudarshan,[4] which also seemed almost universal for those particles known at that time. This raised the possibility that the weak processes might also be mediated by a gauge field. The $V-A$ structure and apparent massless neutrinos also directed our attention to the chirality, or the γ_5 charge, as a possible symmetry which, however, showed up in subtle ways, in spite of the fact that it is violated by the masses of the fermions.

So the problem of mass somehow seemed to be a sticking point in our search for a dynamical symmetry principle for both strong and weak interactions. It is at this point that analogies learned from other branches of physics have turned out to be very fruitful. I will follow this based on my own experiences in the early 50s here and in Japan. There are three topics in this connection.

1. The Debye screening of the Coulomb interaction turns the $1/r$ potential into a Yukawa form $\exp(-\mu r)/r$. In ionized media the screening gives rise to the plasma oscillations, first noted by I. Langmuir and Tonks.[5] It is a mode where the dispersion relation ω versus k has a mass (in particle physics language). I was intrigued when D. Bohm, E. P. Gross, and D. Pines[6] gave a systematic description of this phenomenon, and generalized it to transverse modes. I tried hard, though unsuccessfully, to apply it to the problems of nuclear physics, like the saturation of nuclear forces and the large spin-orbit effects.
2. Tomonaga wrote a paper[7] on a description of collective modes in one-dimentsonal fermionic systems. Stimulated by this, I found that, for bosonic media in general, one could apply Dirac's canonical transformation from the creation and annihilation operators to number and phase

operators, and convert a Hamiltonian for an interacting medium of the type

$$H = \int \frac{1}{2m}|\nabla\psi|^2 + \frac{1}{2}\iint \rho V \rho', \qquad \psi = \sqrt{\rho}e^{i\theta}$$

into an approximate effective form quadratic in ρ and θ

$$H \sim \int \frac{1}{2m}\left(\rho_0|\nabla\theta|^2 + \frac{1}{4\rho_0}|\nabla\rho|^2\right) + \frac{1}{2}\iint \rho V \rho'$$

One immediately obtains the eigenvalues,

$$\omega(k)^2 \sim \frac{k^2}{m}\rho_0\left(V(k) + \frac{k^2}{4m\rho_0}\right)$$

For the Coulomb case $V = e^2/k^2$, we get the plasma "mass"

$$\omega_P^2 = e^2\rho_0/m\,.$$

3. Then there is the more important and mysterious London theory of superconductivity, based on the ansatz $\mathbf{j} = \Lambda\mathbf{A}$. The magnetic field B acquires the "mass" $\Lambda(= \omega_P)$ with an appropriate definition of ρ_0 and m. This was to become the problem to be understood. I learned, among others, of a work by M. R. Schafroth,[8] another former student of Wentzel's, on the Meissner effect of a Bose gas. All these examples gave a hint that the mass of gauge fields in particle physics might also be resolved if one considered the vacuum as a kind of medium. Indeed Dirac had already invoked this analogy in his interpretation of the negative energy solution of his equation.

Now I come to the BCS theory.[9] I first heard about it when our boss Wentzel invited Mr. Schrieffer for a seminar before the work was published. I understood that he was still a graduate student. Subsequently he would join our faculty, and I greatly benefited from our interaction. I remember the excitement I felt at the seminar, but at the same time I became worried about the fact that their Hartree–Fock state was a superposition of different number of charges and did not appear to respect gauge invariance. Soon thereafter N. N. Bogoliubov[10] and J. G. Valatin[11] independently introduced the concept of quasiparticles as fermionic excitations in the BCS medium. The quasiparticles did not carry a definite charge as they were a superposition of electron and hole, with their proportion depending on the momentum.

How can one then trust the BCS theory for discussing the electromagnetic properties like the Meissner effect? As this doubt gradually grew more serious in my mind, I decided to attack it, but it actually took two years before I was able to do so to my satisfaction and publish the result.[12] In the mean time, the problem was addressed by many people,[13] but I wanted to understand it in my own way. Essentially it is the presence of a massless collective mode, now known by the generic name of Nambu–Goldstone (NG) boson, that saves charge conservation or gauge invariance. To describe the Bogoliubov–Valatin (BV) quasiparticle, I introduced a two-component notation, which is similar to the Gor'kov formalism.[14] In this space, the Hamiltonian for the quasiparticle is given as

$$H = \psi^\dagger(\tau_3\varepsilon(k) + \tau_1\Delta)\psi$$

(ε: kinetic energy relative to the Fermi energy; Δ: gap) The charge is diagonal ($\sim \tau_3$) in this space, but the BV state is not. The collective mode is an excitation of the BV pairs $\sim \psi^\dagger\tau_2\psi$. It is not the eigenstate of charge either, but its interaction with the quasiparticle, i.e. its emission and absorption, insures that charge is conserved overall when a quasiparticle undergoes acceleration. In other words, the charge-current operator j_μ is made up of contributions from quasiparticle as well as the collective cloud around it; j_μ is not diagonal in the energy basis, but divergence-free. In addition to the massless NG mode, I found another collective mode in the direction of the gap $\sim \tau_1$, with mass 2Δ. It is a quasiparticle composite with zero binding energy, which may be thought of being similar to the so-called sigma meson and Higgs boson in its relation to the NG mode. I also found that, when the Coulomb interaction between quasiparticles is included, it mixes strongly with the massless collective mode because of their common massless nature, and they turn into the massive plasmon mode. The correct expression for the charge and current is,

$$\rho_{\text{tot}} = \rho_{BV} + \frac{1}{v^2}\frac{\partial\phi}{\partial t}, \qquad v = v_{\text{Fermi}}/\sqrt{3}$$

$$\mathbf{J}_{\text{tot}} = \mathbf{j}_{BV} - \boldsymbol{\nabla}\phi,$$

where ϕ is the NG mode, so that

$$\partial\rho_{\text{tot}}/\partial t + \nabla\cdot J_{\text{tot}} = \frac{1}{v^2}\frac{\partial^2\phi}{\partial^2 t} - \nabla^2\phi = 0$$

The effect of the Coulomb interaction is to screen this longitudinal part of charge-current, but leaves the London term for the transverse part.

Thinking of general characterizations of the BCS mechanism, the idea of spontaneous symmetry breaking (SSB), with the associated massless collective modes arose in my mind. (The word SSB was later introduced by

M. Baker and S. Glashow.[15]) The phenomenon is due to the fact that the lowest energy state of the system is infinitely degenerate, and a Hilbert space of the quasi-particle excitations is built on just one of the vacua. Hilbert spaces built from different vacua are equivalent but are orthogonal to each other in the sense that matrix elements of local observables between them are zero because of the infinite degrees of freedom of the system. They are like different phases of a medium at transition temperature. The NG collective modes correspond to local fluctuations into other vacua, so the excitation energy goes to zero in the long wave length limit.

The broken symmetry under question in the BCS case was that of electric charge conservation, but it immediately reminded me of analogous cases like ferromagnetism and crystal formation. They correspond to a global breaking of symmetry, in these cases that of spin rotation and space of translation and rotation, respectively, and the corresponding collective modes are the magnon and the phonons (longitudinal and tranverse). I thought of collecting further examples systematically before writing a general paper on this problem. I had used the name zeron for the NG boson. Incidentally I also realized that this SSB mechanism was what Heisenberg had invoked in his ill-fated nonlinear unified field theory[16] to generate more degrees of freedom than are in the original Lagrangian, although this would amount to simultaneous existence of different phases, or worlds.

The emergence of the "zeron" and its subsequent transfiguration into a massive gauge field was elucidated by the works of a number of people,[17] starting with J. Goldstone. They made use of a model theory with a scalar field with nonlinear terms, which eventually led to the electroweak unification of Glashow, Salam, and Weinberg and the current Standard Model of particle physics. In hindsight I regret that I should have explored in more detail the general mechanism of mass generation for the gauge field. But I thought the plasma and the Meissner effect had already established it. I also should have paid more attention to the Ginzburg–Landau theory[18] which was a forerunner of the present Higgs description. I had pushed it out of my mind because I had not understood the meaning of their order parameter field. (In writing this, I recall what Wentzel once remarked to me about the Schroedinger equation: "What a bold step to put a Coulomb potential into the de Broglie wave!")

I was actually more intrigued by the similarity of the BV 2-component fermion formalism and the Dirac equation. Only a spinor of left-handed or right-handed chirality is sufficient to describe a massless fermion. The two components of opposite charges in a BV fermion get mixed if a energy gap is formed, and charge conservation is broken. The two spinors of opposite chirality get mixed by a mass term which thus breaks chirality. If this is an SSB, the corresponding NG boson should be a pseudoscalar. I decided to pursue this analogy since I had been always interested in the mass spectrum of the various particles. (By the way, another analog of BCS would be the Majorana mass, which is an issue for neutrinos.) A natural place to start was the nucleon because there was indeed a pseudoscalar boson, i.e. the pion with isospin one, coupled to the isodoublet proton and neutron. The isospin symmetry was not exact, and the pion was not massless, but its mass was small compared to the nucleon masses that were to be dynamically generated by some interaction. The role of the pion in this connection was to restore the conservation of the chirality after it was broken by the nucleon mass M. The isovector current would take the form

$$\mathbf{j}_{5\mu} = \overline{\psi}\gamma_\mu\gamma_5\boldsymbol{\tau}\psi - \frac{2M}{g}\partial_\mu\boldsymbol{\pi}\,,$$

where g is the π–N coupling constant. This is the analog of the London current above. It is conserved, $\partial_\mu j^\mu = 0$ if the pion mass is put to zero. The second term gave the correct value when it was used as the axial current for the $\pi- > \mu\nu$ decay rate. With the coefficient $2M/g$ replaced by an empirical number $f \sim 90$ Mev, it was known as partial conservation of axial current (PCAC). The particular relation between f and $2M/g$ had been noticed by M. Goldberger and S. Treiman.[19] Thus the SSB idea provided theoretical support for it.

I have to mention here some uneasiness that was felt. The mass of the real pion is not zero although it is small compared to that of the nucleon. Unlike the case of a gauge field, there is no natural mechanism to shield the pion field, so to speak, and make it massive. How can I justify my claim when I am following rigorous logic? Here I recalled an edict by S. Sakata. He said that truth is not immutable but will evolve. What we think now as truth is not absolute; it may be modified later as we learn more. Reassured by Sakata's words I decided to go ahead. Indeed we know now that there is a quark substructure to the hadrons, and that the quarks' own base masses are responsible for the pion mass. We also know the so-called chiral anomaly which intrinsically breaks chiral symmetry.

The basic ideas about SSB and mass generation were first presented as a remark at the 1959 High Energy Conference in Kiev,[20] then in the *Physical Review Letters*.[21] A somewhat more detailed exposition was given by my collaborator G. Jona-Lasinio at a regional conference at Purdue University, followed by full papers with him.[22] We considered a model consisting of a nucleon singlet having quadratic interaction terms $(1 \cdot 1, \gamma_5 \cdot \gamma_5)$ $(U(1)_L \times U(1)_R)$, and one consisting of an isodoublet with terms $(1 \cdot 1,\ \boldsymbol{\tau}\gamma_5 \cdot \boldsymbol{\tau}\gamma_5)$ $(\mathrm{SU}(2)_L \times \mathrm{SU}(2)_R)$. As a relativistic theory, it is non-renormalizable and needs a quadratic cut-off, but I chose this BCS-like model for simplicity for displaying the basic principles. The SSB actually generate a scalar meson in addition to the pseudoscalar pion, corresponding to the composites $(\gamma_5, 1) \sim (\pi, \sigma)$ or $(\boldsymbol{\tau}\gamma_5, 1) \sim (\boldsymbol{\pi}, \sigma)$. The σ meson is the analog of the τ_1 mode in the BCS case, with mass $2M$.

When these collective modes were treated as independent effective fields, this model turned out to be equivalent to the so-called sigma model of M. Gell-Mann and M. Lévy,[23] just as the BCS theory gave a dynamical basis to the Ginzburg–Landau theory. The difference from these effective theories is that the fermion and boson masses are related:

$$m_\pi : m_f : m_\sigma = 0 : 1 : 2$$

Translating the NJL model to the quark level, this relation led to the proposition that the observed broad enhancement in the $\pi - \pi$ channel in hadron reactions roughly over the region $\sim 2m_{\text{quark}} \sim 700$ Mev could be interpreted as this σ, although the issue is not settled yet. The above formula was actually vindicated in the case of superconductivity when a collective mode at $\sim 2\Delta$ was found by R. Sooryakumar, M. V. Klein.[24] I could have predicted this explicitly, but I first learned of their result and the relation to my work from C. Varma.[25]

The BCS type system characterized by weak short range interaction yields simple mass relations among fermions and bosons. This is because the Yukawa coupling g and the self-coupling λ of the σ–π field in an effective Gell-Mann–Lévy Hamiltonian are determined from the original theory and are related as $g^2 = \lambda$. In the case of ${}_3He$ B phase there are collective modes with total angular momentum $j = 0.1, 2$ including two NG modes with $j = 0$ and 1 corresponding to the SSB of particle number and spin, respectively. For each j, the mass relations satisfy a generalization of the above form[26]

$$\Delta_{b1}^2 + \Delta_{b2}^2 = 4\Delta_f^2$$

The subsequent developments in particle physics followed the GMLH line because of its generality and renormalizability. I do not have time to elaborate on them, but the efforts to go back to the BCS level of underlying dynamics have also been around from time to time. But it is only a vague hope that we may eventually understand the pattern of the masses of the fermions even if the existence of the Higgs boson is confirmed.

References

1. C. N. Yang and R. L. Mills, *Phys. Rev.* **96**, 191 (1954).
2. J. J. Sakurai, *Ann. Phys.* **11**, 1 (1960).
3. R. P. Feynman and M. Gell-Mann, *Phys. Rev.* **103**, 193 (1958).
4. E. C. G. Sudarshan and R. E. Marshak, *Phys. Rev.* **109**, 1880 (1958).
5. I. Langmuir and L. Tonks, *PR* **33**, 195 (1929).
6. D. Bohm and E. Gross, *Phys. Rev.* **75**, 1851, 1864 (1949); D. Bohm and D. Pines, *Phys. Rev.* **82**, 625 (1951); **85**, 338 (1952); **92**, 609 (1953).
7. S. Tomonaga, *Prog. Theoret. Phys.* **5**, 544 (1950).
8. M. R. Schafroth, *Phys.* **100**, 463 (1955).
9. J. Bardeen, L. N. Cooper and J. R. Schrieffer, *Phys. Rev.* **106**, 162 (1957).
10. N. N. Bogoliubov, *J. Expl. Theoret. Phys. (USSR)* **34**, **58**, **73** (*Soviet Phys. JETP* **34**, **41**, **51**) (1958).
11. J. G. Valatin, *Nuovo Cimento* **7**, 843 (1958).
12. Y. Nambu, *Phys. Rev.* **117**, 648 (1960).
13. J. Bardeen, *Nuovo Cimento* **5**, 1765 (1957); P. W. Anderson, *Phys. Rev.* **110**, 827 (1958); **110**, 1900 (1958); G. Rickaysen, *Phys. Rev.* **111**, 817 (1958); *Phys. Rev. Lett.* **2**, 91 (1959); and others.
14. L. P. Gor'kov, *Soviet Phys. JETP* **9**, 1364 (1959); **10**, 998 (1960).
15. M. Baker and S. L. Glashow, *Phys. Rev.* **128**, 2462 (1962).
16. W. Heisenberg, H. P. Duerr, S. Schlieder and K. Yamazaki, *Zeits. f. Naturf.* **14a**, 441 (1959).
17. J. Goldstone, *Nuovo Cimento* **19**, 154 (1961); F. Englert and R. Brout, *Phys. Rev. Lett.* **13**, 321 (1964); P. Higgs, *Phys. Rev. Lett.* **13**, 508 (1964); G. E. Guralnik, C. R. Hagen and T. W. B. Kibble, *Phys. Rev. Lett.* **12**, 585 (1964); and others.
18. V. L. Ginzburg and L. D. Landau, *Zh. Eksp. Teor. Fiz.* **20**, 1064 (1950).
19. M. L. Goldberger and S. B. Treiman, *Phys. Rev.* **110**, 1178 (1958).
20. *Proc. Int. High En. Conf. in Kiev*, 1959, a remark after a talk by B. F. Toushek.
21. Y. Nambu, *Phys. Rev. Lett.* **4**, 380 (1959).
22. Y. Nambu and G. Jona-Lasinio, *Phys. Rev.* **122**, 345 (1961); **124**, 246 (1961).
23. M. Gell-Mann and M. Lévy, *Nuovo Cimento* **16**, 705 (1958).
24. R. Sooryakumar and M. V. Klein, *Phys. Rev. Lett.* **45**, 660 (1980).
25. C. A. Balseiro and L. M. Falicov, *Phys. Rev. Lett.* **45**, 662 (1980); P. B. Littlewood and C. M. Varma, *Phys. Rev. B* **26**, 4883 (1982).
26. Y. Nambu, *Physica* **15D**, 147 (1985).

BCS AS FOUNDATION AND INSPIRATION: THE TRANSMUTATION OF SYMMETRY

Frank Wilczek
*Center for Theoretical Physics, Department of Physics,
Massachusetts Institute of Technology,
Cambridge, MA 02139, USA*

The BCS theory injected two powerful ideas into the collective consciousness of theoretical physics: pairing and spontaneous symmetry breaking. In the 50 years since the seminal work of Bardeen, Cooper and Schrieffer, those ideas have found important use in areas quite remote from the stem application to metallic superconductivity. This is a brief and eclectic sketch of some highlights, emphasizing relatively recent developments in QCD and in the theory of quantum statistics, and including a few thoughts about future directions. A common theme is the importance of symmetry *transmutation*, as opposed to the simple *breaking* of electromagnetic $U(1)$ symmetry in classic metallic superconductors.

The Bardeen-Cooper-Schrieffer (BCS) theory of superconductivity[1] has been fruitful in many ways. Most obviously, of course, it provided profound and at many points surprising concrete insights into the superconducting state of solids, right from the start. It predicted, for instance, the very different effect of the onset of superconductivity on acoustic versus electromagnetic relaxation, due to the different signs in coherence factors, which is a delicate quantum-mechanical effect. And it provided the intellectual foundation for such wonders as the Josephson effects and Andreev reflection.

The influence of BCS theory on the broader discipline of theoretical physics has been no less profound. Two key ideas abstracted from BCS theory, that have been widely transplanted and borne abundant fruit, are *pairing* and *dynamical symmetry breaking.* Pairing was an essentially new idea, introduced by Cooper and brought to fruition by BCS. The symmetry breaking aspect was mostly implicit in the original BCS work, and in earlier ideas of Fritz London and Landau–Ginzburg; but the depth and success of the BCS theory seized the imagination of the theoretical physics community, and catalyzed an intellectual ferment. The concept of spontaneous symmetry breaking was promptly made explicit, generalized, and put to use by several physicists including Anderson, Josephson, Nambu and Goldstone. The flexibility and transformative power of these ideas revealed itself gradually, in applications to phenomena that at first sight appear to have little or nothing in common with superconductivity.

From a wealth of possible material, I have chosen to discuss some relatively recent developments close to my own work, that I think well illustrate how naturally the basic BCS concepts combine with other ideas of fundamental and emergent symmetry, often with dramatic consequences. A common theme is that symmetry *breaking* forms a special case of a more general phenomenon: symmetry *transmutation.*

1. QCD Meets BCS: Color-flavor Locking, Confinement and Chiral Symmetry Breaking

Quantum chromodynamics or QCD, having run the gauntlet of many exquisite quantitative confrontations with observation, is now established as the fundamental theory of the strong interaction.[2] QCD is a challenging theory to understand, however, and not primarily because of its technical complexity.[a] The real challenge comes in relating the wonderfully "trivial" basic equations to observed reality. The primary ingredients of QCD are massless gluons and nearly massless quarks (u, d, s; the heavy quarks c, b, t are a separate, and much easier, study). The observed hadrons, of course, are neither massless like gluons nor fractionally charged like quarks. Many techniques have been deployed to bridge the chasm separating the theory world from the physical world, but in my opinion none is clearer nor more elegant than the straightforward application of BCS ideas to the regime of high density. (Here by "high density" I intend large baryon number density, at low temperature).[3]

1.1. *QCD meets BCS*

A wise principle states "It is more blessed to ask forgiveness than permission." In that spirit, we consider the possibility of constructing a description

[a]"Technical complexity" is a time-dependent concept. I've heard graduate students accustomed to a diet of high supersymmetry, Calabi–Yau manifolds and intersecting D-branes refer to QCD as "trivial", with no evident ironic intent.

of high-density QCD based on its elementary degrees of freedom, quarks and gluons.

At first sight this approach looks extremely promising. High density means large fermi surfaces. Neglecting interactions, the low-energy excitations are associated with modes near the fermi surface: a mode just above the fermi surface, empty in the ground state, becomes occupied, or a mode just below becomes empty. Since the fermi surface is large, all modes near the fermi surface carry large momentum and energy. Thus scattering among the low-energy excitations will either involve only small angles, and leave the distribution of particles over modes nearly unchanged, or else bring in large momentum transfers, and therefore weak coupling (asymptotic freedom). It appears, therefore, that perturbation theory should be a good approximation.

But when one actually does the calculations, one finds infrared divergences. They arise from two sources:

- The preceding argument only concerns the quarks. Its central point is that Pauli blocking removes the infrared divergences that usually arise through low-virtuality quarks. Gluons, however, are not subject to any such effect. Color electric forces are screened by the quark medium, but color magnetic forces remain long-ranged, and lead to infrared divergences.
- Interacting fermions are subject to the Cooper instability. One has many near-zero energy excitations at zero momentum, associated with particle–particle or hole–hole pairs carrying equal and opposite three-momenta $\pm\vec{p}$. Thus in perturbing around the many-body state which is the ground state of the non-interacting theory, that is the fully occupied fermi sphere, one is engaging in highly degenerate perturbation theory. As a general matter, degenerate perturbation theory can result in significant restructuring of the ground state. In this specific context, Bardeen, Cooper and Schrieffer (BCS) taught us that even a small attractive interaction *will* lead to a drastic re-arrangement of the ground state, by inducing pairing and superfluidity.

In conventional superconductors it is quite subtle to find an effective attractive interaction between electrons. The primary interaction between electrons is the Coulomb interaction, and it is of course repulsive. To find an attractive interaction one must bring in phonons, retardation, and screening, and concentrate on modes within a thin shell around the Fermi surface. For many "unconventional" superconductors, famously including the cuprates, the mechanism of attraction remains unclear at present, despite much effort to identify it. But in all known cases the superconducting transition temperature (which reflects the attractive dynamics) is far below the melting temperature (which reflects the primary dynamics).

In QCD the situation is more straightforward, because the primary interaction — the QCD analogue of the Coulomb interaction — can already be attractive. Two separated quarks, each in the triplet **3** representation, can be brought together in the antisymmetric $\bar{\mathbf{3}}$. The disturbance in the gluon field due to color charge is then half what it was before. Since the energy has decreased, the force is attractive. Nothing like this can happen when there is just one type of charge, of course. The existence of three different color charges is crucial here. (On the other hand, with larger numbers of colors antisymmetrization yields relatively less reduction in flux, and so the attractive force is relatively weaker.)

By zeroing the spin — that is, once again, choosing the antisymmetric channel — we also remove the sources of magnetic disturbance. Thus, on very general grounds we expect a powerful attractive interaction between quarks in the channel where both colors and spins are antisymmetric. This intuition is borne out by calculations using one-gluon exchange, instanton models, and direct numerical simulations, though those simulations could and should be sharpened.

Thus color superconductivity occurs straightforwardly, and should be robust physically, at high density. What does it mean? Here is a list of physical effects we can anticipate, by translating intuition from superconductivity into the language of particle physics:

- Gluons acquire mass — that is a way to state the equations of the Meissner effect. If *all* the gluons acquire mass, their exchange will no longer produce infrared divergences.
- Quarks acquire mass — that is a way to state the equations of the energy gap. If *all* the quarks acquire mass, Cooper's infrared divergence will be removed.
- Thus we construct a new ground state, around which our weak-coupling expansion works.
- This ground state does not contain massless gluons nor exhibit long-range forces. In that sense, it exhibits confinement. We also have the classic phenomenon of confinement — that is, absence of fractional *electric* charge in the spectrum, as I will explain shortly.
- The energy gap for quarks suggests that chiral symmetry, which is associated with massless quarks, may be broken.

In short, we have the prospect of a phase that exhibits the main non-perturbative features of QCD — confinement and chiral symmetry breaking — in a transparent, fully controlled theoretical framework. Let me emphasize that here I am speaking of a phase of QCD itself, not of some idealization of a model of a caricature of QCD. Now let us discuss how it is realized, more concretely.

1.2. *Color-flavor locking*

For concreteness, and because my emphasis here is on QCD rather than astrophysics, I will assume as the initial default that all quarks are massless, that they are subject to a common chemical potential, and that electromagnetism can be treated as a perturbation. I will circle back to revisit these assumptions, and ask your forgiveness, in due course.

Because the most attractive channel for quarks is antisymmetric both in color and spin, Fermi statistics requires another source of antisymmetry. One possibility is antisymmetry in the spatial wave-function of the quark pairs. For example, we might have p-wave pairing. But for simple, purely attractive interaction potentials, s-wave tends to be favored, because it allows pairs from all directions over the Fermi surface to act in phase. So s-wave pairing, if possible, is likely to be favorable.

The remaining possible source of antisymmetry is flavor. Thus we must pair off *different* flavors of quarks to take best advantage of the attractive interaction between quarks. This requirement brings in some significant complications. Obviously, it means that the one-flavor case is not representative, and that we cannot build up the analysis one flavor at a time. The two-flavor case also does not go smoothly. Antisymmetry in flavor and spin (and lack of orbital structure) reduces the quark–quark channel to a single vector in color space. Therefore condensation in this channel can break color symmetry only partially, in the pattern $SU(3) \to SU(2)$. Some gluons remain massless, and some quarks remain gapless, so infrared divergences remain.

1.2.1. *Ground state*

Simplicity and self-consistency (that is, consistent use of weak coupling) first arrive when we consider three flavors.

I will describe the full structure of the condensate momentarily, but since that is a little intimidating, let me begin with a sketch. Since the spin (singlet) and spatial (s-wave) structures are unremarkable I will suppress them, and also chirality. The favored condensate should be antisymmetric in color and in flavor, which suggests the form

$$\langle q_a^\alpha q_b^\beta \rangle \sim \epsilon^{\alpha\beta *} \epsilon_{ab*}$$

where the Greek indices are for color, the Latin indices are for flavor, and * is a wildcard. Now by setting the wild cards equal, and contracting, we maintain as much residual symmetry as possible. Any fixed choices for the wildcards will break both color and flavor symmetries. But by *locking* color to flavor we maintain symmetry under the combined (so-called diagonal) symmetry group. Thus, we arrive at

$$\langle q_a^\alpha q_b^\beta \rangle \sim \epsilon^{\alpha\beta *} \epsilon_{ab*} \rightarrow \epsilon^{\alpha\beta i} \epsilon_{abi} \propto (\delta_a^\alpha \delta_b^\beta - \delta_b^\alpha \delta_a^\beta) \tag{1}$$

This condensate breaks local color times global flavor $SU(3) \times SU(3)$ to a diagonal, "modified flavor" global $SU(3)$.

It also spontaneously breaks baryon number symmetry. To a particle physicist encountering these ideas for the first time, that might sound dramatic — and it is, but not in the sense that it allows the material to decay. With the sample enclosed in a finite volume, outside of which the order parameter vanishes, there is a strict conservation law for the integrated baryon number. As in the theory of liquid helium 4, where one speaks of a condensate of helium atoms, the true implication is that there is easy transport of baryon number within the sample. More specifically, there is a massless Nambu–Goldstone field, which supports the supercurrents characteristic of superfluidity.

Now comes the full structure, in all its glory:

$$\begin{aligned} &\langle \mathbf{1} | (q_a^\alpha)_L^i(\vec{k}) (q_b^\beta)_L^j(-\vec{k}) | \mathbf{1} \rangle \\ &= \epsilon^{ij} \Big(v_1(|\vec{k}|)(\delta_a^\alpha \delta_b^\beta - \delta_b^\alpha \delta_a^\beta) + v_2(|\vec{k}|)(\delta_a^\alpha \delta_b^\beta + \delta_b^\alpha \delta_a^\beta) \Big) \\ &= -(L \leftrightarrow R) \,. \end{aligned} \tag{2}$$

Here some further words of explanation are in order. The mid-Latin indices i, j are for spin. The "L" and "R" are for left and right chirality. The relative sign between left and right condensates reflects conservation of parity. The functions $v_1(|\vec{k}|)$, $v_2(|\vec{k}|)$ are, for weak coupling, peaked near the Fermi surface. Our preceding discussion anticipated the v_1 term, but the v_2 term is also allowed by the residual symmetry. That latter term indeed emerges from calculations based on the microscopic theory, though with $v_1 \gg v_2$.

Tracking chiral flavor symmetry and baryon number together with color, the implied breaking pattern is:

$$SU(3)_{\text{color}} \times SU(3)_L \times SU(3)_R \times U(1)_B \to SU(3)_\Delta \times Z_2 \tag{3}$$

The residual $SU(3)_\Delta$ global symmetry, and the Z_2 of fermion (quark) number, can be used to classify the CFL state's low-energy excitations. There is no residual local symmetry: all the color gluons have acquired mass. A more refined analysis reveals that all the quarks have acquired gaps.

Finally, as a consequence of the underlying — spontaneously broken — baryon number and chiral symmetries we have also generalized ground states, obeying

$$\begin{aligned}
&\langle \mathbf{U}, \theta | (q_a^\alpha)_L^i(\vec{k})(q_b^\beta)_L^j(-\vec{k})|\mathbf{U}, \theta\rangle \\
&\quad = \epsilon^{ij} e^{i\theta}\Big(v_1(|\vec{k}|)(\mathbf{U}_a^\alpha \mathbf{U}_b^\beta - \mathbf{U}_b^\alpha \mathbf{U}_a^\beta) + v_2(|\vec{k}|)(\mathbf{U}_a^\alpha \mathbf{U}_b^\beta + \mathbf{U}_b^\alpha \mathbf{U}_a^\beta)\Big) \\
&\quad = -(L \leftrightarrow R)
\end{aligned} \tag{4}$$

for an any $SU(3)$ matrix $\mathbf{U}$. These generalized ground states are related to one another by global baryon number (phase) or chiral transformations. Low-frequency, long-wavelength modulation of the fields θ and $\mathbf{U}$, which represents slow motion within the vacuum manifold, generates the Nambu–Goldstone bosons.

One more comment about the ground state is in order. Throughout this discussion I have used the language of gauge symmetry breaking and gauge non-singlet order parameters. This is quite familiar and traditional in BCS theory, and also in the standard model of electroweak interactions. Strictly speaking, however, it is based on a lie, for local gauge invariance is never broken. Matrix elements of gauge-variant expectation values always vanish in the physical Hilbert space. Indeed, the physical Hilbert space of a gauge theory is defined by restricting to gauge-invariant states. The usual procedures of "spontaneous symmetry breaking" using gauge-variant operators are a tool — a way of implementing favorable correlations in weak coupling. Their physical content emerges when we use them as a calculational device, in weak coupling, to draw consequences for gauge-invariant quantities such as the physical spectrum or the expectation values of gauge-invariant operators. In the CFL phase, we can identify two non-zero gauge invariant vacuum expectation values that break chiral or baryon number symmetries. They are

$$\begin{gathered}
\langle q_L q_L \bar{q}_R \bar{q}_R \rangle \\
\langle qqqqqq \rangle
\end{gathered} \tag{5}$$

with the color indices suitably contracted. These expectation values arise as powers of the primary, gauge-variant condensates. (After including instanton effects, we also get $\langle q_L \bar{q}_R \rangle$.) By way of contrast, neither conventional s-wave spin-singlet BCS condensation nor doublet condensation in the standard electroweak model support true order parameters.

1.2.2. *Elementary excitations*

We can analyze the elementary excitations from the point of their spin and quantum numbers under the residual $SU(3)_\Delta$ symmetry. There are three types:

1. *Excitations produced by the quark fields*: they are spin-$\frac{1}{2}$ fermions that decompose as $\mathbf{3} \times \bar{\mathbf{3}} \to \mathbf{8} + \mathbf{1}$ under $SU(3)_{\text{color}} \times SU(3)_{\text{flavor}} \to SU(3)_\Delta$. The singlet turns out, at weak coupling, to be significantly heavier than the octet.
2. *Excitations produced by the gluon fields*: they are spin-1 bosons that form an octet.
3. *Collective excitations*: they are a pseudoscalar octet of Nambu–Goldstone bosons, plus the singlet superfluid mode.

Overall, there is a striking resemblance between this calculated spectrum of low-lying excitations and what one might expect for the elementary excitations in the "nuclear physics" of QCD — that is, the nuclear physics of QCD with three massless flavors — based on standard concepts in QCD phenomenology and modeling. The calculated elementary excitations map nicely onto the entries in the expected hadron spectrum. Even the superfluid mode makes sense, because we would expect, in this idealized "nuclear physics", pairing to occur in the dibaryon channel.

Since conventional (heuristic) "nuclear physics" and the asymptotic (calculated) CFL state match so well with regard both to their ground state symmetry and to their low-lying spectrum, it is hard to avoid the conjecture that there is no phase transition separating these states. Consider cranking up the chemical potential, starting from zero. First there is Void. At a critical value nuclear matter appears, with a first-order transition. After that, there is just smooth evolution.

This conjecture of *quark-hadron continuity* is both (superficially) paradoxical and conceptually powerful.

The claim that the baryons of conventional "nuclear physics" are supposed to go over smoothly into excitations produced directly by single quark fields is paradoxical. After all, baryons are famous for containing three quarks, and three cannot evolve smoothly into one! Well, actually it can. When space is filled with a condensate of quark pairs, the difference between three and one is negotiable.

Quark-hadron continuity is a powerful conceptual claim: it implies that the calculable forms of confinement and chiral symmetry breaking we construct by adapting the methods of BCS theory are in the same universality class as confinement and chiral symmetry breaking at low energies, within nuclear (or rather "nuclear") matter.

1.2.3. *Electric charge*

The absence of long-range forces and massless gluons is a rather bloodless characterization of confinement. What we would really like to explain is: why do not fractionally charged particles appear in the spectrum, given that they are in the Lagrangian?

To address that question, we must couple electromagnetism into our theory. Of course, electromagnetism is connected with the photon, which couples as

$$\gamma : e \begin{pmatrix} \frac{2}{3} & 0 & 0 \\ 0 & -\frac{1}{3} & 0 \\ 0 & 0 & -\frac{1}{3} \end{pmatrix} \tag{6}$$

to flavor indices, in an evident notation. The symmetry associated with this generator is broken in the CFL ground state. However, there is a related gluon, which couples as

$$\Gamma : g \begin{pmatrix} \frac{2}{3} & 0 & 0 \\ 0 & -\frac{1}{3} & 0 \\ 0 & 0 & -\frac{1}{3} \end{pmatrix} \tag{7}$$

to color indices. The combination

$$\tilde{\gamma} = \frac{g\gamma + e\Gamma}{\sqrt{g^2 + e^2}} \tag{8}$$

leaves the mixed Kronecker deltas that characterize the CFL ground state invariant, so it defines a massless gauge boson. $\tilde{\gamma}$ is a modified photon, that defines the meaning of electromagnetism to an observer living within the CFL ground state. Formally, for $g \gg e$, it goes over into the ordinary photon. However, in that limit the small part Γ couples much more strongly than the large part γ, and basic properties of the modified photon are, in fact, modified.

Since the electron sees only γ, we read off its effective $\tilde{\gamma}$ charge as $-\frac{eg}{\sqrt{g^2+e^2}}$. Excitations produced by quark fields get a contribution of either $\frac{2}{3}e \times \frac{g}{\sqrt{g^2+e^2}}$ or $-\frac{1}{3}e \times \frac{g}{\sqrt{g^2+e^2}}$ from γ and a contribution of either $-\frac{2}{3}g \times \frac{e}{\sqrt{g^2+e^2}}$ or $\frac{1}{3}g \times \frac{e}{\sqrt{g^2+e^2}}$ from Γ. Adding the two contributions, you see that the quarks are either neutral, or their total $\tilde{\gamma}$ charge is ± 1 times the charge of the electron. The possible charges of quarks in the CFL phase match the observed charges of baryons, which is a nice check on the quark-hadron continuity conjecture. Precisely, these integer charge assignments appeared in the early work of Han and Nambu, who introduced color degrees of freedom together with a non-trivial embedding of electromagnetism to achieve them. Similarly, the gluon and pseudoscalar meson charges are all integral (and match what you find in the particle data tables).

A similar transmutation of the charge spectrum occurs in the standard model of electroweak interactions. In that context, the $SU(2)$ gauge symmetry of weak isospin and the $U(1)$ gauge symmetry of hypercharge are separately broken, but a linear combination survives to become electromagnetism. To reproduce the known electric charges of quarks, leptons, and W bosons one must postulate a very peculiar spectrum of fractional hypercharges. Unified theories of the strong, weak, and electromagnetic interactions, based on symmetry groups such as $SU(5)$ or $SO(10)$, which extend the standard model $SU(3) \times SU(2) \times U(1)$, arrive quite naturally at that very peculiar spectrum. That is perhaps the most compelling evidence that such theories are on the right track.[4]

1.2.4. *Material properties*

What happens to matter, if you keep squeezing? Ultimately — that is, for chemical potentials well above the strange quark mass, but well below the charm quark mass — it goes into the CFL phase, in which

- Hadronic matter forms a transparent insulator. We have discussed how the photon gets modified. Some of the massless Nambu–Goldstone bosons are electrically charged, but once we take into account non-zero quark masses, these bosons (apart from the superfluid mode, which is electrically neutral) acquire mass. Thus all the charged excitations have a gap, and we get an insulator. Note especially that while it is a *color* superconductor, hadronic mater in the CFL phase is not an *electrical* superconductor.
- It is a superfluid.
- It is vastly different from ordinary nuclear matter. It contains an equal

mix of strange quarks, for one thing, and strong trans-baryon correlations among the quarks. One might expect, and model calculations tend to show, that there is a sharp transition between the two phases of hadronic matter, namely nuclear and CFL, including an abrupt jump in density.

The first two items suggest that by squeezing we ultimately arrive at material similar to liquid helium 4, but of course, with vastly higher density. The third item suggests possibilities for astrophysical signatures. (As I will discuss momentarily, at present we cannot preclude the possibility of additional phases of hadronic at intermediate densities.) I expect that eventually observation of gravitational waves from the final infall of neutron star — neutron star or neutron star — black hole binaries, in particular, will bring our knowledge of neutron star interiors to a new, much higher level. Then predictions of this sort will be tested.

1.3. *Beyond color-flavor locking*

As I emphasized earlier, ordinary real-life nuclear matter is quite different from CFL. In practice, the effect of the strange quark's mass on QCD phenomenology is far from negligible. At low density the energetic cost of the strange outweighs its possible advantage in interaction energy, and ordinary nuclear matter has zero strangeness. Because the condensation mechanism at the heart of CFL necessarily connects three different flavors, and must bring in strange quarks, it does not apply to ordinary nuclear matter. CFL will set in when the chemical potential is sufficiently high that the strange quark mass is relatively negligible. That will occur for large enough chemicals potentials — or, equivalently, sufficiently high densities. Unfortunately, at present our calculational ability is not up to the task of predicting what happens subasymptotically. It is possible that nuclear matter transitions directly to CFL; it is also possible that that there are additional intermediate states. Even if the transition is abrupt, as I suspect it is, presently we cannot predict the chemical potential at which it occurs, nor the jump in density that accompanies it. These uncertainties hamstring our ability to make crisp astrophysical applications.

BCS pairing works best when the modes being paired have close to zero (free) energy. Ideally, many pairs should share the same quantum numbers, so that we can get enhancement factors from their coherent contributions. In that case, superconductivity can be triggered by arbitrarily weak interactions. On the other hand, if the Fermi surfaces of the quark species we would like to pair do not match, so that modes at $\vec{k}$ and $-\vec{k}$ cannot both be close to their respective Fermi surfaces simultaneously, then some compromise is necessary. There are several possibilities:

- Meson condensates, involving the Nambu–Goldstone bosons, can soak up some of the unwanted flavor imbalance.
- Less desirable forms of pairing, such as p-wave, that can work with a single flavor, might occur. p-wave pairing (in three dimensions) leaves gapless fermions, which might themselves pair, forming a secondary condensate at a lower energy scale.
- Pairing can occur at one or several non-zero wave-vectors, i.e. involving modes $(\vec{k}+\vec{\kappa},-\vec{k})$. These phases, which break translation invariance, are known as LOFF (Larkin–Ovchinnikov–Ferrell–Fulde) phases, or crystalline superconductivity.
- Pairing can occur between modes that are (nominally) particle and hole; i.e. one can dig into a Fermi ball, or supply a shell, to take advantage of potential energy gains at the cost of kinetic energy. If a vestige of the original Fermi surface remains, one has a breached (gapless) phase, where a superfluid condensate coexists with a normal fluid component.

One bright spot is that cold atom physicists are beginning to explore traps loaded with several fermion species. In that context, it is totally natural to have Fermi surfaces of different sizes and couplings that are not small, so that similar complications arise, but now in systems that are experimentally accessible and allow controlled manipulation of the underlying parameters. There has already been a fruitful cross-migration of ideas and information between these fields, and I expect that will continue.

2. Gauge-Rotation Locking and Quantum Statistics: Anyons[5]

2.1. *Gauge-rotation locking*

Consider a $U(1)$ gauge theory that is spontaneously broken by a condensate associated with a field of ρ of charge mq, with m an integer. Assume the theory also contains particles of charge q, associated with a field η, which does *not* condense. Gauge transformations that multiply ρ by $e^{2\pi ik}$ will multiply η by $e^{2\pi ik/m}$. For integer k such transformations leave the condensate invariant, but not necessarily η. We are therefore left with an unbroken gauge group Z_m, the integers modulo m, that is not entirely trivial, at least mathematically. On the other hand, no conventional long-range gauge interaction survives the symmetry breaking. Does the residual symmetry have any physical consequences?

Indeed it does. They are subtle, but very interesting indeed.

The theory supports vortices with flux quantized in units of

$$\Phi_0 = \frac{2\pi}{mq} \tag{9}$$

(in units with $\hbar \equiv 1$).

The flux is associated with a gauge potential, whose azimuthal piece we can take to have the form

$$\begin{aligned} A_\theta(r,\theta) &= \frac{\Phi}{2\pi} f(r) \\ f(0) &= 0 \\ f(\infty) &= 1 \end{aligned} \tag{10}$$

and a condensate of the form

$$\begin{aligned} \rho(r,\theta) &= g(r) e^{i\frac{\Phi}{\Phi_0}\theta} \\ g(0) &= 0 \\ g(\infty) &= 1 \end{aligned} \tag{11}$$

This condensate is neither rotationally invariant nor gauge invariant. It is, however, invariant under the combined rotation + gauge transformation

$$\tilde{L} = L + \Lambda \equiv -i\frac{\partial}{\partial\theta} - \frac{\Phi}{\Phi_0}\frac{Q}{m} \tag{12}$$

where Q is the charge operator. Here Λ is a generator of spatially constant gauge transformations. Thus, there is a residual modified rotational symmetry, locking naive spatial rotations to appropriate gauge transformations, under which the vortex is invariant. Indeed, from a strictly logical perspective it would be preferable to postulate the symmetry, and use it to motivate the vortex *ansatz*.

For the condensate field ρ, which has charge m, the new contribution to the angular momentum is an integer. Indeed, its "role in life" is to convert the spatial form of the condensation, which whirls in the partial wave with angular momentum $\frac{\Phi}{\Phi_0}$, so that it represents, at spatial infinity, the state of rest. More formally, the *kinetic* angular momentum, which is the gauge invariant version, gets annulled at infinity, by cancellation between the ordinary gradient and vector potential terms in the covariant derivative $D_\theta = \partial_\theta + iQA_\theta$. The square of angular momentum contributes to the energy density, so this cancellation must occur at spatial infinity, in order that the total energy of the vortex (in two dimensions), or the energy per unit length (in three dimensions) remains finite. For broken global symmetries this cancellation is not an option. In that case the vortices in that case have logarithmically divergent energy, and carry angular momentum; these facts underlie the vastly different phenomenology associated with vortices in superconductors versus liquid helium.

For the quanta of fields whose charge is not an integer multiple of the charge Q of ρ, on the contrary, the new contribution to the angular momentum is *not* necessarily an integer, due to the factor Q/m. So composites formed from particles of these kind and vortices will, in general, carry fractional angular momentum.

Since one expects, on very general grounds, that there ought to be a tight connection between the spin of a particle and its quantum statistics, we are led to look into the quantum statistics of these objects, anticipating something unusual.

2.2. *Anyons*

Traditionally, the world has been divided between bosons (Bose–Einstein statistics) and fermions (Fermi–Dirac statistics). Let us recall what these are, and why they appear to exhaust the possibilities.

If two identical particles start at positions (A, B) and transition to (A', B'), we must consider both $(A, B) \to (A', B')$ and $(A, B) \to (B', A')$ as possible accounts of what has happened. According to the rules of quantum mechanics, we must add the amplitudes for these possibilities, with appropriate weights. The rules for the weights encode the dynamics of the particular particles involved, and a large part of what we do in fundamental physics is to determine such rules and derive their consequences.

In general, discovering the rules involves creative guesswork, guided by experiment. One important guiding principle is correspondence with classical mechanics. If we have a classical Lagrangian $L_{\text{cl.}}$, we can use it, following Feynman, to construct a path integral, with each path weighted by a factor

$$e^{i \int dt L_{\text{cl.}}} \equiv e^{iS_{\text{cl.}}} \tag{13}$$

where $S_{\text{cl.}}$ is the classical action. This path integral provides — modulo several technicalities and qualifications — amplitudes that automatically implement the general rules of quantum mechanics. Specifically, it sums over alternative histories, takes products of amplitudes for successive events, and generates unitary time evolution.

The classical correspondence, however, does not instruct us regarding the relative weights for trajectories that are topologically distinct, i.e. trajectories that cannot be continuously deformed into one another. Since only small variations in trajectories are involved in determining the classical equations of motion, from the condition that $S_{\text{cl.}}$ is stationary, the classical equations cannot tell us how to interpolate between topologically distinct trajectories. We need additional, essentially quantum-mechanical rules for that.

Now trajectories that transition $(A, B) \to (A', B')$ respectively $(A, B) \to (B', A')$ are obviously topologically distinct. The traditional additional rule is: For bosons, add the amplitudes for these two classes of trajectories[b]; for fermions, subtract.

Those might appear to be the only two possibilities, according to the following (not-quite-right) argument. Let us focus on the case $A = A', B = B'$. If we run an "exchange" trajectory $(A, B) \to (B, A)$ twice in succession, the doubled trajectory is a direct trajectory. Then the square of the factor we assign to the exchange trajectory must be the square of the (trivial) factor 1 we associate to the direct trajectory, i.e. it must be ± 1.

The argument in the preceding paragraph is not conclusive, however, because there can be additional topological distinctions among trajectories, not captured by the permutation among endpoints. This distinction is especially important in two spatial dimensions, so let us start there. (I should recall that quantum-mechanical systems at low energy can effectively embody reduced dimensionality, if their dynamics is constrained below an energy gap to exclude excited states whose wave functions have structure in the transverse direction.)

The topology of trajectory space is then specified by the *braid group*. Suppose that we have N identical particles. Define the elementary operation σ_j to be the act of taking particle j over particle $j+1$, so that their final positions are interchanged, while leaving the other particles in place. (See Fig. 1.) We define products of the elementary operations by performing them sequentially. Then, we have the obvious relation

$$\sigma_j \sigma_k = \sigma_k \sigma_j ; \quad |j - k| \geq 2 \tag{14}$$

among operations that involve separate pairs of particles. We also have the less obvious Yang–Baxter relation

$$\sigma_j \sigma_{j+1} \sigma_j = \sigma_{j+1} \sigma_j \sigma_{j+1} \tag{15}$$

which is illustrated in Fig. 1. The topologically distinct classes of trajectories are constructed by taking products of σ_js and their inverses, subject only to these relations.

[b]As determined by the classical correspondence, or other knowledge of the interactions.

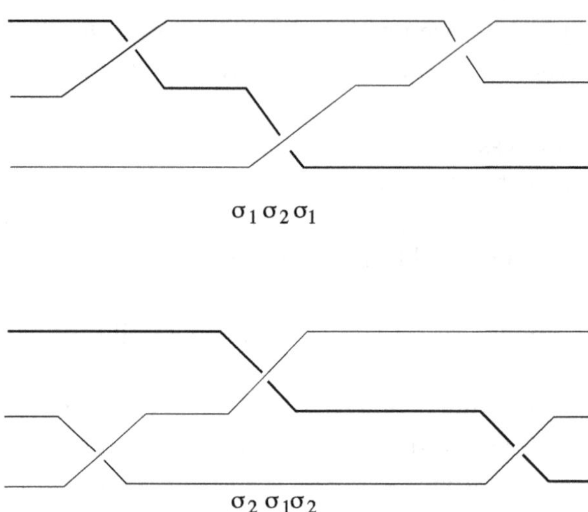

Fig. 1. The elementary acts of crossing one particle trajectory over another generate the braid group. The Yang–Baxter relation $\sigma_1\sigma_2\sigma_1 = \sigma_2\sigma_1\sigma_2$, made visible here, is its characteristic constraint. The top configuration can slide smoothly into the bottom one, with endpoints held fixed.

In three dimensions there is more room to maneuver the strands, and we have an additional relation

$$\sigma_j^2 = 1 \tag{16}$$

When these relations are added to the previous ones, we find that the braid group reduces to the ordinary so-called symmetric group S_N of permutations on N letters. An interesting intermediate possibility is to demand the relation that rotations through 4π are trivial but rotations through 2π might not be, as in the mathematics of spinors. Then, one would impose

$$\sigma_j^4 = 1 \tag{17}$$

in place of Eq. (16). This distinction will come up again shortly.

The defining equations Eqs. (14) and (15) for the braid group allow a continuous range of one-dimensional unitary representations, of the very simple form

$$\sigma_j = e^{i\theta} \tag{18}$$

for all j, with θ an arbitrary real number. One can have any phase, not just the ± 1 characteristic of bosons and fermions. For that reason, I christened particles carrying more general quantum statistics *anyons*.

With this very general background in mind, let us return to our fractional angular momentum vortices. A particle or group of particles with charge bq

moving around a flux Φ acquires, according to the minimal coupling gauge Lagrangian, phase

$$\exp ibq \left(\oint dt\vec{v} \cdot \vec{A} \right) = \exp ibq \left(\oint d\vec{x} \cdot \vec{A} \right) = e^{i\Phi bq} \tag{19}$$

(Note that in two dimensions the familiar flux tubes of three-dimensional physics degenerate to points, so it is proper to regard them as particles.) If the flux is $a\Phi_0$, then the phase will be $e^{2\pi i \frac{ab}{m}}$. Note that this phase does not depend on the velocity, curvature, or any details of the particles' dynamics, other than the topology of how their world-lines interweave. For that reason, we say we have a topological interaction.

Composites with (flux, charge) $=$ $(a\Phi_0, bq)$ will generally be particles with unusual quantum statistics. As we implement the interchange σ_j, each charge cluster feels the influence of the other's flux, and non-trivial phase is required. A close analysis shows that the anomalous statistics is such as to preserve a spin-statistics connection, in the form

$$e^{2\pi i J} = e^{i\theta} \tag{20}$$

Evidently quantum statistics, both conventional and unconventional, can be regarded as a special type of long-range interaction. It is remarkable that this interaction is not associated with the exchange of any massless particle. Indeed, our specific model, with broken gauge symmetry, can be fully gapped. One can also have topological interactions, involving similar accumulations of phase, for non-identical particles. What governs these topological interactions are the quantum numbers, or more formally the superselection sectors, of the particles, excitations or clusters involved, not their detailed internal structure.

The phase factors that accompany winding have observable consequences. They lead to a characteristic "long range" contribution to the scattering cross-section, specifically

$$\frac{d\sigma}{d\phi} = \sin^2 \left(\pi \frac{ab}{m} \right) \frac{1}{2\pi k} \frac{1}{\sin^2 \frac{\phi}{2}} \tag{21}$$

It diverges at small momentum transfer and in the forward direction. A cross-section of this kind was first computed by Aharonov and Böhm[6] in their classic paper on the significance of the vector potential in quantum mechanics. If we could do experiments in the style of high-energy physics, forming beams of quasiparticles and scattering them, we would be in great shape. Unfortunately, as a practical matter the highly characteristic cross-sections associated with anyons may not be easy to access experimentally for

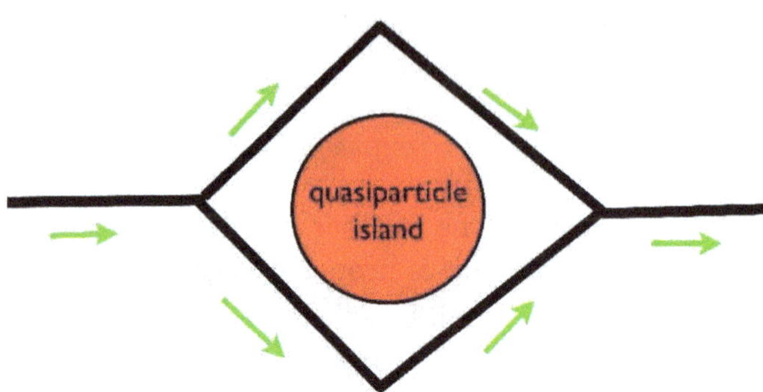

Fig. 2. A schematic interference experiment to reveal quantum statistics. One measures how the combined current depends on the occupation of the quasiparticle island.

the examples that occur as excitations in exotic states of condensed matter (although it could be worth a try!).

Interferometry appears to be more practical. The basic concept is simple and familiar, both from optics and (for instance) from SQUID magnetometers. One divides a coherent flow into two streams, which follow different paths before recombining. The relative phase between the paths determines the form of the interference, which can range from constructive to destructive recombination of the currents. We can vary the superselection sector of the area bounded by the paths, and look for corresponding, characteristic changes in the interference. (See Fig. 2.) Though there are many additional refinements, this is the basic concept behind both Goldman's suggestive experiments[8] and other planned anyon detection experiments.[9]

Elementary excitations in the fractional quantum Hall effect are predicted to be anyons. A proper discussion of that field would require a major digression, which would not be appropriate here. For the central calculation see Ref. 7, for extensive discussion and review, see Ref. 5. I would like to emphasize, in any case, that the general concept of anyons is by no means restricted to the quantum Hall effect; on the contrary, I believe the subject will reach a new level of interest and importance as more robust, user-friendly realizations are discovered. This is a most important area for future research.

2.3. *Nonabelian anyons*

The preceding field-theoretic setting for abelian anyons immediately invites nonabelian generalization. We can have a nonabelian gauge theory broken down to a discrete nonabelian subgroup; vortex-charge composites will then exhibit long range, topological interactions of the same kind as we found in the abelian case, for the same reason.

Though the starting point is virtually identical, when we consider interactions among several anyons the mathematics and physics of the nonabelian case quickly becomes considerably more complicated than the abelian case, and includes several qualitatively new effects. First, and most profoundly, we will find ourselves dealing with irreducible *multidimensional* representations of the braiding operations. Thus by winding well-separated particles[c] around one another, in principle arbitrarily slowly, we can not only acquire phase, but even navigate around a multidimensional Hilbert space. For configurations involving several well-separated particles, the size of the many-body "ground state" Hilbert spaces can get quite large: roughly speaking, they grow exponentially in the number of particles. Since all the states in this Hilbert space are related by locally trivial — but globally non-trivial — gauge transformations, they should be very nearly degenerate. This situation is reminiscent of what one would have if the particles had an internal freedom — a spin, say. However, here the emergent degrees of freedom are not localized on the particles, but more subtle and globally distributed.

The prospect of contructing very large Hilbert spaces that we can navigate in a controlled way using topologically defined (and thus forgiving!), gentle operations in physical space, and whose states differ in global properties not easily obscured by local perturbations, has inspired visions of *topological quantum computing.* (Preskill[10] has written an excellent introductory review). The journey from this exalted vision to real-world engineering practice will be challenging, to say the least, but thankfully there are fascinating prospects along the way.

The tiny seed from which all this complexity grows is the phenomenon displayed in Fig. 3. To keep track of the topological interactions, it is sufficient to know the total (ordered) line integral of the vector potential around simple circuits issuing from a fixed base point. This will tell us the group element a that will be applied to a charged particle as it traverses that loop. (The value of a generally depends on the base point and on the topology of how the loop winds around the regions where flux is concentrated, but not on other details. More formally, it gives a representation of the fundamental group of the plane with punctures.) If a charge that belongs to the representation R traverses the loop, it will be transformed according to

[c]From here on I will refer to the excitations simply as particles, though they may be complex collective excitations in terms of the underlying electrons, or other degrees of freedom.

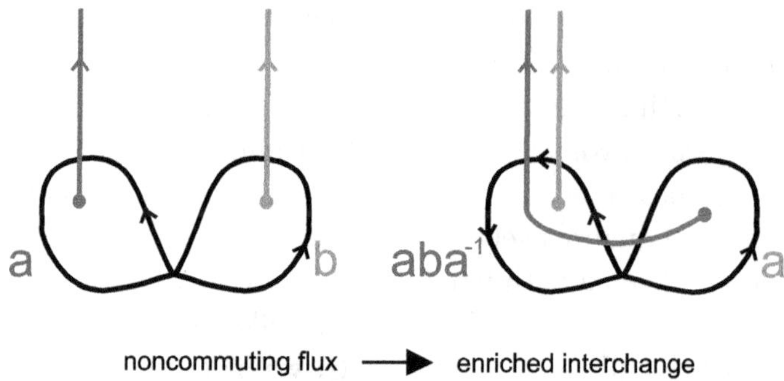

Fig. 3. By a gauge transformation, the vector potential emanating from a flux point can be bundled into a singular line. This aids in visualizing the effects of particle interchanges. Here, we see how nonabelian fluxes, as measured by their action on standardized particle trajectories, are modified by particle interchange.

$R(a)$. With these understandings, what Fig. 3 makes clear is that when two flux points with flux (a, b) get interchanged by winding the second over the first, the new configuration is characterized as (aba^{-1}, a). Note here that we cannot simply pull the "Dirac strings" where flux is taken off through one another, since nonabelian gauge fields self-interact! So motion of flux tubes in physical space generates non-trivial motion in group space, and thus in the Hilbert space of states with group-theoretic labels.

As a small taste of the interesting things that occur, consider the slightly more complicated situation displayed in Fig. 4, with a pair of fluxes (b, b^{-1}) on the right. It is a fun exercise to apply the rule for looping repeatedly, to find out what happens when we take this pair all the way around a on the right. One finds

$$(a, (b, b^{-1})) \rightarrow (a, (aba^{-1}, ab^{-1}a^{-1})) \tag{22}$$

i.e. the pair generally has turned into a different (conjugated) pair. Iterating, we eventually close on a finite-dimensional space of different kinds of pairs. There is a non-trival transformation $\tilde{R}(a)$ in this space that implements the effect of the flux a on pairs that wind around it. But this property — to be transformed by the group operation — is the defining property of charge! We conclude that flux pairs — flux and inverse flux — act as charges. We have constructed, as John Wheeler might have said, Charge Without Charge.

Its abstract realization through flux tubes makes it manifest that nonabelian statistics is consistent with all the general principles of quantum field theory. Practical physical realization in condensed matter is a different issue, for in that context, nonabelian gauge fields are not ready to hand.

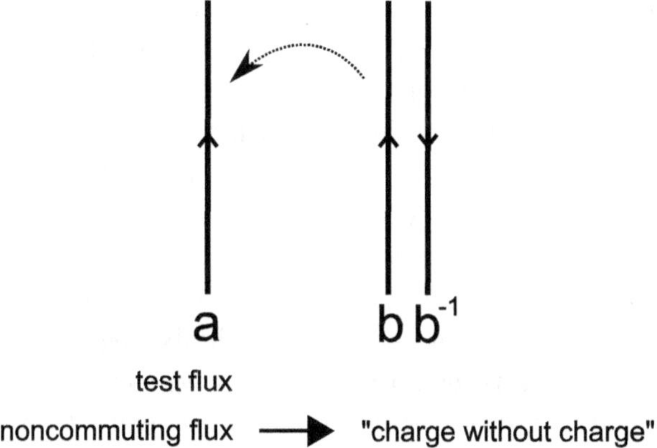

Fig. 4. Winding a flux–antiflux pair around a test flux, and seeing that its elements get conjugated, we learn that the pair generally carries charge.

Fortunately, and remarkably, there may be other ways to get there. At least one state of the quantum Hall effect, the so-called Moore–Read state at filling fraction $\frac{5}{2}$, has been identified as a likely candidate to support excitations with nonabelian statistics.[11]

The nonabelian statistics of the Moore–Read state is closely tied up with spinors.[12,13] I will give a proper discussion of this, including an extension to $3+1$ dimensions, elsewhere.[14] Here, I will just skip to the chase. Taking N γ_j matrices satisfying the usual Clifford algebra relations

$$\{\gamma_j, \gamma_k\} = 2\delta_{jk} \tag{23}$$

the braiding σ_j are realized as

$$\sigma_j = e^{i\pi/4}\frac{1}{\sqrt{2}}(1 + \gamma_j\gamma_{j+1}) \tag{24}$$

It is an easy exercise to show that these obey Eqs. (14) and (15), and $\sigma_j^4 = 1$ [Eq. (17)] but not $\sigma_j^2 = 1$ [Eq. (16)].

2.4. *Pairing, statistical transmutation and zero modes*

Finally, it is appropriate to mention that there is a deep connection between BCS theory and the sorts of quantum Hall states that support nonabelian anyons. They are connected adiabatically — in a conceptual parameter space — through *statistical transmutation.*

The imposition of a constant magnetic field on an two-dimensional electron gas is not a uniformly small perturbation, even if the magnitude of the field B is tiny. That is because in formulating quantum mechanics the Hamiltonian is fundamental, and in the Hamiltonian the vector potential $\vec{A}$ appears. Stokes' law informs us that the vector potential associated with a constant field strength grows linearly with the size of the sample. Thus, large quantities appear in Hamiltonian, and the perturbation associated with a tiny magnetic field is not uniformly small. And indeed we know that such a perturbation can induce a qualitative change in the spectrum, changing (say) the conventional parabolic free-electron spectrum into the quantized Landau levels, which feature a gap above a highly degenerate ground state.

The idea of statistical transmutation is that we can cancel off the growing part of the magnetic vector potential, if we associate to each electron an appropriate change in quantum statistics. Indeed, as I have reviewed above, one can effectively implement changes in the quantum statistics of particles by attaching notional flux and charge to those particles. (In the early days I called this "fictitious flux", to distinguish it from electromagnetic flux.) Now, if we add up the notional gauge potentials from a constant density of electrons, we will get — again according to Stokes' law — notional gauge potentials that grow linearly with the distance. If we add the right amount of flux — in other words, if we make a judicious change in quantum statistics — we can arrange to make a cancellation between the parts of the real and notional gauge potentials which grow with distance. The perturbation implementing this combined operation — a small magnetic field applied *together with* an appropriate small change in quantum statistics — will then be uniformly small.

The required relation between field and statistics can be neatly expressed as a connection between the filling fraction

$$\nu \equiv \frac{\rho}{eB}\,, \tag{25}$$

where ρ is the electron number density, and the quantum statistics parameter θ. We require

$$\Delta\frac{1}{\nu} = \Delta\frac{\theta}{\pi} \tag{26}$$

Since gapped systems which lie along the lines defined by Eq. (26) are related by a sequence of infinitesimal perturbations, we can expect that they lie in the same universality class, and will share universal properties.

$\Delta\theta = 2\pi$ corresponds, on the one hand, to no net change in statistics, and on the other, to $\Delta\frac{1}{\nu} = 2$. In this way our "notional" adiabatic path through anyons can connect proper (fermionic) electron states. A notable example: the fractional Laughlin states at $\frac{1}{\nu} = 2m + 1$ can be connected adiabatically to the integer quantum Hall state at $\nu = 1$ (in other words, $m = 0$). To my mind, this is the most profound[d] way to understand the existence of gapped many-body states at those filling fractions, and their other most distinctive properties.

$\frac{1}{\nu} = 0$ corresponds to zero magnetic field. In zero magnetic field, for appropriate attractive interactions, electrons can form a gapped superconductor, specifically a $p_x + ip_y$ superconductor, through BCS pairing in the $l = 1$ channel. According to Eq. (26), BCS superconductor can be adiabatically connected to $\nu = \frac{1}{2}$ quantum Hall states, which should share its universal properties. (The observed $\nu = \frac{5}{2}$ state plausibly contains two inert Landau levels, so its active dynamics involves $\nu = \frac{1}{2}$.) Prominent among the universal properties we can calculate in the BCS state are: the existence of a gap; the existence of neutral 'pair-breaking' excitations; and the existence of Majorana zero modes on vortices, leading to nonabelian statistics for those vortices. The nonabelian statistics that arises here is of the kind I sketched earlier, in Eq. (24). The Clifford algebra is realized here, concretely, as the algebra of the operator coefficients that multiply the vortex-centered zero modes in the expansion of the electron field. All the aforementioned features should carry over into appropriate $\nu = \frac{1}{2}$ states.

References

1. L. N. Cooper, Bound electron pairs in a degenerate fermi gas, *Phys. Rev.* **104**, 1189 (1956); J. Bardeen, L. N. Cooper and J. R. Schrieffer, Microscopic theory of superconductivity, *Phys. Rev.* **106**, 162 (1957); J. Bardeen, L. N. Cooper and J. R. Schrieffer, Theory of superconductivity, *Phys. Rev.* **108**, 1175 (1957).
2. Reviewed in A. Pich, arXiv:hep-ph/9505231v1 (1995).
3. Reviewed in M. Alford, K. Rajagopal, T. Schäefer and A. Schmitt, *Rev. Mod. Phys.* **80**, 1455 (2008).
4. Reviewed in F. Wilczek, *Nucl. Phys.* **B117** (Proc Suppl.) 410 (2003), hep-ph/0212128.
5. Reviewed in F. Wilczek (ed.), *Fractional Statistics and Anyon Superconductivity* (World Scientific, 1990).
6. Y. Aharonov and D. Bohm, *Phys. Rev.* **115**, 485 (1959).
7. D. Arovas, J. R. Schrieffer and F. Wilczek, *Phys. Rev. Lett.* **53**, 722 (1984).
8. V. Goldman, J. Liu and A. Zaslavsky, *Phys. Rev.* **B71**, 153303 (2005); F. Camino, F. Zhou and V. Goldman, *Phys. Rev. Lett.* **98**, 076805 (2007).
9. See M. Dolev, M. Heiblum, V. Umansky, A. Stern and D. Mahalu, *Nature* **452**, 829 (2008); I. Radu, J. Miller, C. Marcus, M. Kastner, L. Pfeiffer and K. West, *Science* **320**, 899 (2008).
10. J. Preskill, www.theory.caltech.edu/preskill/ph219 (2004).

[d]and the most under-appreciated ...

11. G. Moore and N. Read, *Nucl. Phys.* **B360**, 362 (1991).
12. C. Nayak and F. Wilczek, *Nucl. Phys.* **B479**, 529 (1996).
13. D. Ivanov, *Phys. Rev. Lett.* **86**, 268 (2001).
14. F. Wilczek, *Halflings in Three Dimensions*, paper in preparation.

FROM BCS TO THE LHC

Steven Weinberg
Department of Physics, University of Texas at Austin,
1 University Station C1600, Austin, TX 78712-0264, USA
weinberg@physics.utexas.edu

This article is based on the talk given by Steven Weinberg at BCS@50, held on 10–13 October 2007 at the University of Illinois at Urbana-Champaign to celebrate the 50th anniversary of the BCS paper. For more about the conference see http://www.conferences.uiuc.edu/bcs50/.

It was a little odd for me, a physicist whose work has been mainly on the theory of elementary particles, to be invited to speak at a meeting of condensed matter physicists celebrating a great achievement in their field. It is not only that there is a difference in the subjects we explore, there are deep differences in our aims, in the kinds of satisfaction we hope to get from our work.

Condensed matter physicists are often motivated to deal with phenomena because the phenomena themselves are intrinsically so interesting. Who would not be fascinated by weird things, such as superconductivity, superfluidity, or the quantum Hall effect? On the other hand, I don't think that elementary particle physicists are generally very excited by the phenomena they study. The particles themselves are practically featureless, every electron looking tediously just like every other electron.

Another aim of condensed matter physics is to make discoveries that are useful. In contrast, although elementary particle physicists like to point to the technological spin-offs from elementary particle experimentation, and these are real, this is not the reason we want these experiments to be done, and the knowledge gained by these experiments has no foreseeable practical applications.

Most of us do elementary particle physics neither because of the intrinsic interestingness of the phenomena we study, nor because of the practical importance of what we learn, but because we are pursuing a reductionist vision. All the properties of ordinary matter are what they are because of the principles of atomic and nuclear physics, which are what they are because of the rules of the Standard Model of elementary particles, which are what they are because ... well, we don't know, this is the reductionist frontier, which we are currently exploring.

I think that the single most important thing accomplished by the theory of Bardeen, Cooper, and Schrieffer (BCS) was to show that superconductivity is *not* part of the reductionist frontier.[1] Before BCS this was not so clear. For instance, in 1933 Walter Meissner raised the question whether electric currents in superconductors are carried by the known charged particles, electrons and ions. The great thing showed by Bardeen, Cooper, and Schrieffer was that no new particles or forces had to be introduced to understand superconductivity. According to a book on superconductivity that Leon Cooper showed me, many physicists were even disappointed that "superconductivity should, on the atomistic scale, be revealed as nothing more than a footling small interaction between electrons and lattice vibrations".[2]

The claim of elementary particle physicists to be leading the exploration of the reductionist frontier has at times produced resentment among condensed matter physicists. (This was not helped by a distinguished particle theorist, who was fond of referring to condensed matter physics as "squalid state physics.") This resentment surfaced during the debate over the funding of the Superconducting SuperCollider (SSC). I remember that Phil Anderson and I testified in the same Senate committee hearing on the issue, he against the SSC and I for it. His testimony was so scrupulously honest that I think it helped the SSC more than it hurt it. What really did hurt was a statement opposing the SSC by a condensed matter physicist who happened at the time to be the President of the American Physical Society. As everyone knows, the SSC project was canceled, and now we are waiting for the LHC at CERN to get us moving ahead again in elementary particle physics.

During the SSC debate, Anderson and other condensed matter physicists repeatedly made the point that the knowledge gained in elementary particle physics would be unlikely to help them to understand emergent phenomena like superconductivity. This is certainly true, but I think beside the point, because that is not why we are studying elementary particles; our aim is to push back the reductive frontier, to get closer to whatever simple and general theory accounts for everything in nature. It could be said equally that the knowledge gained by condensed matter physics is unlikely to give us any direct help in constructing more fundamental theories of nature.

So what business does a particle physicist like me have at a celebration of the BCS theory? (I have written just one paper about superconductivity, a paper of monumental unimportance, which was treated by the condensed matter community with the indifference it deserved.) Condensed matter physics and particle physics *are* relevant to each other, despite everything I have said. This is because, although the *knowledge* gained in elementary particle physics is not likely to be useful to condensed matter physicists, or vice versa, experience shows that the *ideas* developed in one field can prove very useful in the other. Sometimes these ideas become transformed in translation, so that they even pick up a renewed value to the field in which they were first conceived.

The example that concerns me here is an idea that elementary particle physicists learned from condensed matter theory — specifically, from the BCS theory. It is the idea of spontaneous symmetry breaking.

In particle physics we are particularly interested in the symmetries of the laws of nature. One of these symmetries is invariance of the laws of nature under the symmetry group of three-dimensional rotations, or in other words, invariance of the laws we discover under changes in the orientation of our measuring apparatus.

When a physical system does not exhibit all the symmetries of the laws by which it is governed, we say that these symmetries are spontaneously broken. A very familiar example is spontaneous magnetization. The laws governing the atoms in a magnet are perfectly invariant under three-dimensional rotations, but at temperatures below a critical value, the spins of these atoms spontaneously line up in some direction, producing a magnetic field. In this case, and as often happens, a subgroup is left invariant: the two-dimensional group of rotations around the direction of magnetization.

Now to the point. A superconductor of any kind is nothing more or less than a material in which a particular symmetry of the laws of nature, electromagnetic gauge invariance, is spontaneously broken. This is true of high temperature superconductors, as well as the more familiar superconductors studied by BCS. The symmetry group here is the group of two-dimensional rotations. These rotations act on a two-dimensional vector, whose two components are the real and imaginary parts of the electron field, the quantum mechanical operator that in quantum field theories of matter destroys electrons. The rotation angle of the broken symmetry group can vary with location in the superconductor, and then the symmetry transformations also affect the electromagnetic potentials, a point to which I will return.

The symmetry breaking in a superconductor leaves unbroken a rotation by 180°, which simply changes the sign of the electron field. In consequence of this spontaneous symmetry breaking, products of any even number of electron fields have non-vanishing expectation values in a superconductor, though a single electron field does not. All of the dramatic exact properties of superconductors — zero electrical resistance, the expelling of magnetic fields from superconductors known as the Meissner effect, the quantization of magnetic flux through a thick superconducting ring, and the Josephson formula for the frequency of the ac current at a junction between two superconductors with different voltages — follow from the assumption that electromagnetic gauge invariance is broken in this way, with no need to inquire into the mechanism by which the symmetry is broken.

Condensed matter physicists often trace these phenomena to the appearance of an "order parameter," the non-vanishing mean value of the product of two electron fields, but I think this is misleading. There is nothing special about *two* electron fields; one might just as well take the order parameter as the product of three electron fields and the complex conjugate of another electron field. The important thing is the broken symmetry, and the unbroken subgroup.

It may then come as a surprise that spontaneous symmetry breaking is nowhere mentioned in the seminal paper of Bardeen, Cooper, and Schrieffer. Their paper describes a mechanism by which electromagnetic gauge invariance is in fact broken, but they derived the properties of superconductors from their dynamical model, not from the mere fact of broken symmetry. I am not saying that Bardeen, Cooper, and Schrieffer did not know of this spontaneous symmetry breaking. Indeed, there was already a large literature on the apparent violation of gauge invariance in phenomenological theories of superconductivity, the fact that the electric current produced by an electromagnetic field in a superconductor depends on a quantity known as the vector potential, which is not gauge invariant. But their attention was focused on the details of the dynamics rather than the symmetry breaking.

This is not just a matter of style. As BCS themselves made clear, their dynamical model was based on an approximation that a pair of electrons interact only when the magnitude of their momenta is very close to a certain value, known as the Fermi surface. This leaves a question: How can you understand the exact properties of superconductors, like exactly zero resistance and exact flux quantization, on the basis of an approximate dynamical theory? It is only the argument from exact symmetry principles that can fully explain the remarkable exact properties of superconductors.

Though spontaneous symmetry breaking was not emphasized in the BCS paper, the recognition of this phenomenon produced a revolution in elementary particle physics. The reason is that (with certain qualification, to which I will return), whenever a symmetry is spontaneously broken, there must exist excitations of the system with a frequency that vanishes in the limit of large wavelength. In elementary particle physics, this means a particle of zero mass.

The first clue to this general result was a remark in a 1960 paper by Yoichiro Nambu, that just such collective excitations in superconductors play a crucial role in reconciling the apparent failure of gauge invariance in a superconductor with the exact gauge invariance of the underlying theory governing matter and electromagnetism. Nambu speculated that these collective excitations are a necessary consequence of this exact gauge invariance.

A little later, Nambu put this idea to good use in particle physics. In nuclear beta decay an electron and neutrino (or their antiparticles) are created by currents of two different kinds flowing in the nucleus, known as vector and axial vector currents. It was known that the vector current was conserved, in the same sense as the ordinary electric current. Could the axial current also be conserved?

The conservation of a current is usually a symptom of some symmetry of the underlying theory, and holds whether or not the symmetry is spontaneously broken. For the ordinary electric current, this symmetry is electromagnetic gauge invariance. Likewise, the vector current in beta decay is conserved because of the isotopic spin symmetry of nuclear physics. One could easily imagine several different symmetries, of a sort known as chiral symmetries, that would entail a conserved axial vector current. But it seemed that any such chiral symmetries would imply either that the nucleon mass is zero, which is certainly not true, or that there must exist a triplet of massless strongly interacting particles of zero spin and negative parity, which isn't true either. These two possibilities simply correspond to the two possibilities that the symmetry, whatever it is, either is not, or is, spontaneously broken, not just in some material like a superconductor, but even in empty space.

Nambu proposed that there is indeed such a symmetry, and it is spontaneously broken in empty space, but the symmetry in addition to being spontaneously broken is not exact to begin with, so the particle of zero spin and negative parity required by the symmetry breaking is not massless, only much lighter than other strongly interacting particles. This light particle, he recognized, is nothing but the pion, the lightest and first discovered of all the mesons. In a subsequent paper with Giovanni Jona-Lasinio, Nambu presented an illustrative theory in which, with some drastic approximations, a suitable chiral symmetry was found to be spontaneously broken, and in consequence the light pion appeared as a bound state of a nucleon and an antinucleon.

So far, there was no proof that broken exact symmetries always entail exactly massless particles, just a number of examples of approximate calculations in specific theories. In 1961 Jeffrey Goldstone gave some more examples of this sort, and a hand-waving proof that this was a general result. Such massless particles are today known as Goldstone bosons, or Nambu-Goldstone bosons. Soon after, Goldstone, Abdus Salam, and I made this into a rigorous and apparently quite general theorem.

This theorem has applications in many branches of physics. One is cosmology. You may know that today the observation of fluctuations in the cosmic microwave background are being used to set constraints on the nature of the exponential expansion, known as inflation, that is widely believed to have preceded the radiation dominated Big Bang. But there is a problem here. In between the end of inflation and the time the microwave background that we observe was emitted, there intervened a number of events that are not at all understood: the heating of the universe after inflation, the production of baryons, the decoupling of cold dark matter, and so on. So how is it possible to learn anything about inflation by studying radiation that was emitted long after inflation, when we don't understand what happened in between? The reason we can get away with this is that the cosmological fluctuations now being studied are of a type, known as adiabatic, that can be regarded as the Goldstone excitations required by a symmetry, related to general coordinate invariance, that is spontaneously broken by the space-time geometry. The physical wavelengths of these cosmological fluctuations were stretched out by inflation so much that they were very large during the epochs when things were happening that we don't understand, so they then had zero frequency, which means that the amplitude of these fluctuations was not changing, so that the value of the amplitude relatively close to the present tells us what it was during inflation.

But in particle physics, this theorem was at first seen as a disappointing result. There was a crazy idea going around, which I have to admit that at first I shared, that somehow the phenomenon of spontaneous symmetry breaking would explain why the symmetries being discovered in strong interaction physics were not exact. Werner Heisenberg continued to believe this into the 1970s, when everyone else had learned better.

The prediction of new massless particles, which were ruled out experimentally, seemed in the early 1960s to close off this hope. But it was a false hope anyway. Except under special circumstances, a spontaneously broken symmetry does not look at all like an approximate unbroken symmetry; it manifests itself in the masslessness of spin — zero bosons, and in details of their interactions. Today we understand approximate symmetries such as isospin and chiral invariance as consequences of the fact that some quark masses, for some unknown reason, happen to be relatively small.

Though based on a false hope, this disappointment had an important consequence. Peter Higgs, Robert Brout and François Englert, and Gerald Guralnik, Dick Hagen, and Tom Kibble were all independently led to look for, and then found, an exception to the theorem of Goldstone, Salam, and me. The exception applies to theories in which the underlying physics is invariant under *local* symmetries, symmetries whose transformations, like electromagnetic gauge transformations, can vary from place to place in space and time. (This is in contrast with the chiral symmetry associated with the axial vector current of beta decay, which applies only when the symmetry transformations are the same throughout space-time.) For each local symmetry there must exist a vector field, like the electromagnetic field, whose quanta would be massless if the symmetry was not spontaneously broken. The quanta of each such field are particles with helicity (the component of angular momentum in the direction of motion) equal in natural units to $+1$ or -1. But if the symmetry *is* spontaneously broken, these two helicity states join up with the helicity-zero state of the Goldstone boson to form the three helicity states of a *massive* particle of spin one. Thus, as shown by Higgs, Brout and Englert, and Guralnik, Hagen, and Kibble, when a local symmetry is spontaneously broken, neither the vector particles with which the symmetry is associated nor the Nambu-Goldstone particles produced by the symmetry breaking have zero mass.

This was actually argued earlier by Anderson, on the basis of the example provided by the BCS theory. But the BCS theory is non-relativistic, and the Lorentz invariance that is characteristic of special relativity had played a crucial role in the theorem of Goldstone, Salam, and me, so Anderson's argument was generally ignored by particle theorists. In fact, Anderson was right: the reason for the exception noted by Higgs *et al.* is that it is not possible to quantize a theory with a local symmetry in a way that preserves both manifest Lorentz invariance and the usual rules of quantum mechanics, including the requirement that probabilities be positive. In fact, there are two ways to quantize theories with local symmetries: one way that preserves positive probabilities but loses manifest Lorentz invariance, and another that preserves manifest Lorentz invariance but seems to lose positive probabilities, so in fact these theories actually do respect both Lorentz invariance and positive probabilities; they just don't respect our theorem.

The appearance of mass for the quanta of the vector bosons in a theory with local symmetry re-opened an old proposal of Chen Ning Yang and Robert Mills, that the strong interactions might be produced by the vector bosons associated with some sort of local symmetry, more complicated than the familiar electromagnetic gauge invariance. This possibility was specially emphasized by Brout and Englert. It took a few years for this idea to mature into a specific theory, which then turned out not to be a theory of strong interactions.

Perhaps the delay was because the earlier idea of Nambu, that the pion was the nearly massless boson associated with an approximate chiral symmetry that is *not* a local symmetry, was looking better and better. I was very much involved in this work, and would love to go into the details, but that would take me too far from BCS. I'll just say that, from the effort to understand processes involving any number of low-energy pions beyond the lowest order of perturbation theory, we became comfortable with the use of effective field theories in particle physics. The mathematical techniques developed in this work in particle physics were then used by Joseph Polchinski and others to justify the approximations made by BCS in their work on superconductivity.

The story of the physical application of spontaneously broken local symmetries has often been told, by me and others, and I don't want to take much time on it here, but I can't leave it out altogether because I want to make a point about it that will take me back to the BCS theory. Briefly, in 1967 I went back to the idea of a theory of strong interactions based on a spontaneously broken local symmetry group, and right away, I ran into a problem: the subgroup consisting of ordinary isospin transformations is not spontaneously broken, so there would be a massless vector particle associated with these transformations with the spin and charges of the rho meson. This of course was in gross disagreement with observation; the rho meson is neither massless nor particularly light.

Then it occurred to me that I was working on the wrong problem. What I should have been working on were the weak nuclear interactions, like beta decay. There was just one natural choice for an appropriate local symmetry, and when I looked back at the literature I found that the symmetry group I had decided on was one that had already been proposed in 1961 by Sheldon Glashow, though not in the context of an exact spontaneously broken local symmetry. (I found later that the same group had also been considered by Salam and John Ward.) Even though it was now exact, the symmetry when spontaneously broken would yield massive vector particles, the charged W particles that had been the subject of theoretical speculation for decades, and a neutral particle, which I called the Z particle, which mediates a "neutral current" weak interaction, which had not yet been observed. The same symmetry breaking also gives mass to the electron and other leptons, and in a simple extension of the theory, to the quarks. This symmetry group contained electromagnetic gauge invariance, and since this subgroup is clearly not spontaneously broken (except in superconductors), the theory requires a massless vector particle, but it is not the rho meson, it is the photon, the quantum of light. This theory, which became known as the electroweak theory, was also proposed independently in 1968 by Salam.

The mathematical consistency of the theory, which Salam and I had suggested but not proved, was shown in 1971 by Gerard't Hooft; neutral current weak interactions were found in 1973; and the W and Z particles were discovered at CERN a decade later. Their detailed properties are just those expected according to the electroweak theory.

There was (and still is) one outstanding issue: just how is the local electroweak symmetry broken? In the BCS theory, the spontaneous breakdown of electromagnetic gauge invariance arises because of attractive forces between electrons near the Fermi surface. These forces don't have to be strong; the symmetry is broken however weak these forces may be. But this feature occurs only because of the existence of a Fermi surface, so in this respect the BCS theory is a misleading guide for particle physics. In the absence of a Fermi surface, dynamical spontaneous symmetry breakdown requires the action of strong forces. There are no forces acting on the known quarks and leptons that are anywhere strong enough to produce the observed breakdown of the local electroweak symmetry dynamically, so Salam and I did not assume a dynamical symmetry breakdown; instead we introduced elementary scalar fields into the theory, whose vacuum expectation values in the classical approximation would break the symmetry.

This has an important consequence. The only elementary scalar quanta in the theory that are eliminated by spontaneous symmetry breaking are those that become the helicity-zero states of the W and Z vector particles. The other elementary scalars appear as physical particles, now generically known as Higgs bosons. It is the Higgs boson predicted by the electroweak theory of Salam and me that will be the primary target of the new LHC accelerator, to be completed at CERN sometime in 2008.

But there is another possibility, suggested independently in the late 1970s by Leonard Susskind and myself. The electroweak symmetry might be broken dynamically after all, as in the BCS theory. For this to be possible, it is necessary to introduce new extra-strong forces, known as technicolour forces, that act on new particles, other than the known quarks and leptons. With these assumptions, it is easy to get the right masses for the W and Z particles and large masses for all the new particles, but there are serious difficulties in giving masses to the ordinary quarks and leptons. Still, it is possible that experiments at the LHC will not find Higgs bosons, but instead will find a great variety of heavy new particles associated with technicolour. Either way, the LHC is likely to settle the question of how the electroweak symmetry is broken.

It would have been nice if we could have settled this question by calculation alone, without the need of the LHC, in the way that Bardeen, Cooper, and Schrieffer were able to find how electromagnetic gauge invariance is broken in a superconductor by applying the known principles of electromagnetism. But that is just the price we in particle physics have to pay for working in a field whose underlying principles are not yet known.

References

1. J. Bardeen, L. N. Cooper and J. R. Schrieffer, *Phys. Rev.* **108**, 1175 (1957).
2. K. Mendelssohn, *The Quest for Absolute Zero* (McGraw-Hill, New York, 1966).

INDEX

www.ingramcontent.com/pod-product-compliance
Lightning Source LLC
Chambersburg PA
CBHW081214220326
41598CB00037B/6774

* 9 7 8 9 8 1 4 3 0 4 6 5 8 *